Palgrave Studies in Cybercrime and Cybersecurity

Series Editors
Thomas J. Holt, Michigan State University, East Lansing, MI, USA
Cassandra Cross, School of Justice, Queensland University of Technology, Brisbane, QLD, Australia

This book series addresses the urgent need to advance knowledge in the fields of cybercrime and cybersecurity. Because the exponential expansion of computer technologies and use of the Internet have greatly increased the access by criminals to people, institutions, and businesses around the globe, the series will be international in scope. It provides a home for cutting-edge long-form research. Further, the series seeks to spur conversation about how traditional criminological theories apply to the online environment. The series welcomes contributions from early career researchers as well as established scholars on a range of topics in the cybercrime and cybersecurity fields. Original series creators and co-founders: Marie-Helen Maras and Thomas J. Holt.

Kristan Stoddart

Cyberwarfare

Threats to Critical Infrastructure

Kristan Stoddart 📍
School of Social Sciences
Swansea Universtiy
Swansea, UK

Palgrave Studies in Cybercrime and Cybersecurity
ISBN 978-3-030-97298-1   ISBN 978-3-030-97299-8   (eBook)
https://doi.org/10.1007/978-3-030-97299-8

© The Editor(s) (if applicable) and The Author(s), under exclusive license to Springer International Publishing AG, part of Springer Nature 2022
This work is subject to copyright. All rights are solely and exclusively licensed by the Publisher, whether the whole or part of the material is concerned, specifically the rights of translation, reprinting, reuse of illustrations, recitation, broadcasting, reproduction on microfilms or in any other physical way, and transmission or information storage and retrieval, electronic adaptation, computer software, or by similar or dissimilar methodology now known or hereafter developed.
The use of general descriptive names, registered names, trademarks, service marks, etc. in this publication does not imply, even in the absence of a specific statement, that such names are exempt from the relevant protective laws and regulations and therefore free for general use.
The publisher, the authors, and the editors are safe to assume that the advice and information in this book are believed to be true and accurate at the date of publication. Neither the publisher nor the authors or the editors give a warranty, expressed or implied, with respect to the material contained herein or for any errors or omissions that may have been made. The publisher remains neutral with regard to jurisdictional claims in published maps and institutional affiliations.

This Palgrave Macmillan imprint is published by the registered company Springer Nature Switzerland AG
The registered company address is: Gewerbestrasse 11, 6330 Cham, Switzerland

Acknowledgments

From 2014–2017 I was involved in a major (£1.2 million) project examining SCADA systems and the Cyber Security Lifecycle funded by Airbus Group and the Welsh Government. This was, and remains, a project in which I will be eternally grateful to have been involved. It was led quite superbly by Dr. Kevin Jones who is now the Chief Information Security Officer at Airbus Group. He was, and remains, inspirational and an active leader in the global cybersecurity field. The genesis of this book was in this project. Through it, I worked with some of the leaders in cybersecurity in locations throughout Europe and North America.

I owe a huge debt of gratitude to my friend and former colleague, Professor Emeritus Len Scott, who read most parts of this when it was a larger two-volume study. Additionally, my long-term collaborator and friend, Professor Emeritus John Baylis, aided me in fundamental discussions of warfare and much else besides. I also need to thank Professor Christian Kaunert at the University of South Wales (where I hold a Visiting Professorship) for all his encouragement with what have become three books (this book, and my related books, *China and its Embrace of Offensive Cyberespionage* and potential for cyber warfare and *Russia's cyber offensives against the West*). I would also like to thank Professor Stuart Macdonald and my colleagues at the Cyber Threats Research Centre (CYTREC) at Swansea University where I am currently an Associate Professor. There are also friends and colleagues I would add too numerous to mention.

Whilst the content I have used produced by United States government agencies is in the public domain I am grateful to be able to reproduce materials from a number of U.S. government agencies. These are the White House, National Security Council, United States Congress and United States Senate, the Department of Energy, Department of Homeland Security, Department of Justice, Federal Aviation Authority, Federal Bureau of Investigation, Federal Emergency Management Agency, the National Institute of Standards, and Technology (NIST)/Department of Commerce. In addition, the Department of Defense and branches of the DOD including the National Security Agency, Central Intelligence Agency, Defense Intelligence Agency, Office of the Director of National Intelligence, U.S. Cybercommand, the Office of the Director of National Intelligence, the Defense Advanced Research Projects Agency, and the Joint Chiefs of Staff.

I am also grateful to NATO's Cooperative Cyber Defence Centre of Excellence (CCDCOE) for permission to quote from James A. Lewis, 'The Role of Offensive Cyber Operations in NATO's Collective Defence', which appeared in Tallinn Paper 8 (2015); Ji Young Kong, Kyoung Gon Kim and Jong In Lim, 'The All-Purpose Sword: North Korea's Cyber Operations and Strategies', in Tomáš Minárik, Siim Alatalu, Stefano Biondi, Massimiliano Signoretti, Ihsan Tolga and Gábor Visky (Eds.), *2019 11th International Conference on Cyber Conflict: Silent Battle* (Tallinn: CCD COE Publications, 2019); and Christian Czosseck, Rain Ottis and Anna-Maria Talihärm, 'Estonia After the 2007 Cyber Attacks: Legal, Strategic and Organisational Changes in Cyber Security', in *Proceedings of the 10th European Conference on Information Warfare and Security at the Tallinn University of Technology Tallinn, Estonia 7–8 July 2011*.

I was fortunate to spend time with the excellent and committed people (civilian and uniformed) who work at the CCDCOE and to live in Tallinn, Estonia for an extended period. I also had (and hope to again) the pleasure of visiting this beautiful city for NATO's CyCon conferences where I met excellent people, felt what is like to be a frontier state of NATO and the EU, and came to appreciate the earnestness they hold for EU and NATO values.

I am grateful to Perry Pederson for granting permission to use 'Aurora revisited—by its original project lead' that was published on langner.com. I am also grateful to Ralph Langer for his help. I am also grateful to 'The Estate of Alexandra Milgram' for permission to quote from Stanley Milgram, 'Behavioral Study of Obedience', which appeared in the *Journal of Abnormal and Social Psychology*, Vol. 67, No. 4 (October 1963). This work and his ideas of 'six degrees of separation' and the small world phenomenon continue to help generate insights. I would also like to thank Yale University Press for permission to quote from *The Vory: Russia's Super Mafia* by Mark Galeotti and to Penguin Press for permission to quote from Joel Brenner, *America the vulnerable: inside the new threat matrix of digital espionage, crime, and warfare*, (New York: Penguin Press, 2011). I also need to thank Eurozine for permission to quote from another work by Mark Galeotti, '(Mis)Understanding Russia's two 'hybrid wars''. Finally, I am very thankful to Verizon for permission to use extracts from their highly informative Data Breach Investigations Reports (DBIR) and to reproduce figures 15 and 16 from the 2021 DBIR.

Whilst best efforts have been made to trace and acknowledge all copyright holders, I would like to apologize should there have been any errors or omissions. I would also like to give my thanks to Josie Taylor, my commissioning editor at Palgrave Macmillan/Springer Nature. This book was begun in earnest during a career break in the summer of 2018. Although it had long been my intention to write up that which I had learned, I came to learn much more still. My knowledge (and hopefully my understanding) grew as I thought about the fundamental issues of cyberwar(fare) and war and warfare itself and then thought in tandem about the myriad of cyberespionage cases and activities I knew of or came to encounter. I came to realize that some separation was necessary but that some acts of cyberespionage were also part of the debate/continuum surrounding cyberwarfare. I hope I have done both subjects justice.

Contents

1	**Introduction**	1
	Cyberwar and Critical Infrastructure	3
	The Threat Actors	6
	The Cyber Context: States as Targets and Attackers	11
	Cybercriminals and Their Usefulness as 'Proxies' and 'Privateers'	13
	The Threat Landscape	15
	Machine Learning, Artificial Intelligence, and High Performance (Quantum) Computing	17
	Critical Infrastructure: ICS and SCADA	19
	Subdue the Enemy Without Fighting	26
	Cyber: The Fifth Domain of Warfare	29
	A Short Guide to Terminology	31
	Malware	31
	Cyber Forensics	33
	Overview	34

Contents

2 **On Cyberwar: Theorizing Cyberwarfare Through Attacks on Critical Infrastructure—Reality, Potential, and Debates** 53
Introduction 53
The Fog of Cyberwar 54
What Is Cyberwar(fare)? 57
Cyberwar Deconstructed 59
Hybrid Warfare 61
International Law: JWT and the LOAC 65
Rules of Engagement 67
The Tallinn Manuals and the Cyberwarfare Debate 70
Cyberwar Against Critical Infrastructure as a War Winner 73
The Failure of Cyber Deterrence and the Attribution Problem 78
Iran 87
North Korea 94
Policy and Debates in the United States 102
The 2018 U.S. National Cyber Strategy: CISA and the Biden Administration 105
The U.S. Military and 'Forward Defense' 106
Conclusion 113

3 **Cyberwar: Attacking Critical Infrastructure** 147
Introduction 147
SCADA Systems and Critical Infrastructure 148
Proof-of-Concept: Aurora and Stuxnet 151
The Implications of Aurora and Stuxnet 152
Real-World Cases 153
  Electricity Generation and Distribution 154
  Electricity Producing Sites Include Nuclear Power Stations 156
  Water Treatment and Sanitation 158
  Dams and Reservoirs 159
  The Oil and Gas Industry: Rigs, Refineries, and Pipelines 160
  Chemical Plants 168
Ports and Logistics 169

	Merchant Shipping	170
	Road and Rail	172
	Civil Aviation	175
	The Good News	182
	The Bad News	186
	Ukraine and Russia's 2022 Invasion	189
	Conclusion	199
4	**Gaining Access: Attack and Defense Methods and Legacy Systems**	**227**
	Introduction	227
	Common Technical Attack Methods	228
	Drive-by Downloads	229
	Watering Hole Attacks	230
	Man-in-the-Middle/Session HIJACKING	231
	Zero-Days	231
	Rootkits	233
	Remote Access Trojans (RATs)	234
	The Use of Mobile/Cellular Devices and Remote Access	236
	Script Kiddies or Nation-States?	237
	Common TTPs	241
	Counters and Defenses	242
	Firewalls	242
	Demilitarized Zones (DMZs)	243
	Intrusion Detection Systems (IDS): HIDS/SIDS/HIPS	243
	Honeypots and Honeytraps	244
	Signature and Behavior-Based Malware Detection	245
	Sandboxing	247
	Packet Sniffers	247
	Application Whitelisting	248
	Security Information and Event Management	249
	Blockchain	249
	Pressing the Reset	250
	The Zero Trust Security Model	251
	Legacy Systems: In-Built Vulnerabilities in Critical Infrastructure	252

Legacy Systems of the U.S. Government	255
Industry and the Costs of 'Keeping the Lights On'	257
Patching	260
Targeting Supply Chains	261
Conclusion	263

5 Hacking the Human — 281

Introduction	281
Social Engineering	282
Examples of Social Engineering	285
Exploiting Cognitive and Behavioral Psychology	287
Hacking the Human	290
Spear Phishing	292
Mitigating Spear Phishing	294
Spear Phishing Attack Tools and Websites	296
State Intelligence: HUMINT Beyond Social Engineering	300
The 'Birds Eye' Macro View and the Micro Level of HUMINT	302
Human Sources and Human Agency	304
Cyber Defense and Offense	306
Defending Insider Threats	307
Mitigation and the Insider Threat	310
Physical Security I	312
Physical Security II: The CIA Triad and 'Full Disclosure'	316
The Cybersecurity Workforce Deficit	318
Computer Emergency Response Teams	322
Cyber Threat Intelligence and the Cybersecurity Community	325
Industry and Government Backed Self-Help Groups	326
Conclusion	328

6 Non and Sub-State Actors: Cybercrime, Terrorism, and Hackers — 351

Introduction	351
Outsider Threats, Insider Threats, and Target Spotting	353
Hackers, Hacking Groups, and Social Engineering	355
Social Network Analysis	357

	SNA as a Law Enforcement and Intelligence Tool	362
	Terrorism	364
	Encryption and the Risk of 'Going Dark'	367
	State-Backed/State-Sanctioned Cybercrime	368
	Cybercriminals and States	370
	'Dark Net' Markets	371
	Organized Crime, Ransomware, and the 'Dark Net'	373
	WannaCry and Petya/NotPetya	376
	The Cloak of Attribution: The Use of Proxy Actors by States	378
	Conclusion	380
7	**Conclusion**	401
	On Cyberwarfare	403
	Attacking Critical Infrastructure	405
	Pinprick Attacks and First Strike	407
	Cybersecurity Defenses: Risk Management and Legacy Systems	408
	Hacking the Human	411
	Reducing Risk	413
	Risk Management and Resilience	414
	States as Advanced Persistent Threats	415
	The U.S. Intelligence Community and a 'Whole of Nation' Effort	418
	Zugzwang	421
Bibliography		431
Index		509

Abbreviations and Concepts[1]

Access privileges	Technical barriers exist that limit access privileges to systems and data.
Active measures	Conducted by state intelligence agencies, active measures include propaganda, destabilization, forgery, assassination, acts of terrorism, hacking of political parties and election interference and dis/misinformation campaigns which generate 'fake news' stories for political effect.
AI	(see Artificial Intelligence).

---

[1] Parts of what follows are drawn from 'Language of Espionage', https://www.spymuseum.org/education-programs/spy-resources/language-of-espionage/, ELINT: A Scientific Intelligence System', https://www.cia.gov/library/center-for-the-study-of-intelligence/kent-csi/vol2no1/html/v02i1a06p_0001.htm, Barry M. Leiner, Vinton G. Cerf, David D. Clark, Robert E. Kahn, Leonard Kleinrock, Daniel C. Lynch, Jon Postel, Larry G. Roberts, Stephen Wolff, 'Brief History of the Internet' (1997), https://www.internetsociety.org/internet/history-internet/brief-history-internet/, all accessed 3 February 2020. 'Espionage', https://www.mi5.gov.uk/espionage, accessed 1 November 2016. '"Covert Action" Memo in report Confrontation or Collaboration? Congress and the Intelligence Community', July 2009, http://belfercenter.ksg.harvard.edu/publication/19149/covert_action.html, accessed 1 November 2016. Julian Richards, *The Art and Science of Intelligence Analysis* (Oxford: Oxford University Press, 2011). Mark Phythian, *Understanding the Intelligence Cycle* (Abingdon: Routledge, 2013).

Agent	A person unofficially employed by an intelligence service, often as a source of information.
Apps	Software applications.
APT	Advanced Persistent Threat. APTs are highly sophisticated and well-resourced threat actors operating in cyberspace and the information and communications environment more generally. They are run by nation-states through their intelligence agencies and can operate for months or years as part of an orchestrated strategy to advance foreign, defense, or economic policies.
APT1	A Chinese cyberespionage group. There are many more.
APT28	Known by a variety of names the most common of which is 'Fancy Bear'. APT28 is a Russian cyberespionage actor run by the GRU.
APT29	Also known by a variety of names or monikers the most common being 'Cozy Bear'. They too are a Russian cyberespionage actor run by the FSB.
Artificial Intelligence	Simulation of human intelligence through computer programming/coding to enable machines to think like humans. This includes learning and problem-solving. A branch of Machine Learning (ML).
Asymmetric warfare	Conflicts between nations or groups which have unequal military capabilities and/or different strategies.
ATC/ATM	Air Traffic Control/Management.
Backdoor	A way in to systems and networks purposefully left or discovered.
BBC	British Broadcasting Corporation.
Big Data	Big data is made up of everything that is created and transmitted online.
BlackEnergy	A cyberespionage campaign run by Russian intelligence against Ukrainian critical infrastructure.

Abbreviations and Concepts   xvii

Black Hat (see also White Hat and Gray Hat)	Black hats are malicious hackers.
Blockchain	An Internet-enabled electronic ledger, based on cryptography, that securely stores and records every transaction between people without third-party involvement such as a bank or payment system like Visa.
Bots/Botnets	Taken from the word robot, these are automated applications that are (among other things) used to generate or spread mis- and disinformation.
Bug	An error, flaw, or fault in software programming leading to an incorrect or unexpected result, causing a program or system to behave in unintended ways.
C^3	Command, Control, and Communications.
CCDCOE	Cooperative Cyber Defence Centre of Excellence (NATO) based in Tallinn, Estonia.
CCP	Chinese Communist Party.
CCTV	Closed-circuit television.
CEO	Chief Executive Officer.
CERT	Computer Emergency Response Team.
CFO	Chief financial officer.
CI	Critical infrastructure—includes public utilities such as electric power generation and distribution, water supplies and water treatment, natural gas and oil production and pipelines, shipping and maritime traffic, hydroelectric dams, traffic lights, and train switching systems.
CIA (Central Intelligence Agency)	The main U.S. foreign intelligence gathering service.
CIA triad	Confidentiality, Integrity, and Availability. A term used in computer science and engineering to prioritize data and systems.
CIO/s	Chief Information Officer/s.
CISA	The U.S. Cybersecurity Information Sharing Act (2015).

CISA	Cybersecurity and Infrastructure Security Agency (U.S.).
CISO	Chief Information Security Officer.
(The) Cloud	Cloud computing is a process of sharing resources to optimize performance using a network of computers to store and process information, rather than a single machine.
CNI (see CI)	Critical National Infrastructure.
CNN	Cable News Network, U.S. television channel.
Code (computer code)	A set of instructions based on rules, written through various programming languages, used to control hardware and build or enable the software.
COIN	Counter Insurgency.
COTS	Commercial-off-the-shelf.
Connectivity (see also hyper-connectivity)	Devices and systems that connect computers and people.
Counterintelligence	The business of thwarting the efforts of foreign intelligence services/agencies. This includes information gathered and activities conducted to counter foreign intelligence threats. This includes but is not limited to spy-catching.
Counterterrorism	Preventing and deterring terrorist threats through all available and permissible means.
Covert action	A wide range of activities that rely on plausible deniability by the perpetrator. This can include an operation designed to affect foreign affairs including the use of lethal force.
Cozy Bear	(See APT29).
Crackers (see also hackers)	People who deliberately set out to compromise software or hardware.

Cryptocurrencies	Created in response to the financial crisis of 2008, these are digital currencies whose value is derived from scarcity through a limited supply. They are created by 'mining', a process of solving complex mathematical problems, in a system regulated by peer-to-peer systems of checks making it a secure way to conduct transactions. They are not physical but virtual and only exist electronically. The system of cryptocurrencies is built on being able to make verifiable transactions outside of the banking or financial services system which (in part) rest on physical holdings like gold but mainly rest on non-physical virtual currency whilst printed or minted notes and coins, whose value is derived from countries their economies and the assets they can trade, circulate. Most commonly associated with Bitcoin.
CSIRT	Computer Security Incident Response Team.
CSIS	Center for Strategic and International Studies (a Washington DC-based think tank).
CTI	Cyber threat intelligence.
CVE	Common Vulnerabilities and Exposures.
Cybercrime	Criminal acts committed primarily through the Internet and the use of technology.
Cyberespionage	Acts beneath the threshold of war defined under international law including 'the use of force'.
CYBERINT	Cyber intelligence.
CyCon	NATO's cybersecurity conferences held in Tallinn, Estonia and organized by their Center of Excellence (CCDCOE).
Cybersecurity	A term that encompasses computer or chip-based technology and its security.
Cyberspace	A generic term that has various meanings encompassing the Internet, intranets, and other computer-based networks and devices.

Cyberwar	A contested term. Throughout the book, this refers to acts defined under international law and the Tallinn Manuals as cyber capabilities capable of critically degrading a nation's capacity to respond to cyber acts and capable of causing physical destruction qualifying as 'armed attack'.
Cyberwarfare	Directly related to cyberwar. Acts of state-on-state war may stand alone or be used in conjunction with cyberespionage or the use of military force.
Cyber privateer	These are hackers who are frequently part of collectives that are used by states for their benefit. They can also be proxy forces.
'Dark Net' (see also TOR)	Part of the Internet that has access privileges and is inaccessible from search engines like Google and requires a TOR browser.
DARPA	(Defense Advanced Research Projects Agency) explores and invests in new technologies for the purposes of national security.
DCEO	Defensive Cyber Effects Operations.
DCO	Defensive cyber operations.
DCS	Distributed Control Systems.
DDoS	Distributed denial-of-service attacks to take websites offline.
Deterrence by denial	Is based on resistant or 'hardened' defenses capable of withstanding attack and capable of responding with a secure 'second-strike' capability.
Deterrence by entanglement	This is based on cooperation through shared national interests which encourage restraint.
Deterrence by punishment	The ability to hold at risk, and be able to strike at, that which an adversary values most or considers sine qua non—especially for the leadership.
DHS	The U.S. Department of Homeland Security founded after 9/11.
DIA	Defense Intelligence Agency (U.S.).

Disinformation	Information that is known to be false, purposely designed to influence and deceive the target.
DMZ	Demilitarized Zone. A term used in computer science/cybersecurity that segregates computers and devices connected to them. A term also applied to geographic locations separating two warring factions (as in the case of North and South Korea).
DNI	Director of National Intelligence (U.S.).
DNS (Domain Name System)	The Internet's addressing system for all devices connected to it.
DoD	Department of Defense (U.S.).
DODIN	Department of Defense Information Network (U.S.).
DOJ	Department of Justice (U.S.).
DoS	Denial of Service
DPRK	Democratic People's Republic of Korea (North Korea).
EASA	European Aviation Safety Agency.
ECA	European Cockpit Association.
ECCSA	European Centre for Cybersecurity in Aviation.
Echo chamber	A place (such as the Internet) where like-minded views are sought or found.
EEA	European Economic Area.
ELINT	Electronic intelligence. The process of electronic intercept and analysis of electronic intelligence.
EMS	Energy Management System/s.
Endpoint security	Securing user devices such as mobile/cellular, laptops, and desktop PCs, as well as servers in a data center referring especially to business/enterprise networks.
ENISA	European Network and Information Security Agency (EU). Now rebranded the European Union Agency for Cybersecurity.
EPCIP	European Programme for Critical Infrastructure Protection (EU).

Espionage	Espionage includes obtaining military, political, commercial, and other secret information through spying (often described as the world's second oldest profession after prostitution), infiltration, or monitoring devices. It is used for intelligence gathering and is illegal under domestic law but an accepted customary norm in international law.
ET	Emerging Technology/Technologies.
ETFMS	Enhanced Tactical Flow Management System (EU).
EU	European Union.
Europol/EUROPOL	The European Union's law enforcement agency.
EW	Electronic Warfare.
FAA	Federal Aviation Administration (U.S.).
Fake news	Mis- and disinformation which is knowingly or unwittingly sown or repeated.
False flag	A tactic used by one group to blame another.
Fancy Bear	(see APT 28).
FBI	Federal Bureau of Investigation. The U.S. domestic counterintelligence service and federal law enforcement agency.
FEMA	Federal Emergency Management Agency (U.S.).
File masquerading	Hiding malicious file types by looking like Microsoft Office documents or PDFs.
Firewall	Most commonly software that allows or disables external access. Can also be hardware.
Firmware	Embedded software in devices installed by the manufacturer.
Five Eyes	An intelligence-sharing partnership between the United State, the United Kingdom, Canada, Australia, and New Zealand facilitated by a series of treaties.
Fog of War (see also friction)	The unforeseen and unforeseeable nature of war that emerges in the conduct of war itself.
FPS	Federal Protective Service (U.S.).

Abbreviations and Concepts   xxiii

Friction	The impediments to battle plans incorporating anything that can go wrong and elements such as weather that create chance and uncertainty.
FSB (Federal'naya Sluzhba Bezopasnosti/ Federal Security Service)	Russia's civilian intelligence service tasked with domestic and foreign operations.
FTSE	Financial Times Stock Exchange
GAO	Government Accountability Office (U.S.).
GCHQ	Government Communications Headquarters (UK). The UK's primary foreign electronic intelligence gathering and analysis center.
GDPR	General Data Protection Regulation. 2016 EU legislation concerning data rights.
GitHub	An Internet-based open source repository for applications and source code.
Globalization	The free movement of goods and services that creates global interdependencies.
GPB	General Political Bureau (North Korea).
GPS	Global Positioning System.
Gray zone	A concept beneath the threshold of war that sees diplomacy, economic coercion, media manipulation, and cyberattacks (including misinformation and disinformation) against a target state. This can include the use of paramilitaries and proxy forces.
Gray hat (see also White hat and Black hat)	Gray hats fall somewhere in between ethical 'white hat' hackers and malicious 'black hat' hackers.
GRU	(Glavnoye Razvedovatel'noye Upravlenie). Russian military intelligence.
GSD	General Staff Department (North Korea).
Hacktivist/hacktivism	Hacking for a political or social cause.
Hard target	A protected site or person which is difficult to access.
HMI	Human Machine Interface. A terminal that allows a human operator to control, monitor, and collect data, which can also be used to program a system.

Honey pot	Used to lure attackers to a false target.
Honey trap	In espionage and intelligence gathering, the use of men or women in sexual situations to entrap, exploit intimidate, or for kompromat (compromising information). In cybersecurity, a term used to ensnare or entrap attackers.
HPC	High Performance Computing.
HUMINT (Human Intelligence)	Intelligence collected from human sources.
HVAC	Heating, Ventilation, Air Conditioning systems.
Hybrid warfare	Mostly associated with Russia (where it is known as *gibridnaya voina*) but with antecedents that stretch back to antiquity. Hybrid warfare can combine information, influence, legal disputes (lawfare), and economic operations, as well as the use of military and paramilitary force and increasing cyber operations.
Hyperconnectivity (see also connectivity)	A highly dependent variant of dependency on the interaction between computers and people.
IAAC	Information Assurance Advisory Council (UK).
IC (Intelligence Community)	Often used singularly or to describe the interactions between national intelligence agencies like the Central Intelligence Agency and National Security Agency. Intelligence communities relate to these as a broad international series of organizations and how national agencies like the CIA interact transnationally and internationally.
ICIJ	Investigative Consortium of Investigative Journalists.
ICS	Industrial Control Systems.
ICS-CERT	Industrial Control Systems Cyber Emergency Response Team.

Abbreviations and Concepts xxv

ICSPA	International Cyber Security Protection Alliance (UK-based).
ICT	Information Communications Technology. A global term to describe software and hardware related to these areas.
IDS	Intrusion Detection System.
IEEE	Institute of Electrical and Electronic Engineers.
Illegal	KGB/SVR operatives infiltrating a target country without the protection of diplomatic immunity, having assumed new identities.
Infiltration	The secret movement of an operative/remote operation/hacking into a target with the intent that the presence will go undetected.
INFOSEC	Information Security.
INL	Idaho National Laboratory.
Internet	A global decentralized and interconnected network of computer systems and information.
Internet Browser	Software that allows users to easily navigate the Internet, examples include Internet Explorer, Google Chrome, or Mozilla's Firefox.
IoT	Internet of Things (see also SMART). All devices that use Internet protocols. This includes everything from home computers and tablets to Internet-connected machinery, wearable technology, and monitoring or sensor-based devices such as electricity meters.
IoT	(Internet of Threats). Related to the Internet of Things, the Internet of Threats is a concern related to the security of devices that increasing connectivity increases risk.
Intranet	A local network that can be wired up or wirelessly connected.

IP (Intellectual Property)	This includes patents and creations where copyright on original designs developed ingeniously is held. These are frequently stolen and counterfeited. IP theft includes hardware and software.
IP (Internet protocols)	A series of distributed computer networking technology standards enabling connections between computers and devices.
IPS	Intrusion Prevention System.
IRGC	Islamic Revolutionary Guard Corps.
ISAC/s	Information Sharing and Analysis Centers (U.S.).
ISACA	Originally the Information Systems Audit and Control Association. It is an international information technology association for professionals founded in 1967 now known just as ISACA.
ISIS	Islamic State of Iraq and the Levant.
ISP	Internet Service Provider.
IT	Information Technology.
IW	Information Warfare (China).
JCPOA	Joint Comprehensive Plan of Action. An agreement with Iran to curb its nuclear program.
JCS	Joint Chiefs of Staff (U.S.)
KGB (Komitet Gosudarstvennoy Bezopasnosti)	The Soviet Union's all-powerful intelligence and security service during the Cold War.
Keylogger/s	Software that records keystrokes on physical keyboards.
KHNP	Korea Hydro & Nuclear Power (a South Korean company).
Kompromat	Russian term for compromising information or evidence.
KPA	Korean People's Army (North Korea).
Legacy systems	These are comprised of hardware and software that can be decades old when the Internet was in its infancy and cybersecurity was not a major concern.

LOAC	Law of Armed Conflict.
Machine Learning (see also AI)	Code/programs that enable machines to learn for themselves.
Malware (Malicious software)	Widely referred to as computer viruses.
Maskirovka	A Russian term for tactics of deception to mask, disguise, or camouflage.
MI6	The British foreign intelligence service; officially known as Secret Intelligence Service (SIS).
MFA	Multi-factor authentication.
Misinformation (see also Disinformation)	Information that is not known to be demonstrable false.
Mitigations	Ways to minimize or prevent.
MITM	Man-in-the-middle.
ML	(see Machine Learning).
Mole	An agent of one organization sent to penetrate an organization or intelligence agency by gaining employment; a term popularized by John Le Carre.
MPAF	Ministry of the People's Armed Forces (North Korea).
MSS	Ministry of State Security (China).
NATO	North Atlantic Treaty Organization.
NCCIC	National Cybersecurity and Communications Integration Center (U.S.).
NCS	National Cyber Strategy (U.S.).
NCSC	National Cyber Security Centre (UK).
NCSC	National Counterintelligence and Security Center (U.S.).
NCW	Network Centric Warfare.
NDS	U.S. National Defense Strategy (U.S.).
NERC	North American Electric Reliability Corporation.
NIDS	Network Intrusion Detection System.
NIS	The European Union's Network and Information Security Directive.
NIST	National Institute of Standards and Technology. NIST is one of the key organizations setting technical standards in and outside the U.S.

NPDO	National Passive Defense Organization (Iran).
NPPD	National Protection and Programs Directorate (U.S.).
NSA	National Security Agency. A branch of the U.S. Department of Defense responsible for ensuring the security of American communications and for breaking into the communications of other nations.
NSC	National Security Council (U.S.).
OCO/OCEO	Offensive cyber operations/ Offensive Cyber Effects Operations.
OECD	Organisation for Economic Co-operation and Development.
OPCW	Organisation for the Prohibition of Chemical Weapons.
Operational Technology (OT)	This is used to control physical processes and devices across the industry and throughout the critical infrastructure. OT includes Industrial Control Systems (ICS) and Supervisory Control and Data Acquisition (SCADA) systems.
OPM	Office of Personnel Management (U.S.).
OSINT	Open source intelligence. Intelligence information is derived from publicly available sources.
OT	(see Operational technology). Packet sniffing technical scanning for pathways into networks and devices such as routers which are used to transmit and receive data packets.
Patches	Fixes or updates to programs/applications and especially important against known threats.
Pattern	The behavior and daily routine of an individual that makes their identity unique.
Pattern of life	Part of Pattern but enhanced with Open Source Intelligence, Social Media Intelligence and electronic footprints to deduce psychological and social behavior patterns.

Abbreviations and Concepts   xxix

Phishing	Contacts by e-mail, telephone, or text message by someone posing as legitimate to obtain sensitive information. This can include directing an end user via e-mail to click on a link hosting malware designed to extract information.
Pivoting	The ability to move from one part of a network to another.
PLA	People's Liberation Army of the People's Republic of China.
PLCs	Programmable Logic Controllers. These are used in industry to control processes such as temperature, pressure, timings and so on.
PPD	Presidential Policy Directive (U.S.).
Pre-positioning	A strategy used to move to a position of best advantage and be able to maintain that position. This includes being able to control or direct an engagement from a winning or advantageous position.
Pretexting	Pretexting is an example of a 'confidence trick'. At a base level, it is simply someone pretending to be someone they are not or inventing a scenario designed to give access rights such as passwords or to give out confidential details.
PRISM	Planning Tool for Resource Integration, Synchronization, and Management. An NSA program disclosed by Edward Snowden in 2013.
Privilege escalation	Moving through a network to gather data by bypassing or hacking access controls, user permissions, or privileges that restrict access to certain users.
Proxy server (see also Server)	A server that acts as a middleman in the flow of Internet traffic, so that the Internet activities appear to come from another location (this could be in the same country or a different one or used to route through many hiding the original location).

Quantum computing	Using quantum mechanical phenomenon to process information and a generational leap from current computing which leverages the quantum mechanical phenomena of superposition and entanglement.
Quantum encryption/ quantum cryptography	Utilizing quantum mechanics to send messages that are technically unhackable.
R & D	Research and development.
Ransomware	Encrypting files to extort money from organizations and individuals.
RAT (Remote Access Trojan/Remote Access Tools)	Legitimate or malware tools to enable computer or systems access remotely.
RCE	Remote Code Execution.
RDP	Remote Desktop Protocol.
RFID	Radio Frequency Identification. A technology common to many modern swipe cards (including credit cards) and ID tags.
RGB	Reconnaissance General Bureau (North Korea).
RMA	Revolution in Military Affairs.
RoE	Rules of Engagement.
ROK	Republic of Korea (South Korea).
SAC	State Affairs Commission (North Korea).
SACCS	Strategic Automated Command and Control System (U.S. military).
Sandbox	A test environment with no external connections (such as to the Internet).
SCADA	Systems Control and Data Acquisition. SCADA systems enable the management and remote access to remote control devices, or field devices through a central system. They are the underlying control system of most critical infrastructures.
Screen capture	A method of 'photographing' an image on a computer screen/s at any given time mirroring the print screen function on keyboards.
Script kiddies	Young hackers often thought of as hoodie-wearing teenagers.

SE (Social engineering)	Exploiting human psychology and social norms for covert ends through cyber means and in face-to-face or spoken interactions.
Segmentation	The separation of systems and data to improve security.
Server	A computer used in a network that provides a service/s to a client computer/s (which are requesting data held by servers). Servers usually have more processing power, memory, and storage than clients.
SIEM	Security Information and Event Management.
SIGINT	Signals intelligence; consists of COMINT (communications intelligence) and ELINT (electronic intelligence).
Signature	A unique identifier used to ensure the authenticity of a user or request.
SIS	Secret Intelligence Service; the official name of Britain's MI6.
SMART (see also IoT)	Self-Monitoring Analysis and Reporting Technology.
SME/s	Small to Medium-Sized Enterprise/s.
SNA	Social network analysis.
SOCMINT	Social Media Intelligence.
SOC/s	Security Operations Centers.
Spear phishing (see also Social engineering)	A targeted and tailored cyberattack through e-mail or electronic communications (phish/scam) tailored towards a specific individual and/or organization with the aim of defrauding the target, extracting information for malicious purposes, installing malware and/or providing a gateway for deeper activity towards individuals and their associates or organizations. Social engineering is being increasingly used in spear phishing.
Spyware (see also Malware)	This is a form of malware that is designed to steal sensitive information or spy on a computer/computer user.

STRATCOM	U.S. Strategic Command based in Omaha, Nebraska.
SWIFT	Society for Worldwide Interbank Financial Telecommunication.
TCP	Transmission Control Protocol (see also IP) This transfers data (broken up as packets) through the Internet Protocol (IP), which routes communications. TCP/IP are used alongside one another to transfer data ensuring reliable and synchronized communication. 90% of data traffic uses TCP/IP, mostly over the Internet.
TOR (The Onion Router)	TOR is an encrypted technology that routes data through multiple layers and was originally developed by the U.S. Navy and Defense Advanced Research Projects Agency (DARPA) in the 1990s before being made open source in 2004 and is now run by a nonprofit organization. It is still funded by U.S. government agencies and offers anonymity to its users which 'hides' U.S. military users in a sea of other TOR users. These users include cybercriminals, state actors, and whistleblowers. Anyone can use a TOR browser and gain access to the 'Dark Net'.
Tradecraft	The methods developed by intelligence operatives to conduct their operations.
Trojan (see also Malware)	In cybersecurity, a Trojan horse or Trojan is a type of malware that is disguised as legitimate to infect a host.
TTP	Tactics, Techniques, and Procedures. Sometimes referred to as Tools, Techniques, and Procedures.
UN	United Nations.
UNCHR	United Nations Commission on Human Rights.
UNSC	United Nations Security Council.

URL (Uniform Resource Locator)	Colloquially termed a web address (e.g., google.com).
USAF	United States Air Force.
USB	Universal Serial Bus (a type of connector for hardware).
US-CERT	United States Cyber Emergency Response Team.
USCYBERCOM	United States Cyber Command is primarily run out of the NSA in Fort Meade but has capabilities across the U.S. military for offensive and defensive cyber operations.
VERIS	Vocabulary for Event Recording and Incident Sharing.
VPN	Virtual Private Networks are designed to be secure, encrypted tunnels for online data between systems. As in most other areas of encryption, VPNs add a layer of security but do not guarantee anonymity or complete protection.
Walk-in	A defector who declares his or her intentions by walking into an official installation and asking for political asylum or volunteering to work in-place.
Watering hole/Watering hole attack	These hijack legitimate websites to spread malware.
White Hat (see also Black hat and Gray hat)	White hats are ethical hackers who identify and flag security vulnerabilities.
Wikileaks	An Internet repository for leaked information founded by Julian Assange.
Wiper	A wiper is a form of malware designed to delete information from computer systems.
W.O.P.R.	War Operations Planned Response (a fictional computer system from the 1983 movie War Games).

World Wide Web (www)	Commonly thought of as coterminous with the Internet, the World Wide Web is a global information communication and retrieval system providing universal access to web pages containing text, graphics, video, and sound.
Worm (see Malware)	A worm is a malware that enters a computer through a vulnerability in the system and takes advantage of file-transport or information-transport features on the system, allowing it to travel unaided.
WPK	Workers' Party of Korea (North Korea).

List of Figures

Fig. 3.1	Offshore oil fields (circa 2010) (Author diagram with data from https://www.bbc.co.uk/news/10298342, Infield Systems, and Petroleum Economist)	164
Fig. 3.2	Gulf of Mexico: producing oil and gas fields (By US Energy Information Administration - http://www.eia.gov/pub/oil_gas/natural_gas/data_publications/crude_oil_natural_gas_reserves/current/pdf/gomwaterdepth.pdf, Public Domain, https://commons.wikimedia.org/w/index.php?curid=29929461)	165
Fig. 4.1	Top threat actor motive over time in breaches (*Source* 2021 Data Breach Investigations Report, http://verizon.com/dbir/, p. 12)	239
Fig. 4.2	Top threat actor varieties in breaches (n = 2,277)	240
Fig. 6.1	Simple SNA	360

1

Introduction

Probably the most vivid example of the arguments put forward in this book is fictional. It is the 2004 pilot of the reimagined television series *Battlestar Galactica*. The android Cylons, having rebelled against their programming and their human masters, install malware through seducing a trusted government insider which shuts down all their networked defenses except those on the oldest unnetworked battlestar, *Galactica*. This means when the Cylons attack, humanity's twelve colonies are defenseless. The underlying premise is realistic. A former Obama-era U.S. government cybersecurity official, drawing directly on *Battlestar Galactica*, worried this scenario would render "A trillion dollars of hardware…worthless if you can't get the first shot off".[1]

This book concentrates not on military networks but on civilian critical infrastructure/critical national infrastructure (CI/CNI) attacks.[2] They underpin a developed society. More than any other area, attacks on CI demonstrate the potential for cyberwarfare. CI encompasses a wide variety of sectors. This includes public utilities such as electric power generation and distribution, water supplies and water treatment, natural gas and oil production and pipelines, shipping and maritime traffic, hydroelectric dams, traffic lights and train switching systems. They are the backbones of every industrialized state.[3]

At the far end of the scale, and one of the most difficult to envisage outside of a third world war involving China and/or Russia and America, is a countrywide cyberattack on critical infrastructure. This has been postulated. An attack on the electric power sector would be the likely first target as electricity affects all other sectors. This could begin with cyberattacks on the generation of electricity at power plants, transmission from the power plants onto regional and national grids and could also be used against high-voltage transformers. This could produce cascading failures throughout the grid and long-lasting power outages. Modern societies like the United States have become reliant on electricity.

This would likely immediately impact water treatment and pumping stations and large dams, taking hydropower production offline, and produce a threat to life. Human life (indeed all organic life) needs water to survive. Transport would grind to a standstill as pumping stations run out of gasoline with oil and gas transport inoperative as fuel distribution also largely relies on electrical pumping and transportation. Effects would ripple outwards and cascade immediately as backup electric supplies begin to fail as batteries run out of charge and generators run out of fuel. In time, there would be no one to call for help. Telephone lines would likely be dead and the Internet inaccessible once their backup power systems fail as telecommunications runs on electricity.[4]

Until this point, public utilities would likely be flooded with calls and Internet queries and complaints. At home, batteries or fire will provide some heat and light but clean drinking water would not come from grid-connected taps as the pumping stations will have no power once the backup generators run out of fuel. Whilst you hope that someone is fixing the problems, many have stayed at home—unable or unwilling to leave for work. Effectively this would revert a modern developed society from the year 2022 back to 1900 when the first electric grids began to be installed. This plausibility is finely detailed in this book but, if you can remember, think of the last time you experienced a lengthy power cut to begin to comprehend the immediate impact this has…

Robert Knake reported to the Council on Foreign Relations in 2018 that "The U.S. power grid has long been considered a logical target for a major cyberattack. Besides the intrinsic importance of the power grid to a functioning U.S. society, all sixteen sectors of the U.S. economy deemed to make up the nation's critical infrastructure rely on electricity. Disabling or otherwise interfering with the power grid in a significant way could thus seriously harm the United States".[5] Knake added, "Rapid digitization combined with low levels of investment in cybersecurity and a weak regulatory regime suggest that the U.S. power system is as vulnerable—if not more vulnerable—to a cyberattack as systems in other parts of the world".[6]

A well-orchestrated series of attacks could significantly impact the functioning of the economy temporarily or for longer periods, it will likely create social harms and induce panic or crisis and will necessitate remedial actions—which incur economic costs across many affected industries.[7] How any nation copes, and how individuals respond, will depend on resiliency and preparedness. 'Cybergeddon' scenarios such as this are neither science fiction nor far-fetched, but they do lie at the far end of the scale of cyber threats.[8] A concerted attack of this nature has been described as achieving "rapid dominance" or "strategic paralysis" by simultaneously attacking the government, military, and people in parallel and greatly reducing the ability to counterattack.[9]

Cyberwar and Critical Infrastructure

A concerted attack on one sector of critical infrastructure alone, especially electric power generation and distribution, could well be life-threatening and produce mass casualty events. It is also likely to produce unanticipated and unpredictable cascade effects to other sectors.[10] This includes militaries. As a 2017 U.S. Department of Defense (DoD) taskforce concluded: "Due to the centrality of electrical power generation in supporting military strike capabilities, the cyber security and resilience of electrical power deserves particular attention".[11]

The possibilities and likelihood are open for debate, but we live in an increasingly interdependent world driven by technology and global communications embodied by the Internet which considerably aids the movement of people, goods, and services. These global tools and hyperconnectivity enable 'globalization' to happen and develop, and the Internet continues to expand into the 'Internet of things' (IoT) through the advent of a wide variety of sensor-based (SMART/ 'smart') devices.[12] Information Communications Technology (ICT) has brought far-reaching benefits to business and society whilst breeding high-dependency on technology, with the IoT also referred to as the 'Internet of Threats' can extend to social control and political manipulation.[13]

Whilst the Internet is the main vehicle for global connectivity it is one of many forms of connectivity that includes intranets in a wide variety of facilities and home networks. In industry this includes Operational Technology (OT). This sits alongside IT/ICT systems and networks. These "play a critical role in manufacturing, defense, emergency services, food and agriculture, financial systems, and critical infrastructure, just to name a few". Indeed, "many of today's OT systems are transited or tunneled over corporate networks, leverage common internet protocols, run on general-purpose hardware and mainstream operating systems, and are increasingly connected via wireless technologies".[14]

National and global dependencies on ICT, whilst not always evenly distributed within and between sectors and states, "will continue to spread as long as there are sufficient political, social or economic benefits to be gained from interconnection".[15] High dependency is accompanied by long-term security risks through exploitable vulnerabilities in hardware and software; the architectural spine and enabler of this business, governmental and social interactions. Individual and group interactions are also exploited. People are widely seen as the weakest link in the cybersecurity fence.

Around 80 percent of critical infrastructure is owned and operated by the civilian private sector in liberal democracies. These sectors are under constant cyberattack but there are no declarations of war. Instead, these and many other events, are grouped together under the umbrella term 'cyberwar'. What this means, what has been demonstrated, and what the potential of cyber is for conflict in our digitally interconnected world is the main focus of this book. Its concentration is on threats to the Western world, particularly the United States as the lead nation, but many of the threats it describes are common to all industrialized nations. Vulnerabilities are exploited on a minute-by-minute basis, but this does beg several questions:

- Why have attacks we have seen been limited in scope?
- Why haven't we seen widespread power blackouts or explosions caused by cyberattack?
- Could we see a 'cyber 9/11'?
- Is a cyber Pearl Harbor possible?

One of the most vexing questions is why was Russia unable to fully utilize cyberattacks against Ukraine before, during, and after it invaded in February 2022 to target CI? The answers to these questions are tackled and explored systematically. What cyberwar might look like is one of the most difficult.

This book highlights the interdependencies described above and provides a detailed examination of the threats and dangers facing the West through current and future attacks on hardware, software, and end users (people). It also outlines the threats posed by the two most potent adversaries Russia and China. Russia and China are unequivocally the two biggest threats facing the West. North Korea and Iran, represent lower order threats. North Korea and Iran each have a section devoted to them. China and Russia are the subjects of two separate books, *China and its embrace of offensive cyber espionage* and *Russia's cyber offensive against the West*. Each carefully examines their different strategies and goals and provides an in-depth analysis of each based on wide-ranging case studies. They complement this book on cyberwarfare.

In this book, through looking at underlying vulnerabilities and analyzing the two main threat actors, many common tactics, techniques, and procedures (TTPs) emerge alongside a series of mitigations. Identifying them will demonstrate the vulnerabilities as well as the resiliency and mitigation measures that exist. This is empirically driven but is also driven by the need to evolve the nascent debate on cyberwarfare and conceptions of 'cyber war'.

To evolve this debate, Chapter 2, 'On Cyberwar', draws on Carl von Clausewitz's classic text, *On War*, and international legal definitions of war to firmly make the case that vulnerabilities throughout CI make war through cyber means alone possible. This argument engages with debates initiated and sustained by NATO's Cooperative Cyber Defence Centre of Excellence (CCDCOE) based in Tallinn, Estonia. This argues and demonstrates that cyberwar is possible not only by "war on [or over] the Internet".[16] Instead, it is more in line with the thinking of Jon R. Lindsay, who sees "the pragmatic value of rules of engagement that distinguish reversible damage to code versus irreversible damage to equipment".[17] What is not debatable is that cyber defenders are under continuous bombardment from cyberattackers, and the most advanced and persistent are those in China and Russia.

The Threat Actors

The U.S. National Counterintelligence Strategy for 2020–2022 outlined that "Foreign intelligence actors—to include nation-states, organizations, and individuals—are employing innovative combinations of traditional spying, economic espionage, and supply chain and cyber operations to gain access to critical infrastructure and steal sensitive information, research, technology, and industrial secrets".[18] It also warned:

> **The number of actors targeting the United States is growing.** Russia and China operate globally, use all instruments of national power to target the United States, and have a broad range of sophisticated intelligence capabilities. Other state adversaries such as Cuba, Iran, and North Korea; non-state actors such as Lebanese Hizballah, ISIS, and al-Qa'ida; as well as, transnational criminal organizations and ideologically motivated entities such as hacktivists, leaktivists, and public disclosure organizations, also pose significant threats. Additionally, foreign nationals with no formal ties to foreign intelligence services steal sensitive data and intellectual property.[19]

It added that these foreign intelligence actors can exploit, disrupt, and damage U.S. and allied critical infrastructure and military operations and capabilities (especially during a crisis). These could place critical infrastructure at risk or diminish current advantages held by the U.S. and their allies across multiple domains. This includes satellites in space and command, control, and communications (C^3) over them.[20]

Probing attacks keep critical infrastructure at risk and latent capabilities could be activated in a crisis or used as leverage. It expressed genuine concerns which could undermine national security, economic wellbeing, public health, or safety. In particular, "adversaries seeking to cause societal disruption in the United States could attack the electrical grid causing a large-scale power outage that affects many aspects of daily life".[21] Moreover, supply chains that enable and support CI are also targets for foreign adversaries. From concept to design through to deployment and integration, supply chains are vulnerable to compromise. Including downstream patches and maintenance, malware can potentially be inserted in one or more of these stages to disrupt, damage, or provide future exploits into information technology networks, including military networks and communications systems. This can include IoT devices, weapons platforms, and the C^3 of assets, especially in a crisis.[22]

These and other cyberthreats are enduring despite growing awareness alongside growing numbers and development of cybersecurity products and services by companies that have become major players in the ICT industry. Cyberthreats are reflected in public awareness campaigns and European Union (EU) initiatives like the Budapest Convention on Cyber Crime which seeks to coordinate policing and cooperation within and beyond Europe,[23] and NATO's Tallinn process which analyzes how existing international law applies to cyberspace in two Tallinn Manuals.[24]

Fundamentally there are two categories of threat: external attackers and insider threats.[25] External or outsider threats can fall into several categories ranging from 'script kiddie' recreational hackers, crackers, and virus creators (now commonly seen as malicious software/malware 'authors') whose primary motivation is the challenge and interest provided by technology and coding, through to Advanced Persistent Threats (APTs) representing nation-states and their intelligence agencies. In 2011 the cybersecurity provider McAfee believed "that 120 countries had, or were developing, cyber espionage or cyber war capabilities".[26] In 2016, it was suspected that there were 29 states with offensive cyber capabilities; 16 of them declared.[27] The 2018 U.S. Intelligence World Threat

Assessment indicated that number has risen to over 30 and continues to rise.[28] This was the last assessment to offer a number.[29] As importantly, disaggregating the demonstrable potential for cyberwar and cyberespionage goals and gains are intimately linked. It is too much ground to cover in a single book and part of the reason why Chinese and Russian cyberespionage campaigns are tackled in two separate books.

In terms of 'cyber war', "Perhaps the most persistent concern…is the idea that it deepens asymmetries of power between strong states and weaker states, and between all states and some 'super-empowered' non-state actors…as David proved against Goliath, strength can be beaten".[30] These 'David and Goliath' arguments can be found in the writings of two of the grand masters of strategic thought, Sun Tzu and Clausewitz.[31] How asymmetrical state-on-state cyberattacks can and have been used against CI is the subject of Chapter 3, 'Cyberwar: Attacking Critical Infrastructure'.

'Script kiddies', commonly thought of as young, self-taught, hoodie-wearing teenagers have, as with criminal malware authors, support networks through online hacking and cracking communities and, although there are 'lone wolves', they often work in collectives. Hacktivists and hacktivist groups such as 'Anonymous', who conduct publicity-seeking attacks and campaigns against targets they deem fit, are mainly young and impressionable 'script kiddies' and mainly politically driven towards causes. Both can also harbor inquisitive, thrill-seeking, or hostile intentions. The fictional TV series, *Mr. Robot*, is a good technically informed illustration of this community.

Through online forums and one-to-one communications hackers and crackers routinely exchange information and knowledge through a self-organizing meritocracy founded on coding ability and other skills, competences, and proficiency, alongside forms of barter. This knowledge exchange includes vulnerabilities, techniques, targets, and stolen data. Through the hacker/cracker community, entry-mid level hackers gain further knowledge and skills from more proficient members through 'how to' guides, knowledge exchanges and coaching. This aids specialization in these areas.[32] A number go onto become cybercriminals to satisfy their curiosity and because there is 'easy money' to be made. There is also a vibrant hacker and cybersecurity conference scene, including the major DEFCON and Black Hat conferences.[33]

The telecommunications giant Verizon reported in 2018 that around three-quarters of data breaches were financially motivated and came from outside. Most cyberattacks upon organizations were "opportunistic" targeting "the unprepared".[34] Half of these are perpetrated by organized crime groups. Organized crime is a major problem. Although direct evidence is hard to come by, Robert Hannigan, the former Director of Britain's GCHQ warned in 2017, "Countries that mean us harm are co-operating with each other, sharing expertise, and using wider criminal groups. The overlap of crime and state actors is one of the most alarming developments of the past few years".[35] The U.S. cybersecurity firm, CrowdStrike, also reported "the blurring of lines between...highly skilled nation-state adversaries and their criminally motivated counterparts...".[36]

Meanwhile, most insider threats come from grudges, or for financial or personal gain but there are cases of whistleblowers. Insiders like these might have virtuous motives but even these can be exploited by criminals, organized crime gangs, or for corporate and state espionage. These spy for competitive advantage and state goals and are used by states for economic advantage, intelligence gathering, and national security goals. Verizon also records that over a quarter of attacks are from insiders and these are difficult to defend. Their 2019 Data Breach Investigations Report said categorically that "There's definitely a feeling in InfoSec [Information Security] that the attackers are outpacing us...[and] even the best security departments can only do so much".[37]

Terrorists, who so far have used the Internet to transmit their ideas and radicalize their followers, are an area of concern against targets that own and operate critical infrastructure. Terrorists can also have criminal links. Terrorists can mobilize directly or act by proxy by being or by hiring cyber 'guns for hire' recruited through the hidden 'Dark Net' (websites that are not indexed by search engines and have access privileges through encryption and peer moderation/access authorization).[38] They too are proving difficult to stop, especially if they are a 'lone wolf' attacker like Anders Breivik, the far-right Norwegian terrorist who killed 77 people in 2011, or the Islamic State-inspired, Sayfullo Saipov, who killed eight with a truck in New York in 2017.[39] Sadly, there are many more examples of 'lone wolves', terrorist cells, and coordinated activities.

Both Breivik and Saipov sought out 'echo chamber' sites on the Internet that supported their existing beliefs and helped radicalize them. This 'echo chamber' effect is also used to spread propaganda on the Internet. Milton Mueller, a professor at Georgia Tech, wrote in 2013 that:

> Any sentient user of the internet knows by now that creating a gigantic, globalized sphere for social interaction means that the activity there will exhibit all the classical problems of human society: bullying, fads and follies, propaganda, political domination, rumor-mongering, theft, fraud, and inter-group conflicts ranging from nationalism to racism. We have been accepting and responding to that reality for the past decade, instituting various technical, legal and behavioral controls on Internet use while seeking to preserve the freedom and openness that made the Internet such a valuable resource. That is why we are having dialogue about internet governance and security at the global, national and local levels.[40]

These are important debates; especially as national governments use many of the same technologies found in private industry and across critical infrastructure.[41] All attackers make use of generic threats such as worms (that burrow into networks), trojans (programs that mask themselves as trustworthy), and malware. These are largely spread through the Internet to infect systems but there are other means. These will be outlined in Chapter 4 'Gaining Access'.

They can especially affect systems using commercially available software and protocols, particularly those which have not been updated with patches that deal with known threats. Although this might not be the result of a deliberate attack, systems could become infected as a result of an Internet-facing connection or through outside infection imported onto a secure site. None of these threats and actors are going to go away in the near future. These are issues that cannot be fully insured against and there is always likely to be a residual risk.[42] No matter how high the barriers to entry, one of these residual risks are people. They are the invariably weakest link in the cybersecurity fence. This is the subject of Chapter 5 'Hacking the human'.

The Cyber Context: States as Targets and Attackers

The Federal Bureau of Investigation (FBI) currently ranks cyber-based attacks third on its list of priorities after counterterrorism and counterintelligence. In 2012 Robert S. Mueller, as Director of the FBI, anticipated that cyber threats would surpass terrorism in the coming years.[43] In the intervening years, cyber offense and defense are increasingly part of a counterterrorism and clearly evident in counterintelligence activities (a focal point of Mueller's investigation into Russian interference into U.S. elections, especially the 2016 presidential election).[44]

A wholly defensive approach to cybersecurity is unlikely to be anything other than reactive and places governments on the back foot. It is known, partly through the Snowden revelations which disclosed PRISM and associated programs run by the National Security Agency (NSA) with help from their partners in the United Kingdom, Canada, Australia, and New Zealand—collectively the 'Five Eyes', that both the NSA and GCHQ have offensive cyber capabilities against both state and non-state actors. What is less clear is where the balance lies between cyber defense and cyber offense especially when this includes intelligence gathering and espionage practices. This includes surveillance and reconnaissance by cyber means.[45]

These are areas in which conceptual arguments of deterrence add value.[46] This is not a straightforward calculation and is further exacerbated by the speed, range, sophistication, and diversity of attacks. This means that the range of passive and active cybersecurity measures not only has to contend with Stuxnet-type attacks at the highest (state-based) level but also lower level threats. Relatedly, is there a point at which a cyberattack will lead to a kinetic (military) response? Under current international law, cyberattacks would have to lead to violent "real-world" deeds before a kinetic (military) response could be contemplated.[47] As

Jaak Aaviksoo, then Estonian defense minister, asked after the 2007 Russian cyberattacks on Estonia:

> Do we have a proper legal code that defines the cyber attacks in detail—where does cyber crime stop and terrorism or war begin? Should NATO, for example, safeguard and defend not only its communications and information systems but also some national critical physical infrastructures? And what to make of collective defense in case of cyber war against one of the allies?[48]

Although this line of reasoning helped initiate the *Tallinn Manual* (discussed in detail below) this set of issues remains unresolved. They remain viable, difficult, and complex issues for NATO and national governments.[49] This is part of the reason a third iteration of the Tallinn Manual is being undertaken.[50]

Partly for these reasons, several former policymakers have called for an international treaty under the United Nations to mitigate or penalize cyberattacks by nation-states or individuals and groups within states.[51] Currently, preparations against actual and potential cyberattacks continue to be conducted in what remains a largely anarchic environment. Considerable resources are plowed into protecting state security assets, with advanced economies aiming to employ only 'assured' hardware and software which have undergone stringent security testing in their critical infrastructure. These high-value 'hard targets' are extremely difficult or impossible to access remotely and can also be subjected to physical surveillance and attempted infiltration.

This calculation would have formed part of the Natanz nuclear facility breach in Iran and the insertion of the U.S.-Israeli Stuxnet worm, first discovered in 2010, and the failed breach of the Organization for the Prohibition of Chemical Weapons (OPCW) by Russian intelligence officers in 2018. Both cases point to the importance of traditional spycraft in a world where intelligence gathering is increasingly dominated by cyberespionage. Cyberespionage is directly related to cyberwarfare. Although many believe cyberspace should not be used as a battleground, states remain both targets and attackers. McAfee and CSIS assessed in 2011 that:

Cyberspace today most resembles what Hobbes called a state of nature—a "war of every man against every man." Hobbes thought that only government and law could end that war. But in cyberspace, the role of government is more complicated. Globally, a majority of critical infrastructure is in the hands of private companies, which often operate in more than one country. For these companies, governments are partners; they are regulators and policemen; they are owners, contractors and customers; but they are also seen as aggressors, infiltrators and adversaries. Even when governments assume the role of defender, seeking to prevent attacks and improve security, many IT and security executives are skeptical about their ability to deter or protect against cyberattacks— although attitudes vary from country to country. One area where government is seen as having a generally positive impact is in regulation. Audit and enforcement rates and the impact of regulation on security vary widely from country to country, as do perceptions about its effectiveness. Many governments have sponsored cybersecurity cooperation among owners and operators of critical infrastructure, but widely varying levels of participation are reported.[52]

Because states are targets and attackers, this poses a conundrum for international law and is now a more visible issue due to the 2013 disclosures of the U.S. National Security Agency's PRISM program by Edward Snowden. The activities of Russian election interference have also raised awareness along with the stakes. Russia remains an active player and not only uses tactics honed during the Cold War but has expanded them. Russian campaigns and strategies against CI are part of a separate book as is China's use of cyberespionage, which is even more far-reaching and comprehensive than Russia's.

Cybercriminals and Their Usefulness as 'Proxies' and 'Privateers'

The usefulness of cybercriminals for states like Russia to hide behind is also evident.[53] The majority of organized crime gangs operate out of Eastern Europe, especially former Soviet states, many with Russian ties.[54] Alina Polyakova of the Atlantic Council sees this as operating in the gray zones. This means they are not necessarily directed centrally or controlled from the top. Nigel Inkster, a former Deputy Director of Britain's Secret Intelligence Service (SIS), sees instead a sort of symbiosis between Russia's organized criminal gangs and the Russian state. This *modus vivendi* means that so long as their activities are directed outside of Russia and (if instructed) their skills and connections are able to be leveraged on demand, they are free from significant disruption.[55]

Inkster is one of a number who now believes this effectively means these sub-state hackers have become state-protected 'privateers.'[56] Tim Maurer also makes a persuasive case that "states' proxy relationships range across a spectrum: from delegation to orchestration and sanctioning".[57] Delegation means proxies effectively operate under state control and a tight leash. Orchestration is characterized as "sharing strong ideational bonds with the state" which can involve funding or tools but where they are tethered on a looser leash. Sanctioning is the form of tacit or active approval or permitting proxy activities without penalties, and where governments like Russia do little or nothing to stop them so long as it believes it benefits state interests.[58] Cases include the 2021 Colonial Pipeline ransomware attack in the U.S.[59]

For Russia, as well as China, this 'privateer' status lends a useful cloak of plausible deniability.[60] This intermix is not restricted to Russia or China and one has to be careful of 'false flags' through misattribution. Similar methodologies can also be (or are simultaneously) used by industrial competitors and foreign intelligence services or simply hackers or hacktivists who draw on political or ideological rationales with no formal affiliation to a state.[61] A year after Russian cyberattacks on Estonia in 2007, Aaviksoo spoke of, "the Paris Declaration Respecting Maritime Law…to abolish privateering, which basically was seen as state-sponsored piracy" which "represented the first multilateral attempt to codify in times of peace rules, which were to be applicable in the event of war".[62] Then or now, a cyber Paris Declaration is not on the agenda. Instead, the situation has got worse.

The Threat Landscape

Threats can come from the logical layer of data, through physical equipment and at the social layer. CrowdStrike found in their 2018 analysis of "1 trillion security events a week across 176 countries", 48 percent of identified intrusions involved "adversaries with a nation-state nexus", with 19 percent "conducted by eCrime actors".[63] For the private sector, national governments, and their security services, these threats are growing faster than their ability to act.[64] State-backed cyberespionage APT campaigns pose the most significant threat to critical infrastructure.

These 'security events' include phishing e-mails that entice us to click on links or disclose information and distribute malware. They are often a first step to ID and password theft, leading to compromise and data breaches. Distributed Denial of Service (DDoS) attacks are also common, but they do not seek to breach systems, instead flooding website servers with so many requests websites become inaccessible. System intrusions and data breaches are much more serious. This includes websites and Internet browsers.[65]

It is also worth noting that "Most compromises took minutes, or less" and "Two-thirds went undiscovered for months or more".[66] These encompass APTs which can operate for months or years as part of an orchestrated strategy to advance foreign, defense, or economic goals. Mandiant was one of the first major cybersecurity firms to draw on frontline industry experience to suggest the methodologies of APTs. Its 2010 M-Trends report outlined seven steps APTs took: (1) Reconnaissance; (2) Initial intrusion into the network; (3) Establish a 'backdoor' into the network; (4) Obtain user credentials; (5) Install various utilities; (6) Privilege escalation/Lateral movement/Data Exfiltration; (7) Maintain persistence.[67]

Three years later Mandiant modified these seven steps to: (1) Initial compromise; (2) Establish foothold (through software allowing them to establish 'backdoors'); (3) Escalating privileges—most commonly through obtaining usernames and passwords or cracking passwords; (4) Internal reconnaissance—this collects information on the target environment, such as users, groups, affiliations, and responsibilities. Through internal reconnaissance data of interest can be detected by file extensions, keywords or by date modified, with file servers, e-mail servers or domain controllers the main targets. (5) Move laterally—this increases the chances for collecting, and being able to exploit, data, systems, and devices. (6) Maintain presence—this ensures continued remote access or control in a network using the above or by creating further backdoors. (7) Complete mission—exfiltrate stolen data from victim's network.[68] These methods are explored in more detail in Chapter 4, 'Gaining Access'.

These phases, especially the initial compromise, use Social Engineering (SE) and Open Source Intelligence (OSINT) gathering and exploitation to breach their target. This forms part of Chapter 5, 'Hacking the Human'. A multiplying myriad of known and unknown ('zero-day') software vulnerabilities also means that ideas of perimeter defenses through antivirus software and firewalls is not going to help by itself. Hardware widely used in CI and across businesses (including products used at home) often has passwords hardwired into factory settings (firmware) from when they were manufactured. These passwords, alongside instruction manuals and help guides are widely available on the Internet. There are many gateways and backdoors into systems and even viruses thought to be dormant can resurface and be repurposed.[69]

These issues were raised in a presentation by Kevin Mitnick, an industry celebrity who gained repute as the FBI's most wanted hacker and has since acted as an FBI (and Fortune 500) security consultant. Mitnick said "In my experience, everything has been hackable". He added, "Can they be defended? You can raise the bar extremely high and make it extremely difficult, but at the end of the day, everything that I've seen out there has been broken. It just depends on time, money and resources."[70] This 'hacker ethos' has been applied against individuals, private industry, governments, and military networks with the majority of today's defense communications reliant upon civilian-owned and operated networks which use commercial off-the shelf hardware and software.[71]

These threats arrive in several guises, varying in magnitude and tempo, and might be basic or sophisticated. They often emanate from groups that are well organized and based in hard-to-reach jurisdictions. 'Lone wolves', as well as organized groups, have stolen intellectual property, commercially sensitive data, accessed government and defense-related information, disrupted government and industry services, and exploited information security weaknesses through the targeting of partners, subsidiaries, and foreign and domestic supply chains.

Machine Learning, Artificial Intelligence, and High Performance (Quantum) Computing

At the same time Machine Learning (ML) and Artificial Intelligence (AI) are also impacting society. Both are growth areas with several branches of research and significant private industry and government funding. This includes artificial neural networks that try to reproduce human brain functions.[72] Early on ML was being designed in part to aid large and complex tasks and heuristics.[73] ML can now "accurately predict a wide variety of complex phenomena" and "ML models are capable of producing knowledge about domain relationships contained in data, often referred to as interpretations".[74]

AI, as a development of ML, aims to advance research from 'narrow AI' which tackles specific applications or domains such as Natural Language Processing (NLP) into learning and adaptation to new and untaught areas. It is in use in everyday life by millions for "planning, problem solving, speaking, understanding language, recognising objects (face, image, voice), or performing social or business transactions", designed to "learn human behaviour".[75] Amazon's Alexa, Google's Assistant, and Apple's Siri each use AI. Further evolutions of AI aim to pass the Turing Test, where a machine cannot be distinguished from a human being in a spoken language test.[76] Ultimately the goal is to achieve 'Singularity' where a machine surpasses human cognitive abilities.[77]

In an increasingly networked world of Big Data, fueled by the Internet, 4G and 5G cellular technology, and the IoT, ML and AI could significantly impact the potential to engage in cyberwarfare, especially when allied to quantum computing and quantum encryption.[78] Cyberwarfare exercises, like NATO's 'Locked Shields', "where the attack and defender try to anticipate the opponent's strategy with complete and incomplete information with learning", are unlikely to be able to defend against an ML/AI enabled adversary or produce potential defensive responses without ML/AI capabilities of their own.[79]

This future also points to increasing automation across current and emerging technologies (ET). This has already raised concerns over the reduction or removal of human politico-military decision-making across a widening range of systems and technologies. This has raised a series of ethical and legal questions with significant implications for the Law of Armed Conflict (LOAC), International Humanitarian Law (IHL), and military Rules of Engagement (RoE).[80]

In terms of cyberwarfare, the risks of diminishing or removing people from the decision-making loop are provided by another vivid example from popular culture. At the finale of the 1983 film *War Games* the automated War Operations Planned Response (W.O.P.R.) computer system that had replaced human decision-making narrowly averted launching the U.S. nuclear arsenal and starting World War III by self-learning that there was no way to win.[81] Whether such a zero-sum game can be applied to cyberwarfare is difficult to assess as it can be conceived in a variety of symmetric or asymmetric contexts between states. This could be coercive or threatening, an attempt to execute a 'Cyber Pearl Harbor', escalatory, decisively damaging, and/or an essential adjunct to military intervention. Would people behave differently to machines that are programmed to achieve desired ends or would an AI behave like W.O.P.R.? Some of this would depend on computing power and available data.

Computing power is "The fundamental key to AI is the learning processing, adaptation, and consumption of data".[82] Supercomputers are probably the most well-known example of High Performance Computing (HPC). These are purpose-built machines that have hitherto been used for dedicated tasks, including weather and climate change modeling; infectious disease modeling and vaccine development; quantum physics, astrophysics; and nuclear weapons modeling and analysis.[83] HPC can also involve sequences of computers operating in parallel which includes the use of remote cloud computing power and storage.[84] There is also an evolutionary pathway to HPC in the form of quantum computing. According to Intel, "Quantum computing offers the extraordinary potential to decrease the time needed to solve problems". More fundamentally, "Quantum computing reimagines the binary on–off encoding of data fundamental to present-day computing, and replaces bits with qubits that can simultaneously manifest multiple states. That could enable computing at unprecedented levels of massive parallelism".[85]

Quantum computing also poses a threat to present-day encryption. Although still in early-stage development, "Cryptography is one of the areas that is most seriously impacted by the potential of quantum information processing" with "the impact of quantum computers on cryptanalysis…tremendous".[86] Should the United States be beaten to quantum computing, one of the main concerns is that the winner will "gain unfettered access to any state secret of the losing government". However, even if this should happen it is argued there will be an ongoing need to maintain "the long-term security of secret information secured by contemporary cryptographic protections", without which, current protection "will fail against an attack by a future quantum computer…".[87]

This is not only a race between nation-states. Should there be a breakthrough, code breaking will likely be far quicker, the cryptography potentially unbreakable, and detection of assets that seek to remain hidden made easier. However, the move "from digital to quantum" comes with uncertainty and "we are unlikely to be able to predict exactly what those impacts will be". This said, "for government leaders in national security who face significant stakes for getting things wrong, doing nothing is not an option".[88] The same could be said of critical infrastructure.

Critical Infrastructure: ICS and SCADA

Although nations define critical infrastructure differently, they almost always encompass the following: energy; water; food; health; finance; government; transport; communications; and emergency services. The definition adopted by the United States specifically lists sixteen sectors as critical infrastructure sectors: chemicals; commercial facilities; communications; critical manufacturing; dams; the defense industrial base; emergency services; energy; financial services; food and agriculture; government facilities; healthcare and public health; information technology; nuclear power; transportation; and water.[89]

Disruption or damage to these sectors would have profound effects on developed societies like the United States that are used to electricity on demand, traffic systems that are safe, and water supplies that are uncontaminated and always available. Nations derive the hierarchy of importance of each based on perceived national priorities. Each is interconnected and vulnerable across the cyber security lifecycle from the time products enter the marketplace until they are replaced.

IT and ICT are familiar terms to many people but Operational Technology (OT) less so. OT is used to control physical processes and devices across industry and throughout critical infrastructure. OT includes Industrial Control Systems (ICS) and Supervisory Control and Data Acquisition (SCADA) systems. These are in widespread use around the world and across all major industries and all sectors of the economy. ICS and SCADA can be found:

> inside an automated manufacturing floor, outside a chemical processing plant managing valves and switches, on a rig in the middle of the ocean, or out in the arctic monitoring oil and gas pipelines. OT systems often perform simple yet essential tasks, such as monitoring a valve and shutting it off when a certain value is triggered. As a result, they can perform their tasks with little change for years. Which also means they sometimes run on aging operating systems and obsolete hardware using home grown applications. Since the goal for an OT system is to run exactly as designed, even patches are only applied if they do not hinder the process of the OT system.[90]

SCADA systems enable the management and access to "remote station control devices, or field devices. They are the underlying control system for most critical infrastructure including power, energy, water, transportation, telecommunication".[91] SCADA and the devices that make up Industrial Control Systems are detailed in Chapter 3.

Prior to the rise of the Internet, "Remote locations and proprietary industrial networks used to give SCADA system[s] a considerable degree of protection through isolation".[92] Now the Internet can be used to locate and target "public control interfaces for heating systems, geo-thermal energy plants, building control systems and manufacturing plants…treatment systems, power plants and traffic control systems".[93] An official U.S. government report stated that in 2014 the U.S. Cyber Emergency Response Team (US-CERT) "was aware of 82,000 cases of industrial hardware or software directly accessible from the public Internet".[94] Hitherto:

> These systems have traditionally been kept separate from IT networks…because OT systems are often tasked with monitoring and managing the highly sensitive processes associated with critical infrastructure. The other is that these systems can be notoriously delicate. Something as benign as an active system scan can cause these devices to fail. And any failure or compromise can have serious if not catastrophic results. However, new requirements being driven by digital transformation, such as smart environmental control systems, just in time manufacturing, and interactive systems tied to Big Data have begun to change all of that. In addition, companies are looking for productivity improvements and cost savings…that requires real time online data.[95]

This OT ecosystem is now beginning to converge with Information Technology (IT) networks made up of switches, routers, wireless access points, and the use of 4G (and 5G) communications standards.

This means that traditional ideas of hermetically sealed Local Area Networks (LAN), including home networks, as well as Wide Area Networks (WAN) are no longer walled off and are increasingly Internet facing. LAN is frequently used in homes and small-to-medium sized enterprises (SMEs) through routers and modems and the use of ethernet cables and Wi-Fi. LAN and WAN are replete in businesses as the adoption of computer technology increased due to the commercial benefits. This has meant increasing computing connectivity within buildings and between locations in the same organization. This is known as an Enterprise Network.

Business needs, including the requirements for increased connectivity and the ability to connect remotely, coupled with the uses of the Internet itself, are leading to a lack of separation between IT/Enterprise networks and the OT environment. Mo Cashman of McAfee believes that "Information technology (IT) and operational technology (OT) convergence is increasing the cyber risk. Digital transformation projects are connecting historically isolated production systems with enterprise IT infrastructure".[96]

In a 2016 government-commissioned study into cyber vulnerabilities in the U.S. electric sector, Idaho National Laboratory (INL) saw that an attack resulting from this convergence could "cause an operator loss of view or control, and/or directly manipulate operational performance to affect dire consequences, including physical damage".[97] IHL also found that "these unpredictable elements interact in ways so complex they can never be fully comprehended by us, let alone fully accounted for or protected". This means "attackers may manipulate digital components to cause unintended physical consequences to real equipment, such as falsifying sensor signals, causing temperature shifts to destroy electronics, over- or under-pressurizing valves, among many other possibilities in a complex control system".[98] Details on Internet-connected devices are searchable on specialized Internet search engines like Shodan.[99] Many ICT specialists are already well aware of the types of attack vectors search engines such as Shodan can generate. YouTube is also an extremely useful repository for hackers and crackers many of whom are self-taught with online hacker groups, the most infamous being 'Anonymous', growing in number and sophistication. These groups also populate and frequent the hidden 'Dark Net'.

Shodan is further enabled by the emerging IoT which "encompasses everything connected to the Internet, but it is increasingly being used to define objects that ""talk" to each other" from sensors, wearable technologies such as fit bits and watches, to an increasing number of smart devices.[100] In 2017 there were 13 billion Internet-connected devices. By 2023 there will be around 30 billion, all of which contribute to Big Data.[101] The IoT represents "one aspect of a Future Internet", for which "many application areas have been postulated—not only in industrial domains such as manufacturing, logistics, retail, service management, energy, public security, and insurance, but also for the life of every citizen – where IoT can bring significant improvements, leading even to new business models."[102] However, a significant proportion of IoT devices are weakly secured, and some cannot even accept software updates. This means that IoT vulnerabilities are apparent and around 30 percent of these effect SCADA and IoT systems.[103] This is why the IoT has also been dubbed the 'Internet of Threats'.

The potential threat level increases with developments in ML, AI, and quantum computing. Big Data and data mining can be leveraged by these technologies in ways politico-military decision-makers would struggle with because of heuristic constraints. Vast amounts of unstructured and seemingly disconnected data, allied to OSINT and that gathered by APTs through cyberespionage "spanning distributed databases of critical nature present in several public, private clouds, and government owned cyber infrastructures" make CI "highly vulnerable in the cyber ecosystem and become the weakest link in the entire chain which can compromise security of the entire system".[104]

The growing numbers of threats and threat actors in what is an ever-expanding information ecosystem mean individuals and organizations around the globe are vulnerable despite heavy investment in ICT security. Private industry is a highly profitable target in an already target-rich environment. Here the security barriers vary widely, partly depending on the size of the business, but even the largest multinationals, including global ICT multinationals like Apple, have been breached.[105]

This cybersecurity nexus means that even the largest multinationals cannot be complacent regarding cyber dependency and vulnerability and the risks of reputational damage, ransomware attacks, or physical harm resulting from cyberattacks.[106] Due to the potential for economic and reputational damage there have been cases where researchers/bug hunters have faced prosecution in Europe and the United States for disclosing vulnerabilities. The current legal framework, combined with differing company perspectives on disclosing vulnerabilities, can have "a chilling effect on security research, with several experts reporting holding back discoveries due to concerns over legal action".[107] At the same time, as Chapter 4 explains, "'bug bounties' are also offered to 'White Hat' ethical hackers by a number of companies.[108]

Insecurity within the cyber ecosystem also can directly affect governments with major and sustained implications for the international system itself. A brutal appraisal of this present and future reality was provided by Chris Demchak of the U.S. Naval War College in her testimony to the U.S. government in 2017:

> Cyberspace is widely misunderstood and has been from its outset. It is now a deeply intertwined 'substrate' connecting all the critical components of every nation's domestic 'socio-technical-economic system' (STES), built with fault-tolerant programming and insecure hardware routinely sent too quickly to market with overblown promises of fast returns. Three interrelated cognitive blinders in western approaches to the spread of cyberspace hindered accurate assessments of the emerging reality. These were unrealistic optimism in early utopian cyber visions blended with security-blind IT capital goods business models, and endured far longer than reason would suggest due to deeply institutionalized Western societies' hubris about the permanency and moral superiority of their Cold War legacy control of the international system despite the overwhelming demographic and eventual economic scale of the rest of the non-western world. The 'winners' of the Cold War ignored the reality of their cultural uniqueness. The result was insufficient security concerns for the national wealth in their own IT capital goods manufacturing, and of the possibility that the international system they created could be contested and bested by the scale of dedicated, rising adversaries.[109]

Others have also recognized that from the 1990s into the first half of the 2000s "the cyber security focus was on criminal and espionage attacks (if recognised as a national security issue at all)" and then "a surge in politically motivated cyber attacks". In this period cyber criminality and espionage not only increased but metastasized. Now, alongside financially motivated cybercrime are politically motivated attacks "from high-profile strikes against critical infrastructure, to millions of pinprick attacks that can weaken the state over a long period of time". This "shift in targets requires everyone to reassess their risks and security requirements…[which] can cover the entire spectrum of cyber attack".[110]

Politically motivated cyberattacks continue to evolve, increase, and morph to include mis- and disinformation campaigns ('fake news') to also be part of these million pinprick attacks against a target. Attacks against power grids and other parts of critical infrastructure have demonstrated that no targets, including civilian infrastructure like hospitals, are out of bounds. High-profile attacks upon critical infrastructure have occurred, most notably in Ukraine, but as yet we have not seen a 'cyber Pearl Harbor' or 'cyber 9/11' type attack.[111] Governments and commentators might see this possibility as unrealistic but the 9/11 Commission Report recorded that part of the reason 9/11 happened was due to a "'failure of imagination" and a mind-set that dismissed possibilities".[112] Understanding the potential that exists is one way of guarding against threats to critical infrastructure. In 2017 the U.S. Defense Department said:

> Conducting detailed advance planning for responses to every plausible cyber attack, with every potential adversary in every conceivable scenario, is neither possible nor necessary. Nor is it feasible to have in hand the "optimal" response to each hypothetical attack scenario. However, it is both possible and essential to conduct systematic planning and wargaming, to establish clear employment and declaratory policies, and to establish priorities for the development of a range of potential cyber and non-cyber (and military and non-military) responses to cyber attacks.[113]

What we know about the strategies and activities of our adversaries is complicated by a range of non- and sub-state actors in a fast-moving and fast-changing cyber environment which reflects national development and international relations. Moreover, the complexity of attacks has seen exponential growth through improving capabilities, practice, and the reuse of malware and tactics, techniques, and procedures (TTPs). The barriers to entry continue to fall and "What was considered a sophisticated cyber attack only a year ago might now be incorporated into a downloadable and easy to deploy internet application, requiring little or no expertise to use".[114] Our increasing dependency on ICT, and its architectural underpinnings, has increased our vulnerability to cyberattack. At the far end of this security spectrum there lies the potential for cyberwar through systematic attacks on critical infrastructure.[115]

Subdue the Enemy Without Fighting

Sun Tzu, in perhaps his most famous dictum said, "The supreme art of war is to subdue the enemy without fighting".[116] To attempt to accomplish this, one sector of critical infrastructure stands out: attacking national or regional electrical supplies, as happened in Ukraine in 2015 and 2016 resulting in temporary blackouts.[117] Most essential services have only a very limited backup of alternative electricity supplies in the form of generators and fuel. Other parts of the energy sector also rely on mains electricity supplied regionally onto national grids. This also applies to water which needs to be sanitized and pumped. Transport requires electricity for fuel pumps, pipelines, and air, rail, road, and maritime traffic. Hospitals, and supporting healthcare services, rely on electricity and computerization to function in the developed world. Large-scale heating too, including that of fossil fuels, is mostly dependent on electricity.

The world of finance is also dependent on power and communications and an extended power cut to Wall Street, the Tokyo Stock Exchange, or the Square Mile in London would likely have national and global implications. Normal communications could be disrupted or cease entirely for a period of time without electrical power.[118] Although redundancy measures are in place, even a medium term and regional cyberattack on the electrical grid would place a strain on the local and national government and hamper or prevent emergency services from responding and coordinating, including police and the military. Militaries will be degraded by well-orchestrated cyberattacks on CI by affecting their ability to coordinate and deploy. Agriculture too in developed states has seen an increased reliance on mechanization and the distribution of food requires transportation and coordination.

There is a degree of resilience to shock, including preparedness for and responses to natural disasters such as hurricanes, tornadoes, earthquakes, tsunamis, flooding, and disease control but this varies widely.[119] However, akin to a natural disaster, a cyberattack on the electric grid alone could come with no warning and leave many, including policymakers, unprepared and unsure, and (at worst) unable to respond.[120] A cyber first strike of this nature could be followed up by further cyberattacks and might be accompanied by denials or obscuration of the perpetrator/s. These dangers are recognized by governments around the world, including by the United States.[121]

To mount such an attack would be highly likely to require a major state actor. Russia has a demonstrated potential to attack the West's critical infrastructure, but it is likely their capacity and capability do not extend beyond a local or regional scale. Russia has used cyberattacks against a number of its former Soviet-era satellite states and a concerted attack of this nature could weaken or even cripple unprepared nations in the immediate aftermath. China by weight of sheer numbers, combined with a highly advanced and capable technology base, has the human resources and capabilities to conduct a much wider, possibly countrywide, cyberattack. These capabilities and capacities are backed by theoretically informed stratagems and technical studies.

Still, a peer-to-peer attack from comparable nations, perhaps one which is more cyber-enabled but less cyber dependent than the other, is problematic to assess. A direct conflict involving Russia and America, or a Sino-U.S. conflict would fit this category. This is but one example of a 'cyber Pearl Harbor' moment. Whether this is a forerunner of a conflict to follow will help determine the responses and the capability to recover and respond. The 2019 U.S. Worldwide Threat Assessment judged that:

> *China and Russia pose the greatest espionage and cyber attack threats, but we anticipate that all our adversaries and strategic competitors will increasingly build and integrate cyber espionage, attack, and influence capabilities*...In the last decade, our adversaries and strategic competitors have developed and experimented with a growing capability to shape and alter the information and systems on which we rely. For years, they have conducted cyber espionage to collect intelligence

and targeted our critical infrastructure to hold it at risk. They are now becoming more adept at using social media to alter how we think, behave, and decide. As we connect and integrate billions of new digital devices into our lives and business processes, adversaries and strategic competitors almost certainly will gain greater insight into and access to our protected information.[122]

To this was added: "***China presents a persistent cyber espionage threat and a growing attack threat to our core military and critical infrastructure systems***...China has the ability to launch cyber attacks that cause localized, temporary disruptive effects on critical infrastructure—such as disruption of a natural gas pipeline for days to weeks—in the United States..."[123] Russia:

> continues to be a highly capable and effective adversary, integrating cyber espionage, attack, and influence operations to achieve its political and military objectives. Moscow is now staging cyber attack assets to allow it to disrupt or damage US civilian and military infrastructure during a crisis and poses a significant cyber influence threat...Russia has the ability to execute cyber attacks in the United States that generate localized, temporary disruptive effects on critical infrastructure—such as disrupting an electrical distribution network for at least a few hours—similar to those demonstrated in Ukraine in 2015 and 2016. Moscow is mapping our critical infrastructure with the long-term goal of being able to cause substantial damage.[124]

The 2022 Annual Threat Assessment of the U.S. Intelligence Community recognized China as a "near-peer competitor", and Russia, as an increasing threat. China's ability to conduct cyberattacks against CI was again restated, including rail alongside oil and gas pipelines. Russia too was focusing on CI, including ICS, as well as threatening the undersea cables which provide Internet and other communications access. It also continues to use cyber influence operations for politico-strategic effects and to advance its aims.[125] These threats also extend *to* Russia and China as well as from them.[126]

One of the challenges in risk assessing and evaluating cyberattacks is that they emanate from a wide number of sources. They could be from a state; a state-based actor; a state-sponsored actor; a private sector company engaged in espionage; a group or individuals involved in cybercrime, or a state sponsoring for their own ends (including for terrorist purposes). Their reasons might range from the curious to malicious and dangerous with cases where it is problematic (often deeply problematic from a legal standpoint) to identify the perpetrator. This is the so-called 'attribution problem' and it leads to difficulties when it is a 'nation-state like attack' as it is known in the cybersecurity community.[127] Another challenge is that cyberespionage by states is permissible under international law, even when wrongful under national laws.[128] This permissibility has led to violations largely without penalties and rather than persuading states to disengage from these activities has instead encouraged wider use. This state of affairs is revisited in the conclusion.

Cyber: The Fifth Domain of Warfare

Cyberattacks can be a force multiplier for weaker nations against stronger, more developed nations and help offset hard military capabilities and the economic and industrial capacity that underpin them. This also encompasses highly protected military systems and assets, many of which are profoundly dependent on ICT and increasingly rely on global positioning systems to perform to their best.[129] This means that although "Cyber in its most general sense is heralded as a force-multiplier in the arsenal of both State and non-State actors" it can also be a force inhibitor.[130] As John Arquilla, the originator of the concept of cyberwar,[131] has argued, "this new way of war-possibly quite potent on the battlefield…[is] also able to strike at others' homelands without the need to defeat their military forces first."[132] This is not lost on Western military planners and political decision-makers.

At NATO's meeting in Warsaw in July 2016, the United States, together with their allies in NATO, declared offensive cyber operations at the operational/tactical and strategic levels part of military planning and cyber the fifth domain of warfare.[133] As James A. Lewis highlights, "cyber operations are increasingly embedded into military operations…Offensive cyber capabilities will shape the battlefields of the future."[134] They are also shaping present-day conflicts and concepts with new forms of hybrid conflicts and warfare that utilize multiple attack surfaces beyond hard military-backed power and the conventional use of force.[135]

However, military forces and the power projection they provide depend on the civilian infrastructure they are tasked to defend. This includes the information infrastructure embodied by the Internet, which can house any Internet-facing military networks. Compromise of these systems and networks has the capacity to degrade, disrupt, or even cripple military interventions and the command and control on which they depend before, during, or after deployment. In addition, the intelligence and information on which they depend can also be compromised from the outside through external hacking or from trusted insiders run by foreign intelligence agencies. As Lewis argues, this can "disrupt data and services, sow confusion, damage networks and computers (including software and computers embedded in weapons systems) [and] machinery". He adds, "Offensive cyber operations would strike military, government and perhaps civilian targets such as critical infrastructure in the opponents homeland used to support war efforts."[136]

Even before this declaration cyberspace has become a domain next to land, air, and sea, in which nation-states attempt to assert power over each other. As a response, national and international organizations are setting up defensive mechanisms, commands, and renewing doctrine. The United States Department of Defense declared cyberspace an operational domain in 2011 "for purposes of organizing, training, and equipping U.S. military forces."[137] NATO too sees cyberspace "as a domain of operations in which NATO must defend itself as effectively as it does in the air, on land, and at sea" whilst reaffirming their mandate is defensive not offensive.[138] This said, the recognition of the level and types of threats emanating from cyberspace to national defense, security, social and economic structures varies across nations and agencies. For the West and their regional friends and allies, it is imperative that national governments bolster their ability to account for, monitor, and respond to these cyber threats. In the meantime, state attackers often benefit from playing in the gray zone below the threshold of the use of force or armed attack. This too is examined in Chapter 2 and in the conclusion.

A Short Guide to Terminology

Prior to entering the complex world of cybersecurity and detailed threats to critical infrastructure some explanation of the terminology might help guide readers unfamiliar with this world. Although there are attempts to standardize the language used, like the Vocabulary for Event Recording and Incident Sharing (VERIS) system of classification and taxonomy, terminology frequently gets conflated and a variety of names applied to the same actor or acts.[139] Malware includes viruses, worms, trojans, and automated bots. Malicious bots mainly generate and spread spam or dis/mis and malinformation. Malware is a graver concern in terms of cybersecurity.

Malware

In a compressive assessment of the wide range of cybersecurity terminology, Cisco Systems notes that viruses, worms, and trojans are designed to gain access to, disrupt, steal, or damage systems, hosts (including servers, clients, and networks), and data. Some of these can self-propagate or bundle with other software or affixed to software through macroinstructions (macros) which enable or request functionality. Other methods of infection include exploiting vulnerabilities (especially known vulnerabilities) in software, including operating systems and network devices through design errors (bugs) in legitimate code.[140] Internet browsers too can be infected, and websites can contain malware and even legitimate sites can become unwitting hosts; a method known as a man-in-the-middle (MITM) attack.

Most are spread by end users. This can include something as innocuous as clicking an e-mail attachment or downloading files from the Internet. Adware and spyware are common and increasingly this type of attack leads to ransomware or the unknowing installation of backdoors into systems and networks. Unless quarantined, the effects can vary from mild irritation (such as pop-up ads) to the theft of confidential information, the encryption, destruction or wiping of data, to major compromises or disabling systems and networks.[141] Two types are prevalent, especially through phishing e-mails. These are 'droppers' and 'malware downloaders'. Both do similar jobs, but 'dropper's' have malware pre-packaged within them. Malware downloaders need an Internet connection.[142]

Viruses are passed from one computer to another. They spread by copying themselves into other programs and into systems. Worms are similar to viruses and similarly replicate themselves. The damage they can inflict is also similar. However, worms can operate as standalone software and do not need a host program or people to propagate. Instead, they spread by exploiting known or newly discovered vulnerabilities or bugs. Worms also often infected by first Social Engineering the target using 'confidence tricks' which exploit human behavior. They can travel unaided by taking advantage of file-transport or information-transport features. More sophisticated worms also employ encryption, wipers (wiperware), and ransomware in their designs.[143]

Encryption is used to make files unusable unless a ransom is paid in the form of an electronic cryptocurrency like Bitcoin or Monero. These and other cryptocurrencies are secured through Blockchain, an Internet-enabled electronic ledger, that securely stores and records every transaction between people without third-party involvement such as a bank or payment system like Visa or SWIFT.[144] Ransomware encrypts files to extort money from organizations and individuals whilst wipers will delete information from computer systems and make it almost impossible to recover. Another common group of malware are trojans.

Named eponymously from Greek myth, trojans are malicious software which masquerade as legitimate. Trojans trick users to execute them. When executed trojans can perform a wide variety of attacks. These again can vary from irritation (such as pop-up windows or changing a desktop background) to deleting files, data theft, or by activating and spreading other malware, including viruses. They can create backdoors into systems and through this into organizations. In contrast to viruses and worms, trojans do not self-replicate or infect other files. Instead, they disseminate by targeting end users (people) through e-mail attachments or Internet files.[145]

Infections can also be accomplished by requests to run macros, used widely in Microsoft Office products and propagation of malware can be accomplished through botnets. Bots (from the word robot) are automated programs spread via the Internet. Bots collect information that can be used for both positive and malicious intent. For example, bots are used by search engines like Google to index websites. They are also used by cybercriminals to "infect a host and connect back to a central server or servers that act as a command and control (C&C) center for an entire network of compromised devices, or "botnet." With a botnet, attackers can launch broad-based, "remote-control," flood-type attacks against their target(s)".[146] Botnets are also used to spread mis and disinformation and have even been used to rig a children's talent show on Russian TV.[147] They are also widely known for Distributed Denial of Service (DDoS) attacks which can take websites offline.

Malware can also be used to log keyboard strokes. Known as a keylogger this can harvest passwords and personal and financial information, and open backdoors on an infected host. They can also be used to send spam (junk) e-mails and "hackers take pride in offering a complete attack with just a few mouse clicks".[148] Like all forms of malware, the infection can be by downloading from the Internet, by e-mail, and through physical media like USB drives. They "are truly a low-effort attack that knows no boundaries and brings attackers either direct revenue through financial account compromise or infrastructure to work from".[149]

Cyber Forensics

There have also been considerable advancements in the field of cyber forensics. This is being used alongside socio-technical research to better understand and determine state-based and cybercriminal attackers and aid the 'attribution problem'. Multinational research on cyber forensics alongside that conducted by the cybersecurity industry/cybersecurity providers has helped them, in conjunction with law enforcement and government agencies, to identify attackers with a higher degree of certainty than was previously possible.

The expanding profusion of Open Source Intelligence (OSINT) aids these efforts and clandestine intelligence gathering, leveraging area and subject matter experts, can be brought to bear by governments. Additionally, common or newly discovered TTPs used by cyber attackers add to the body of evidence that help expose the motives and actors behind intrusions, breaches, and cyberespionage or cybercrime campaigns. Indicators include malware types, codebase, and level of capabilities, identification of a common language or alphabet, the times of day attackers are active, and their selection of targets.[150] This has bred confidence in publicly naming those considered responsible with the Federal Bureau of Investigation's 'Cyber Most Wanted' page the most visible example.[151] This is being employed against cybercriminals, as well as cyberespionage actors and acts—the precursors to cyberwarfare—with cybercriminals also in play.

Overview

The potential of cyberwarfare and the debates surrounding it are the subject of Chapter 2, 'On Cyberwar: Theorizing cyberwarfare through attacks on critical infrastructure—Reality, Potential and Debates'. Moving beyond theory, real-world cases are examined in Chapter 3, 'Cyberwar: Attacking Critical Infrastructure'. This includes an examination of Russia's invasion of Ukraine which began in February 2022. Technical barriers exist that limit access privileges to systems and data but Chapter 4, 'Gaining Access: Attack and defense methods and legacy systems' will show how and why the protective barriers that are currently in place are permeable. In addition, it will detail underlying vulnerabilities by analyzing legacy systems that were put in place prior to the rise of the Internet when cybersecurity was barely considered and now require bolt-on solutions and a future that prioritizes 'security-by-design'.

With people widely considered the weakest link in the cybersecurity fence, Chapter 5, 'Hacking the Human' will look at attacks at the social layer of the cybersecurity strata. It concentrates on human behavioral factors behind cybersecurity breaches in the areas that make up critical infrastructure both from the outside and by current or former insiders from within organizations. Finally, Chapter 6 'Non and sub-state actors: Cybercrime, terrorism and hackers', looks at each in turn, where and how they collaborate, and what and how they target—particularly using social engineering as a gateway and means of exploitation. It goes on to examine social networks and their use of encryption to cover their tracks and whether this is impeding police, law enforcement, and intelligence agencies. It then moves on to discuss linkages between states hostile to the West (and their friends and allies in the wider world) before analyzing 'Dark Net' markets, organized crime and ransomware, and the WannaCry and Petya/NotPetya cases which demonstrate the linkages between cybercrime and states and the use of 'false flags' which helps mask attribution. It finally discusses China's use and cooperative agreements with Western universities which are being used to enhance Chinese development and how these collaborations are being used to boost their sub-state private sector and national goals.

The conclusion will digest the findings of the book and question what can be done to reduce risk and the possibility of interstate cyberwarfare. It draws from experience, the case studies discussed throughout the book, as well as the discourse and conceptual debates that influence decision-making in business and government. It will seek to synthesize one of the main themes of this book, that increased cybersecurity and the avoidance of cyberconflict and cyberwar are best achieved by increased cooperation and adherence to the Rule of Law to deepen the benefits of ICT and reduce the threats and hazards within critical infrastructure.

Notes

1. Emily Dreyfuss, 'US Weapons Systems Are Easy Cyberattack Targets, New Report Finds A new report says the Department of Defense "likely has an entire generation of systems that were designed and built without adequately considering cybersecurity."' (October 10 2018), https://www.wired.com/story/us-weapons-systems-easy-cyberattack-targets/, accessed 5 April 2020.
2. Both terms are widely used and refer to the same thing.
3. Pierluigi Paganini, 'Improving SCADA System Security' (December 6 2013), http://resources.infosecinstitute.com/improving-scada-system-security/, accessed 26 July 2015.

4. This is because although telephone exchanges usually operate an uninterrupted power supply (UPS) policy which includes "(arrays of lead-acid batteries) for backup and also a socket for connecting a generator during extended periods of outage" but once battery and generator fuel is exhausted their systems will fail. They too are dependent on electricity from regional and national grids. Abdullah M. Shaalan, 'Adopting measures to reduce power outages', *Electrical & Computer Engineering: An International Journal (ECIJ)*, Vol. 7, Issue 1/2 (June 2018), p. 11.
5. Robert Knake, 'A Cyberattack on the U.S. Power Grid Contingency Planning Memorandum No. 31' (April 3 2018), https://www.cfr.org/report/cyberattack-us-power-grid, accessed January 20 2018.
6. Knake, 'A Cyberattack on the U.S. Power Grid Contingency Planning Memorandum No. 31'.
7. This will be unevenly distributed depending on the state/s or sector/s involved. For example, an economy such as the U.S. might be more resistant whilst an economy such as Latvia will be harder hit. Moreover, some sectors such as financial services, central government and militaries tend to be highly protected. The same cannot be said for other parts of the economy or local governments which will need to coordinate emergency responses in the case of blackouts.
8. Patrick J. Kiger, "American Blackout': Four Major Real-Life Threats to the Electric Grid' (October 25 2013), https://www.nationalgeographic.com/environment/great-energy-challenge/2013/american-blackout-four-major-real-life-threats-to-the-electric-grid/, 'Blackout: The Power Outage That Left 50 Million W/o Electricity | Retro Report | The New York Times' (November 11 2013), https://www.youtube.com/watch?v=nd3teNgUq8E, both accessed 13 February 2020.
9. Government, the military, and the people are often, as Sharma has, been mischaracterized as the Clausewitzian trinity (discussed in chapter 2). Both "rapid dominance" and "strategic paralysis" draw on the writings of Sun Tzu and Carl von Clausewitz. Amit Sharma, 'Cyber Wars: A Paradigm Shift from Means to Ends', in Christian Czosseck and Kenneth Geers (eds.), *The Virtual Battlefield: Perspectives on Cyber Warfare* (Amsterdam: IOS Press, 2009), pp. 3–17.
10. Cyber Security Building Resilience Reducing Risk conference, Chatham House, London (19–20 May 2014) attended by author. See also Chris Keeling, Waking Shark II: Desktop Cyber Exercise: Report

11. to Participants, 12 November 2013, https://www.bba.org.uk/wp-content/uploads/2014/02/Banking_3192106_v_1_Waking-Shark-II-Report-v1.pdf.pdf, accessed 6 April 2020.
11. Department of Defense Defense Science Board Task Force on Cyber Deterrence February 2017, p. 19, https://www.armed-services.senate.gov/imo/media/doc/DSB%20CD%20Report%202017-02-27-17_v18_Final-Cleared%20Security%20Review.pdf, accessed 1 December 2018.
12. Self-Monitoring Analysis and Reporting Technology. See also Elnur Hasan Mikail, Cavit Emre Aytekin, 'The Communications and Internet Revolution in International Relations', *Open Journal of Political Science*, Issue 6 (September 2016), pp. 345–350.
13. Philip N. Howard, *Pax Technica: How the Internet of Things May Set Us Free or Lock Us Up* (New Haven CT: Yale University Press, 2015).
14. Derek Manky, 'Securing OT Networks Against Rising Attacks' (March 13 2018), https://www.csoonline.com/article/3261448/security/securing-ot-networks-against-rising-attacks.html, accessed August 13 2018.
15. Dave Clemente, 'Cyber Security and Global Interdependence: What Is Critical?', https://www.chathamhouse.org/publications/papers/view/189645 (February 2013), p. 7.
16. This is a phrase repeated throughout his article. See Erik Gartzke, 'The Myth of Cyberwar Bringing War in Cyberspace Back Down to Earth', *International Security*, Vol. 38, Issue 2 (Fall 2013), pp. 41–73.
17. Jon R. Lindsay and Lucas Kello, 'Correspondence: A Cyber Disagreement', *International Security*, Vol. 39, Issue 2 (Fall 2014), pp. 181–192, at p. 186. See also Lucas Kello, *The Virtual Weapon and International Order* (New Haven, CT: Yale University Press, 2017).
18. National Counterintelligence Strategy of the United States of America 2020–2022, p. 2. https://www.dni.gov/files/NCSC/documents/features/20200205-National_CI_Strategy_2020_2022.pdf, accessed 2 September 2021.
19. Original emphasis. National Counterintelligence Strategy of the United States of America 2020–2022, p. 2.
20. National Counterintelligence Strategy of the United States of America 2020–2022, p. 3.
21. National Counterintelligence Strategy of the United States of America 2020–2022, p. 6.

22. National Counterintelligence Strategy of the United States of America 2020–2022, p. 7.
23. Council of Europe, 'Chart of signatures and ratifications of Treaty 185', https://www.coe.int/en/web/conventions/full-list/-/conventions/treaty/185/signatures?p_auth=Q4ufyD6q, and Council of Europe, 'Details of Treaty No.185 Convention on Cybercrime', http://conventions.coe.int/Treaty/en/Treaties/Html/185.htm, both accessed 11 September 2018.
24. 'Tallinn Manual 2.0', https://ccdcoe.org/research/tallinn-manual/, accessed 22 August 2019.
25. On insider threats see Markus Kont, Mauno Pihelgas, Jesse Wojtkowiak, Lorena Trinberg, and Anna-Maria Osula, *Insider Threat Detection Study* (Tallinn: CCD COE Publications, 2015).
26. McAfee/Center for Strategic and International Studies (CSIS), 'In the Crossfire: Critical Infrastructure in the Age of Cyberwar' (2011), p. 5, https://www.govexec.com/pdfs/012810j1.pdf, accessed 26 July 2019.
27. Remarks made during presentations at the 8th International Conference on Cyber Conflict, Tallinn, Estonia, 1–3 June 2016 attended by author.
28. Worldwide Threat Assessment of the US Intelligence Community Daniel R. Coats 13 February 2018, p. 5, https://www.dni.gov/files/documents/Newsroom/Testimonies/2018-ATA---Unclassified-SSCI.pdf, accessed 13 December 2018.
29. There was no public assessment in 2020. Zachary Cohen, 'Intelligence officials ask Congress not to hold threats hearings after angering Trump last year' (January 16 2020), https://edition.cnn.com/2020/01/16/politics/us-intelligence-officials-world-wide-threats-hearing-testimony/index.html, Annual Threat Assessment of the US Intelligence Community, Office of the Director of National Intelligence (April 9, 2021), https://www.dni.gov/files/ODNI/documents/assessments/ATA-2021-Unclassified-Report.pdf, Annual Threat Assessment of the US Intelligence Community, Office of the Director of National Intelligence (February 2022), https://www.dni.gov/files/ODNI/documents/assessments/ATA-2022-Unclassified-Report.pdf, all accessed 6 August 2022.
30. David J. Betz and Tim Stevens, *Cyberspace and the State Towards a Strategy for Cyber-Power,* (New York: Routledge, 2011), p. 90.
31. Sun Tzu, *The Art of War* translated and annotated by Samuel B. Griffith (Oxford: Oxford University Press, 1963) and Carl von Clausewitz,

On War, edited and translated by Michael Howard and Peter Paret (Ware: Wordsworth Editions Limited, 1997). See also Kenneth Geers, *Strategic Cyber Security* (Tallinn: CCD COE Publications, 2011), pp. 95–110 for his excellent discussion 'Sun Tzu: Can Our Best Military Doctrine Encompass Cyber War?'.

32. Ahmed Abbasi, Weifeng Li, Victor Benjamin, Shiyu Hu and Hsinchun Chen, 'Descriptive Analytics: Examining Expert Hackers in Web Forums', 2014 IEEE Joint Intelligence and Security Informatics Conference, p. 56, https://ieeexplore.ieee.org/stamp/stamp.jsp?arnumber=6975554, accessed 21 July 2019. See also Thomas J. Holt and Erik Lampke, 'Exploring stolen data markets online: Products and market forces', *Criminal Justice Studies*, Vol. 23, Issue 1 (January 2010), pp. 33–50, and Thomas J. Holt, Deborah Strumsky, Olga Smirnova and Max Kilger, 'Examining the Social Networks of Malware Writers and Hackers', *International Journal of Cyber Criminology*, Vol. 6, Issue 1 (June 2012), pp. 891–903.
33. https://www.blackhat.com/us-22/defcon.html, accessed 4 June 2022.
34. Verizon, 2018 Data Breach Investigations Report Executive Summary, p. 2, https://enterprise.verizon.com/resources/reports/DBIR_2018_Report_execsummary.pdf, accessed 13 April 2020.
35. Thomas Mackie, 'Revealed: New era of state sponsored hacking can turn oil rigs into 'bomb' that can kill' (February 18 2018), https://www.express.co.uk/news/world/920437/computer-hacker-cyber-hack-saudi-arabia-cyber-criminals-oil-rigs, accessed 7 November 2018.
36. CrowdStrike, 'Observations from the front lines of threat hunting' (October 2018), https://go.crowdstrike.com/rs/281-OBQ-266/images/Report2018OverwatchReport.pdf, p. 6 and '2018 CrowdStrike Global Threat Report', https://www.crowdstrike.com/resources/reports/2018-crowdstrike-global-threat-report-blurring-the-lines-between-statecraft-and-tradecraft/, both accessed 19 January 2018.
37. Verizon, 2019 Data Breach Investigation Report, p. 27, https://enterprise.verizon.com/resources/reports/2019-data-breach-investigations-report.pdf, accessed 13 July 2019.
38. Accessing 'Dark Net' sites is less of a problem than identifying users for law enforcement and intelligence agencies. Those who frequent 'Dark Net' sites, including paedophiles, organized criminals, purchasers of drugs etc. know that authorities are looking for them and take precautionary measures to hide their activities. This includes the use of Virtual Private Networks (VPNs) which act as a secure communication tunnel

from computer to computer/computer server, encrypted communications and files, coded language, as well as nom-de-plumes. This poses a range of practical and ethical problems for police and law enforcement agencies who are in some instances posing as fellow travellers signed up to their beliefs and causes. Problems include producing new material, purchasing illegal goods or services (to prove they are legitimate) and legal issues of entrapment. See for example Julia Buxton and Tim Bingham, 'The Rise and Challenge of Dark Net Drug Markets', Policy Brief 7 (January 2015), Swansea University, https://core.ac.uk/download/pdf/34722885.pdf, accessed 17 January 2020.

39. Jacob Aasland Ravndal, 'Anders Behring Breivik's use of the Internet and social media', *Journal EXIT-Deutschland*, 2 (2013), pp. 172–185. Ines Von Behr, Anaïs Reding, Charlie Edwards, Luke Gribbon, 'Radicalisation in the digital era The use of the internet in 15 cases of terrorism and extremism', RAND Europe (2013), available from https://www.rand.org/pubs/research_reports/RR453.html, and David Patrikarakos, 'Social Media Networks Are the Handmaiden to Dangerous Propaganda' (November 2 2017), https://time.com/5008076/nyc-terror-attack-isis-facebook-russia/, both accessed 7 September 2018.

40. Milton Mueller, 'What is Evgeny Morozov Trying to Prove? A review of "The Net Delusion"', https://www.internetgovernance.org/2011/01/13/what-is-evgeny-morozov-trying-to-prove-a-review-of-the-net-delusion/ (January 13 2011), accessed 9 September 2018.

41. Sean Gallagher, '"EPIC" fail—how OPM hackers tapped the mother lode of espionage data' (22 June 2015), https://arstechnica.com/information-technology/2015/06/epic-fail-how-opm-hackers-tapped-the-mother-lode-of-espionage-data/, accessed 7 September 2018.

42. Lee Hazell, 'SCADA system vulnerabilities so common, insurers won't insure', http://cybersecuritynews.co.uk/scada-system-vulnerabilities-so-common-insurers-wont-insure/, accessed 27 July 2015.

43. Audit of the Federal Bureau of Investigation's Implementation of Its Next Generation Cyber Initiative (July 2015), p. i, https://oig.justice.gov/reports/2015/a1529.pdf, accessed 6 April 2020.

44. The United States Department of Justice Special Counsel's Office, https://www.justice.gov/sco, accessed 6 August 2018.

45. This includes the use of biometrics.

46. Martin Libicki, 'The Nature of Strategic Instability in Cyberspace', *Brown Journal of World Affairs*, Vol. 18, Issue 1 (Fall/Winter 2011), pp. 71–79.
47. This point is emphasized in Thomas Rid, *Cyber War Will Not Take Place* (London: Hurst. 2013), pp. 11–34.
48. Defence Minister Jaak Aaviksoo: Cyber Defense—The unnoticed Third World War (8 May 2008), http://www.kaitseministeerium.ee/en/news/defence-minister-jaak-aaviksoo-cyber-defense-unnoticed-third-world-war, accessed 18 October 2016.
49. Michael N. Schmitt (ed.), *Tallinn Manual on the International Law Applicable to Cyber Warfare* (Cambridge: Cambridge University Press, 2013), Michael N. Schmitt (ed.), *Tallinn Manual 2.0 on the International Law Applicable to Cyber Operations* Second edition (Cambridge: Cambridge University Press, 2017). Some of the issues that the expert legal group encountered in drafting the *Tallinn Manual* are discussed in Michael N. Schmitt, 'The Law of Cyber Warfare: Quo Vadis?', *Stanford Law & Policy Review*, Vol. 25 (2014), pp. 269–300. A cyber "code of conduct" was proposed by China and Russia to the United Nations General Assembly in 2011. Timothy Farnsworth, 'China and Russia Submit Cyber Proposal', *Arms Control Today*, 2 November 2011, https://www.armscontrol.org/act/2011_11/China_and_Russia_Submit_Cyber_Proposal, accessed 20 July 2014.
50. The Tallinn Manual, https://ccdcoe.org/research/tallinn-manual/, accessed 22 June 2022.
51. See, for example, Richard A. Clarke and Robert K. Knake, *Cyber War The Next Threat to National Security and What To Do About It* (New York: HarperCollins, 2010), pp. 220–228, 235–242, 268–269; and David Omand, *Securing the State* (London: Hurst, 2010), pp. 11, 66–72, 80–84.
52. McAfee/CSIS, 'In the Crossfire Critical Infrastructure in the Age of Cyber War' (2011), p. 25. Their report also found deep-seated scepticism of the ability of governments around the world to positively legislate and regulate cybersecurity.
53. In the West a number of these become 'poacher turned gamekeeper' and have gone onto work for intelligence agencies, the cybersecurity industry, become freelance 'white hats' or journalists like Kevin Poulsen, 'Kingpin: How One Hacker Took Over the Billion-Dollar Cybercrime Underground', https://www.kingpin.cc/about/, accessed 17 February 2020.

54. Paul Peachey, 'Director of Europol: 'Top computer graduates are being lured into cybercrime' (December 29 2014), https://www.independent.co.uk/news/uk/crime/director-of-europol-top-computer-graduates-are-being-lured-into-cybercrime-9948990.html, accessed 4 September 2018.
55. Sam Jones, 'Licensed to hack: the rise of the cyber privateer Russia is increasingly using criminal proxies, say western intelligence officials' (March 16 2017), https://www.ft.com/content/21be48ec-0a48-11e7-97d1-5e720a26771b, accessed January 21 2018.
56. Nigel Inkster, *China's Cyber Power* (Abingdon: Routledge, 2016), p. 68. See also Tim Maurer, *Cyber Mercenaries: The State, Hackers, and Power* (Cambridge: Cambridge University Press, 2018).
57. Maurer, *Cyber Mercenaries*, p. xii.
58. Maurer, *Cyber Mercenaries*, p. xii.
59. Stephen Collinson, 'Ransomware attacks saddle Biden with grave national security crisis' (June 7 2021), https://edition.cnn.com/2021/06/07/politics/president-joe-biden-cyber-attacks-russia-putin-trump-economy/index.html, accessed 8 July 2021.
60. Wider perspectives can be found in R.B. Andrew, 'State-sponsored militias are coming to a server near you' (2013), *Foreign Policy*, http://foreignpolicy.com/2013/02/12/cyber-gang-warfare/, accessed 3 April 2017.
61. On hacktivism, see Jonathan Diamond, 'Early Patriotic Hacking', in Jason Healey (ed.), *A Fierce Domain: Conflict in Cyberspace 1986–2002* (Vienna, VGN: CSSA/Atlantic Council, 2013), pp. 136–151.
62. Defence Minister Jaak Aaviksoo, Cyber defense—the unnoticed third world war (May 8 2008).
63. 'CrowdStrike Report Reveals Cyber Intrusion Trends from Elite Team of Threat Hunters' (October 9 2018), https://www.crowdstrike.com/resources/news/crowdstrike-report-reveals-cyber-intrusion-trends-from-elite-team-of-threat-hunters/, accessed 19 January 2019.
64. Issues relating to economic loss from cybercrime activities are being dealt with by the World Economic Forum on risk and resilience. Their 2020 report, conducted jointly with Marsh & McLennan and Zurich Insurance Group, is available from https://www.weforum.org/reports/the-global-risks-report-2020, accessed 27 March 2020.
65. Anonymous, 'British Airways: Suspect code that hacked fliers 'found" (11 September 2018), https://www.bbc.co.uk/news/technology-45481976, 'Browser Extensions: Are They Worth the Risk?' (September

5 2018), https://krebsonsecurity.com/2018/09/browser-extensions-are-they-worth-the-risk/, both accessed 11 September 2018.
66. Verizon Data Breach Investigations Report 11th edition (2018), p. 6, https://enterprise.verizon.com/resources/reports/DBIR_2018_Report.pdf, accessed 17 January 2020.
67. Mandiant, M-Trends (2010) p. 3, https://www.fireeye.com/current-threats/annual-threat-report/mtrends/rpt-2010-mtrends.html, accessed 17 January 2019.
68. Mandiant, APT1 Exposing One of China's Cyber Espionage Units (2013), pp. 63–65, https://www.fireeye.com/content/dam/fireeye-www/services/pdfs/mandiant-apt1-report.pdf, accessed 17 January 2019.
69. Mark Ward, 'Why some computer viruses refuse to die' (August 14 2018), https://www.bbc.co.uk/news/technology-44564709, accessed 1 October 2019.
70. Alex Choros, 'World's most famous hacker says metadata is an "extremely attractive" target, everything is hackable',
   http://www.cybershack.com.au/news/worlds-most-famous-hacker-says-metadata-extremely-attractive-target-everything-hackable (dead link), last accessed 4 April 2017. These are issues Mitnick tackled in his books including Kevin D. Mitnick and William L. Simon, *The Art of Intrusion: The Real Stories Behind the Exploits of Hackers, Intruders and Deceivers* (Indianapolis, IN: Wiley, 2005).
71. 'Cyber Primer', Ministry of Defence, December 2013. This document is no longer officially accessible online.
72. 'Deep Neural Network' (undated), https://www.sciencedirect.com/topics/computer-science/deep-neural-network, accessed 8 January 2021.
73. Avrim L. Blum and Pat Langley, 'Selection of relevant features and examples in machine learning', *Artificial Intelligence*, 97 (1997), pp. 245–271.
74. W. James Murdoch, Chandan Singh, Karl Kumbiera, Reza Abbasi-Aslb, and Bin Yu, 'Definitions, methods, and applications in interpretable machine learning', *Proceedings of the National Academy of Sciences (PNAS)*, Vol. 116, Issue 44 (October 2019), p. 22071.
75. Bil Hallaq, Tiia Somer, Anna-Maria Osula, Kim Ngo, and Timothy Mitchener-Nissen, 'Artificial Intelligence Within the Military Domain and Cyber Warfare', in Mark Scanlon and Nhien-An Le-Khac (eds.),

Proceedings of 16th European Conference on Cyber Warfare and Security (Reading: Academic Conferences Publishing International, 2017), p. 153.

76. Ayse Pinar Saygin, Ilyas Cicekli, and Varol Akman, 'Turing Test: 50 Years Later', *Minds and Machines*, Vol. 10, Issue 4 (2000), pp. 463–518.

77. Vernor Vinge, 'The Coming Technological Singularity, *Whole Earth Review*, Issue 81 (Winter 1993), pp. 88–95. Ray Kurzweil, *The Singularity Is Near: When Humans Transcend Biology* (New York: Penguin, 2005). Toby Walsh, 'The Singularity May Never Be Near', *AI Magazine*, Vol. 38, Issue 3 (Fall 2017), pp. 58–62.

78. Matteo E. Bonfanti, Artificial intelligenceand the offense-defense balance in cyber security', and Jon R. Lindsay, 'Quantum computing and classical politics The Ambiguity of advantage in signals intelligence', both in Miriam Dunn Cavelty and Andreas Wenger (eds.), *Cyber Security Politics Socio-Technical Transformations and Political Fragmentation* (New York: Routledge, 2022), pp. 64–79, 80–94.

79. Vasisht Duddu, 'A Survey of Adversarial Machine Learning in Cyber Warfare', *Defence Science Journal*, Vol. 68, Issue 4 (July 2018), p. 363. 'Locked Shields', https://ccdcoe.org/exercises/locked-shields/, accessed 8 January 2021.

80. Noel E. Sharkey & Jan N.H. Heemskerk, 'The Neural Mind and the Robot', in Anthony Browne (ed.), *Neural Network Perspectives on Cognition and Adaptive Robotics* (Bristol: IOC Publishing, 1997), pp. 1–25, Noel E. Sharkey, 'The evitability of autonomous robot warfare', *International Review of the Red Cross*, Vol. 94, Issue 886 (Summer 2012), pp. 787–799, David Danks & Joseph H. Danks, 'The Moral Permissibility of Automated Responses during Cyberwarfare', *Journal of Military Ethics*, Vol. 12, Issue 1 (April 2013), pp. 18–33, Anzhelika Solovyeva, Nik Hynek, 'Going Beyond the "Killer Robots" Debate: Six Dilemmas Autonomous Weapon Systems Raise', *Central European Journal of International and Security Studies*, Vol. 12, Issue 3 (September 2018), pp. 166–208, Tim Stevens, 'Knowledge in the grey zone: AI and cybersecurity', *Digital War*, Vol. 1, Issue 1 (December 2020), pp. 164–160. 'Risks from Artificial Intelligence', https://www.cser.ac.uk/research/risks-from-artificial-intelligence/ (regularly updated), accessed 8 January 2022.

81. For scholarly debate on these issues using the example of *War Games* see Randy Eshelman and Douglas Derrick, 'Relying on the Kindness

of Machines? The Security Threat of Artificial Agents', *Joint Forces Quarterly*, No. 77 (April 2015), pp. 70–75,
82. Hallaq, Somer, Osula, Ngo, and Mitchener-Nissen, 'Artificial Intelligence Within the Military Domain and Cyber Warfare', pp. 153–156.
83. 'Supercomputing and Exascale Computing Overview' (undated), https://www.intel.com/content/www/us/en/government/exascale-supercomputing.html, accessed 9 January 2022.
84. Priyanshu Srivastava and Rizwan Khan, 'A Review Paper on Cloud Computing', *International Journals of Advanced Research in Computer Science and Software Engineering*, Vol. 8, Issue 6 (June 2018), pp. 17–20.
85. 'Supercomputing and Exascale Computing Overview'.
86. Gorjan Alagic, Anne Broadbent, Bill Fefferman, Tommaso Gagliardoni, Christian Schaffner, Michael St. Jules, 'Computational Security of Quantum Encryption', in Anderson C.A. Nascimento and Paulo Barreto (eds.), *9th International Conference, ICITS 2016 Tacoma, WA, USA, August 9–12, 2016 Revised Selected Papers* (Cham: Springer, 2016), pp. 47–48.
87. Jake Tibbetts, 'Quantum Computing and Cryptography: Analysis, Risks, and Recommendations for Decisionmakers', Center for Global Security Research, Lawrence Livermore National Laboratory, August 2019, p. 1, https://cgsr.llnl.gov/content/assets/docs/QuantumComputingandCryptography-20190920.pdf, accessed 9 January 2021.
88. Scott Buchholz, Joe Mariani, Adam Routh, Akash Keyal, Pankaj Kamleshkumar Kishnani, 'What nontechnical government leaders can do today to be ready for tomorrow's quantum world' (6 February 2020), https://www2.deloitte.com/us/en/insights/industry/public-sector/the-impact-of-quantum-technology-on-national-security.html, accessed 9 January 2021.
89. 'Critical Infrastructure Sectors', https://www.dhs.gov/critical-infrastructure-sectors, accessed 9 September 2018. For further information see Presidential Policy Directive 21 (PPD-21), 'Critical Infrastructure Security and Resilience', signed by President Obama in February 2013. Available from https://obamawhitehouse.archives.gov/the-press-office/2013/02/12/presidential-policy-directive-critical-infrastructure-security-and-resil, accessed 9 September 2018.
90. Manky, 'Securing OT Networks Against Rising Attacks' (March 13 2018).

91. Bonnie Zhu, Anthony Joseph and Shankar Sastry, 'A Taxonomy of Cyber Attacks on SCADA Systems', 2011 IEEE International Conferences on Internet of Things, and Cyber, Physical and Social Computing, p. 380, https://ieeexplore.ieee.org/stamp/stamp.jsp?tp=&arnumber=6142258, accessed 21 July 2019.
92. Zhu et al., 'A Taxonomy of Cyber Attacks on SCADA Systems', p. 380.
93. Mark Ward, 'How to hack a nation's infrastructure' (May 20 2013), https://www.bbc.co.uk/news/technology-22524274, accessed 15 July 2014.
94. Cybersecurity Reference and Resource Guide 2018, Department of Defense August 22, 2018, p. 44. https://ics-cert.us-cert.gov/, accessed 4 October 2018.
95. Manky, 'Securing OT Networks Against Rising Attacks' (March 13 2018).
96. Harry Menear, 'McAfee: The shifting threat landscape in the manufacturing sector' (June 21 2019), https://www.businesschief.com/leadership/8209/McAfee:-the-shifting-threat-landscape-in-the-manufacturing-sector and 'Protecting productivity: Industrial Security for the Digital Enterprise with Claroty and McAfee' (12 April 2019), https://www.youtube.com/watch?v=ZUkkI3S8JNc, accessed 24 July 2019.
97. Cyber Threat and Vulnerability Analysis of the U.S. Electric Sector Prepared by: Mission Support Center Idaho National Laboratory August 2016, p. 7, https://www.energy.gov/sites/prod/files/2017/01/f34/Cyber%20Threat%20and%20Vulnerability%20Analysis%20of%20the%20U.S.%20Electric%20Sector.pdf, accessed 12 August 2019.
98. Cyber Threat and Vulnerability Analysis of the U.S. Electric Sector, p. 4.
99. 'Shodan is the world's first search engine for Internet-connected devices', https://www.shodan.io/, accessed 6 September 2018.
100. Matt Burgess, 'What is the Internet of Things? WIRED explains From hairbrushes to scales, consumer and industrial devices are having chips inserted into them to collect and communicate data' (February 16 2018), https://www.wired.co.uk/article/internet-of-things-what-is-explained-iot, accessed 30 August 2018.
101. 'Cisco Annual Internet Report (2018–2023) White Paper', https://www.cisco.com/c/en/us/solutions/collateral/executive-perspectives/annual-internet-report/white-paper-c11-741490.html, accessed 8 July 2021.

102. Stephan Haller, Carsten Magerkurth, 'The Real-time Enterprise: IoT-enabled Business Processes', IETF IAB Workshop on Interconnecting Smart Objects with the Internet, March 2011, https://www.iab.org/wp-content/IAB-uploads/2011/03/Haller.pdf, p. 1.
103. Joss Meakins, 'A zero-sum game: The zero-day market in 2018', *Journal of Cyber Policy*, Vol. 4, Issue 1 (January 2019), p. 67.
104. Duddu, 'A Survey of Adversarial Machine Learning in Cyber Warfare', p. 356.
105. Rebecca Opie, 'Adelaide teen hacked into Apple twice hoping the tech giant would offer him a job' (May 27 2019), https://www.abc.net.au/news/2019-05-27/adelaide-teenager-hacked-into-apple-twice-in-two-years/11152492, accessed 13 April 2020.
106. Paul Cornish, David Livingstone, Dave Clemente, and Claire Yorke, 'Cyber Security and the UK's Critical National Infrastructure A Chatham House Report', p. vii. https://www.chathamhouse.org/sites/default/files/public/Research/International%20Security/r0911cyber.pdf, accessed 21 July 2014.
107. Meakins, 'A zero-sum game', p. 67.
108. Marleen Weulen Kranenbarg, Thomas J. Holt and Jeroen van der Ham 'Don't shoot the messenger! A criminological and computer science perspective on coordinated vulnerability disclosure', *Crime Science*, Vol. 7, Issue 16 (December 2018), pp. 1–9.
109. Key Trends across a Maturing Cyberspace affecting U.S. and China Future Influences in a Rising deeply Cybered, Conflictual, and Post-Western World Dr. Chris C. Demchak Testimony before Hearing on China's Information Controls, Global Media Influence, and Cyber Warfare Strategy Panel 3: Beijing's Views on Norms in Cyberspace and China's Cyber Warfare Strategy U.S.-China Economic and Security Review Commission.

    Washington, DC 4 May 2017, p. 4, https://www.uscc.gov/sites/default/files/Chris%20Demchak%20May%204th%202017%20USCC%20testimony.pdf, accessed 3 April 2020.
110. Christian Czosseck, Rain Ottis and Anna-Maria Talihärm, 'Estonia After the 2007 Cyber Attacks: Legal, Strategic and Organisational Changes in Cyber Security', *International Journal of Cyber Warfare and Terrorism*, Vol. 1, Issue 1 (January-March 2011), pp. 61–62.
111. Clarke and Knake outlined a cyber Pearl Harbor scenario in Clarke and Knake, *Cyber War*, pp. 64–68, as has John Carlin in 'A Farewell to

Arms' (May 1 1997), http://archive.wired.com/wired/archive/5.05/netizen.html, accessed 3 February 2015.
112. The 9/11 Commission Report, p. 336, https://www.9-11commission.gov/report/911Report.pdf, accessed 13 January 2019. This 'failure of imagination' was a feature of the report with a section devoted to foresight and hindsight (pp. 339–360).
113. Department of Defense Defense Science Board Task Force on Cyber Deterrence February 2017, p. 9, https://www.armed-services.senate.gov/imo/media/doc/DSB%20CD%20Report%202017-02-27-17_v18_Final-Cleared%20Security%20Review.pdf, accessed 1 December 2018.
114. Kathleen Hall, 'Cyber security is a board-level issue, warn government spooks' (September 5 2012), https://www.computerweekly.com/news/2240162676/Companies-must-tackle-cyber-threats-warn-government-spooks, accessed 27 March 2020.
115. There are competing and contested views of what cyberwar means which are discussed in the opening chapter.
116. Tzu, *The Art of War*, p. xxvii.
117. 'Cyber-Attack against Ukrainian Critical Infrastructure', Industrial Control Systems Cyber Emergency Response Team, 25 February 2016, https://ics-cert.us-cert.gov/alerts/IR-ALERT-H-16-056-01, accessed 28 June 2016. The activities of ICS-CERT and related organizations provide a valuable protective barrier. See also Dan Gooding, 'Hackers trigger yet another power outage in Ukraine For the second year in a row, hack targets Ukraine during one of its coldest months' (11 January 2017), https://arstechnica.com/information-technology/2017/01/the-new-normal-yet-another-hacker-caused-power-outage-hits-ukraine/, accessed January 19 2018.
118. And a reason that the U.S. Department of Defense is studying power and communication restoration in the event of a major cyberattack. Defense Science Board Task Force on Cyber Deterrence February 2017, p. 20.
119. See for example Ted G. Lewis, *Critical Infrastructure Protection in Homeland Security: Defending a Nation Second Edition* (Hoboken, NJ: Wiley, 2015), pp. 3–63.
120. For examples of responses to natural disaster and emergency relief see for example Brenda D. Phillips, *Disaster Recovery* Second Edition (Boca Raton FL: Taylor and Francis, 2016), George Haddow, Jane Bullock

121. National Security Strategy of the United States of America (December 2017), pp. 12–14, https://www.whitehouse.gov/wp-content/uploads/2017/12/NSS-Final-12-18-2017-0905.pdf, accessed 13 September 2018.
122. Original emphasis. Daniel R. Coats, Statement for the Record Worldwide Threat Assessment of the US Intelligence Community, 29 January 2019, p. 5, https://www.dni.gov/files/ODNI/documents/2019-ATA-SFR---SSCI.pdf, accessed 25 August 2019.
123. Original emphasis. Coats, Statement for the Record Worldwide Threat Assessment, p. 5.
124. Coats, Statement for the Record Worldwide Threat Assessment, pp. 5–6.
125. Annual Threat Assessment of the U.S. Intelligence Community, Office of the Director of National Intelligence, February 2022, pp. 4–13, https://www.odni.gov/files/ODNI/documents/assessments/ATA-2022-Unclassified-Report.pdf, accessed 8 May 2022.
126. Bronte Cullum, 'China seeks public comment on critical infrastructure cybersecurity review rules' (29 May 2019, https://globaldatareview.com/article/1193471/china-seeks-public-comment-on-critical-infrastructure-cybersecurity-review-rules, accessed 23 August 2019. Patrick Tucker, 'Russia Will Build Its Own Internet Directory, Citing US Information Warfare' (November 28 2017), http://www.defenseone.com/technology/2017/11/russia-will-build-its-own-internet-directory-citing-us-information-warfare/142822/?oref=d-river, accessed 27 December 2017 and Sasha Baranovskaya, 'Moscow's cyber-defense How the Russian government plans to protect the country from the coming cyberwar' (19 July 2017), https://meduza.io/en/feature/2017/07/19/moscow-s-cyber-defense, accessed 28 December 2017.
127. See for example Rid, *Cyber War will not take place*, pp. 139–162.
128. Schmitt (ed.), *Tallinn Manual 2.0*, pp. 168–176.
129. Remarks made under Chatham House Rules at the NATO Intelligence Fusion Centre, RAF Molesworth, 2–5 November 2015 attended by author. For a conceptual discussion of military thinking on these issues drawing on the prisoner's dilemma, see Libicki, 'The Nature of Strategic Instability in Cyberspace', pp. 71–79.
130. Paul Ducheine and Jelle van Haaster, 'Fighting Power, Targeting and Cyber Operations' in Pascal Brangetto, Markus Maybaum and Jan

Stinissen, *2014 6th International Conference on Cyber Conflict* (Tallinn: CCD COE Publications, 2014), p. 303.
131. John Arquilla and David Ronfeldt, 'Cyberwar Is Coming!', *Comparative Strategy*, Vol. 12, Issue 2 (Spring 1993), pp. 141–165.
132. John Arquilla, 'The Computer Mouse That Roared: Cyberwar in the Twenty-First Century', *Brown Journal of World Affairs*, Vol. 18, Issue 1 (Fall/Winter 2011), pp. 39–48.
133. Remarks made during presentations at the 8th International Conference on Cyber Conflict, Tallinn, Estonia, 1–3 June 2016 attended by author.
134. James A. Lewis, 'The Role of Offensive Cyber Operations in NATO's Collective Defence', Tallinn Paper 8 (2015), p. 3, https://ccdcoe.org/uploads/2018/10/TP_08_2015_0.pdf, accessed 7 June 2016.
135. See for example Ofer Fridman, *Russian 'hybrid warfare': Resurgence and politicisation* (London: Hurst, 2018). Alexander D. Chekov, Anna V. Makarycheva, Anastasia M. Solomentseva, Maxim A. Suchkov and Andrey A. Sushentsov, 'War of the Future: A View from Russia', *Survival*, Vol. 61, Issue 6 (December 2019-January 2020), pp. 25–48.
136. Lewis, 'The Role of Offensive Cyber Operations in NATO's Collective Defence', p. 4.
137. DoD Cyber Strategy (April 2015), https://www.hsdl.org/?abstract&did=764848, accessed 10 January 2018.
138. Warsaw Summit Communiqué, Issued by the Heads of State and Government participating in the meeting of the North Atlantic Council in Warsaw 8–9 July 2016, https://www.nato.int/cps/en/natohq/official_texts_133169.htm#cyber, accessed January 10, 2018.
139. 'VERIS Overview', http://veriscommunity.net/veris-overview.html, accessed 13 July 2019.
140. This can be inadvertent. Anonymous, 'Copycat coders create 'vulnerable' apps' (October 7 2019), https://www.bbc.co.uk/news/technology-49960387, accessed 3 November 2019.
141. 'What Is the Difference: Viruses, Worms, Trojans, and Bots?' (undated), https://www.cisco.com/c/en/us/about/security-center/virus-differences.html, accessed 11 July 2019.
142. Verizon Data Breach Investigations Report 2022, pp. 33–35, https://www.verizon.com/business/resources/reports/2022/dbir/2022-data-breach-investigations-report-dbir.pdf, accessed 11 June 2022.
143. 'What Is the Difference: Viruses, Worms, Trojans, and Bots?'.

144. There are many explanations of cryptocurrencies and the underlying Blockchain technology that underpins it. Much of it can be baffling to the uninitiated. Among the more accessible is Mohit Kaushal and Sheel Tyle, 'The Blockchain: What It Is and Why It Matters' (January 13 2015), https://www.brookings.edu/blog/techtank/2015/01/13/the-blockchain-what-it-is-and-why-it-matters/, accessed 14 February 2020.
145. 'What Is the Difference: Viruses, Worms, Trojans, and Bots?'.
146. 'What Is the Difference: Viruses, Worms, Trojans, and Bots?'.
147. Automated tackling of disinformation Study Panel for the Future of Science and Technology European Science-Media Hub (March 2019), http://www.europarl.europa.eu/RegData/etudes/STUD/2019/624278/EPRS_STU(2019)624278_EN.pdf, accessed 16 July 2019. Anonymous, 'Russian bots rigged Voice Kids TV talent show result' (16 May 2019), https://www.bbc.co.uk/news/world-europe-48293196, accessed 17 July 2019.
148. Noa Bar-Yosef, 'The Structure of a Cybercrime Organization—Hackers Have Supply Chains Too!' (September 23 2010), https://www.securityweek.com/structure-crybercrime-organization-hackers-have-supply-chains-too, accessed 11 July 2019.
149. Verizon, 2019 Data Breach Investigation Report, p. 26.
150. Brenden Kuerbis, Farzaneh Badiei, Karl Grindal, and Milton Mueller, 'Understanding transnational cyber attribution Moving from "who-dunit" to who did it', in Cavelty and Wenger (eds.), *Cyber Security Politics*, pp. 220–238.
151. 'Cyber's Most Wanted', https://www.fbi.gov/wanted/cyber, accessed 2 May 2022.

2

On Cyberwar: Theorizing Cyberwarfare Through Attacks on Critical Infrastructure—Reality, Potential, and Debates

Introduction

The title of this opening chapter borrows from Carl von Clausewitz's classic nineteenth-century text *On War* which explored "the phenomenon of war" and its "manifestations".[1] This included its nature and character.[2] This chapter utilizes Clausewitz to explore the present and future potential for cyberwar to better understand its nature and character with war, and conflict more generally, viewed as the continuation of politics by other means.[3] This is "where the original German term *politik* means both politics and policy combined" and the motives for undertaking war are applied.[4]

This is the locus where Clausewitz saw war as a political act which can also be understood by looking at "the enemy's character, from his institutions, the state of his affairs, and his general situation…[which] forms an estimate of its opponent's likely course and acts accordingly".[5] Clausewitz recognized that the character of the state and society itself also provides indicators of capability and intentions.[6] These remain avenues of analyses that help conceptualize cyberwar. This book draws on this rationale for as a member of India's Ministry of Defence described, "cyber warfare has assumed the shape of an elephant assessed by a group of blind people, with every one drawing different meanings based upon their perceptions".[7] The aim of this chapter is to improve our sight and assign more meaning.

It is first worth considering that as a general schooled in the Napoleonic Wars of the early nineteenth century, Clausewitz has limitations in advancing conceptions of cyberwar. A 2018 account published by the U.S. Army War College on the meaning and influence of *On War* suggests that:

> the fragmented, networked enemy, produced by today's information revolution—which might well turn out to be just as transformative as the French Revolution, or the Industrial Revolution— fits badly into Clausewitz's abstract account of war as a two-way military fight between unified entities. This enemy is not new but well-known to the Western tradition of strategic thought in the imperial and small-wars context—though historically known more at a local or a regional level than at the global-level networks of today's Information Age.[8]

The state-centrism of *On War* is a valid criticism. Cyberattacks can originate at the sub-state level or from non-state actors, particularly in cyberespionage campaigns, and forms part of the 'fog of war'.

The Fog of Cyberwar

The 'fog of war' is the unforeseen and unforeseeable nature of strategizing that emerges in the conduct of war itself and friction the impediments to battle plans incorporating anything that can go wrong and elements such as weather. Here each part of the war effort, down to the role of individuals, are contributors to friction and interact to increase and create chance and uncertainty.[9] Clausewitz understood in his earlier treatise *Principles of War* that trusting subordinates to carry out orders carries uncertainty.[10] Unless directly subordinated to and directed by a state, sub-state actors such as cybercriminals used as proxy forces, not only add to what Clausewitz described as the 'fog of war' they also magnify the role of uncertainty and chance and are one of Clausewitz's 'contributors to friction'.

In militaries around the world cyber is being incorporated into tactical operations for in-area operations and for 'preparation of the battlespace'. This incorporates localized jamming or disruption of signals and communications traffic and to generate and provide intelligence and strategic level situational awareness of the battlespace. These fit into broader political and military strategies and cyber, which is now the fifth domain of warfare alongside land, sea, air, and space, is a vital component of these operations.[11] Insofar as critical infrastructure is concerned, state-on-state intelligence-led cyberspace operations known as Advanced Persistent Threats (APTs) are the main concern and an increasingly important part of the foreign and defense policy tools of major states. APTs are defined by the widely respected US-based National Institute of Standards and Technology (NIST) as:

> An adversary that possesses sophisticated levels of expertise and significant resources which allow it to create opportunities to achieve its objectives by using multiple attack vectors (e.g., cyber, physical, and deception). These objectives typically include establishing and extending footholds within the information technology infrastructure of the targeted organizations for purposes of exfiltrating information, undermining or impeding critical aspects of a mission, program, or organization; or positioning itself to carry out these objectives in the future. The advanced persistent threat: (i) pursues its objectives repeatedly over an extended period of time; (ii) adapts to defenders' efforts to resist it; and (iii) is determined to maintain the level of interaction needed to execute its objectives.[12]

These APTs present themselves as not only a clear and present danger but also a direct threat to national security and state cohesion through cyberattacks and cyberespionage campaigns. These are being actively pursued as a zero-sum game by nations like China and Russia, mainly (but not exclusively) against the United States and its Western and non-Western allies. Thus far, Western responses have been limited (including diplomatic responses) and have not engaged in symmetrical escalation through the return use of their own offensive cyber capabilities—at least not in widely visible or publicly reported manner. Offensive cyber operations the U.S. Department of Defense (DoD) conducts operate under strict legal boundaries of national and international law. This was seen as an impediment under the Trump administration.[13]

Whether or not we can (or should) define current uses of offensive cyber as 'war' (in particular an adversarial 'cold' or non-shooting war),[14] cyber adds elements of risk or chance into the international security environment. In *On War*, Clausewitz came to understand that "only the element of chance is needed to make war a gamble, and that element is never absent".[15] This is part of Clausewitz's famous trinity of violence (the use of force), chance and probability with war a political action.[16] Although these retain contemporary relevance, Clausewitz is of limited use in forming an improved understanding of the *conduct* of twenty-first-century war, in particular 'cyberwar'. Nevertheless, he is still informative about the *character* of war itself. Currently, cyberwar is conceptually contested, and its meaning confused.[17] Its *character* is conceptually informed by the practices we are witnessing but defining its *character* is inchoate at present. Instead, descriptions have been built up by the *conduct* of 'cyberwar'.

They are headline grabbers: 'Russia Ukraine conflict could trigger cyberwar' (VOA News, April 20 2014). 'Cyber WAR ALERT: Britain prepares for Russian retaliation as Putin FUMES' (*Daily Express*, April 16 2018). 'US is waking up to the deadly threat of cyber war' (CNN, May 6 2018). 'Iran prepares for cyberwar amid rising tensions' (Fox News, October 4 2019). 'As Cyberwar Rages in Ukraine, Demand for Israeli Cyber Firms Soars' (Haaretz, March 2 2022). There have been many more headlines that securitize these and other cyber activities as cyberwar and there will be many more to come.[18] Currently, the use of the term *war* in cyberwar does not fit the conduct we are seeing. At its worst we are seeing orchestrated strategic campaigns of cyberespionage, disinformation, and sabotage. This is part of Thomas Rid's argument that 'Cyber War Will Not Take Place' but this does not paint the whole picture.[19] Newspaper headlines are probably guilty of too much imagination, others not enough.

Cyber is different to all other classes of war-fighting tools for two reasons. First, cyber alone can generate kinetic effects by manipulating code to produce physical effects in civilian and military systems. Any weapons system or communications asset that relies on computer code can be manipulated to generate unintended effects be they kill switches that degrade or render systems inoperable, make them volatile, or able to be turned upon those wielding them. So far this is largely theoretical but there have been a wide variety of proof-of-concept demonstrations that with enough scalability and will could make this a reality in the coming years. Second, unlike in all other classes of weapon, computer code has no visibility unless used. This makes capabilities difficult to assess and does not require large-scale and monitorable facilities. Policymakers have not yet fully grasped the potential of cyber. The following chapters will demonstrate in detail that the capacity for cyberwar—far greater that hitherto seen—exists by attacking critical infrastructure, why it exists and the forms it could take. First, this requires a discussion of cyberwar as a phenomenon and its manifestations.

What Is Cyberwar(fare)?

Debates over what constitutes cyberwar and cyberwarfare remain contested and frequently misapplied to wider areas of cybersecurity. In Western discourse, the terms 'cyberwar' and 'cyberwarfare' are seen as interchangeable.[20] The difference might well be semantics but acts of cyberwarfare are possible without military conflict. This debate began with Arquilla and Ronfeldt's seminal 1993 article 'Cyberwar is Coming!' which offered a militaristic definition of cyberwar as "conducting, and preparing to conduct, military operations according to information-related principles".[21] Since then militaries have become secondary players to intelligence agencies, adding to contestations between cyberwar and cyberespionage.

In a study of cyberwar(fare) definitions published in 2017, it was concluded that "a majority of articles[22] do not offer explicit definitions of either cyber war or cyber warfare from which to base their analysis…characterized by both intra and interdisciplinary competition between dozens of definitions".[23] General Michael Hayden and Richard Clarke, both long-serving U.S. national security officials, also recognize this definitional problem.[24] In part, this helps explain why many remain skeptical of cyberwar as a potential reality.[25] The eminent political scientist Joseph S. Nye, wrote in 2018, "maybe we are looking in the wrong place, and the real danger is not major physical damage but conflict in the gray zone of hostility below the threshold of conventional warfare".[26] This line of reasoning is best summed up by Thomas Rid and his belief that 'Cyber War Will Not Take Place'.

Rid takes the definition of warfare propounded by Clausewitz in *On War* and bases his argument that cyberattacks are making political interactions less violent around three core arguments. First, the use of cyberattacks is being used as a tool of non-violent sabotage. Second, the use of cyber for espionage operations decreases risk. Third, that subversion is decreasing the resort to armed force. According to this line of argument, "cyber war has never happened in the past, it does not occur in the present, and it is highly unlikely that it will disturb our future".[27] For Rid, "All past and present political cyber attacks-in contrast to computer crime-are sophisticated versions of three activities that are as old as human conflict itself: sabotage, espionage and subversion".[28] Whilst these are undoubtedly present, he does not completely dismiss the potential for catastrophic uses of cyber against critical infrastructure but for Rid "all such scenarios… remain fiction, not to say science fiction".[29]

Meanwhile, the counterpoise to Rid's argument was articulated by the aforementioned Richard Clarke, a former U.S. National Security Adviser, who along with Robert Knake in their book *Cyber War: The Next Threat to National Security and What to Do About It* called it a 'wakeup call' to policymakers. They warned, "a hand can reach out from cyberspace…causing things to shut down or blow up, things like the power grid, or a thousand other critical systems, things like an opponent's weapons".[30] They argue that this is producing an incentive to strike first with "military and intelligence organizations…preparing the cyber battlefield with things called "logic bombs" and "trapdoors"".[31] The prime targets for these peacetime preparations reside in the civilian private sector whilst "The speed at which thousands of targets can be hit, almost anywhere in the world, brings with it the prospect of highly volatile crises".[32]

These two sets of arguments effectively bookend the cyberwar(fare) debate but its examination remains highly active.[33] This is found in a growing body of literature on a range of cybersecurity issues.[34] Cybersecurity, including cybercrime concerns, has emerged in the mainstream media, in policymaking debates and amongst civil society. To varying degrees, they draw expertise from the disciplines of (computer) science, technology, engineering, and math (STEM) subjects alongside political science/international relations, international law, and military theory and are driven by events.

While this book draws on each of these disciplines as well as events, it also draws significantly on the technical world at the frontline of the civilian-run, private sector industries Clarke and Knake were referring to. Scholarship and debate have evolved and play an inquiring and informing role for policymakers and wider society. However, 'cyberwar' remains under-conceptualized and overused and is frequently conflated with wider cybersecurity issues, especially cyberespionage. So long as this terminology remains ill-defined the lack of boundaries and separation between them will cause confusion.[35]

Cyberwar Deconstructed

To better understand the meanings attributed to cyberwar, it is first important to disaggregate the term. As the first principle to a better understanding of what cyberwar means we need to first question what constitutes war (as accepted by the international community). This helps establish a baseline to better understand the current and future cyber threat landscape and the uses of cyber as war-fighting weapons or tools. Moreover, if escalation is to be avoided, this is all the more important. If not, "with attention diverted elsewhere, we may be laying the groundwork for cyber war", especially if uncontrolled.[36]

War begins with aggression.[37] This is defined by the United Nations (UN) in Resolution 3314 (XXIX) as "the use of armed force by a State against the sovereignty, territorial integrity or political independence of another State".[38] Although cyberspace is essentially borderless and inherently transnational, this state-centric view of 'armed force' against territory and sovereignty is used as one of the baselines for the UN as well as the Tallinn process (described below) in defining international norms determining not only what constitutes a cyberattack but also how states are legally entitled to respond.[39]

'The use of force' is a problematic starting point for arriving at a definition of cyberwar or describing cyberwarfare. Its use and usefulness as a baseline for defining traditional conceptions of war is limited to cyber-physical attacks. The political scientist Kalevi Holsti argues, "The 'meaning' of war refers to the policy-makers' conceptualizations of war—what type of activity and ethical connotations it involves, for example—and their attitudes toward the use of force".[40] This book makes the case that unless applied to destructive and kinetic forms of attack, especially upon critical infrastructure but also directed at military sites and systems (including as a separate act), cyberwar(fare) does not fit conventional and legally derived definitions of war or warfare.

This said, maybe a whole new paradigm of understanding needs to be built up and evolved because legally, only a sovereign state can declare war but there have been many, and arguably increasing, cases of undeclared wars between nations. This includes American intervention in the Vietnam War.[41] A state of war can exist without a declaration of war or might, as in the case of the Japanese strike on Pearl Harbor in 1941, followed by a pre-emptive attack. Pearl Harbor also points to the value of intelligence gathering, diplomacy, and espionage prior to attack. This has led some international law scholars to argue "that the use of espionage is allowable as both a necessary part of statecraft and a means of peremptory self-defense under both the U.N. Charter and customary international law".[42]

According to Darien Pun, "This view claims espionage serves as a form of either arms control or conflict prevention by making aware the capabilities of other states, serving as a mechanism to promote stability and peace".[43] Preemption through cyberespionage by pre-positioning for preemptive attack against critical infrastructure (CI) may challenge this state of affairs depending on whether these activities are seen to enable mutual deterrence or undercut stability. This is demonstrable and underestimated. An attack on CI would almost certainly violate the principles of the Law of Armed Conflict (LOAC), described below, by targeting non-combatants deliberately or through collateral damage. An attack on military systems such as forcing an aircraft to crash on a populated area or a destroyer to career into a port could also be used to cause civilian casualties or far worse if, for example, nuclear targeting coordinates were changed and launch orders were able to be faked.[44]

One of the important dimensions of this discussion of the 'war' aspect of cyberwar concerns escalation. For Clausewitz, escalation develops from 'a mere state of armed observation' all the way through to 'wars of extermination'. Drawing on this Clausewitzian definition, the American political philosopher, Michael Walzer remarked, "War is not usefully described as an act of force without some specification of the context in which the act takes place and from which it derives its meaning".[45] War is a social construct as is cyberwar. As Walzer frames it, "What is war and what is not-war is in fact something that people decide…[and] in a variety of cultural settings they have decided, that war is limited war – that is, they have built up certain notions about who can fight, what tactics are acceptable, when battle has to be broken off, and what prerogatives go with victory into the idea of war itself".[46] What is war and what is not war in terms of cyberwar is a vexing question, displaying many of these same characteristics. The status of combatants can be blurry and tactics controversial, the idea of a 'battle' in cyberspace limited to computer code (which can affect kinetic weapons and computer-controlled machinery). And the idea of territorial defense in cyberspace or securing national cyberspace within a state—as many national strategy documents make mention—holds little meaning in practice.

Hybrid Warfare

Neutral civilians and those who align with various factions are also on these battlefields as are terrorists who use violence for political ends. This is part of the dynamic of Clausewitz's remarkable trinity. This is not a pitched battle of two clearly identifiable enemies and in state-on-state conflicts and open war, it is also where counterinsurgency campaigns, nationalism, and terrorism can be combined to blur the lines. This is part of the definition of 'hybrid warfare'—a series of activities in that gray zone of hostility below the threshold of conventional warfare that Nye talked about. These blur the distinction between combatants and non-combatants and state, non-state, and sub-state actors.

Despite its contemporary re-emergence, hybrid warfare's "historical pedigree goes back at least as far as the Peloponnesian War in the fifth century BC".[47] In 1991 Martin van Creveld was among the first to refocus on these dynamics as low-intensity warfare seen in the aftermath of the Cold War as the "last gasp" of "Large scale, conventional war"[48] and the rise of 'non-trinitarian' Clausewitzian warfare.[49] Frank G. Hoffman was perhaps the first Western military scholar to redefine what others were characterizing as 'Fourth Generation Warfare (4GW)'. 4GW is where state and non-state actors utilize means such as "terrorism and information, to undermine the will of the existing state, to de-legitimize it, and to stimulate an internal social breakdown".[50] This is certainly in evidence, especially in Russian political interference campaigns.

Hoffman saw merit in 4GW but preferred to style this blurring of combatants, non-combatants, conflict types, definitions of war and peacetime, and use of geographic and cyber ('informational') terrain as a "convergence into multi-modal or Hybrid Wars".[51] Hoffman also saw this convergence against the background of history with "Thucydides' ponderous tome on the carnage of the Peloponnesian War…an extended history of the operational adaptation of each side as they strove to gain a sustainable advantage over their enemy".[52]

The theories underlying 4GW, 'hybrid warfare' and the subsequent Russian articulation of this as the 'Gerasimov doctrine' are not new inventions. Instead, they were responses to Counter Insurgency (COIN) campaigns in Afghanistan, the Caucasus, the Middle East, and international counterterrorism operations that spilled out of these conflicts and the 2011 Arab Spring.[53] This lens helps illuminate Russia's take on hybrid war, which for them was honed in COIN operations during the two-decade-long Chechen Wars which began in 1999.[54] Whilst it is certainly true that COIN have been features of conflict throughout recorded history what is different about this re-emergence is the cyber context.[55] The reach and speed cyber provides to combatants and non-combatants and state and non- or sub-state actors into geographically distant states and societies (connected nationally and globally through technology such as the Internet and 4G mobile) has no historical parallel. Whether state-on-state or at the non- or sub-state level, unconventional regular and irregular threats are offering asymmetrical challenges to Clausewitzian orthodoxy. Mark Galeotti suggests that:

> Putin and his government have concluded that the West began this campaign of *gibridnaya voina* [voyna] – hybrid war – not least through, in its eyes, organizing a whole string of regime-changing insurrections, from the 'colour revolutions' of the post-Soviet space, through the 'Arab Spring' risings and eventually the 2013–14 Euromaidan revolt, which brought down Ukrainian president Ukrainian president Victor Yanukovych.[56]

Galeotti's analysis also draws on Russia's politico-military interventions in Chechnya, Georgia in 2008, and Ukraine in 2014 which saw Crimea annexed. Prior to Ukraine, military reforms had been initiated by Defence Minister Anatoly Serdyukov, but especially the Chief of the General Staff Nikolai Makarov. Both fell from grace in 2012 leading to the appointment of General Valery Gerasimov as Chief of the General Staff. Galeotti points out that "their successors continued the process, hence the irony of Gerasimov getting the credit".[57] In addition, at least three other Russians are worthy of mention. The first is Igor Korobov, a career intelligence officer and graduate of Russia's primary military intelligence arm, the GRU (*Glavnoye Razvedovatel'noye Upravlenie*). Appointed in 2016 as director of the Military Intelligence Directorate, and trained by the GRU's 'Conservatory', he died in 2018. The second is Sergei Shoigu, the Minister of Defence appointed in 2012. The third is Pyotr Ivashutin, who is "credited with establishing the "GRU empire"" from its headquarters—'The Aquarium'. Ivashutin spent his career leading military intelligence from the early 1960s until his death in 2002.[58]

Following the Russia-Georgia conflict in 2008, Shoigu saw a need for Russia to militarize cyber and to modernize its increasingly outdated military with a need to recruit and train in the field of cybersecurity for *informatsionnaya voyna* (information war). This gained traction and greater focus prior to the 2014 'Euro Maidan' revolution, where

Yanukovych (widely seen as a pro-Russian proxy[59]) was ousted and Crimea annexed. Keir Giles argues this was formative because:

> the techniques visible in and around Ukraine represent the culmination of an evolutionary process in Russian information warfare theory and practice, seeking to revive well-established Soviet techniques of subversion and destabilisation and update them for the internet age…current Russian approaches have deep roots in long-standing Soviet practice.[60]

According to Galeotti, in Ukraine in 2014 cyber and influence operations were used to lay the groundwork for a traditional aerial offensive followed by ground forces. Like Giles, he adds:

> Russia has reached back and re-learned a particular Soviet lesson, that political effects are what matters, not the means used to achieve them. Instead of trying to contest NATO where it is strongest, on the battlefield – it is worth remembering that NATO has a combined defence budget of $946 billion to Russia's $47.4 billion (2.84 trillion roubles) – it is instead an example of asymmetric warfare, using gamesmanship, corruption and disinformation instead of direct force.[61]

This helps set Russian dis- and misinformation campaigns (aided and abetted by state-controlled media outlets), election interference and even state-sanctioned or state-controlled large-scale cybercrime into context.[62] But is it also a part of *maskirovka*; a tactic of deception to mask, disguise, or camouflage (described by both Sun Tzu and Clausewitz) to serve politico-military ends.[63] These are used to mislead, confuse, distract, and interfere.[64] This is referred to by Costinel Anuta of Romania's intelligence service, as "feeding the current Russian 4D (dismiss, distort, distract and dismay) approach of information war".[65] What *maskirovka* does not explain is Russian pre-positioning in Western critical infrastructure, especially in the United States (an area where the Chinese are even more advanced). These are all active measures by Russia's intelligence agencies and Russia's 'hybrid warfare' strategy takes many forms.

Relatedly, Buchanan and Sulmeyer make the following observation: "Holding key targets at risk is an important part of many modern military strategies. The idea is that, in the event of hostilities or other contingencies, one has the ability to destroy or disable key targets. This can be done either as a means of denying an adversary the capacity for action or of imposing costs on an adversary".[66] These fit with Russian and Chinese activities against critical infrastructure which are more than just cyberattacks or cyberespionage. There is a growing fear that this is part of a sustained long-term strategy to pre-position and hold key sectors such as the electrical grid at risk. This also fits the framework of operations enunciated in the 'Gerasimov doctrine' and in Chinese strategy.

International Law: JWT and the LOAC

Another area that is problematic, especially to international law scholars, is found in Just War Theory or Tradition (JWT) which is a basis for the LOAC in the Western world.[67] Especially troublesome is the meaning of 'armed attack' in cyberspace. JWT is generally considered to have two elements: *jus ad bellum* (the justice of war) and *jus in bello* (justice in war). There is plausibly also a third that needs consideration, *jus post bellum* (justice after war). Walzer sees *jus ad bellum* as requiring "us to make judgements about aggression and self-defense" and *"jus in bello* about the observance or violation of the customary and positive rules of engagement".[68]

Jus in bello has helped to instigate some rules of war, such as the prohibition of the use of certain kinds of weapons to ideas of proportionality and the non-deliberate targeting of civilians.[69] These have become widely accepted international norms underpinned by treaty obligations, especially following World War II and the establishment of the United Nations.[70] In the years after World War II, great thought has been given over to defining non-combatants (not that this has always made a practical or humanitarian difference in conflict situations). Nowadays, cyberspace already contains an array of sub and non-state civilian 'combatants'.

Russia in particular is blurring the definition of a state combatant in cyberspace through its direct and indirect use of cybercriminals and sub-state proxies in state operations and by providing sanctuary for criminal activities. China has also been accused of leveraging sub-state actors.[71] In these and other states, civilians can also be separately tasked or can take it upon themselves to be 'patriotic hackers' as some Ukrainians have in the war against Russia. Collectives, including 'hacktivists' such as 'Anonymous', can also mobilize. At the non-state level, terrorist groups can also be leveraged for state use and plausible deniability. These groups not only affect conventional ideas of what constitutes combatants and non-combatants but, like the insertion of covert special forces in disputed territories or conflict zones, also complicate state-based resolutions.

International law, which is accepted by a majority of states, includes the LOAC which adds both context and complexity to conflict in cyberspace.[72] International law reflects realism and international liberalism in the search for enforceable norms and rules of behavior through ongoing political and legal discourse.[73] Although the LOAC only operates under conditions of armed conflict it may also affect national strategies and operational planning for the cyber elements of conflict and one of a number of good reasons to arrive at a "deeper comprehension of what war is in the cyber-age".[74]

The LOAC is not a single piece of international law but has been gradually codified under the Hague Conventions (1899 and 1907), the Geneva Protocol (1925), and the Geneva Conventions (1949). It is largely seen as coterminous with International Humanitarian Law (IHL), also sometimes referred to in the United States as the Law of War.[75] Although it "has never been authoritatively defined as a matter of treaty law…it [conflict] has now replaced the term 'war' for most international law purposes".[76] The United States still uses the term 'war' when describing its responsibilities under the LOAC which "rests on fundamental principles of military necessity, proportionality, distinction (discrimination), and avoidance of unnecessary suffering".[77]

The LOAC is used as a legally binding instrument to establish some 'rules of the game' in conflict situations. This includes not deliberately targeting civilian populations and the avoidance 'collateral damage' on non-military targets. With most cyberwarfare targets owned and operated by private industry—including the civilian networks that form most of the backbone architecture—this poses a considerable problem for the LOAC and traditional war-fighting principles. This includes military

Rules of Engagement (RoE) which "are driven by a mix of legal, military, and political factors" in the nation concerned and in military alliances.[78]

International laws including the LOAC are broken but without them, peace and security would be more problematic (as would domestic order without civil law).[79] Relatedly, international bodies, most especially the UN, operate through the gift of power and the prism of power politics and national interests.[80] This includes Article 2 of the UN Charter which centers on restraining "the threat or use of force against the territorial integrity or political independence of any state, or in any other manner".[81] The UN Charter did not restrain Russia, a UN Security Council (UNSC) member, in February 2022 when it invaded Ukraine nor has the LOAC restrained its military during the invasion.[82]

The presence of the LOAC has not prevented humanitarian tragedy, especially in civil wars, and the extent to which it acts as a restraining mechanism varies depending on the aggressor and combatants involved. Cicero, the ancient Roman statesman and philosopher, argued from experience "In times of war, the law falls silent" (*Inter arma enim silent lēgēs*).[83] This has seen "Breaches of the law of armed conflict [that] are frequent and dreadful". This includes "Ethnic cleansing, murder, and widespread breaches of international law in the former Yugoslavia and in Rwanda".[84] To this unhappy list, Ukraine can now be added. Nevertheless, the presence and extent of humanitarian considerations in interpreting the LOAC and determining military Rules of Engagement are important considerations. This extends to wider state-based actors including intelligence agency-led cyberespionage.[85]

Rules of Engagement

The RoE for militaries varies from state to state but is devised to act as guidelines that operate through a chain of command for military warfighting, disengagement, and restraining conflict under the LOAC. This limits the potential for unintended escalation and enables the cessation of hostilities and re-establishing peace. There are only a very small number of publicly known policies on RoE for cyberwar. These are ascribed to militaries, but cyberwarfare has state intelligence and civilian components within the cyber battlespace. This is where the British Ministry of Defence's (MoD) 2013 'Cyber Primer' document offers some useful guidance. It notes:

> When observing changes in cyberspace, timescales vary from days or months to milliseconds. Individuals and groups operating in cyberspace leave digital trails but these can be disguised, thus making accurate identification, geo-location and attribution difficult. While there are no international treaties specifically governing cyber activity, cyber operations must be conducted in accordance with existing domestic law. The international law that applies to military cyber operations will depend on whether an armed conflict is in existence, be it an international armed conflict or a non-international armed conflict. Where there is no armed conflict, military cyber activities are governed by domestic and international law applicable in peacetime.[86]

The Second Edition of the 'Cyber Primer' published in 2016 shows how this is interpretive as:

> Armed attack is not defined in international law, but it is generally accepted that it must be an act of armed force of sufficient gravity, having regard to its scale and effects. A cyber operation may constitute an armed attack if its method, gravity and intensity of force is such that its effects are equivalent to those achieved by a kinetic attack which would reach the level of an armed attack… The inherent right of individual and collective self-defence is customary international law and is also recognised by Article 51 of the United Nations Charter. An armed attack or imminent armed attack triggers the right of self-defence or anticipatory self-defence. Any response under self-defence must be necessary and proportionate…- Cyber operations conducted during an armed conflict to which the UK is a party, and which are related to that conflict, are governed by the existing rules of the Law of Armed Conflict (LOAC) including the prohibition on perfidy (inviting the confidence of an adversary as to protection under the LOAC) and principles of neutrality.[87]

In addition, the implications of the law of self-defense turn on three practical issues: attribution; the speed with which an attack can be conducted, which greatly reduces the ability to respond to an imminent attack; and the difficulty of determining intent, even if actions are provable and actors identifiable.[88] This could potentially lead to greater use of preemption or 'going first'.

The 'attribution problem' leads to difficulties when it is a 'nation-state like attack' and a reason why the U.S. Department of Defense has examined "whether to declassify intelligence based, for example, on human sources or cyber exploitation" in addition to computer forensics.[89] The attribution problem is well recognized already and one of the reasons many have escaped judicial proceedings (it is also quite likely that most private sector intrusions lay undetected or unreported). Whether acts such as an attack on critical infrastructure committed by cyber 'guns for hire' through the 'Dark Net' can be deterred or prosecuted is a major problem. In this way, both state and non-state actors such as al Qaeda or Islamic State can also have a force multiplier effect and become 'David' to 'Goliath'.[90]

The International Group of Experts behind *Tallinn 2.0* could not agree whether the LOAC applied only in the context of a conflict that had broken out under the conditions set out in the Geneva Conventions.[91] Depending on the operation and context, neutrality of any kind could also prove to be extremely difficult during 'the fog of cyberwar' with NATO's Secretary General, Jens Stoltenberg, acknowledging in 2018 "Nowhere is the 'Fog of War' thicker than it is in cyberspace".[92] As *Tallinn 2.0* debated, the lines set out by the LOAC in terms of territoriality and neutrality have already been blurred by counter-terrorist operations since 9/11.[93] More fundamentally still, Kalevi J. Holsti argues, "To many, war is an irrational activity, representing a rejection of politics for an entirely different domain of behavior. It must be, therefore, structures and processes that lead to war and not the deliberate calculations of policy-makers who might be bent on conquest".[94]

NATO was faced with these sets of issues when Russia invaded Ukraine. With respect to cyberwarfare, since 2014, NATO's position is that "a cyber-attack could trigger Article 5" where an *armed attack* on one NATO member is considered an attack against all.[95] By Fall 2019 Stoltenberg was publicly recognizing that "In just minutes, a single cyberattack can inflict billions of dollars' worth of damage to our economies, bring global companies to a standstill, paralyse our critical infrastructure, undermine our democracies and cripple our military capabilities".[96] Perhaps wisely, the scenarios or 'red lines' under which NATO could invoke Article 5 and the seriousness of that cyberattack remain ill-defined and this position provides latitude in terms of responses and decision-making.

Michaela Pruckovà of NATO's Cooperative Cyber Defence Centre of Excellence (CCDCOE) based in Tallinn, Estonia agues, this reasoning invites "evaluation on a case-by-case basis" built on the rationale that uncertainty can enhance deterrence.[97] To enhance deterrence over Ukraine, NATO strengthened its military deployments and readiness levels. Alongside, NATO has made public pronouncements regarding the sanctity of Article 5 and has "provide[d] significant support to Ukraine both to bolster their cyber defenses, to share information, and to share best practices".[98] As yet, the Ukraine war has left open the question of under what conditions a cyberattack on NATO might invoke Article 5. Michael N. Schmitt, the distinguished law professor who leads the International Group of Experts behind the Tallinn process at the CCDCOE, noted in a February 2022 legal opinion, "whether Russia will conduct hostile cyber operations against NATO States remains to be seen".[99] *If* that should happen, NATO has latitude in deciding its collective response or that of a coalition of the willing.

The Tallinn Manuals and the Cyberwarfare Debate

International Law debates regarding cyberwar(fare), are centered around the *Tallinn Manual*; the second edition of which was published in 2017.[100] The Tallinn Manuals "represents only the views of its authors, and not of NATO, the CCDCOE, [or] its Sponsoring Nations" but "It is an influential resource for legal advisers dealing with cyber issues".[101] The Tallinn Manuals are the culmination of a process and what this process has set out to do is to show that existing international law applies to cyberspace and also "how such law applies in the cyber context".[102] Schmitt noted in his introduction to *Tallinn 2.0*: "There are very few treaties that deal directly with cyber operations and those that have been adopted are of limited scope". Schmitt added, "because State cyber practise is mostly classified and publicly available expressions of *opinio juris* are sparse, it is difficult to definitively identify any cyber specific customary international law".[103]

There are a wide array of difficulties in applying current international (as well as national) law to the cyber domain. These include: What measures constitute self-defense to cyberattack? What would be a proportionate response? To what extent do variances in domestic law inhibit joint operations and inter-operability?[104] On the law of self-defense *Tallinn 2.0* signposts some of the difficulties. It reasoned that in cases where "the gap between less grave uses of force and those that qualify as an armed attack for the purposes of the law of self-defence…will remain an open question until uncertainty as to the use of force and armed attack thresholds is resolved".[105]

Tallinn 2.0 draws distinctions between *jus ad bellum* and *jus in bello* in terms of military necessity and proportionality.[106] Moreover, anticipatory self-defense is clearly a difficult area as a "victim state will lose its opportunity to effectively defend itself unless it acts" when "the attacker is clearly committed to launching an armed attack".[107] In international law this has been framed as the Caroline test which holds anticipatory self-defense as only justified by showing that its necessity was "instant, overwhelming, leaving no choice of means, and no moment of deliberation".[108] This has clear implications for a potentially disarming cyber first strike, especially one against an unprepared adversary or a foe "unable to defend itself when that attack actually starts".[109]

For the *Tallinn 2.0* experts, they drew a "distinction between actions that constitute an initial phase of an armed attack and those that are merely preparatory"…but "the lawfulness of any defensive response will be determined by the reasonableness of the victim State's assessment of the situation".[110] This leaves room for maneuver not only in how a state can distinguish preparedness with immanency but also that that preparedness to engage in cyber operations against a victim state can leave a victim state to conclude that it is vulnerable to the point where it is incapable of self-defense or is incapacitated leading to coercion or blackmail.[111]

Of all the main disciplines that have addressed themselves to the issue of cyberwar(fare), in the Western world at least, the law has seen the greatest level of the discussion centered around "the implications that the emergence of cyber war and warfare will have on the existing Law of Armed Conflict, particularly the conditions under whether cyber war or warfare can be considered as a 'use of force', or 'armed attack'".[112] The consensus view in *Tallinn 2.0* was that "the application of the law of armed conflict to cyber operations can prove problematic".[113] Even among the technical cybersecurity community there is no clear definition of cyberwar or what cyberwarfare constitutes. Cyberwarfare might not emphasize "the ultimate goal of destruction" as Knapp and Langill wrote in 2015.[114] Instead, the goals might be to bring about economic dislocation, disruption to a region, societal panic, or breakdown as a precursor to physical invasion of territory.[115]

Robinson, Jones, and Janicke offer the following definition based on a thorough critique of the discourse on cyberwar and cyberwarfare terminologies.[116] Based on an 'Actor and Intent Definition Model' they define cyberwarfare as "The use of cyber attacks with a warfare-like intent", whilst cyberwarfare is defined as "a situation [where]…cyber warfare is the only type of warfare used. If a kinetic attack is used during the war, such as an air strike, the situation should not be classified as cyber war - it should simply be seen as war where cyber warfare was used".[117] Although their 'Actor and Intent Definition Model' is thought-provoking, and worthy of discussion, tighter definitions are required.[118]

The terms cyberwar and cyberwarfare are used interchangeably every day around the world. Media headlines provide a snapshot but beneath the headlines, the security discourse has been polluted by non-warlike activities, especially by cyberespionage, cybercrime, and straightforward criminality. The media, academia, militaries, lawmakers, policymakers, and citizens who read the headlines and follow the debates are focusing more on the espionage and criminal dimensions of the much wider cybersecurity landscape. The focus should also be squarely fixed on the war-making potential and demonstrations of cyberattacks that are both part of ongoing political and military conflicts and signs of future potential.

Cyberwar Against Critical Infrastructure as a War Winner

The strategic uses of cyber represent a Revolution in Military Affairs (RMA)[119] but cyber itself and the malicious use of cyberspace extends far beyond militaries and into the fabric of society itself across the globe.[120] Colin S. Gray wrote that "Historically viewed, strategic thought, as a practical subject, tends to slumber between episodes of security alarm".[121] The alarm is ringing. Pressing the snooze button will not make it go away. Should there be a 'cyber 9/11' or 'cyber Pearl Harbor' Clarke and Knake's 'wakeup call' will ring loudly. Neither is hyperbole.

In May 2017 Admiral Mike Rogers, then Director of the U.S. National Security Agency (NSA) and U.S. Cybercommand testified to the Senate Armed Services Committee that there continued to be increasing activities from state and non-state actors in cyberspace. Ranging from the government to private sector targets, they included network intrusions into the Joint Chiefs of Staff unclassified network to those controlling U.S. critical infrastructure. Rogers believed, "These threat actors are using cyberspace… to shape future operations with a view to limiting our options in the event of a crisis". Rogers ominously added, "The worst case scenario in my mind…[includes] outright destructive activity focused on some aspects of critical infrastructure".[122]

In a written statement to the committee, Admiral Rogers recorded that alongside the multitude of other threats the United States experiences in cyberspace, are persistent remote intrusions by state adversaries and attempts to penetrate CI and the systems that control these services. Iran was mentioned as was the 2015 BlackEnergy attack (an attack that had Russian fingerprints). According to the U.S. Attorney General, Iran's attacks led to 46 separate companies suffering tens of millions of dollars in losses with the Justice Department indicting seven Iranians for cyber disruptions in the U.S. financial sector alone. BlackEnergy "identified in energy-sector systems worldwide" was primarily aimed at Ukraine's electrical power systems.[123] Its third incarnation, BlackEnergy 3.0, was used to precision-target three regional power distribution companies via external attacks causing power cuts at Christmas 2015.[124] Despite lasting only 3–6 hours, 225,000 customers were affected.[125] Rogers noted:

> The Department of Homeland Security has been warning systems administrators at critical infrastructure sites in the United States and abroad about sophisticated cyber threats from malicious actors employing Black Energy…Infiltrations in US critical infrastructure—when viewed in the light of incidents like these—can look like preparations for future attacks that could be intended to harm Americans, or at least to deter the United States and other countries from protecting and defending our vital interests.[126]

BlackEnergy could have had a longer-term impact by conceivably alerting Ukraine and the West to the dangers cyberattacks pose from a military, public safety, and public health perspective and preparing them for future attacks when they came in 2022.[127] These and other issues raised in this chapter will be explored further in Chapter 3 when examining Russia's use of cyberattacks against Ukraine prior to and during its invasion.

More is likely to emerge in respect to Ukraine, but there are broader underlying issues worthy of attention. Shodan and other tools can provide 'discoverability' to remotely accessible ICS and SCADA devices. These are backbone technologies that enable computer control over local and national critical infrastructure. Moreover, "Some vendors of ICS equipment unintentionally generate cyber security problems due to vendor maintenance policies, such as creating intentional or unintentional "backdoors" for access to devices or software, or by threatening to void equipment warranties if reconfigured from factory settings, i.e., changing passwords or installing unapproved security packages".[128]

With these issues in mind, the then head of Britain's National Cyber Security Centre (NCSC), Ciaran Martin, was unequivocal. In a public address in January 2018, he foresaw that a major cyberattack on UK infrastructure is a case of "when, not if". Martin added, "we will be fortunate to come to the end of the decade without having to trigger a category one attack".[129] This is defined as "A cyber attack which causes sustained disruption of UK essential services or affects UK national security, leading to severe economic or social consequences or to loss of life".[130] He also indicated that although "thus far, the UK has avoided a Category 1 – most of our foremost international partners have not".[131] Although Martin offered no further details, there is growing recognition that cyberattacks can have grave consequences.

Attacks upon critical infrastructure was recognized by *Tallinn 2.0* as an area of growing concern for states and qualifying as grave and imminent peril.[132] *Tallinn 2.0* documented that the temporal nature of these threats or actions might not be immediate but over time can constitute 'grave peril' with immanency when the 'window of opportunity' for preventive measures is about to close.[133] What form of cyberattack amounted to an armed attack divided *Tallinn 2.0*'s legal experts. Stuxnet, which was used to destroy centrifuges in the Natanz nuclear processing plant in Iran, was used as a test case. Opinion was divided on whether it constituted an armed attack, but all saw it as constituting 'the use of force'.[134] This is a narrow test case to go on with specific characteristics.[135] BlackEnergy and Industroyer are better test cases. Attacks that fall beneath the threshold of an armed attack but form part of a series of 'cyber incidents' that when aggregated "meet the requisite scale and effects" were considered to be a "composite armed attack".[136] More pointedly, when considering attacks on critical infrastructure, "severe effects" which would not necessarily be destructive would also qualify as an armed attack.[137]

This ties into considerations about cyberespionage and whether international norms and attempts at consensus building are needed. As Darien Pun describes, "malware…can be changed on-the-fly to achieve a destructive capacity rather than mere surveillance (and often does so to wipe its traces from the hardware). For the victim state in both cases, there is a fear that the vehicle is intended for combat rather than surveillance. A counterargument is that this is no different than a traditional spy who can cause immense physical harm, and thus cyber espionage should be treated the same as traditional espionage".[138] These legal opinions matter. They are relied upon by policymakers, especially in liberal democracies, and help formulate policy and inform responses.

Bearing in mind current legal opinion, contemporary uses of cyber means alone—including in a standalone cyberwar or cybered conflict that does not involve military use of force—might remain beneath the threshold of 'armed attack'. If it is deemed to cross that line, cyberwar would qualify as an armed attack permitting individual and collective self-defense under Article 51 of the UN Charter. Such attacks could be localized, regional, or on a countrywide scale depending on the actor/s involved. Indeed, it might not even be possible to fully engage in a countrywide (total) cyberwar at present, but it will be demonstrated how the groundwork is being laid by "'pre-positioning' for a significant attack in the future".[139] One hopes that such a test case does not materialize but there was widespread feeling that Russia's invasion of Ukraine would be that test case.[140] Chapter 3 details that cyberwarfare was attempted by Russia against Ukraine, but many attacks were blunted or prevented. Some were near misses.

As mentioned in the introduction, for cyberwar to be fully effective much would depend on the size and capacity of the states involved, their ability to absorb attacks and recover, their resources, their allies, and their cyber dependencies and vulnerabilities. What cyber cannot do is to take and hold territory. What it can do is to degrade, damage or weaken a state in a matter of hours and days or over time as part of a campaign, much like through a million pinprick attacks or a frog boiling in water. It could also currently be accomplished by a persistent barrage aimed mainly at industrial, economic, and political targets. This is particularly acute in areas comprising critical infrastructure. This would enable an opponent "to subdue the enemy without fighting".[141]

This does not need to involve physical destruction, although that is certainly possible on a scale far greater than that seen hitherto with Stuxnet in Iran and BlackEnergy and Industroyer in Ukraine. Cyberattacks are already being aimed at striking at the heart and ability of a state and the society on which it depends to mobilize, unite, and fight back. Much like the Maginot Line, a 940-mile string of defenses built by the French government in the run-up to World War II, which "held up superbly against several direct assaults" but which the German blitzkrieg sidestepped by invading Belgium, a defense-in-depth model can be circumvented or burrowed into.[142] This has been a feature of cyber campaigns conducted by Russia and China whilst the United States and its allies have also developed and used offensive cyber weapons for the purposes of espionage and, through Stuxnet, for sabotage.

A future cyberwar might potentially have the ability to cause a state to sue for peace before a bullet has been fired by being able to subdue an enemy and succumb to the political will of the belligerent. Within Clausewitz's trinity of violence, chance, probability, and war as a political action lies the tendency to escalate and to rouse passions leading to previous constraints being removed. This has clear dimensions for cyberwar. Friction and chance also play into the 'fog of war'.[143] As Joel Brenner, a former NSA Inspector General has argued:

> the current and foreseeable state of cyber technology "enables numerous instances of friction to emerge below the threshold of violence" in "the gray space between war and peace…If this environment is showing signs of strategic stability, it is partly…because mutual vulnerability is creating mutual restraint among nation-states. But the vulnerabilities remain, and they could be exploited by China or Russia in a crisis and by a growing number of second-tier cyber players that are not so constrained.[144]

An offensive-led 'shooting war' between major states—including a 'West versus the rest' scenario—risks escalation and all that entails. Whatever the end goals might be, uncertainty plays a significant part in weighing up the decision-making process of whether to escalate hostilities and cross the armed attack threshold. For the time being, "incentives for restraint in cyberspace make it better suited for intelligence operations than for coercive diplomacy or strategic attack. When options for inflicting real harm are limited by operational barriers and strategic deterrence, then espionage and harassment become attractive, if less effective, alternatives".[145]

Cyberattacks and cyber campaigns and wider politico-military strategies do not form in a vacuum. They are shaped by the political values and aims of the states conducting them, reflecting their history and strategic culture.[146] However, for states, and potentially for non-state terrorist groups, war is but one of the tools that can be used for political ends. This is also a feature of Machiavelli's *The Prince*; a treatise on gaining and maintaining power that Clausewitz greatly admired.[147] War is one of these end states. In crises, where the risk of escalation is always present, war should be avoided unless a last resort. Relatedly, cyber arms control has not been afforded the attention it should. Whether or not this is desirable or achievable, deterrence will continue to play a role.

Returning fire through counter cyberattacks threatens escalation and can deepen the anarchic use of cyberspace through the use of proxy non- and sub-state actors to produce a measure of plausible deniability and produce local–regional tactical effects and national-level strategic effects. Resort to military use of force could be impaired by the precursor use of cyberattacks over the medium to long term and military systems themselves could be vulnerable to cyberattacks.[148] Civilian societies which enable militaries are already vulnerable and in battlefield conditions cyber is planned to achieve short-term tactical advantages and planned strategic effects.[149]

The Failure of Cyber Deterrence and the Attribution Problem

Cyber deterrence is being explored by a number of national governments and in NATO forums. It is also being explored by China. One of China's most prominent thinkers on this issue was Dai Qingmin, a key strategist in the mid-late 1990s and early 2000s when ideas of Information Warfare (IW) and Network Centric Warfare (NCW) were in vogue and Dai headed the Fourth Department of the PLA's General Staff. Dai wrote in 2002 that network warfare "could develop in another direction and work to create 'network deterrence' or 'network containment'" by complying with a rulebook where if peer or near peer adversaries could inflict equivalent network damage there was mutual deterrence.[150]

For national governments as well as NATO, which remains the primary military vehicle for the vast majority of the European Union (EU), it is contended that "Cyberdeterence strategy remains largely unexplored and underdeveloped, due to a limited understanding of how the principles of deterrence can be applied to the cyber domain".[151] Grappling with these issues has however taken place, notably through NATO's valuable CCDCOE CyCon conferences.[152] Clarke and Knake are categorical: "deterrence, does not work well in cyber war".[153] Cyber deterrence exists so long as none of the major belligerents launches a full-blown cyberattack because of fears of cyber-retaliation but these questions of deterrence have to be extended to concerning cases of cyberespionage which have risen in the 2010s. What are the proportionate responses to cyberespionage? What is a proportionate response to a cyber-physical attack made possible by cyberespionage campaigns years in advance?

Cyberespionage has made for difficulties in embedding the Rule of Law into the cyber domain and has led to a drip feed of attacks and escalations. This is storing up the prospect of far worse things to come than has hitherto been seen. Raising the barriers to entry as high as feasible is part of the pathway towards a successful deterrence strategy but an argument can also be made to employ offensive cyber means and to 'shoot back'. The 2013 Snowden revelations strongly indicate that offensive means are available *by* the West and Stuxnet provides one clear indicator of use.

Offensive use risks escalation and the risk of escalation (intended or unintended) is high and "Since espionage itself is not definitively illegal, states are uncertain what countermeasures, or perhaps even acts of self-defense, are proportional and what would violate international law". Furthermore, "Without clear ways to retaliate as a means of deterrence, the calculus becomes extremely asymmetrical where actors have very little to lose by constantly deploying remote cyber espionage" and "greatly favors offense over defense".[154] These are also considerations for large multinational corporations and critical infrastructure owner-operators.

Deterrence has not worked against cyberespionage and signaling capabilities also reveal them. It is difficult to come to any other conclusion with the most powerful nation on earth, the United States, attacked with impunity short of cyber-kinetic attacks on critical infrastructure or its military. As the late John McCain, then Chair of the Senate Armed Services Committee, lamented at a May 2017 hearing on U.S. cyber policy, the "lack of a strategy and policy continues to undermine the development of a meaningful deterrence in cyberspace. The threat is growing. Yet, we remain stuck in a defensive crouch, forced to handle every event on a case-by-case basis and woefully unprepared to address these threats".[155]

This is not to say that Western intelligence agencies are passive actors. The Snowden disclosures detailed that the NSA, Great Britain's Government Communications Headquarters (GCHQ), and their three other

'Five Eyes' partners (Canada, Australia, and New Zealand) which were each implicated in foreign and domestic cyber surveillance activities have taken a variety of active measures against foreign actors.[156] Despite the capabilities unveiled by Snowden, which may or may not still be operating in different forms, mass surveillance has not deterred attacks upon the United States and its allies.[157] A two-year-long study on cyber deterrence by a DoD Science Board Task Force reported its findings in February 2017. Its co-chair, Craig Fields, stated at the outset of the report:

> The Task Force notes that the cyber threat to U.S. critical infrastructure is outpacing efforts to reduce pervasive vulnerabilities, so that for the next decade at least the United States must lean significantly on deterrence to address the cyber threat posed by the most capable U.S. adversaries. It is clear that a more proactive and systematic approach to U.S. cyber deterrence is urgently needed.[158]

Echoing the findings of the Task Force, the Senate Armed Services Committee, repeated that there was now a ten-year gap "during which we will not have the defensive capability to defeat our peer adversaries' offensive capabilities".[159]

The spectrum of cyberattacks the United States and its friends and allies currently experience, and could potentially face, does not bode well but the Task Force report does offer some steps in the right direction. Naming Russia and China as major powers who "have a significant and growing ability to hold U.S. critical infrastructure at risk via cyber attack, and an increasing potential to also use cyber to thwart U.S. military responses to any such attacks" is a heads-up to U.S. policymakers and society at large.[160] It is also a heads-up to America's friends and allies.

Concerns were also raised in relation to Iran and North Korea as able to hold the U.S. hostage whilst sub and non-state actors were part of a pinprick series of attacks that are beginning to be felt *en masse*. America's dependence on ICT across both civilian infrastructure and military make the attack surface simply too large to be defended, with many systems increasingly networked. The innovation, integration, and networks that have helped propel America into economic and military ascendency, and which support American political power, have now become a source of bodily vulnerability.

Three measures were put forward by the Task Force as a remedy. One: tailored deterrence campaigns geared directly to each potential adversary. Two: bolstering cyber resilience of military assets and across critical infrastructure with the ability to 'impose unacceptable costs' in retaliation to cyberattacks ranging up to large-scale attacks by major powers. This included creating a hardened cyber capacity resilient to attack and able to conduct a second strike against major powers. This would serve as a 'thin line' underwriting cyber deterrence. Third: foundational capabilities needed to be enhanced in areas of attribution, cyber resilience, and support for technological innovation.

For tailored deterrence to succeed requires in their view, "a wider range of military cyber options, and a clear policy and legal framework for their employment…to add essential rungs to the U.S. escalation ladder".[161] Without these, the United States would be defaulting to diplomatic expulsions, criminal prosecutions, and economic sanctions. If ineffective this could lead to 'next level' "sustained campaigns to undermine U.S. economic growth, financial services and systems, political institutions (e.g., elections), and social cohesion".[162]

In terms of tailored deterrence, the report outlines two well-known Cold War deterrence strategies. The first is 'deterrence by denial', which is based on resistant or 'hardened' defenses capable of withstanding attack and capable of responding with a secure 'second-strike' capability. The second is 'deterrence by punishment' (or 'cost imposition' as the report terms it). This means holding at risk, and able to strike at, that which the adversary values most or considers *sine qua non*—especially for the leadership. This invites a menu of options invoking an 'escalation ladder' to be able to respond to cyberattacks. As the report frames it: "Because "massive retaliation" to limited cyber attacks by nuclear-capable adversaries such as Russia and China is not credible, the United States must develop cyber and non-cyber proportional (although not necessarily symmetrical) response capabilities to attacks ranging from low-level disruption to catastrophic destruction and loss of life".[163] This was not limited to offensive cyber but also should include economic, diplomatic, and law enforcement responses.

Scott Jasper argues in his 2017 study *Strategic Cyber Deterrence* that "in the current threat landscape, cyber attacks seriously challenge the strategic option of deterrence by denial".[164] Jasper is skeptical deterrence by denial can be made to work with "Current security mechanisms and practices…simply inadequate to achieve deterrence and likely will always be".[165] These deterrence strategies make sense coming from a position of strength but not from one of weakness. It is difficult to countenance how the resort to the Cold War logic of deterrence will work in the cyber domain when the United States is at a disadvantage.

Jasper instead suggests a strategy of 'deterrence by entanglement'. Deterrence by entanglement is based on cooperation through shared national interests which encourage restraint. This is founded on the work of Robert O. Keohane and Joseph S. Nye and models of complex interdependency.[166] This seeks to alter the cost–benefit calculations for state actors by clearly signaling unacceptable behavior and having the will and capability to do something about it. Deterrence by entanglement is also difficult to achieve and requires norm building and compellence to be able to hold uncooperative states to account or to desist.

It draws on principles established multilaterally through treaty-based international law "which echo the basic values of international society that are intrinsic to international order".[167] Jasper invokes the norms that were established for other classes of weapons that produced a Revolution in Military Affairs (RMA) such as chemical and biological weapons, strategic bombing platforms, and for nuclear weapons. In order to promote the curtailment of development and deployment as well as use, a majority of states have to recognize that it serves their national self-interest. This is related to arms control which will be returned to later, but Jasper's contention is that the lack of convergence between Russia, China, and the United States on issues of cybersecurity make norms building challenging.[168] As Singer and Friedman argue, "Cyber deterrence may play out on computer networks, but it's all about a state of mind".[169]

The second measure of 'imposing costs' recommended by the Task Force did not specify what those costs might be. However, in discussing the potential for norms building, they believed it would be possible to establish 'red lines' of behavior. They believed this would also act as a disincentive for arms racing in offensive cyber capabilities and for preemption or 'going first'. This logic is part of the security dilemma through the development of offensive and defensive capabilities which produce insecurity between states.[170] Differentiating what qualifies as cyberespionage and what constitutes cyberwar(fare) forms part of the security dilemma. This is a vexing problem.

Wargaming, scenario planning, and studies of escalation dynamics were recommended alongside 'playbooks' "to provide flexible response options…in response to cyber attacks and costly cyber intrusions during peacetime – as well as to support operations in crisis and war".[171] These 'playbooks' would be informed by legal opinion and by the Chairman of the Joint Chiefs of Staff and included evaluations of potential cascading effects.[172] Deterrence founded on the use of hard power/military force is ill-suited to the cyber domain but not without its merits.

The first two measures in particular replicate the Cold War nuclear weapons strategy adopted by the United States and subsequently incorporated into NATO strategy as the doctrine of 'flexible response'. A posture built on adversarial deterrence, war-fighting, and escalation control threatens to further militarize cyberspace. These would be familiar to Cold War warriors as concepts that for many guided a successful long-term strategy that brought strategic stability and a peaceful end to the Cold War.[173] However, in the cyber domain civilian private industry is in the crosshairs far more than militaries. This kind of posture will also likely lead to cyber weapons development and deployment which, it is argued, kept the peace during the Cold War. This includes ideas of 'extended deterrence' to allies which the Task Force wants to extend into cyberspace.

Although deterrence, through the threat of nuclear escalation, did not fail during the Cold War it *could* have. There were a number of crises, most notably the 1962 Cuban Missile Crisis and 1983 Able Archer exercise where the prospect of a nuclear attack hung in the air and it was always a palpable background fear. In addition, the large-scale use of nuclear weapons, even if confined to the superpowers of America and the Soviet Union, would have had global effects. Deterrence strategies based on deployed forces and declaratory strategies of potential use, No

First Use, limiting follow-on use and so on were as difficult and troubling to politico-military policymakers during the Cold War as they are in the cyber context in the twenty-first century.[174] Also, unlike in the Cold War, there are many other actors who could mount damaging attacks on the United States and its friends and allies with undeclared capabilities—visible only when used.

This raises questions of whether the United States is capable of generating extended cyber deterrence or whether deterrence is better served through cooperative efforts with its friends and allies?[175] Moreover, offensive use of cyber in defense of an ally raises questions related to extended deterrence as not only does it reveal capabilities and risk escalation, but also return fire. What might or might not be targeted and whether a current or future U.S. administration sees this as crossing a threshold into 'armed attack' raises difficult questions.

It should be noted that the co-chairs of the Task Force, James N. Miller and James R. Gosler are both Cold War veterans with backgrounds in nuclear strategy. The remaining twenty-three members were a mix of U.S. cybersecurity experts with most, if not all, either former or serving government and military officials. The language and concepts in the report borrow heavily from nuclear strategy. There are also a number of other borrowings from nuclear strategy alongside proposals for 'Non-Nuclear Long-Range Strike Capability' which would see conventional weapons and platforms be 'cyber resilient' and able to provide asymmetric options against major cyberattacks.[176] There were some notable steps forward, not least in the frank recognition of American vulnerability to large-scale cyberattacks. The Task Force report stated early on:

> known cyber attacks on the United States to date do not represent the "high end" threats that could be conducted by U.S. adversaries today – let alone the much more daunting threats of cyber attack that the Nation will face in coming years as adversary capabilities continue to grow rapidly. A large-scale cyber attack on civilian critical infrastructure could cause chaos by disrupting the flow of electricity, money, communications, fuel, and water. Thus far, we have only seen the virtual tip of the cyber attack iceberg.[177]

As a result, "Cyber-resilient electrical power, water, waste-water and communications systems should be particular priorities".[178] The Task Force called for an urgent, proactive, and systematic approach to cyber deterrence as the current policy of sanctions and diplomatic expulsions was not deterring.

The report warned that "if the United States were in a major war with another nation, we should not expect to be able to deter even debilitating cyber attacks on U.S. military capabilities that produced little or no collateral damage to civilian society".[179] This indicates that they see potential attacks by major powers potentially as being confined to military-on-military in the first instance and not a 'first strike' against civilian critical infrastructure. Such an attack would purposely degrade and disrupt military and policymaking responses and weaken America's military overmatch. Indeed, this could already be having an effect as the Task Force recommended ways of using 'retro tech' and de-networking critical military assets which, whilst potentially degrading performance, would improve cyber resilience.[180]

They also warned of the 'boiling frog' mentality as "A range of state and non-state actors' growing capacity for persistent cyber attacks and costly cyber intrusions against the United States, which individually may be inconsequential (or be only one element of a broader campaign) but which cumulatively subject the Nation to a "death by 1,000 hacks"".[181] They also recognized that the roll-out of Self-Monitoring Analysis and Reporting Technology (SMART/'smart') and deepening of the Internet of Things was increasing still further the already large attack surface.[182]

The U.S. Government, especially the DoD, has previously sought to bolster the U.S. cyber deterrence posture. However, the Task Force warned, "it will take many more years of effort, consistent senior-leader attention, and a sufficient budget for ongoing and planned steps to come to fruition…additional steps are urgently needed".[183] Consistent senior-leader attention is certainly needed, and cybersecurity has seen significant funding, but more thought needs to be given over to lessening the threat of large-scale attacks.

The Task Force grappled with how to respond to cyberattacks and what constituted attacks where responses were merited. Those constituting responses included China's 2015 breach of the U.S. Office of Personnel Management and Russia's 'Dragonfly' and BlackEnergy campaigns. 'Dragonfly' was a campaign of malware attacks that began around 2010 that initially targeted aviation and defense companies in the United States and Canada before migrating into the energy sector early in 2013.[184] It is reported to have spread to more than 1,000 energy companies in North America and Europe and between 2,500 and 2,800 in total.[185] 'Dragonfly' has also been found in industries including pharmaceuticals and academia with Russia looking for gateway (and persistent) access into facilities. The BlackEnergy campaign has evolved into a campaign spanning over a decade from the early 2010s through to the present. In December 2015, 'Dragonfly' was used to precision-target three regional power distribution companies in Ukraine through an external attack leading to power cuts in Kyiv at Christmas 2015.[186]

The Task Force reasoned, "U.S. leaders must not be paralyzed into inaction by fear of escalation" as not responding invites follow-on attacks even if it carries the risk of an 'intelligence loss'.[187] The Task Force was dismissive of treaty-based cyber arms control because they believed verification would not be feasible. Its lack of exploration is well worth addressing in future studies as are ideas of 'cyber peacekeeping' and cyberwar termination should deterrence fail. Norms building was believed to be achievable and this will be returned to at various points.[188]

Unlike in almost all other areas of international relations, the United States is facing the threat of a major cyberattack from a position of vulnerability and able to be coerced.[189] Here it is not in a position of strength where it can set the field and lead. The Task Force did not pull its punches when it recognized:

> it will not be possible (for the foreseeable future) to deny highly capable actors the ability to conduct catastrophic cyber attacks on the United States. This is primarily because the limited U.S. efforts to defend U.S. information systems to date are unlikely to accelerate (in the near- to mid-term at least) to the point where they can offset the combination of major powers' technical wherewithal, vast supply of resources (including a supporting intelligence apparatus), and the ability to influence supply chains and exploit vulnerabilities at scale.[190]

This was squarely directed at Russia and China. The two other states directly referred to by the Task Force, Iran, and North Korea, it felt could be deterred and could be denied the ability to attack.[191]

Iran

Iran is the subject of much attention, as is the Middle East more generally, but the cyber threat they pose is of a far lower order than either China or Russia. They are far less sophisticated than either despite the fact that Iran was an early adopter of the Internet and Iranians have a relatively high degree of 'Internet Freedom'.[192] There are three main limiting factors effecting Iran's use of offensive cyber. The first is the 'brain drain' from Iran that sees around 150,000 Iranians leave every year. Approximately 2.6 million have departed since the 1979 Islamic revolution and around 130,000 students are enrolled at foreign universities. Many do not return, including those working on cybersecurity or in the ICT sector, and "When Iranian engineers leave for Silicon Valley and Europe, the country's capacity for effective offensive and defensive cyber operations goes with them".[193]

The second is that "Unlike the cyber operations of the United States and Israel, which are conducted by professional intelligence services supported by billion dollar budgets, Iran's offensive and defensive capabilities are disorganized and modestly funded".[194] This is despite a 12-fold budget increase for cyber under President Hassan Rouhani (2013–2021).[195] Third, whilst many of Iran's state-level cyber threat actors, including some groups defined as APTs, have been widely attributed to a multitude of external offensive cyberespionage campaigns (including intrusions into critical infrastructure) many of these same groups have also been directed to internal political repression or external critics.[196] These practices were given impetus after mass protests occurred following the disputed presidential election between Mahmoud Ahmadinejad and Mir-Hossein Mousavi in 2009 dubbed the Green Movement (after the color adopted by Mousavi and worn by his supporters).[197] The Arab Spring beginning in 2011, which toppled regimes in North Africa and spread to the Middle East before being suppressed, also have added to concerns. Against this background, there have been at least two waves of significant domestic protests in 2019–2020 for the regime to contend with, with another following in Fall 2022.[198]

Domestically, "Iran has shown little success in fostering a mature cybersecurity industry and lags behind both developed economies and key regional rivals in terms of investing in defense or formulating national policies to secure critical infrastructure". Moreover, "Unlike China, Iran has limited use for commercial espionage given its lack of an industrial production sector that could utilize stolen intellectual property. Iran's industrial espionage activities serve to boost its commodities industries and military technological prowess rather than its domestic manufacturing sector. Nor has Iran attempted to offset the impact of economic sanctions through large-scale financial crime, as North Korea appears to do".[199] Part of the reason Iran has lagged behind is the effect of U.S.-led economic (and travel) sanctions. Largely imposed due to concerns over its nuclear program, these far-reaching sanctions have hit hard. This includes its ability to sell oil and repatriate revenue frozen in foreign banks.[200] So long as sanctions remain in place, the 'brain drain' from Iran is also likely to continue.

Despite Iran's limitations and a domestic focus on maintaining the Islamic regime against internal and external dissidents, there have been several international cyberespionage campaigns that are causes for concern. Most notably Iran is suspected of developing Shamoon, a 'wiper' malware campaign that "removed and overwrote the information on the hard drives of 30,000 to 55,000…workstations of Saudi Aramco". Although it "did not disrupt an industrial process"[201] Shamoon "resulted in millions of dollars in damages", even though "the malware was unsophisticated and the attack did not require significant resources".[202]

Shamoon was subsequently upgraded and used against targets in the energy sector, specifically the oil refining and petrochemicals industry and aviation sector in countries with ties to Saudi Arabia—a key regional rival.[203] This has been attributed to a group designated APT33 which FireEye suspects "works at the behest of the Iranian government".[204] APT33 was also implicated in unsuccessful cyberespionage/phishing campaigns in June 2019.[205] In 2020–2021 Iran also stood accused by Israel of cyberattacks against a hospital as well as part of Israel's water supply (by specifically trying to raise the level of chemicals including chlorine used for water treatment). In return, Israel hit back with a cyberattack on Iranian petroleum distribution.[206] Both Saudi Arabia and Israel have been prime targets for Iran. Both are regional rivals and a supporter and ally of the United States, respectively.[207]

Attempts have been made to penetrate government, military, and scientific institutions in many other regions across the Middle East, and in the U.S., Europe, and Africa. The sectors include aviation (both military and commercial), wider defense and security industries, energy (especially petrochemicals, including oil companies and engineering firms), finance and financial institutions, government agencies, human rights organizations and journalists, technology companies, telecommunications, travel, universities, and think tanks (relating to policy research, diplomacy, and all aspects of international affairs that concern Iran). Most appear very limited intrusions. However, Iran has also been implicated in attacks on sectors that comprise critical infrastructure. This has seen Iranian hackers gain knowledge of operations and also engage in ransomware attacks against owner-operators.[208]

At least half a dozen APTs have been attributed to Iran. However, these are not static entities with a clearly identifiable base of operations. Instead, as Marie Baezner suggests, "Iranian APTs are fluid entities that usually disappear once a cybersecurity company reports on them". After unmasking, these "APT groups are dissolved, and their members are reallocated to other groups".[209] Iran's government also sub-contracts to sub-state hackers who act as proxy forces for the regime and have been known to freelance. They are not regime loyalists. Consequently, managerial oversight from either the central government or the Islamic Revolutionary Guard Corps (IRGC) is needed to maintain alignment with state goals and religious ideology through the Basij Cyber Council.[210]

As of 2018, there were believed to be around fifty such groups all competing for government contracts, but they have struggled to find groups with the right technical skillset. This might explain why there appears to be a division of labor at times with groups instructed to work with one another (e.g., one to develop or repurpose malware and another to carry out the operation or campaign). Some of these groups also appear to have freelanced into cybercrime, which although not state-sanctioned has been linked back to the Iranian government. In addition, Shahid Beheshti University and Imam Hossein University have also been implicated in sub-contracted cyberattacks.[211]

There are several known or suspected Iranian APTs. Not all remain active. These include APT33 ('Refined Kitten'), APT34 ('Helix Kitten'), APT35 ('Charming Kitten'), APT39 ('Remix Kitten'). Other groups include 'Clever Kitten', 'Charming Kitten', 'Domestic Kitten', 'Flying Kitten', 'Rocket Kitten', and 'Twisted Kitten'.[212] These are the monikers given to them by CrowdStrike and other companies in the cybersecurity industry. They, like other APTs, are also known by a variety of other names. All use tactics, techniques, and procedures (TTPs) common to other well-organized groups including spear phishing and pretexting. APT34 is the most mature of Iran's APT groups, although most are probably not deserving of the title of Advanced Persistent Threats.

APT34's primary targets have been entities in the Middle East, but it has also targeted African nations and the United States[213] APT34 was initially identified in 2015, focused primarily on Saudi Arabian financial and technology companies. It has also been implicated in attacks on other sectors including energy, aviation, universities, and government and critical infrastructure sites more widely across the Middle East using Russian-developed malware that was possibly sold on by the developers.[214]

'Charming Kitten' (APT35), believed active since at least 2014, has concentrated on long-term cyberespionage is aimed at governments and diplomacy, defense technology, and militaries. Whilst it is persistent it lacks advanced skills and capabilities. However, they are well-resourced with overlaps between them, APT33, and other Iranian hacker groups.[215] APT33 ('Refined Kitten') has also been associated with APT34 ('Helix Kitten') since around 2017. APT33's target sectors include aviation (both military and commercial), and energy (especially petrochemicals). Geographically, its targets have been the United States, Saudi Arabia, as well as South Korea.[216]

APT39 ('Remix Kitten'), which has not been clearly linked to the Iranian government, focuses on the Middle East and telecommunications, travel, and IT companies that work in high technology. Their interest in the travel industry is to enable them to locate persons of interest.[217] 'Clever Kitten' is similarly suspected of targeting individuals as is 'Domestic Kitten' (which targeted Iranians supporting ISIS and those of Kurdish and Turkish heritage).[218] 'Fox Kitten', first identified in 2017, was a widespread cyberespionage campaign through to 2020 which targeted multiple companies and organizations concentrating on IT, telecommunications, oil and gas, aviation, government, and the security sector in Israel and other parts of the world, whose goal was to exfiltrate valuable information/intelligence. 'Fox Kitten' was linked to APT33 but with overlaps with APT34 and a connection with APT39.[219]

'Rocket Kitten', which was first identified in 2011, conducts cyberespionage on individuals working in academic institutions (especially those working on physics and nuclear sciences), defense industries, well-placed government and defense officials, embassies, and media outside Iran. It is one of the groups that has diversified into cybercrime and was active through to 2017.[220] According to a Trend Micro report from 2015, "Their favorite targets seem to be involved in policy research, diplomacy, all aspects of international affairs, defense, security, journalism, human rights, and others". Those targeted were mainly located in the Middle East and although specific goals are unclear, what "is clear that it has to do with espionage".[221] 'Pioneer Kitten', active from around 2017–2020, was another intelligence gathering operation mainly directed at North American and Israeli targets.[222]

Another called 'Flash Kitten'/'Leafminer'/'Raspite' targeted electric utilities in the United States and a broader range of targets in the Middle East, Europe, and East Asia using methodologies associated with Russia's APT 'Berserk Bear' or 'Energetic Bear' and the 'Dragonfly 2.0' campaign.[223] Whether this is part of a rapprochement or *modus vivendi* between Russia and Iran remains open to interpretation.[224] According to Dragos, from 2017 to date 'Raspite' has concentrated on ICS owner-operators but "has not demonstrated an ICS-specific capability to date…[and] there is no current indication the group has the capability of destructive ICS attacks including widespread blackouts like those in Ukraine". However, the potential to try and hold CI 'at risk' remains because the "targeting focus and methodology are clear indicators of necessary activity for initial intrusion operations into an IT network to prepare the way for later potential ICS events".[225]

There are some other warning indicators, including the case of 'Flying Kitten'. 'Flying Kitten' is one of Iran's 'patriotic hacking' groups and the Iranian government has "openly encouraged these groups to launch cyberattacks against enemies of Iran". Similar groups such as the 'Iranian Cyber Army' and 'The Cyber Fighters of Izz ad-Din al-Qassam' have also conducted low-level attacks.[226] Seven members of Izz ad-Din al-Qassam were indicted by the Department of Justice for DDoS attacks on U.S. banks. More importantly, one member was also accused of the cyberattack on the Bowman Dam in Rye, New York in 2013 where he learned "critical information about the dam's operation, including details about gates that control water levels and flow rates".[227] In 2011/2012 individuals working on critical infrastructure projects were also targeted by an unnamed group from Iran.[228]

Despite Interpol Red Notices which could see them arrested and extradited, the seven accused remain free whilst in Iran. Instead, these are attempts to 'name and shame' Iran. It is worth bearing in mind that "Tehran frequently masks its involvement in such operations using cutouts (intermediaries) to avoid attribution and provide it plausible deniability".[229] Still, who authorizes these operations is unclear and it is not publicly known how these are directed or agreed upon within government, or whether the president, or the power behind the throne, the Supreme Leader, Ali Khamenei, inputs or directs operations or campaigns.

Relatedly, Anderson and Sadjadpour make the case that whilst Iran is a rational actor it is not a unitary actor and they speculate that "Iran's security apparatus can easily conduct hostilities in cyberspace without the consent or awareness of the rest of the government".[230] The use of proxy hacking groups, some of whom have been known to freelance, lends credence to this view. At the same time, it also raises diplomatic concerns over unintentional risk-taking and unintended consequences, and the risk of mistaken signaling and worries regarding intentions and government control over proxy forces. This is a cause for concern with several Iranian government institutions having cyber responsibilities.

According to a 2020 Congressional Service Report, the most important of these is the Supreme (or High) Council of Cyberspace which "coordinates cyberspace policy for the Iranian government and coordinates between offensive and defensive cyber operations". Next is the IRGC which oversees cyber offense and manages sub-contracted hacker groups/proxies and works in tandem with Iran's Cyber Defence Command/Cyber Headquarters.[231] A thorough report by the British journalist Deborah Haynes for Sky News, based on classified documents obtained from a covert offensive cyber unit called Shahid Kaveh, run by the IRGC, also detailed intelligence collection against companies working in critical infrastructure as well as planning to exploit the intelligence they have gathered.[232]

The Ministry of Intelligence and Security appears to play little part in external cyber operations but it is suspected of running 'Magic Kitten', a group mostly directed at internal repression and the targeting of external critics of Iran's hardliner Islamic regime but also believed to be an active and widespread external cyberespionage actor in the early 2010s.[233] It also engages in counterintelligence and between 2018 and 2019 allegedly identified 290 CIA agents, a number of whom were likely Iranian moles, with dozens killed. This followed the compromise of the CIA's covert communications system (COVCOM) by Chinese intelligence in the early 2010s.[234]

Iran has also been a target, especially through Stuxnet (Operation Olympic Games). Stuxnet derailed Iran's nuclear program sufficiently to buy time to negotiate the Joint Comprehensive Plan of Action (JCPOA) signed in 2015, which aimed to divert Iran from developing nuclear weapons (a deal the Trump administration withdrew the United States from in 2018).[235] There followed Operation Nitro Zeus, developed by the NSA. This operation went much wider than Stuxnet by targeting Iran's critical infrastructure.[236] Israel is also alleged to have conducted cyberattacks on Natanz in July 2020 and again in April 2021.[237] In June 2022 a steel plant was also hit by a cyber-kinetic attack that caused molten steel to spill and start a fire. A group known as 'Predatory Sparrow' claimed responsibility and released hacked video footage to prove its claims.[238] To raise barriers against further attacks Iran established the National Passive Defense Organization (NPDO) in October 2020.[239]

In summation, according to Anderson and Sadjadpour "Having been the target of sustained cyber espionage and destructive attacks, Iran is bound to seek the same capabilities used against it. These capabilities provide Tehran opportunities to impose costs during potential hostilities". However, "While Iran may not appear able to perform synchronized multistage attacks wherever it would like, it can repeatedly hammer away at soft targets".[240] Iran remains an active threat (and Iranian capabilities and intentions are discussed elsewhere in this book) but currently, this is the extent of Iranian capabilities against critical infrastructure.[241] The NSA has recognized that capabilities exist "to take down control systems that operate U.S. power grids, water systems and other critical infrastructure".[242] However, these capabilities do not extend to either Iran or North Korea.

North Korea

North Korea (the Democratic People's Republic of Korea, DPRK) has also faced sanctions because of its nuclear and missile programs. Unlike Iran, which is attempting to develop nuclear weapons, North Korea is a fully fledged nuclear weapons state. It is an authoritarian communist regime ruled by Kim Jong-un who succeeded his father, Kim Jong-il, who in turn had succeeded his father, Kim Il-sung (who was part of the resistance to the Japanese occupation of the Korean peninsula, 1910–1945). After World War II North and South Korea (the Republic of Korea, ROK) were separated at the 38th parallel, with the Soviet Union and China backing the North, and Western allies the South. Under the leadership of his grandfather, Kim Il-sung, North Korea invaded the South in 1950. In 1953 a truce was negotiated and both sides retreated. No peace treaty was signed leaving both technically at war, separated by a demilitarized zone (DMZ) along the 38th parallel.

The DPRK's hereditary and authoritarian regime has led to a system of patronage with domestic cohesion enforced by the military. Since the truce, the DPRK has become one of the most closed and secretive countries on earth leaving its people blinded to the outside world to a large degree with Internet access tightly controlled by Kim Jon-un and his inner circle. Out of an estimated population of 25 million, nearly 1.3 million active-duty personnel serve in the military.[243] This is "the world's highest ratio of military personnel to the general population".[244] From the time of Kim Jong-il, the North has also been implicated in terrorism.[245] Despite a state policy of *Juche* (self-reliance), it is far from self-reliant and depends on food and fuel from abroad. Kim Jong-il, who ruled from 1994–2011, "lived a life of luxury" whilst "the overwhelming majority of the population lived on the edge of starvation".[246] Nothing has changed under his son. A 2019 UNCHR report stated "10.9 million people (over 43 per cent of the total population) are undernourished and suffer from food insecurity" with "unmet health, water, sanitation and hygiene needs" or access to "safe drinking water".[247]

Famines in the 1990s and 2000s saw between 580,000 and 1.1 million people starve (3–5% of the population) but "privileged groups—especially members of the government, the military, and the Korean Worker's Party" were protected from the worst.[248] The UNCHR also reported that "in the midst of the famine, the Government had allocated "disproportional amounts of resources on its military, on the personality cult of the Supreme Leader, related glorification events and the purchase of luxury goods for the elites".[249]

It also detailed widespread bribery, corruption, and extortion; a lack of food, clothing, and other basic necessities; human rights violations including arrests, detentions, torture, and denying fair trials; and economic destitution and disparities between urban and rural areas. Almost 30,000 have fled across the northern border with China and still more try to flee. Those fleeing the south are unlikely to make it past the DMZ.[250] Given all of this, it is both tragic and remarkable that North Korea is so technologically advanced. North Korea's leadership might one day have to answer for the path they have chosen but how they have accomplished this is the subject of widespread reports, and it includes cyberespionage.[251]

Gaining understanding and insight into the structure of the DPRK's cyberespionage operations "is extremely difficult at the unclassified level".²⁵² What is accepted is that North Korean cyberespionage is directly controlled by the leadership headed by Kim Jong-un and although the army has cyber units and missions which include signals intelligence (SIGINT), electronic warfare, information warfare, and psychological operations (PSYOPS), cyberespionage is largely directed by the high-level Reconnaissance General Bureau (RGB).²⁵³ Within the RGB are subdivisions comprised of: First Bureau Operations; Second Bureau Reconnaissance; Third Bureau Foreign Intelligence; Fifth Bureau Inter-Korean Affairs; Sixth Bureau (technology/technical)²⁵⁴; Seventh Bureau (support/rear services); and Bureau 121. There is no fourth Bureau for cultural reasons.²⁵⁵

The RGB was established in 2009 after a large-scale reorganization. Until then "the DPRK's general cyber capabilities were strewn about numerous departments, bureaus, and offices".²⁵⁶ The RGB is headed by General Kim Yong-chol, a 30-year veteran of North Korea's intelligence community who is "reported to have a heavy hand in overseeing Bureau 121" (described below).²⁵⁷ He is alleged to have ordered the attack on Sony Entertainment in 2014 and is the equivalent of the U.S. Director of National Intelligence and reports directly to the State Affairs Commission (SAC) headed by Kim Jong-un. Kim is also Supreme Commander of the Korean People's Army (KPA), and Chairman of the Central Military Commission of the Workers' Party of Korea (WPK). Furthermore:

> As the Supreme Commander of the KPA, Kim commands the General Political Bureau (총정치국, GPB), General Staff Department (총참모부, GSD) and the Ministry of the People's Armed Forces (인민무력성, MPAF). Specifically, the GPB oversees party organs within the military and is responsible for issues related to political ideology; while the GSD is responsible for conducting military operations; and the MPAF for administering military diplomacy, military logistics, procurement, and finance. Furthermore, as the Chairman of the Central Military Commission of the WPK, Kim deliberates and decides what measures are necessary for implementing military policy and provides guidance for overall defense affairs at a party level.²⁵⁸

According to a 2015 report by CSIS, the RGB draws its personnel from across the government as well as the Workers [Communist] Party.

The RGB is an intelligence organization that has been responsible for "black/special operations force that can spy, abduct, and provoke during peacetime". More important still, "with a direct channel to senior leadership and an independent budget, [it] is probably not beholden to the operational planning of the GSD, and [has] a certain freedom….to act independently from the larger military strategy and doctrine".[259] Within the GSD Kong, Lim, and Kim believe that are three Units (31, 32, and 56) which "are responsible for malware development" along with an Enemy Collapse Sabotage Bureau directed at South Korea "tasked with information and psychological warfare".[260]

The DPRK has followed the cyber pathway since the First Gulf War in 1991 which demonstrated the advantages of a networked military. The groundwork for the embrace of cyber, electronic warfare, and nuclear weapons were laid by Kim Jong-il.[261] According to a United Nations Panel of Experts, the selection process for any of North Korea's cyber groups is stringent and recruits are chosen at a very young age and given rigorous training by the military and secret services.[262] North Korea is estimated to have around 7,000 'cyber soldiers'. 6,000 of these work in Bureau 121. Many are said to operate outside the DPRK including in China, India, Bangladesh, Nepal, Malaysia, Thailand, Russia, Belarus, Kenya, and Mozambique.[263] North Korean defectors indicate that members of Bureau 121 live in luxury apartment complexes near their purpose-built (and possibly subterranean) headquarters in northern Pyongyang (constructed in 2013).[264] From 2017 to 2019 there was a 300% increase in state-run cyberespionage and cybercrime operations by North Korea that was partly enabled through a deal with Russia's TransTelekom. This provided increased capacity and bandwidth in addition to the main provider, China Unicom.[265]

The RGB is "the linchpin of the DPRK's cyber operations, as well as its clandestine operations", but within the RGB, Bureau 121 coordinates most of North Korea's cyberespionage operations and campaigns with indications that the Bureau is "organizationally significant" and "personally overseen by the supreme commander".[266] There are believed to be subordinate units again beneath Bureau 121. This series of groups, some of which are designated APTs, use TTPs common to other threat actors including watering hole attacks, social engineering and spear phishing, and the use of macros. These include 'Andariel', APT37, APT38, 'Bluenoroff', 'Labyrinth Chollima', 'Stardust Chollima', 'Velvet Chollima'/ 'Kimsuky', 'TEMP.Hermit', 'TEMP.Firework', the 'Lazarus Group', and an electronic warfare regiment.[267] Groups, including at Kim Il Sung University in Pyongyang, are also involved in cryptojacking which covertly steals computing power to mine cryptocurrencies.[268]

Thailand's Computer Emergency Response Team lists 11 North Korean hacking groups and the overlaps between them.[269] It is difficult to draw clear distinctions between some of these groups and the intrusions and campaigns they are associated with, with these groups containing "hackers, cryptologists, and software developers". Whilst the U.S. government refers to all DPRK cyber operations as 'Hidden Cobra' most others draw distinctions.[270] The problems delineating these groups is discussed below, but the UN's Panel of Experts articulated that through them the DPRK engages in "widespread and increasingly sophisticated" methods of cybercrime and money laundering as a means of circumventing the economic sanctions placed on it. This includes theft from financial institutions, online cryptocurrency exchanges, and money laundering using cryptocurrencies which can be converted into fiat. The attacks on cryptocurrency exchanges "are harder to trace and subject to less government oversight and regulation than the traditional banking sector" and North Korea uses front companies as well as operatives overseas working "under diplomatic cover".[271]

This all adds to a complex and confusing picture, aiding the DPRK's attempts at "deception and concealment" which also includes using alternate names or cover designations for their cyber teams.[272] Spending time and energy analyzing what many believe are specific groups based on their TTPs might be counterproductive because "North Korea with its very small decision-making circle, can pursue a strictly object-oriented strategy and redirect various resources toward that objective as needed, rather than being organizationally confined". Given the tight control from the leadership, and the close involvement of both the intelligence and military, they all work for common purpose with little division of responsibilities. The stranglehold of the leadership on North Korea means that "Activities such as cyber crime and hacktivism are not confined to non-state actors, nor are the functions of espionage and military action strictly delineated" as they are in other states.[273]

Relatedly, as Pierluigi Paganini questioned at the time of the Sony hack, "is very difficult to attribute the cyber attack to a specific threat actor or to understand the nature of the offense. Is it an act of cybercrime or a state-sponsored attack run by a foreign government?"[274] The Insikt Group more broadly recorded that "Attribution of specific cyber activity to the North Korean state or intelligence organizations is difficult".[275] Adding to this confusion, there might also have been a series of reorganizations around 2012/3 concerning Bureau 121 that led to the establishment of several new organizations including 'Lab 110' (the 110th Research Center, Third Bureau of the RGB).[276]

Of these groups, 'Andariel'targets "South Korean government and military, foreign businesses, government agencies, financial services infrastructure, private corporations, and businesses" and has been active since at least 2009. They concentrate on the defense sector but have also been implicated in "malicious cyber operations on foreign businesses, government agencies, financial services infrastructure, private corporations, and businesses" as well as "financially-motivated computer intrusions".[277] 'Bluenoroff' has been accused of the same and according to the U.S. Treasury, both 'Andariel' and 'Bluenoroff' are part of, or work with, the 'Lazarus Group'.[278] 'Labyrinth Chollima' and 'Stardust Chollima' have also been linked to the 'Lazarus Group'.[279] 'Velvet Chollima'/'Kimsuky', is a cyberespionage group directed against South Korea.[280] Its targets have included South Korea's Ministry of National Defense and Korea Hydro & Nuclear Power (KHNP) which supplies around 27% of the South's electricity. It is also tasked with stealing information for the DPRK's nuclear weapons program.[281]

APT37 is a cyberespionage group with counterintelligence priorities that has been active since at least 2012. It has targeted chemicals, electronics, manufacturing, banking, aerospace, automotive, and healthcare sectors. Its primary focus has been on South Korea, but it has also been identified as going after targets in the U.S., Japan, Vietnam, and the Middle East. Like other APTs and other threat actors, they are adept users of social engineering (detailed in Chapter 5). APT38 targets banks and financial institutions, including online cryptocurrency exchanges and since it was first identified in 2014, APT38 has targeted over 16 organizations in at least 13 countries.[282] FireEye believe that APT38 concentrates on large-scale cybercrime focusing on financial institutions which they describe as "a large, prolific operation with extensive resources".[283] APT38's activities are global and its "primary mission is targeting financial institutions and manipulating inter-bank financial systems to raise large sums of money for the North Korean regime".[284] To give an idea of the lengths DPRK agents have gone to, in an attack on Redbanc, an interbank network in Chile, a target employee was first enticed through a job opportunity on LinkedIn. There followed "an entire interview over Skype in Spanish to build trust before asking the target to download malware".[285]

'TEMP.Hermit' has been attributed to Lab 110 and is another of the DPRK's cyberespionage groups. Its primary targets are government, defense, energy, and financial institutions in South Korea, but it has worldwide operations (including against U.S. targets). It has been active since around 2013 and has been linked to APT38 and is suspected of using front companies in northeast China to mask its activities.[286] 'TEMP.Firework' has targeted government agencies, universities, and think tanks. The locations of these targets are primarily in South Korea and the United States, but it has also been found in Australia, Italy, India, Japan, Hong Kong, Hungary, and the Philippines. It is particularly interested in nuclear security, nonproliferation, and the U.S.-South Korea alliance and military capabilities. It also conducted cyberespionage during the 2020 U.S. presidential election to gain insight into any potential policy changes relating to the DPRK. In addition, it and another DPRK team targeted multiple pharmaceutical companies and medical centers (primarily for data on Covid-19 and the development of vaccines). 'Kimsuky', which primarily focuses on intelligence gathering, has gone after government organizations and think tanks as well as individual experts working in multiple fields (including biomedicine) and has been active since around 2012.[287]

In late 2018 McAfee warned of "a new global campaign targeting nuclear, defense, energy, and financial companies" which it attributed to the 'Lazarus Group'.[288] This was the same group that was responsible for WannaCry (discussed in Chapter 6).[289] 87 organizations spread across the globe were targeted in a spear phishing campaign (many with the lure of job opportunities). Most were in the United States with "the majority of targets...defense and government-related organizations", including a prime defense contractor in Lockheed Martin.[290] Once the two-stage attack was executed, reconnaissance was performed on the victim's network. Drive scanning was then performed which located documents of interest and date stamps. The data (intelligence) they gathered was then encrypted and exfiltrated back to North Korea (or servers controlled by them and then rerouted).[291] Three members of this group were indicted by the FBI in February 2021.[292]

'Lazarus Group' is associated with APT37 and has been linked to APT38 as well as Lab 110. It was also deemed responsible by the FBI for the attack on Sony Pictures in 2014 following the release of *The Interview* (a black comedy parodying Kim Jong-un). It was also accused of an $81 million 'cyber heist' from Bangladesh Bank which involved fraudulently authenticated Society for Worldwide Interbank Financial Telecommunication (SWIFT) requests for fund transfers (part of a much wider and longer-term campaign that could have netted at least $1 billion). In addition, the FBI also accused them of "numerous other attacks or intrusions on the entertainment, financial services, defense, technology, and virtual currency industries, academia, and electric utilities".[293] Between 2016 and 2020, the UN's Panel of Experts has estimated that the DPRK has acquired as much as $2 billion from cybercrime from at least 35 reported cases spread across 17 countries.[294]

As far as cyberwarfare against critical infrastructure is concerned, even when one takes into account the activities of 'Velvet Chollima'/'Kimsuky' against KHNP and the 2018 'Lazarus Group' campaign, there have not been any other (publicly reported) serious ventures or moves to disrupt critical infrastructure, let alone damaging attacks, to ends other than intelligence gathering and ransomware. This leads to the conclusion that for Kim Jong-un, regime survival and the continuance of North Korea are best served by a "desire to remain within the gray zone".[295]

The U.S. government maintains that the "DPRK has the capability to conduct disruptive or destructive cyber activities affecting U.S. critical infrastructure".[296] Disruptive attacks for financial gain through ransomware are widely evidenced. There are no public reports that intrusions and breaches into critical infrastructure aimed at destructive attacks qualifying as 'armed attack', especially against South Korea, are being planned or conducted. This said it has gained insights into worldwide and country-specific critical infrastructure, as well as the global defense-industrial base and supply chains. Moreover, its groups have been successful against their targets. Neither these attacks nor attackers should be taken lightly, especially if used as a prelude or part of attacking the ROK or an invasion (moves that could draw the United States and China into direct or proxy conflict). Should North Korea continue to remain risk-averse to destructive attacks CI whether in the ROK or elsewhere, it could nevertheless become a proxy actor for others. Policy changes at the top, human error and miscalculation, and the potential for escalation remain clear and present dangers especially if the regime believed it faced existential threats domestically or externally. If the regime collapsed there would be many takers for North Korean hackers, including in China on their northern border.[297] These are concerns for the United States and its allies, including the ROK.

Policy and Debates in the United States

A CNN report from November 2017 observed there is "no explicit definition or legal framework in the United States for what constitutes cyberwar".[298] Defining the term through informed debate would help determine policy in this internationally contested area. The Cyber Act of War Act proposed in 2016 would have necessitated some deep thinking around the issue and helped define and determine a political and legal classification of what constitutes cyberwar.[299] Instead, the debate has centered around mis and disinformation campaigns under the banner of 'fake news' and election interference in the United States (interference which has also been seen widely across Europe).[300] Economic and sanctions and diplomatic expulsions are neither a like-for-like response or sufficient.[301]

In a May 2017 written statement to the Senate Armed Services Committee Admiral Mike Rogers, then the serving Director of the U.S.

National Security Agency (NSA) and U.S. Cybercommand, did not define either cyberwar or cyberwarfare. Nevertheless, he asserted that:

> 'Cyber war' is not some future concept or cinematic spectacle, it is real and here to stay. The fact that it is not killing people yet, or causing widespread destruction, should be no comfort to us as we survey the threat landscape. Conflict in the cyber domain is not simply a continuation of kinetic operations by digital means, nor is it some Science Fiction clash of robot armies. It is unfolding according to its own logic, which we are continuing to better understand.[302]

The December 2017 National Security Strategy also made no mention of cyberwar(fare). Previously, the doctrine for cyberwar(fare) was laid out in a Presidential Policy Directive (PPD-20), a classified document that was disclosed by Edward Snowden in 2013. PPD-20 offered the following 'Guiding Principles' for Defensive Cyber Effects Operations (DCEO) and Offensive Cyber Effects Operations (OCEO):

> DCEO and OCEO may raise unique national security and foreign policy concerns that require additional coordination and policy considerations because cyberspace is globally connected. DCEO and OCEO, even for subtle or clandestine operations, may generate cyber effects in locations other than the intended target, with potential unintended or collateral consequences that may affect U.S. national interests in many locations.[303]

PPD-20 emphasized the adherence to international law in both DCEO and OCEO "consistent with its obligations under international law, including with regard to matters of sovereignty and neutrality, and, as applicable, the law of armed conflict".[304]

For offensive cyber use, it called for the identification of "potential targets of national importance where OCEO can offer a favorable balance of effectiveness and risk as compared with other instruments of national power" and to "establish and maintain OCEO capabilities integrated as appropriate with other U.S. offensive capabilities".[305] During the Obama administration Rules of Engagement were tightly defined by PPD-20 and tightly controlled by the White House given the blurry distinctions between military uses of offensive and defensive cyber capabilities and their uses (and usefulness) to the intelligence community. Resultantly, PPD-20 sought to maintain political control.[306] The capabilities and capacity to respond are clearly there. Some of the extant capabilities of the NSA, unveiled by the Snowden revelations of 2013, point to a wide range of tools that could be used as both preventive measures and for offensive 'hack-back'.

President Trump's Executive Order (EO) 13800, *Strengthening the Cybersecurity of Federal Networks and Critical Infrastructure*, of May 2017 was directed to these ends. EO 13800 built on President Obama's Presidential Policy Directive-21 (PPD-21) of February 2013 dealing with how the government could improve the resiliency of private sector owner-operators of critical infrastructure. Specifically, the Department of Energy and DHS were tasked with jointly assessing:

i. the potential scope and duration of a prolonged power outage associated with a significant cyber incident, as defined in Presidential Policy Directive 41 of July 26, 2016 (the United States Cyber Incident Coordination), against the United States electric subsector;
ii. the readiness of the United States to manage the consequences of such an incident; and
iii. any gaps or shortcomings in assets or capabilities required to mitigate the consequences of such an incident.[307]

The first of a series of reports requested by Executive Order (EO) 13800 was published in May 2018 by the Office of Management and Budget (OMB). Using the widely respected NIST Framework of Identify, Protect, Detect, Respond, and Recover of the 96 federal agencies surveyed 74% were identified as either 'At Risk' or 'High Risk'.[308] Resultantly, the U.S. government will need to try and shore up its own systems before it points the finger at the private sector. Coordinated counterintelligence against foreign threats is aided by the National Counterintelligence and Security Center (NCSC), created in December 2014, which reports to the Office of the Director of National Intelligence (DNI).[309]

The 2018 U.S. National Cyber Strategy: CISA and the Biden Administration

The 2018 National Cyber Strategy (NCS), the "first fully articulated cyber strategy in 15 years", recognized that cyber deterrence had failed and that critical infrastructure was vulnerable and attacks were taking place beneath the threshold of war.[310] To better manage cybersecurity risks the Trump administration planned to centralize a number of federal authorities and promote American interests under four pillars 'Protect the American People, the Homeland, and the American Way of Life', 'Promote American Prosperity', 'Preserve Peace through Strength' and 'Advance American Influence'.

Under Pillar I, supply chain risks to government, infrastructure, and defense procurement are intended to be much more stringently monitored, including sub-contracted firms. It also laid out the need to improve threat and vulnerability information when 'sharable'—a long-standing debate and an area of discord between the government and private industry in the United States and elsewhere. Critical Infrastructure Protection (CIP) was vaguely defined by the NCS with risk-management being used to improve base levels of cybersecurity. Deterrence would attempt to be improved by prosecutions and economic sanctions with equally vague language about other forms of deterrence strategy with no mention of a cyber offensive strategy or Rules of Engagement. This was found instead in National Security Presidential Memorandum 13 which indicated a more reflexive posture and greater willingness to integrate offensive cyber into operations.[311]

Notably, the Cybersecurity and Infrastructure Security Agency (CISA) was established in November 2018. Its priorities include infrastructure resilience and field operations, emergency communications, and federal network protection. Previously CISA was the National Protection and Programs Directorate (NPPD) within the Department of Homeland Security (DHS) including the National Cybersecurity and Communications Integration Center (NCCIC). CISA now integrates functions previously run independently by the U.S. Computer Emergency Readiness Team (US-CERT) and the Industrial Control Systems Cyber Emergency Response Team (ICS-CERT). CISA, like its CERT predecessors, is responsible for issuing advisories, reports, and alerts to critical infrastructure providers.[312]

In May 2021 President Joe Biden signed EO 14028 'Improving the Nation's Cybersecurity'. Its stated aim is to improve the protection and security of both IT and OT systems with top priority accorded to the prevention or remediation of cyber incidents. Intelligence sharing barriers were ordered to be removed between government agencies and more rapid reporting of incidents initiated. Legacy software was also highlighted as an area of vulnerability and software testing improvements requested. This included IoT devices coming to market and consideration whether a consumer software labelling program was feasible. A public–private Cyber Safety Review Board aims to be established under the Department of Homeland Security (DHS). The Secretary of Homeland Security alongside the Attorney General also would establish a Cyber Safety Review Board. This board would be convened in the event of a significant cyber incident. In turn, this would trigger the formation of a Cyber Unified Coordination Group (UCG). Both will operate under Presidential Policy Directive 41 of July 2016, and both appear to be ad hoc rather than standing groups. A 'playbook' of operational responses to incidents based on NIST standards was also planned as is a National Cyber Director (NCD).[313] This top-down drive for improvement is welcome but leaves many unanswered questions including outside of the Federal Government, what is expected of the private sector? And will the Federal Government share more actionable intelligence with private industry and in particular the owner-operators of critical infrastructure?

The U.S. Military and 'Forward Defense'

For the U.S. military, U.S. Cyber Command (USCYBERCOM), founded in 2009, is part of the Department of Defense and like the NSA based at Fort Meade in Maryland. It is the military branch of U.S. cyber operations. Their policy is found in the 2018 National Defense Strategy (NDS) and USCYBCERCOM's *Achieve and Maintain Cyberspace Superiority Command Vision for US Cyber Command* doctrine. Drawing on the

National Defense Strategy, USCYBCERCOM pointed to the restrictions of PPD-20 in arguing:

> We cede our freedom of action with lengthy approval processes that delay US responses or set a very high threshold for responding to malicious cyber activities. Our adversaries maneuver deep into our networks, forcing the US government into a reactive mode after intrusions and attacks that cost us greatly and provide them high returns. This reactive posture introduces unacceptable risk to our systems, data, decision-making processes, and ultimately our mission success.[314]

The issue of legal restrictions was addressed by the Trump administration which outlined a deterrence model which involved:

> Defending forward as close as possible to the origin of adversary activity extends our reach to expose adversaries' weaknesses, learn their intentions and capabilities, and counter attacks close to their origins. Continuous engagement imposes tactical friction and strategic costs on our adversaries, compelling them to shift resources to defense and reduce attacks.[315]

Forward defense is a long-standing part of U.S. strategy and a reflection of U.S. strategic culture.[316] It permits engagements far from the U.S. mainland which is protected by the Atlantic and Pacific Oceans; its Eastern and Western seaboards. Mexico and Canada on its two land borders are friends or allies. Forward bases in Europe, Asia–Pacific, and the Middle East combined with satellite reconnaissance, long-range missiles and the force projection enabled by aircraft carriers and global reach stealth bombers, next-generation hypersonic platforms and conventional heavy bombers like the B-52 give the United States a global reach that no other nation possesses.

The United States is also, across many classes of weapons systems, the lead nation, and its aircraft carriers dwarf those of other nations in size, number, and capabilities. The two nations with comparable carriers are Great Britain with its two *Queen Elizabeth* class carriers and France with the *Charles de Gaulle* and both are NATO allies. Russia has a single carrier, the *Admiral Kuznetsov*, which is a late Soviet design with another upgraded carrier undergoing sea trials. China operates the *Lioaoning*, the sister ship of the *Admiral Kuznetsov*, which it purchased from Ukraine with another indigenously designed and built carrier, the *Shandong*, entering service in December 2019. India also operates a Soviet-built carrier, the *INS Vikramaditya* with another undergoing sea trials.[317] In short, the United States remains the dominant military power in the world. Its only peer competitors are other nuclear weapons powers, particularly Russia, and peer adversaries in the cyber domain. Aside from China, America's adversaries do not seek to compete at a symmetrical level, whilst Russia takes the path of 'hybrid warfare' in 'gray zones' beneath the level of armed conflict with the West.

Lt. Gen. Paul Nakasone, the current director of the NSA and commander of United States Cyber Command who supported the recommendations of the Task Force during his Senate confirmation hearing in March 2018, was questioned, "Is U.S. Cyber Command and the military services actively developing capabilities to threaten the critical infrastructure of peer adversaries?" Nakasone replied with a simple, "Yes".[318] Nakasone had stated in advance:

> We face a challenging and volatile threat environment, and cyber threats to our national security interests and critical infrastructure rank at the top of the list. Cyber threats are already challenging public trust and confidence in global institutions, governance and norms, and are imposing significant costs on the U.S. and global economies. Cyber threats also pose an increasing risk to public health, safety and prosperity as cyber technologies are integrated with critical infrastructure in key sectors.[319]

The National Guard could also step into to try and secure civilian (private sector) critical infrastructure from attack. Nakasone also outlined campaign plans built on deterrence or offensively developed "to go after key assets of our opponents, such as the wealth of Putin oligarchs, their financial transactions, corruption in Iran, information issues with respect to Russia, China, and North Korea".[320]

Nakasone's testimony "was the first time U.S. cyber attack capabilities against foreign infrastructure were discussed in public".[321] This was one of the recommendations of the Task Force in calling for "declaratory policy relating to U.S. responses to cyber attack and use of offensive cyber capabilities".[322] There is evidence that undeclared capabilities exist, and that Russia has already been targeted. Russia is the most active of the Advanced Persistent Threat (APT) actors in cyberspace but could itself be experiencing similar pre-positioning for future attacks on its critical infrastructure that they are inflicting on others. There are few publicly available examples of cyberattacks against Russia but these are likely to be the tip of the iceberg. One such cyberattack or campaign was through a 'cyber-weapon' called 'Sputnik' in 2013. This was part of a cyberespionage operation that gathered intelligence on military agencies, institutes, and diplomatic consulates.

In 2016 the FSB (*Federal'naya Sluzhba Bezopasnosti*/Federal Security Service): Russia's civilian intelligence service tasked with domestic and foreign operations[323] also "discovered several trojan viruses in the IT infrastructure of Russia's state, scientific, and defense institutions (in total, about two dozen facilities). The FSB indicated that the attack was planned carefully and carried out by trained professionals".[324] It bore the hallmarks of a nation-state with different exploits for each enterprise and also used 'zero day' vulnerabilities which is usually another marker of a nation-state APT. It began with a phishing expedition designed to infect target computers through modular infections that installed a keylogger, which records keystrokes, access to any connected webcams and microphones and allowed network traffic interception. There was a Kremlin enquiry, and the Russian Duma pointed the finger at the Americans.[325] If true, this was almost certainly the NSA.

Supporting this theory is a well-sourced article from the *Washington Post* in June 2017. It reported that President Obama ordered a "covert measure that authorized planting cyberweapons in Russia's infrastructure" which could be triggered remotely "if the United States found itself in an escalating exchange with Moscow".[326] They were designed by the NSA, CIA, and U.S. Cyber Command to be capable of a lawful proportional response to future Russian actions permitted under international law.[327] They were being "designed to be detected by Moscow but not cause significant damage".[328] This was in the planning stages when the Obama administration left office and was, at least in part, a response to Russian U.S. election interference in 2016. This was part of an ongoing campaign which unless countermanded by the Trump or Biden administrations, remains an active covert program.[329]

Given these offensive cyber activities Russia has been instituting defensive cyber measures. In 2013 the FSB began to develop GosSOPKA which is a central system of detection, warning, and mitigation for Russian government networks, intended as a hub for warnings and information sharing for critical infrastructure. In 2015 GosSOPKA was divided into regional response centers and government departments with a Computer Emergency Response Team (RU-CERT) with a 24/7 response center CERT-GIB whilst Fin-CERT operates for the financial services sector.[330]

Federal Law No. 187-FZ 'On the Security of the Russian Federation's Critical Data Infrastructure' also came into force in January 2018.[331] In January 2017 hearings on this bill, Nikolay Murashov of the FSB testified to the Duma that Russia had faced 70 million cyberattacks during 2016 alone.[332] Global attack statistics through to 2020 show similar numbers.[333] The FSB now has prime responsibility for the protection of Russia's CI through GosSOPKA.[334]

Part of the renewed public discussion in American politics, and with its friends and allies, is finding means to deter peer adversaries, which include Russia and China. For Nakasone, this includes U.S. military forces having "the ability and will to retaliate effectively against the critical infrastructure of peer adversaries" as a deterrent to first-strikes on U.S. infrastructure—or in retaliation.[335] This could be construed as an admittance of the Obama-era operation.

Plans are prepared for retaliatory attacks against Russian and Chinese critical infrastructure and plans made to make this capability public as part of a wider, more focused, cyber deterrence strategy. The same network intrusions and mapping of CI chokepoints and vulnerabilities that both Russia and China have separately conducted are part of this strategy. This would amount to the same pre-positioning that both Russia and China have engaged and intelligence-led by the DoD. Nakasone stated: "To be operationally effective in cyberspace, U.S. forces must have the ability to conduct a range of preparatory activities which may include gaining clandestine access to operationally relevant cyber systems or networks".³³⁶ One of the recommendations of the Task Force was for CYBERCOM to develop "specific capabilities to support approved "playbook" options, including capabilities that do not require "burning" intelligence accesses (sources and methods) when exercised".³³⁷

Part of the counterintelligence efforts of USCYBERCOM, supported by the NSA and National Counterintelligence and Security Center,³³⁸ is found in the assignment of National Mission Teams (NMTs) commanded by U.S. Cyber Command to defend the United States against significant cyberattacks. Their function "is to identify and surveil potential adversary cyber forces, learn as much as possible about those forces and their networks, plans, capabilities, and tactics, and be prepared to disrupt their activities if or when so ordered".³³⁹

With concerns that DHS lacks the ability to defend against a major attack on American CI, Nakasone also revealed that if the DoD had spare capability and capacity to support DHS, that would fall under the (longstanding) Defense Support of Civil Authorities (DSCA) process.³⁴⁰ However, given the waves of cyberattacks America and its allies have faced there have been criticisms of the DoD, of which USCYBERCOM and the NSA are part. During hearings of the Senate Subcommittee on Cybersecurity in March 2018, Senator Bill Nelson (D-FL) complained that the DoD was disorganized in the way it was conducting information warfare and information operations.³⁴¹ Nakasone had noted in his confirmation hearing that the DoD, FBI, and DHS with their separate remits, capabilities and capacity needed better coordination with "90 percent of our networks in the private sector" and "likely to be the first indicator of…intrusion or attack".³⁴²

Nelson went on to criticize the lack of progress in these areas and U.S. responses to Russian and North Korea cyberattacks or cyberattacks against Islamic State. This included counter information operations. He also criticized constraints on cyber operations imposed by "outmoded concepts of sovereignty, attribution, and intelligence gain/loss calculations".³⁴³ Major General Christopher P. Weggeman of the US Air Force, however, testified in DoD's defense that U.S. policy should be guided by the Rule of Law and should not seek "to act like the irresponsible actors" but to continue to "operate under the Law of Armed Conflict…[and]

rules of engagement".[344] These legally set limits on counter-information operations such as political interference.

Cyber-physical attacks have been technically demonstrated in Ukraine and by Stuxnet, but we have not yet seen them used more widely. Russia has not conducted cyber-physical attacks in the West—or farmed out the capability to Russian cybercriminals or to mercenaries or terrorists. These remain potential scenarios and Nakasone is aware that actors are not currently being deterred. Current norms mean instead, "they do not think much will happen to them", "They don't fear us", he said.[345] Against Russia, a number of former officials have called for aggressive 'shoot-back' policies rather than just diplomatic expulsions and economic sanctions.[346]

The use of economic and travel sanctions alongside expulsions might fail, pressuring deterrence calculations, and leading to greater use of offensive cyber operations. This risks escalation and further cyber arms racing. This was not lost on the 2017 DoD Task Force:

> The United States and Russia, and the United States and China, share extremely strong stakes in avoiding major war, including through misperception and inadvertent escalation. The dynamics of cyber offensive weapons will increase challenges to crisis stability, as each side is likely to perceive significant advantages and relatively low risks (no direct casualties, no visible damage) to going first with offensive cyber against the other side's military. At the same time, one side's assessment of imminent/underway offensive cyber attacks against its offensive cyber capabilities or military more broadly could be viewed as a compelling indicator of imminent conflict – and create real fears of "use or lose." Thus, as offensive cyber capabilities continue to grow, and are likely to outpace cyber defense and resilience, there are likely to be growing risks of misperception that could lead to rapid cyber escalation – and the potential for rapid escalation to armed conflict.[347]

Inadvertent or intended escalation might not be restricted to militaries. Critical infrastructure is in the crosshairs. This could well entail direct casualties, including mass casualty events, and significant and widespread damage. Russian and Chinese goals differ but both appear to be gearing up through increasing levels of pre-positioning in American and Western critical infrastructure. If Western intelligence agencies and their militaries have corresponding programs this will be a non-zero-sum game if the triggers are pushed. At the very least national economies will suffer, and very likely the global economy.

Conclusion

The human geographer, Robert Kaiser, questioned in 2015: "Hardly a week goes by when cyberwar is not a featured news story. Yet only a few years ago it was barely acknowledged as a realistic security threat, and its imaginative production was limited largely to sci-fi novels and films. What happened to bring about such a fundamental change in western security discourse?".[348] The answer lies in developments, events, and elevation into the mainstream through popular culture. Mainstream media reporting of real-world developments and events has also been raising its visibility. This has all led to the securitization of cyber, its politicization, scholarly study, and military adoption of cyber as the fifth domain of warfare. Hughes and Colarik argue in their 2017 study of cyberwar definitions, "the discourse peaks in response to what we consider to be the three most notable cyber incidents in the international domain; the cyber conflicts between Russia and Estonia in 2007, between Russia and Georgia in 2008, and the Stuxnet attack [in Iran] in 2011".[349] To this, we can now add Ukraine and U.S. election interference.

Whilst both have spiked interest, the latter has dominated the Western security discourse even more than the destruction of Islamic State in the Middle East. Election interference by Russia is part of a far broader campaign. The attacks on Ukraine's power grid and other parts of CI are part of that. These proof-of-concept attacks are a harbinger of what could come. Cyberdeterrence is also (at best) ineffective under current conditions. David E. Sanger, a veteran journalist and author wrote in 2018 that Pentagon wargaming for future confrontations with Russia and China (as well as Iran and North Korea) is predicated on cyberattacks against civilians because of expected targets being part of critical infrastructure. Sanger cautions, "It would fry power grids, stop trains, silence cellphones and overwhelm the internet. In the worst-case scenarios, food and water would begin to run out; hospitals would turn people away. Separated from their electronics, and thus their connections, Americans would panic, or turn against one another".[350]

Pentagon planning now weaves in planning for this scenario. Its own war plans also begin this way. These are ends Clausewitz and Sun Tzu would have understood, as they would the underlying roles of friction, chance, and uncertainty, but the means belong to the cybered environments of the twentieth and twenty-first centuries. Clausewitz and Sun Tzu still provide lessons to learn from. This is especially so with too little in-depth debate "on how this new revolution is reshaping global power".[351] Let us begin to improve our handle on this revolution in the next chapter by looking in detail at vulnerabilities throughout critical infrastructure which are the backbones of developed states.

Notes

1. Carl von Clausewitz, *On War*, edited and translated by Michael Howard and Peter Paret (Oxford: Oxford University Press, 1997), p. xiii.
2. Also characterized by Clausewitz as its 'essence' (*wesen* in the original high German).
3. von Clausewitz, *On War*, pp. 20–21, 27–29.
4. András Rácz, 'Russia's Hybrid War in Ukraine: Breaking the Enemy's Ability to Resist', Finnish Institute of International Affairs (2015), p. 19, http://www.fiia.fi/en/publication/514/russia_s_hybrid_war_in_ukraine/, accessed 28 October 2018.
5. von Clausewitz, *On War*, p. 20.
6. von Clausewitz, *On War*, pp. 45–47.
7. Amit Sharma, 'Cyber Wars: A Paradigm Shift from Means to Ends', in Christian Czosseck and Kenneth Geers (eds.), *The Virtual Battlefield: Perspectives on Cyber Warfare* (Amsterdam: IOS Press, 2009), p. 3.
8. Emile Simpson, 'Clausewitz's Theory of War and Victory in Contemporary Conflict', *Parameters*, Vol. 47, Issue 4 (Winter 2017–2018), pp. 17–18.

9. von Clausewitz, *On War*, pp. 65–68. How the U.S. technological advantage (in particular, advances in surveillance and reconnaissance) is believed to be able to clear the 'fog of war' can be found in Barry D. Watts, *Clausewitzian Friction and Future War* Revised Edition, McNair Paper 68 (Washington, DC: National Defense University, 2004), esp. pp. 1–4.
10. Carl von Clausewitz, *Principles of War*, translated by Hans Wilhelm Gatkze (Harrisburg, PA: Military Service Publishing, 1942), p. 27.
11. Andrew Foxall, 'Putin's Cyberwar: Russia's Statecraft in the Fifth Domain', Russia Studies Centre Policy Paper No. 9 (2016), pp. 1–15.
12. Computer Security Resource Center (CSRC), 'Glossary', https://csrc.nist.gov/Glossary/?term=2856, accessed 3 September 2018.
13. Joint Publication 3-12, 'Cyberspace Operations', Joint Chiefs of Staff (June 8 2018), p. III-11. https://www.jcs.mil/Portals/36/Documents/Doctrine/pubs/jp3_12.pdf?ver=2018-07-16-134954-150, accessed 8 March 2020.
14. For a breakdown of the conceptual underpinnings of war see Quincy Wright, *A Study of War* (Chicago: Chicago University Press, 1942). See also Waqar H. Zaidi, 'Stages of War, Stages of Man: Quincy Wright and the Liberal Internationalist Study of War', *The International History Review*, Vol. 40, Issue 2 (March 2018), pp. 416–425.
15. von Clausewitz, *On War*, p. 26.
16. The trinity has also been appropriated to mean "government, military, and population (society)". von Clausewitz, *On War*, p. xxviii.
17. 'Cyber Defence in the EU Preparing for Cyber Warfare?' (October 2014), p. 3, http://www.europarl.europa.eu/EPRS/EPRS-Briefing-542143-Cyber-defence-in-the-EU-FINAL.pdf, accessed 28 October 2018.
18. On securitization see Barry Buzan, Ole Wæver and Jaap de Wilde, *Security: A New Framework for Analysis* (London: Lynne Rienner, 1997) and Jan Ruzicka, 'Failed securitization: Why It Matters', *Polity*, Vol. 51, Issue 2 (April 2019), pp. 365–377.
19. Thomas Rid, *Cyber War Will Not Take Place* (London: Hurst, 2013), pp. 3–4. The 9/11 Commission Report, p. 336, https://www.9-11commission.gov/report/911Report.pdf, accessed 13 January 2019.
20. Daniel Hughes and Andrew Colarik, 'The Hierarchy of Cyber War Definitions', in G. Alan Wang, Michael Chau and Hsinchun Chen (eds.), *Pacific Asia Workshop on Intelligence and Security* (2017),

p. 20, https://pdfs.semanticscholar.org/34c5/8f3a28f836bd78352381e9f6054dd78f374d.pdf, accessed 13 September 2018. See also Keir Giles and William Hagestad II, 'Divided by a Common Language: Cyber Definitions in Chinese, Russian and English' and Scott D. Applegate and Angelos Stavrou 'Towards a Cyber Conflict Taxonomy', both in Karlis Podins, Jan Stinissen and Markus Maybaum (eds.), *5th International Conference on Cyber Conflict Proceedings* (Tallinn: CCD COE Publications, 2013), pp. 413–430, 431–450.

21. John Arquilla and David Ronfeldt, 'Cyberwar Is Coming!', in John Arquilla and David Ronfeldt, *In Athena's Camp: Preparing for Conflict in the Information Age* (Santa Monica: RAND Corporation, 1997), p. 30. The original article is in *Comparative Strategy*, Vol. 12, Issue 2 (Spring 1993), pp. 141–165.
22. This includes books concerning cyberwar(fare). There are too many to list but, beyond those already cited, include Jeffrey Carr, *Inside Cyber Warfare: Mapping the Cyber Underworld* (Sebastopol, CA: O'Reilly Media, 2010), Peter W. Singer and Allan Friedman, *Cybersecurity and Cyberwar: What Everyone Needs to Know* (Oxford: Oxford University Press, 2014).
23. Hughes and Colarik, 'The Hierarchy of Cyber War Definitions', p. 30.
24. 'Michael Hayden, Richard Clarke on Greatest Cyberthreats Facing America', Washington Post Live (October 6 2017), https://www.youtube.com/watch?v=FdiAQBXGsMg, accessed 17 October 2018.
25. Kello, 'The Meaning of the Cyber Revolution,' pp. 9–14, 22. In addition to the ongoing examples, theoretical and conceptual debate abound. See, for example, David Betz, 'Cyberpower in Strategic Affairs: Neither Unthinkable nor Blessed', *Journal of Strategic Studies*, Vol. 35, Issue 5 (October 2012), pp. 689–711, Timothy J. Junio, 'How Probable Is Cyber War? Bringing IR Theory Back in to the Cyber Conflict Debate', *Journal of Strategic Studies*, Vol. 36, Issue 1 (February 2013), pp. 125–133, Adam P. Liff, 'The Proliferation of Cyberwarfare Capabilities and Interstate War, Redux: Liff Responds to Junio', *Journal of Strategic Studies*, Vol. 36, Issue 1 (February 2013), pp. 134–138; and Erik Gartzke, 'The Myth of Cyberwar Bringing War in Cyberspace Back Down to Earth', *International Security*, Vol. 38, Issue 2 (Fall 2013), pp. 41–73.

26. Joseph S. Nye, 'Is Cyber the Perfect Weapon?' (July 5 2018), https://www.project-syndicate.org/commentary/deterring-cyber-attacks-and-information-warfare-by-joseph-s--nye-2018-07, accessed 13 September 2018.
27. Rid, *Cyber War Will Not Take Place*, p. xiv.
28. Rid, *Cyber War Will Not Take Place*, p. xiv.
29. Rid, *Cyber War Will Not Take Place*, p. 4.
30. Richard A. Clarke and Robert K. Knake, *Cyber War: The Next Threat to National Security and What to Do About It* (New York: HarperCollins, 2010), p. 101.
31. Clarke and Knake, *Cyber War*, p. xi.
32. Clarke and Knake, *Cyber War*, p. xi.
33. See for example John Arquilla, 'Twenty Years of Cyberwar', *Journal of Military Ethics*, Vol. 12, Issue 1 (April 2013), pp. 80–87, Julian Richards, *Cyber-War: The Anatomy of the Global Security Threat* (London: Palgrave Macmillan, 2014), Lucas Kello, *The Virtual Weapon and International Order* (New Haven CT: Yale University Press, 2017), Alexander Klimburg, *The Darkening Web: The War for Cyberspace* (New York: Penguin, 2017), Tobias Kliem, 'You Can't Cyber in Here, This Is the War Room! A Rejection of the Effects Doctrine on Cyberwar and the Use of Force in International Law', *Journal on the Use of Force and International Law*, Vol. 4, Issue 2 (June 2017), pp. 344–370, Damien Van Puyvelde and Aaron F. Brantly, *Cybersecurity Politics, Governance and Conflict in Cyberspace* (Cambridge: Polity Press, 2019), Elizabeth Van Wie Davis, *Shadow Warfare: Cyberwar Policy in the United States, Russia and China* (Lanham: MD: Rowman & Littlefield, 2021).
34. Lucas Kello, 'The Meaning of the Cyber Revolution Perils to Theory and Statecraft', *International Security*, Vol. 38, Issue 2 (Fall 2013), pp. 7–40.
35. Jack Goldsmith, 'How Cyber Changes the Laws of War', *European Journal of International Law*, Vol. 24, Issue 1 (February 2013), pp. 129–138, Michael Robinson, Kevin Jones and Helge Janicke, 'Cyber Warfare: Issues and Challenges', *Computers & Security*, Vol. 49 (March 2015), pp. 70–94, Martin C. Libicki, *Conquest in Cyberspace: National Security and Information Warfare* (New York: Cambridge University Press, 2007), and Martin C. Libicki, 'Cyberspace Is Not a Warfighting Domain', *Journal of Law and Policy for the Information Society*, Vol. 8, Issue 2 (Fall 2012), pp. 325–340.
36. Clarke and Knake, *Cyber War*, p. 257.

37. This can be preceded by posturing (which complicates how aggression is defined), a buildup of forces which create insecurity and raise tensions, or by miscalculation, insubordination and accident. The latter especially being contributors to friction.
38. 3314 (XXIX). Definition of Aggression, https://documents-dds-ny.un.org/doc/RESOLUTION/GEN/NR0/739/16/IMG/NR073916.pdf?OpenElement, accessed 9 October 2018.
39. Anna-Maria Osula and Henry Rõigas (eds.), *International Cyber Norms Legal, Policy & Industry Perspectives* (Tallinn: CCD COE Publications, 2016). Katharina Ziolkowski (ed.), *Peacetime Regime for State Activities in Cyberspace. International Law, International Relations and Diplomacy* (Tallinn: NATO CCD COE Publications, 2013).
40. Kalevi J. Holsti, *Major Texts on War, the State, Peace, and International Order* (London: Springer Nature, 2016), p. 21.
41. Danny Sjursen, 'America Is Addicted to Fighting Undeclared Wars' (May 7 2017), https://nationalinterest.org/feature/america-addicted-fighting-undeclared-wars-20535, accessed 10 October 2018.
42. Darien Pun, 'Rethinking Espionage in the Modern Era', *Chicago Journal of Internal Law*, Vol. 18, Issue 1 (Summer 2017), p. 363.
43. Pun, 'Rethinking Espionage in the Modern Era', p. 363.
44. This is at the very far end of possibilities. As someone who has written extensively on nuclear weapons strategy, and interviewed decision-makers and military personnel at all levels of the firing chain, there is hope and expectation that the protection accorded to nuclear weapon command and control is more robust than *any* other area. Given levels of classification this cannot be proven and Chapter 4 will indicate there is cause for concern.
45. Michael Walzer, *Just and Unjust Wars: A Moral Argument with Historical Illustrations* Fifth Edition (New York: Basic Books, 2015), p. 24.
46. Walzer, *Just and Unjust Wars*, p. 24.
47. Peter R. Mansoor, 'Introduction: Hybrid Warfare in History', in Williamson Murray and Peter R. Mansoor (eds.), *Hybrid Warfare: Fighting Complex Opponents from the Ancient World to the Present* (New York: Cambridge University Press, 2012), p. 3.
48. Martin van Creveld, *The Transformation of War* (New York: Free Press 1991), p. 2. See also William S. Lind, Keith Nightengale, John F. Schmitt, Joseph W. Sutton, Gary I. Wilson, 'The Changing Face of War: Into the Fourth Generation', *Marine Corps Gazette* (October 1989), pp. 22–26.

49. Martin van Creveld, 'The Transformation of War Revisited', *Small Wars & Insurgencies*, Vol. 13, Issue 2 (Summer 2002), pp. 3–15.
50. Frank G. Hoffman, *Conflict in the 21st Century: The Rise of Hybrid Wars* (Arlington VA: Potomac Institute for Policy Studies, 2007), p. 18.
51. Original emphasis. Hoffman, *Conflict in the 21st Century: The Rise of Hybrid Wars*, p. 28.
52. Frank G. Hoffman, 'Hybrid Warfare and Challenges', *Joint Forces Quarterly*, Issue 52 (Spring 2009), p. 34.
53. Robert Wilkie, 'Hybrid Warfare: Something Old, Not Something New', *Air & Space Power Journal*, Vol. 23, Issue 4 (Winter 2009), pp. 13–17, Gjorgji Veljovski, Nenad Taneski and Metodija Dojchinovski, 'The Danger of "Hybrid Warfare" from a Sophisticated Adversary: The Russian "Hybridity" in the Ukrainian Conflict', *Defense and Security Analysis*, Vol. 33, Issue 4 (December 2017), pp. 292–307.
54. William J. Nemeth, 'Future War and Chechnya: A Case for Hybrid Warfare', MA thesis, Naval Postgraduate School, Monterey, California (June 2002), https://calhoun.nps.edu/bitstream/handle/10945/5865/02Jun_Nemeth.pdf;sequence=3, accessed 23 December 2018.
55. Klimburg, *The Darkening Web*, pp. 205–252.
56. Mark Galeotti, '(Mis)Understanding Russia's Two 'Hybrid Wars'', *Critique & Humanism*, Vol. 59, Issue 1 (2018), p. 1. Reprinted in https://www.eurozine.com/misunderstanding-russias-two-hybrid-wars/?pdf, accessed 3 January 2018.
57. Galeotti, '(Mis)Understanding Russia's Two 'Hybrid Wars'', p. 4.
58. Daniil Turovsky, 'What Is the GRU? Who Gets Recruited to Be a Spy? Why Are They Exposed so Often? Here Are the Most Important Things You Should Know About Russia's Intelligence Community' (6 November 2018), https://meduza.io/en/feature/2018/11/06/what-is-the-gru-who-gets-recruited-to-be-a-spy-why-are-they-exposed-so-often, accessed 26 December 2018.
59. Miriam Lanskoy and Dylan Myles-Primakoff, 'Power and Plunder in Putin's Russia', *Journal of Democracy*, Vol. 29, Issue 1 (January 2018), p. 81.
60. Keir Giles, 'The Next Phase of Russian Information Warfare', p. 4, https://www.stratcomcoe.org/next-phase-russian-information-warfare-keir-giles, accessed 3 January 2018.
61. Galeotti, '(Mis)Understanding Russia's Two 'Hybrid Wars'', p. 5.
62. Ion Mihai Pacepa and Ronald J. Rychlak, *Disinformation: Former Spy Chief Reveals Secret Strategies for Undermining Freedom, Attacking*

Religion, and Promoting Terrorism (Washington, DC: WND Books, 2013). See also Booz Allen Hamilton, 'Bearing Witness: Uncovering the Logic Behind Russian Military Cyber Operations' (2020), https://www.boozallen.com/content/dam/boozallen_site/ccg/pdf/publications/bearing-witness-uncovering-the-logic-behind-russian-military-cyber-operations-2020.pdf, accessed 25 July 2021.

63. Daniel W. Kruger, 'Maskirovka—What's in it for Us?', School of Advanced Military Studies, Fort Leavenworth Kansas (December 4 1987), https://apps.dtic.mil/dtic/tr/fulltext/u2/a190836.pdf, accessed 13 January 2019.
64. Timothy C. Shea, 'Post-Soviet Maskirovka, Cold War Nostalgia, and Peacetime Engagement', *Military Review*, Vol. 82, Issue 3 (May–June 2002), pp. 63–67.
65. Costinel Anuta, 'Old and New in Hybrid Warfare', in Nicolai Iancu, Andrei Fortuna, Cristian Barna and Mihaela Teodor (eds.), *Countering Hybrid Threats: Lessons Learned from Ukraine* (Amsterdam: IOS Press: 2016), p. 48.
66. Ben Buchanan and Michael Sulmeyer, 'Russia and Cyber Operations: Challenges and Opportunities for the Next U.S. Administration' (December 13 2016), https://carnegieendowment.org/2016/12/13/russia-and-cyber-operations-challenges-and-opportunities-for-next-u.s.-administration-pub-66433, accessed 28 January 2018.
67. Also known as the Just War Tradition, JWT has its roots in Roman ethics and morality and Western Christian theology. As it relates to cyber, see for example Rex Hughes, 'Towards a Global Regime for Cyber Warfare', in Christian Czosseck and Kenneth Geers (eds.), *The Virtual Battlefield: Perspectives on Cyber Warfare* (Amsterdam: IOS Press, 2009), pp. 106–117.
68. Walzer, *Just and Unjust War*, p. 21.
69. For example, outlawing the use of chemical weapons after their widespread use in World War I (eventually resulting in the Chemical Weapons Convention and the Organisation for the Prohibition of Chemical Weapons). Despite the CWC, stockpiles continue to be held by a number of states and were used in the Syrian civil war and for targeted assassinations.
70. Hugo Slim, *Killing Civilians: Method, Madness, and Morality in War* (Oxford: Oxford University Press, 2010).
71. Clarke and Knake, *Cyber War*, pp. 54–56 and Maurer, *Cyber Mercenaries*, pp. 108–109.

72. Carr, *Inside Cyber Warfare*, pp. 31–76. Michael N. Schmitt (ed.), *Tallinn Manual 2.0 on the International Law Applicable to Cyber Operations* (New York: Cambridge University Press, 2017), pp. 373–400.
73. Jutta Brunnée and Stephen J. Toope, *Legitimacy and Legality in International Law* (Cambridge: Cambridge University Press, 2010), esp. pp. 1–19.
74. Massimo Durante, 'Violence, Just Cyber War and Information', in Ludovica Glorioso and Anna-Maria Osula (eds.), *1st Workshop on Ethics of Cyber Conflict Proceedings* (Tallinn: CCD COE, 2014), p. 59, https://ccdcoe.org/uploads/2018/10/2013ethics-workshop-proceedings.pdf, accessed 19 September 2019.
75. Page Wilson, 'The Myth of International Humanitarian Law', *International Affairs*, Vol. 93, Issue 3 (May 2017), pp. 563–579, 'What Is International Humanitarian Law?', https://www.icrc.org/en/doc/assets/files/other/what_is_ihl.pdf. See also related discussions on Customary International Law such as 'Customary Law', https://www.icrc.org/en/war-and-law/treaties-customary-law/customary-law, 'Customary International Law', https://www.law.cornell.edu/wex/customary_international_law, all accessed 29 February 2020. Maurizio D'Urso, 'The Cyber-Combatant: A New Status for a New Warrior', in Glorioso and Osula (eds.), *1st Workshop on Ethics of Cyber Conflict*, pp. 41–48.
76. Schmitt (ed.), *Tallinn Manual 2.0*, p. 375.
77. Joint Publication 3-12, 'Cyberspace Operations', Joint Chiefs of Staff (June 8 2018), p. III-11.
78. Rick Evertsz, Frank E. Ritter, Simon Russell and David Shepherdson, 'Modeling Rules of Engagement in Computer Generated Forces', in *Proceedings of the 16th Conference on Behavior Representation in Modeling and Simulation* (Orlando, FL: University of Central Florida, 2007), pp. 123–134.
79. 'International Law and Justice', https://www.un.org/en/sections/issues-depth/international-law-and-justice/, accessed 18 February 2020.
80. Ian Hurd, *After Anarchy: Legitimacy and Power in the United Nations Security Council* (Princeton, NJ: Princeton University Press, 2007).
81. Article 2(1)–(5), https://legal.un.org/repertory/art2.shtml, accessed 29 February 2020.
82. https://www.icc-cpi.int/ukraine, accessed 14 May 2022.
83. Quoted in L. C. Green, 'Cicero and Clausewitz or Quincy Wright: The Interplay of Law and War', *United States Airforce Academy Journal of Legal Studies*, Vol. 9 (1998–1999), p. 59. See G. A. Harrer, 'Cicero

on Peace and War', *The Classical Journal*, Vol. 14, Issue 1 (October 1918), pp. 26–38 for a fuller appreciation and Durante, 'Violence, Just Cyber War and Information', in Glorioso and Osula (eds.), *1st Workshop on Ethics of Cyber Conflict*, pp. 59–72 for wider a still wider historical appreciation.
84. Bill Boothby, 'Law, Ethics and Cyber Warfare', in Glorioso and Osula (eds.), *1st Workshop on Ethics of Cyber Conflict*, p. 22.
85. Kubo Mačák, 'From the Vanishing Point Back to the Core: The Impact of the Development of the Cyber Law of War on General International Law' and Peter Z. Stockburger, 'Control and Capabilities Test: Toward a New *Lex Specialis* Governing State Responsibility for Third Party Cyber Incidents' both in Henry Rõigas, Raik Jakschis, Lauri Lindström and Tomáš Minárik (eds.), *2017 9th International Conference on Cyber Conflict Defending the Core* (Tallinn: CCD COE Publications, 2017), pp. 135-148, 149-162.
86. Ministry of Defence, Cyber Primer, December 2013, 1-23/1-26, http://www.securethecyber.uk/wp-content/uploads/2015/10/201 40716_DCDC_Cyber_Primer_Internet_Secured-VERSION-TO-BE-USED.pdf, (dead link), last accessed 18 October 2016. In the second edition of the Cyber Primer the relevant passages relating to timescales and international law can be found on pp. 12–14, 26–27. Cyber Primer Second Edition, July 2016. https://www.gov.uk/gov ernment/uploads/system/uploads/attachment_data/file/549291/201 60720-Cyber_Primer_ed_2_secured.pdf, accessed 18 October 2016.
87. Cyber Primer Second Edition, July 2016, p. 13.
88. These operations are governed by four principles: military necessity, distinction, proportionality, and humanity. Cyber Primer Second Edition, July 2016, pp. 13–14.
89. Department of Defense, Defense Science Board, 'Task Force on Cyber Deterrence' (February 2017), p. 25, https://www.armed-services.sen ate.gov/imo/media/doc/DSB%20CD%20Report%202017-02-27-17_ v18_Final-Cleared%20Security%20Review.pdf, accessed 1 December 2018.
90. See, for example, Myriam Dunn Cavelty, 'Cyber-Terror—Looming Threat or Phantom Menace? The Framing of the US Cyber-Threat Debate', *Journal of Information Technology and Politics*, Vol. 4, Issue 1 (April 2008), pp. 19–36, David J. Betz and Tim Stevens, *Cyberspace and the State Towards a Strategy for Cyber-Power* (New York: Routledge, 2011), pp. 134–139, Jason Rivera, 'Achieving Cyberdeterrence

and the Ability of Small States to Hold Large States at Risk', in Markus Maybaum, Anna-Maria Osula and Lauri Lindström (eds.), *2015 7th International Conference on Cyber Conflict: Architectures in Cyberspace* (Tallinn: NATO Cooperative Cyber Defence Centre of Excellence, 2015), 7–24.

91. Schmitt (ed.), *Tallinn Manual 2.0*, pp. 375–378.
92. Jens Stoltenberg, 'Stoltenberg Provides Details of NATO's Cyber Policy' (May 16 2018), https://www.atlanticcouncil.org/blogs/natosource/stoltenberg-provides-details-of-nato-s-cyber-policy/, accessed 23 May 2022.
93. Schmitt (ed.), *Tallinn Manual 2.0*, pp. 378–379.
94. Holsti, *Major Texts on War, the State, Peace, and International Order*, p. 14.
95. 'Stoltenberg Provides Details of NATO's Cyber Policy' (May 16 2018).
96. 'NATO Will Defend Itself' (August 27 2022), https://www.nato.int/cps/en/natohq/news_168435.htm?selectedLocale=en, accessed 23 May 2022.
97. Michaela Prucková, 'Cyber Attacks and Article 5—A Note on a Blurry But Consistent Position of NATO' (undated, 2022), https://ccdcoe.org/incyder-articles/cyber-attacks-and-article-5-a-note-on-a-blurry-but-consistent-position-of-nato/, accessed 23 May 2022.
98. https://www.atlanticcouncil.org/news/transcripts/transcript-nato-head-jens-stoltenberg-on-russian-aggression-ukraines-capabilities-and-expanding-the-alliance/ (January 28 2022), accessed May 22 2022.
99. Michael Schmitt, 'Expert Backgrounder: NATO Response Options to Potential Russia Cyber Attacks Understanding the Legal Framework' (February 24 2022), https://www.justsecurity.org/80347/expert-backgrounder-nato-response-options-to-potential-russia-cyber-attacks/, accessed 23 May 2022.
100. Schmitt (ed.), *Tallinn Manual 2.0*.
101. 'Tallinn Manual 2.0', https://ccdcoe.org/research/tallinn-manual/, accessed 28 February 2020.
102. Schmitt (ed.), *Tallinn Manual 2.0*, p. 3.
103. Schmitt (ed.), *Tallinn Manual 2.0*, p. 3.
104. This largely applies to the military dimensions but can also apply to intelligence activities. The latter can be stand-alone but the former will be intelligence-led where practicable. Military RoE can also be different even amongst close allies. See for example Joint Publication

3-12, 'Cyberspace Operations', Joint Chiefs of Staff (June 8 2018), pp. IV-24–25.
105. Schmitt (ed.), *Tallinn Manual 2.0*, p. 126.
106. Schmitt (ed.), *Tallinn Manual 2.0*, pp. 348–350.
107. Schmitt (ed.), *Tallinn Manual 2.0*, p. 351.
108. Pun, 'Rethinking Espionage in the Modern Era', pp. 364–365. Martin A. Rogoff and Edward Collins Jr., 'The Caroline Incident and the Development of International Law', *Brooklyn Journal of International Law*, Vol. 16, Issue 3 (1990), pp. 493–507. Michael N. Schmitt, '21st Century Conflict: Can the Law Survive?', *Melbourne Journal of International Law*, Vol. 8, Issue 2 (October 2007), pp. 443–476.
109. Schmitt (ed.), *Tallinn Manual 2.0*, p. 351.
110. Schmitt (ed.), *Tallinn Manual 2.0*, p. 352.
111. Schmitt (ed.), *Tallinn Manual 2.0*, pp. 350–354.
112. Hughes and Colarik, 'The Hierarchy of Cyber War Definitions', p. 20.
113. Schmitt (ed.), *Tallinn Manual 2.0*, p. 377. A valuable critique of the strengths and limitation of Tallinn 2.0 can be found in David A. Wallace, Amy H. McCarthy and Mark Visger, 'Peeling Back the Onion of Cyber Espionage After Tallinn 2.0', *Maryland Law Review*, Vol. 78, Issue 2 (2019), pp. 205–246.
114. Eric D. Knapp and Joel Thomas Langill, *Industrial Network Security: Securing Critical Infrastructure Networks for Smart Grid, SCADA, and Other Industrial Control Systems* Second Edition (New York: Elsevier, 2015), p. 50.
115. Chris C. Demchak, *Wars of Disruption and Resilience: Cybered Conflict, Power and National Security* (Athens: University of Georgia Press, 2011).
116. Robinson, Jones and Janicke, 'Cyber Warfare: Issues and Challenges', pp. 70–94.
117. Robinson, Jones and Janicke, 'Cyber Warfare: Issues and Challenges', p. 14.
118. Rid has also criticized the quality of the discourse around cyber. Rid, *Cyber War Will Not Take Place*, p. ix.
119. On the RMA debate, see for example MacGregor Knox and Williamson Murray, *The Dynamics of Military Revolution 1300–2050* (Cambridge: Cambridge University Press, 2001), Colin S. Gray, *Strategy for Chaos: Revolutions in Military Affairs and the Evidence of History* (London: Frank Cass, 2002), and Dima Adamsky, *The Culture*

of *Military Innovation: The Impact of Cultural Factors on the Revolution in Military Affairs in Russia, the US, and Israel* (Stanford: Stanford University Press, 2010).
120. Whilst digital divides exist within and between states.
121. Colin S. Gray, *Irregular Enemies and the Essence of Strategy: Can the American Way of War Adapt?* (Carlisle Barracks, PA: Strategic Studies Institute, U.S. Army War College, 2006), p. 3.
122. 'Cybersecurity Threats and Defense Strategy', https://www.c-span.org/video/?428023-1/nsa-director-rogers-russia-poses-threat-congressional-elections&start=1166, accessed 27 September 2018.
123. 'Cybersecurity Threats and Defense Strategy'.
124. Natalia Zinets, 'Ukraine Charges Russia with New Cyber Attacks on Infrastructure' (February 15 2017), http://www.reuters.com/article/us-ukraine-crisis-cyber/ukraine-charges-russia-with-new-cyber-attacks-on-infrastructure-idUSKBN15U2CN, accessed 7 September 2017. BlackEnergy is further detailed in my companion book, *Russia's Cyber Offensive Against the West*.
125. 'Cyber-Attack Against Ukrainian Critical Infrastructure', Industrial Control Systems Cyber Emergency Response Team, 25 February 2016, https://ics-cert.us-cert.gov/alerts/IR-ALERT-H-16-056-01, accessed 28 June 2016. The activities of ICS-CERT and related organizations are discussed in Chapter 3.
126. 'Cybersecurity Threats and Defense Strategy'.
127. 'Russia Behind Cyber-Attack with Europe-Wide Impact an Hour Before Ukraine Invasion' (May 10 2022), https://www.gov.uk/government/news/russia-behind-cyber-attack-with-europe-wide-impact-an-hour-before-ukraine-invasion, CISA, 'Russian State-Sponsored and Criminal Cyber Threats to Critical Infrastructure', https://www.cisa.gov/uscert/sites/default/files/publications/AA22-110A_Joint_CSA_Russian_State-Sponsored_and_Criminal_Cyber_Threats_to_Critical_Infrastructure_4_20_22_Final.pdf, both accessed 17 May 2022.
128. 'Cyber Threat and Vulnerability Analysis of the U.S. Electric Sector', Prepared by: Mission Support Center Idaho National Laboratory (August 2016), p. 15, https://www.energy.gov/sites/prod/files/2017/01/f34/Cyber%20Threat%20and%20Vulnerability%20Analysis%20of%20the%20U.S.%20Electric%20Sector.pdf, accessed 12 August 2019.

129. Anonymous, 'Major Cyberattack on UK Infrastructure Is 'When, Not If'' (January 23 2018), https://news.sky.com/story/major-cyberattack-on-uk-infrastructure-is-when-not-if-11219026, accessed 27 September 2018.
130. 'New Cyber Attack Categorisation System to Improve UK Response to Incidents' (April 12 2018), https://www.ncsc.gov.uk/news/new-cyber-attack-categorisation-system-improve-uk-response-incidents, accessed 27 September 2018.
131. National Cyber Security Centre a part of GCHQ, 'Annual Review 2018 Making the UK the Safest Place to Live and Work Online', p. 10, https://www.ncsc.gov.uk/annual-review-2018/docs/ncsc_2018-annual-review.pdf, accessed 17 October 2018.
132. Schmitt (ed.), *Tallinn Manual 2.0*, pp. 25–26, 37–38, 135–141,
133. Schmitt (ed.), *Tallinn Manual 2.0*, pp. 135–142. Stuxnet has been variously described as a worm/malware/or a weapon. The loose use of this terminology is not helpful.
134. Schmitt (ed.), *Tallinn Manual 2.0*, pp. 342–345.
135. Discussed in John Richardson, 'Stuxnet as Cyberwarfare: Applying the Law of War to the Virtual Battlefield', *Journal of Information Technology and Privacy Law*, Vol. 29, Issue 1 (Fall 2011), pp. 1–27.
136. Schmitt (ed.), *Tallinn Manual 2.0*, pp. 342–345.
137. Schmitt (ed.), *Tallinn Manual 2.0*, p. 343.
138. Pun, 'Rethinking Espionage in the Modern Era', p. 375.
139. And attributed to Russia by GCHQ and the NSA. National Cyber Security Centre a part of GCHQ, *Annual Review 2018*, p. 10.
140. Christopher Krebbs and Robert Chesney, 'Gray Zone, Twilight Zone, or Danger Zone? Russian Cyber and Information Operations in Ukraine' (March 18 2022), https://warontherocks.com/2022/03/gray-zone-twilight-zone-or-danger-zone-russian-cyber-and-information-operations-in-ukraine/, Sean Lyngaas, 'Biden Administration Remains 'on Guard' for Russian Cyberattacks Amid War in Ukraine' (March 2 2022), https://edition.cnn.com/2022/03/02/politics/blinken-russia-cyberattacks-ukraine/index.html, both accessed 17 May 2022.
141. Tzu, *The Art of War*, p. xxvii.
142. FireEye and Mandiant, 'Cybersecurity's Maginot Line: A Real-World Assessment of the Defense-in-Depth Model', p. 13, http://www2.fireeye.com/rs/fireye/images/fireeye-real-world-assessment.pdf, accessed 27 September 2014.
143. von Clausewitz, *On War*, pp. 88–89.

144. Joel Brenner, 'Correspondence: Debating the Chinese Cyber Threat', *International Security*, Vol. 40, Issue 1 (July 2015), p. 193.
145. Lindsay, 'Correspondence: Debating the Chinese Cyber Threat', p. 194.
146. Alastair Iain Johnston, 'Thinking About Strategic Culture', *International Security*, Vol. 19, Issue 4 (Spring 1995), pp. 32–64. Colin S. Gray, 'Strategic Culture as Context: The First Generation of Theory Strikes Back', *Review of International Studies*, Vol. 25, Issue 1 (January 1999), pp. 49–69. Alexander Wendt, 'Anarchy Is What States Make of It: The Social Construction of Power Politics', *International Organization*, Vol. 46, Issue 2 (Spring 1992), pp. 391–425 and Alexander Wendt, *Social Theory of World Politics* (Cambridge: Cambridge University Press, 1999) for social constructivist arguments.
147. von Clausewitz, *On War*, p. xv.
148. A point made by the Department of Defense, Defense Science Board, 'Task Force on Cyber Deterrence' (February 2017), pp. 10, 17–24.
149. Joint Publication 3-12, 'Cyberspace Operations', Joint Chiefs of Staff (June 8 2018), p. IV-3.
150. Dai Qingmin, *Wangdian Yiti zhan Yinlun (Introduction to Integrated Network and Electronic Warfare)* (Beijing: PLA Press, 2002), p. 58. Quoted in Timothy L. Thomas, *Decoding the Virtual Dragon Critical Evolutions in the Science and Philosophy of China's Information Operations and Military Strategy* (Fort Leavenworth, KS: Foreign Military Studies Office, 2007), p. 106.
151. Rivera, 'Achieving Cyberdeterrence and the Ability of Small States to Hold Large States at Risk', pp. 8–9.
152. Kenneth Geers, *Strategic Cyber Security* (Tallinn: CCD COE Publications, 2011), pp. 111–121, Michael P. Fischerkeller and Richard J. Harknett, 'Deterrence Is Not a Credible Strategy for Cyberspace', *Orbis*, Vol. 61, Issue 3 (Summer 2017), pp. 381–393 and Brandon Valeriano, Benjamin Jensen and Ryan C. Maness, *Cyber Strategy The Evolving Character of Power and Coercion* (New York: Oxford University Press, 2018). Aaron F. Brantly, 'The Cyber Deterrence Problem', Max Smeets and Herbert S. Lin, 'Offensive Cyber Capabilities: To What Ends?', Quentin E. Hodgson, 'Understanding and Countering Cyber Coercion', Daniel Moore, 'Targeting Technology: Mapping Military', all in Tomáš Minárik, Raik Jakschis and Lauri Lindström (eds), *2018 10th International Conference on Cyber Conflict CyCon X: Maximising Effects* (Tallinn: CCD COE Publications, 2018),

pp. 31–54, 55–72, 73–88, 89–108. Jason Healey and Neil Jenkins, 'Rough-and-Ready: A Policy Framework to Determine if Cyber Deterrence Is Working or Failing', in Tomáš Minárik, Siim Alatalu, Stefano Biondi, Massimiliano Signoretti, Ihsan Tolga and Gábor Visky (eds.), *2019 11th International Conference on Cyber Conflict: Silent Battle* (Tallinn: CCD COE Publications, 2019), pp. 123–142.
153. Clarke and Knake, *Cyber War*, p. xi.
154. Pun, 'Rethinking Espionage in the Modern Era', pp. 375–376.
155. Committee on Armed Services, 'United States Senate: Hearing to Receive Testimony on Cyber Policy, Strategy and Organization' (Thursday May 11 2017), p. 3, https://www.armed-services.senate.gov/imo/media/doc/17-45_05-11-17.pdf, accessed 16 December 2017.
156. Glenn Greenwald, *No Place to Hide: Edward Snowden, the NSA and the Surveillance State* (New York: Hamish Hamilton, 2014), David P. Fidler, *The Snowden Reader* (Bloomington: Indiana University Press, 2015).
157. There are signs of increased intelligence sharing beyond the 'Five Eyes' including France, Germany and Japan. Noah Barkin, 'Exclusive: Five Eyes intelligence Alliance Builds Coalition to Counter China' (October 12 2018), https://www.reuters.com/article/us-china-fiveeyes-idUSKCN1MM0GH, accessed 14 March 2020.
158. Department of Defense, Defense Science Board, 'Task Force on Cyber Deterrence' (February 2017), p. 4.
159. United States Senate, 'Cyber Posture of the Services' (Tuesday March 13 2018), p. 5, https://www.armed-services.senate.gov/imo/media/doc/18-25_03-13-18.pdf, accessed 3 December 2018.
160. Department of Defense, Defense Science Board, 'Task Force on Cyber Deterrence' (February 2017), Memorandum for the Chairman, Defense Science Board.
161. Department of Defense, Defense Science Board, 'Task Force on Cyber Deterrence' (February 2017), p. 9.
162. Department of Defense, Defense Science Board, 'Task Force on Cyber Deterrence' (February 2017), p. 9.
163. Department of Defense, Defense Science Board, 'Task Force on Cyber Deterrence' (February 2017), p. 6.
164. Scott Jasper, *Strategic Cyber Deterrence* (Lanham, MD: Rowman & Littlefield, 2017), p. 111.
165. Jasper, *Strategic Cyber Deterrence*, p. 130.

166. Robert O. Keohane and Joseph S. Nye, *Power and Interdependence: World Politics in Transition* (Boston: Little, Brown, 1977) and Robert O. Keohane and Joseph S. Nye, 'Power and Interdependence Revisited', *International Organization*, Vol. 41, Issue 4 (Autumn 1987), pp. 725–753.
167. Jasper, *Strategic Cyber Deterrence*, p. 143.
168. Jasper, *Strategic Cyber Deterrence*, p. 143.
169. Singer and Friedman, *Cybersecurity and Cyberwar What Everyone Needs to Know*, p. 147.
170. Shiping Tang, 'The Security Dilemma: A Conceptual Analysis', *Security Studies*, Vol. 18, Issue 3 (October 2009), pp. 587–623. This is influenced by, amongst other things, regime type, ethnocentrism, worst-case forecasting, and enemy imaging.
171. Department of Defense, Defense Science Board, 'Task Force on Cyber Deterrence' (February 2017), pp. 12–13.
172. Department of Defense, Defense Science Board, 'Task Force on Cyber Deterrence' (February 2017), p. 14.
173. However, this also included inflated (as well as accurate) intelligence estimates producing arms racing as well as real and unfounded fears of capabilities and intentions requiring a degree of luck as well as judgement. What we think we know is not reality. It is a judgement which for the purposes under discussion is framed and informed by the intelligence cycle.
174. Kristan Stoddart, *Losing an Empire and Finding a Role: Britain, the USA, NATO and Nuclear Weapons 1964–1970* (Palgrave Macmillan, 2012), Kristan Stoddart, *The Sword and the Shield: Britain, America, NATO and Nuclear Weapons 1970–1976* (Palgrave Macmillan, 2014), Kristan Stoddart, *Facing Down the Soviet Union: Britain, the USA, NATO and Nuclear Weapons 1976–1983* (Palgrave Macmillan, 2014).
175. Barkin, 'Exclusive: Five Eyes intelligence Alliance Builds Coalition to Counter China' (October 12 2018).
176. Department of Defense, Defense Science Board, 'Task Force on Cyber Deterrence' (February 2017), pp. 18–19.
177. Department of Defense, Defense Science Board, 'Task Force on Cyber Deterrence' (February 2017), p. 2.
178. Department of Defense, Defense Science Board, 'Task Force on Cyber Deterrence' (February 2017), p. 28.
179. Department of Defense, Defense Science Board, 'Task Force on Cyber Deterrence' (February 2017), p. 4.

180. Department of Defense, Defense Science Board, 'Task Force on Cyber Deterrence' (February 2017), pp. 21–22.
181. Department of Defense, Defense Science Board, 'Task Force on Cyber Deterrence' (February 2017), p. 4.
182. This was also Idaho National Laboratory's assessment. 'Cyber Threat and Vulnerability Analysis of the U.S. Electric Sector', p. 15.
183. Department of Defense, Defense Science Board, 'Task Force on Cyber Deterrence' (February 2017), p. 4.
184. 'Energetic Bear—Crouching Yeti Kaspersky Lab Global Research and Analysis Team', https://media.kasperskycontenthub.com/wp-content/uploads/sites/43/2018/03/08080817/EB-YetiJuly2014-Public.pdf, accessed 5 October 2018, p. 40. Stephen Blank and Younkyoo Kim, 'Economic Warfare a la Russe: The Energy Weapon and Russian National Security Strategy', *The Journal of East Asian Affairs*, Vol. 30, Issue 1 (Spring/Summer 2016), pp. 1–39.
185. Joel T. Langhill, 'Defending Against the Dragonfly Cyber Security Attacks' (December 10 2014), p. 3, https://www.controlglobal.com/assets/15WPpdf/150311-Belden-DragonflyCybersecurity.pdf, accessed 5 October 2018 and 'Energetic Bear—Crouching Yeti', accessed 5 October 2018, p. 3.
186. 'Cyber-Attack Against Ukrainian Critical Infrastructure' (February 25 2016).
187. Department of Defense, Defense Science Board, 'Task Force on Cyber Deterrence' (February 2017), p. 7.
188. Department of Defense, Defense Science Board, 'Task Force on Cyber Deterrence' (February 2017), p. 8.
189. In attempting to wall themselves off from the main Internet Russia, China and other nations are indicating there is a shared sense of vulnerability through methods of electronic territorial defense and domestic control. Leonid Kovachich, 'Russia Flirts with Internet Sovereignty China Specialist Leonid Kovachich on How Russia Might Overtake China in Internet Censorship' (February 1 2019), https://www.themoscowtimes.com/2019/02/01/russia-flirts-with-internet-sovereignty-op-ed-a64369, accessed 18 February 2020.
190. Department of Defense, Defense Science Board, 'Task Force on Cyber Deterrence' (February 2017), p. 12.
191. On the thinking of North Korea's leadership see Ji Young Kong, Kyoung Gon Kim and Jong In Lim, 'The All-Purpose Sword: North Korea's Cyber Operations and Strategies', in Minárik, Alatalu, Biondi,

Signoretti, Tolga and Visky (eds.), *2019 11th International Conference on Cyber Conflict: Silent Battle*, pp. 143–162.
192. Altug Yalcintas and Naseraddin Alizadeh, 'Digital Protectionism and National Planning in the Age of the Internet: The Case of Iran', *Journal of Institutional Economics*, Vol. 16, Issue 4 (August 2020), pp. 519–536.
193. Colinn Anderson and Karim Sadjadpour, 'Iran's Cyber Threat: Espionage, Sabotage, and Revenge' (2018), p. 52, https://carnegieendowment.org/files/Iran_Cyber_Final_Full_v2.pdf. See also PA/MEHR News Agency, 'Iran Loses $150 Billion a Year Due to Brain Drain' (January 8, 2014), https://en.mehrnews.com/news/101558/Iran-loses-150-billion-a-year-due-to-brain-drain. Pooya Azadi, Matin Mirramezani and Mohsen B. Mesgaran, 'Migration and Brain Drain from Iran', Working Paper No. 9, Stanford Iran 2040 Project, Stanford University (April 2020), https://stanford.app.box.com/s/zv18ed560o38q0sefkxx4leikz5q2cpw, all accessed 26 December 2021.
194. Anderson and Sadjadpour, 'Iran's Cyber Threat: Espionage, Sabotage, and Revenge' (2018), p. 13. See also pp. 5, 10–14.
195. Gabi Siboni, Léa Abramski and Gal Sapir, 'Iran's Activity in Cyberspace: Identifying Patterns and Understanding the Strategy', *Cyber, Intelligence, and Security*, Vol. 4, Issue 1 (March 2020), p. 22.
196. Anderson and Sadjadpour, 'Iran's Cyber Threat: Espionage, Sabotage, and Revenge' (2018), pp. 39–48.
197. Siboni, Abramski and Sapir, 'Iran's Activity in Cyberspace', pp. 22–23.
198. Afshin Shahi and Ehsan Abdoh-Tabrizi, 'Iran's 2019–2020 Demonstrations: The Changing Dynamics of Political Protests in Iran', *Asian Affairs*, Vol. 51, Issue 1 (January 2020), pp. 1–41.
199. Anderson and Sadjadpour, 'Iran's Cyber Threat: Espionage, Sabotage, and Revenge' (2018), pp. 13, 37.
200. Abigail Ng, 'These 6 Charts Show How Sanctions Are Crushing Iran's Economy' (March 22 2021), https://www.cnbc.com/2021/03/23/these-6-charts-show-how-sanctions-are-crushing-irans-economy.html, accessed 29 December 2021.
201. Heather Mackenzie, 'Shamoon Malware and SCADA Security—What Are the Impacts?' (October 25 2012), https://www.tofinosecurity.com/blog/shamoon-malware-and-scada-security-%E2%80%93-what-are-impacts, accessed 4 March 2020.
202. Anderson and Sadjadpour, 'Iran's Cyber Threat: Espionage, Sabotage, and Revenge' (2018), p. 26.

203. Charlie Osborne, 'Shamoon Data-Wiping Malware Believed to Be the Work of Iranian Hackers' (December 20 2018), https://www.zdnet.com/article/shamoons-data-wiping-malware-believed-to-be-the-work-of-iranian-hackers/, accessed 4 March 2020. Kim Ghattas, *Black Wave: Saudi Arabia, Iran, and the Forty-Year Rivalry That Unraveled Culture, Religion, and Collective Memory in the Middle East* (New York: Henry Holt, 2020).
204. Jacqueline O'Leary, Josiah Kimble, Kelli Vanderlee and Nalani Fraser, 'Insights into Iranian Cyber Espionage: APT33 Targets Aerospace and Energy Sectors and Has Ties to Destructive Malware' (September 20 2017), https://www.fireeye.com/blog/threat-research/2017/09/apt33-insights-into-iranian-cyber-espionage.html, accessed 4 March 2020. In July 2021 an Iranian cyberespionage campaign dubbed Operation GhostShell first seen in 2018 was identified. This also targeted aerospace and telecommunications (largely in the Middle East but also hit organizations in the U.S., Russia, and Europe) and was linked to APT39. 'APT Group: MalKamak', https://apt.thaicert.or.th/cgi-bin/showcard.cgi?g=MalKamak&n=1, accessed 28 December 2021.
205. Andy Greenberg, 'Iranian Hackers Launch a New US-Targeted Campaign as Tensions Mount: Three cybersecurity firms have identified phishing attacks stemming from Iran—that may lay the groundwork for something more destructive' (June 20 2021), https://www.wired.com/story/iran-hackers-us-phishing-tensions/, accessed 29 December 2021.
206. Hadas Gold, 'We Know Who Is Attacking Us and We Know How to Get Even, Says Israel's Cyber Defense Chief' (December 4 2021), https://edition.cnn.com/2021/12/04/middleeast/israel-cyberattack-intl-cmd/index.html. See also Mark Stone, 'Iran Claims It Has 'Missile Cities' as 6 Incidents Prompt Theories It Is Under Attack' (July 6 2020), https://news.sky.com/story/coincidence-or-attack-what-is-behind-the-six-curious-incidents-in-iran-12021907, both accessed 4 December 2021.
207. Anderson and Sadjadpour, 'Iran's Cyber Threat: Espionage, Sabotage, and Revenge' (2018), pp. 35–37.
208. Sean Lyngaas, 'US Warns That Iranian Government-Sponsored Hackers Are Targeting Key US Infrastructure' (November 17 2021), https://edition.cnn.com/2021/11/17/politics/us-iran-hackers-warning/index.html, accessed 29 December 2021.

209. Marie Baezner, 'Hotspot Analysis: Iranian Cyber-Activities in the Context of Regional Rivalries and International Tensions' (May 2019), p. 9, https://www.research-collection.ethz.ch/bitstream/handle/20.500.11850/344841/1/20190507_MB_HS_IRNV1_rev.pdf, accessed 26 December 2021.
210. Baezner, 'Hotspot Analysis: Iranian Cyber-Activities', p. 9 and Catherine A. Theohary, 'Iranian Offensive Cyberattack Capabilities' (January 13 2020), https://crsreports.congress.gov/product/pdf/IF/IF11406, accessed 29 December 2021.
211. Anderson and Sadjadpour, 'Iran's Cyber Threat: Espionage, Sabotage, and Revenge' (2018), pp. 30–31, 37–38. Baezner, 'Hotspot Analysis: Iranian Cyber-Activities', p. 9.
212. Others include 'CopyKittens', 'Cutting Kitten', 'DarkHydrus', 'DNSpionage', 'Ferocious Kitten', and 'Group5', https://apt.thaicert.or.th/cgi-bin/listgroups.cgi, accessed 27 December 2021.
213. Like most APTs the cybersecurity industry has many monikers for APT34. These include 'Twisted Kitten', 'OilRig', 'Cobalt Gypsy', 'Crambus', and 'IRN2'. 'OilRig', https://malpedia.caad.fkie.fraunhofer.de/actor/oilrig. accessed 27 December 2021.
214. Itay Kozuch, 'APT Group OilRig: Who They Are and What You Need to Know' (April 10 2018), https://intsights.com/blog/apt-group-oilrig-who-they-are-and-what-you-need-to-know, Robert Falcone and Bryan Lee, 'OilRig' (undated), https://attack.mitre.org/groups/G0049/. Another group named 'Lyceum' and 'Hexane', which is also interested in oil and gas companies in the Middle East has also been linked to APT33, https://apt.thaicert.or.th/cgi-bin/showcard.cgi?g=Hexane&n=1, all accessed 27 December 2021. Siboni, Abramski and Sapir, 'Iran's Activity in Cyberspace', p. 25.
215. 'Charming Kitten', https://malpedia.caad.fkie.fraunhofer.de/actor/charming_kitten, 'APT Group: Magic Hound, APT 35, Cobalt Gypsy, Charming Kitten', https://apt.thaicert.or.th/cgi-bin/showcard.cgi?g=Magic%20Hound%2C%20APT%2035%2C%20Cobalt%20Gypsy%2C%20Charming%20Kitten&n=1, both accessed 26–28 December 2021.
216. 'APT Group: APT 33, Elfin, Magnallium', https://apt.thaicert.or.th/cgi-bin/showcard.cgi?g=APT%2033%2C%20Elfin%2C%20Magnallium&n=1, accessed 27 December 2021.

217. 'APT Group: Chafer, APT 39', https://apt.thaicert.or.th/cgi-bin/showcard.cgi?g=Chafer%2C%20APT%2039&n=1, accessed 27 December 2021.
218. 'APT Group: Clever Kitten', https://apt.thaicert.or.th/cgi-bin/showcard.cgi?g=Clever%20Kitten&n=1, 'APT group: Domestic Kitten', https://apt.thaicert.or.th/cgi-bin/showcard.cgi?g=Domestic%20Kitten&n=1, both accessed 27 December 2021.
219. 'APT Group: Parisite, Fox Kitten, Pioneer Kitten', https://apt.thaicert.or.th/cgi-bin/showcard.cgi?g=Parisite%2C%20Fox%20Kitten%2C%20Pioneer%20Kitten&n=1, accessed 28 December 2021.
220. 'Rocket Kitten', https://malpedia.caad.fkie.fraunhofer.de/actor/rocket_kitten, 'APT Group: Rocket Kitten, Newscaster, NewsBeef', https://apt.thaicert.or.th/cgi-bin/showcard.cgi?g=Rocket%20Kitten%2C%20Newscaster%2C%20NewsBeef&n=1, both accessed 29 December 2021.
221. Cedric Pernet and Eyal Sela, 'The Spy Kittens Are Back: Rocket Kitten 2', p. 8, https://documents.trendmicro.com/assets/wp/wp-the-spy-kittens-are-back.pdf, accessed 29 December 2021.
222. Alex Orleans, 'Who Is PIONEER KITTEN?' (August 31 2020), https://www.crowdstrike.com/blog/who-is-pioneer-kitten/, accessed 29 December 2021.
223. 'APT Group: Leafminer, Raspite', https://apt.thaicert.or.th/cgi-bin/showcard.cgi?g=Leafminer%2C%20Raspite&n=1, 'Leafminer: New Espionage Campaigns Targeting Middle Eastern Regions Active Attack Group Is Eager to Make Use of Available Tools, Research, and the Work of Other Threat Actors' (July 25 2018), https://symantec-enterprise-blogs.security.com/blogs/threat-intelligence/leafminer-espionage-middle-east, both accessed 28 December 2021. US-CERT, 'Alert (TA17-293A) Advanced Persistent Threat Activity Targeting Energy and Other Critical Infrastructure Sectors' (October 23 2017), https://www.us-cert.gov/ncas/alerts/TA17-293A, accessed 20 December 2017. See also Yonathan Klijnsma, 'New Insights into Energetic Bear's Watering Hole Attacks on Turkish Critical Infrastructure' (November 2 2017), https://www.riskiq.com/blog/labs/energetic-bear/, accessed 20 December 2017.
224. Omree Wechsler, 'The Iran-Russia Cyber Agreement and U.S. Strategy in the Middle East' (March 14 2021), https://www.cfr.org/blog/iran-russia-cyber-agreement-and-us-strategy-middle-east, accessed 20 December 2021.

225. 'RASPITE' (undated), https://www.dragos.com/threat/raspite/, accessed 29 December 2021.
226. Baezner, 'Hotspot Analysis: Iranian Cyber-Activities in the Context of Regional Rivalries and International Tensions' (May 2019), pp. 9–12. Levi Gundert, Sanil Chohan and Greg Lesnewich, 'Iran's Hacker Hierarchy Exposed How the Islamic Republic of Iran Uses Contractors and Universities to Conduct Cyber Operations' (May 9 2018), https://www.recordedfuture.com/iran-hacker-hierarchy/. Claudio Guarnieri and Collin Anderson, 'Flying Kitten to Rocket Kitten, a Case of Ambiguity and Shared Code' (December 5 2017), https://iranthreats.github.io/resources/attribution-flying-rocket-kitten/. 'APT Group: Flying Kitten, Ajax Security Team', https://apt.thaicert.or.th/cgi-bin/showcard.cgi?g=Flying%20Kitten%2C%20Ajax%20Security%20Team&n=1, all accessed 26 December 2021.
227. 'International Cyber Crime Iranians Charged with Hacking U.S. Financial Sector' (March 24 2016), https://www.fbi.gov/news/stories/iranians-charged-with-hacking-us-financial-sector, accessed 26 December 2021.
228. 'APT Group: Madi', https://apt.thaicert.or.th/cgi-bin/showcard.cgi?g=Madi&n=1, accessed 28 December 2021.
229. Anderson and Sadjadpour, 'Iran's Cyber Threat: Espionage, Sabotage, and Revenge' (2018), p. 5. Other Iranian threat groups include 'Boss Spider'/'Gold Lowell', 'Cadelle', 'Aggah', 'Infy', 'Imperial Kitten', 'Iridium', 'ITG18', 'Mabna Institute', 'Nazar', 'Static Kitten', and 'xHunt'. https://apt.thaicert.or.th/cgi-bin/listgroups.cgi, accessed 28 December 2021.
230. Anderson and Sadjadpour, 'Iran's Cyber Threat: Espionage, Sabotage, and Revenge' (2018), p. 52.
231. Theohary, 'Iranian Offensive Cyberattack Capabilities'.
232. Deborah Haynes, 'Iran's Secret Cyber Files' (undated), https://news.sky.com/story/irans-secret-cyber-files-on-how-cargo-ships-and-petrol-stations-could-be-attacked-12364871, accessed 30 December 2021.
233. Anderson and Sadjadpour, 'Iran's Cyber Threat: Espionage, Sabotage, and Revenge' (2018), pp. 20–21.

234. Radio Farda, 'Iran's Intelligence Minister Boasts of Wide-Ranging Successes' (April 20 2019), https://en.radiofarda.com/a/iran-s-intelligence-minister-boasts-of-wide-ranging-successes/29892972.html, Julian E. Barnes and Adam Goldman, 'Captured, Killed or Compromised: C.I.A. Admits to Losing Dozens of Informants' (October 5 2021), https://www.nytimes.com/2021/10/05/us/politics/cia-informants-killed-captured.html, both accessed 30 December 2021.
235. Ardavan Khoshnood, 'The Attack on Natanz and the JCPOA', BESA Center Perspectives Paper No. 1,997, April 14, 2021, https://besacenter.org/the-attack-on-natanz-and-the-jcpoa/. 'Status of Iran's Nuclear Programme in Relation to the Joint Plan of Action', Report by the Director General (July 20 2015), https://www.iaea.org/sites/default/files/gov-inf-2015-15.pdf, both accessed 27 December 2021.
236. David E. Sanger, *The Perfect Weapon War, Sabotage, and Fear in the Cyber Age* (London: Scribe, 2018), pp. 43–47. Thomas Rid, 'An Imperfect Weapon', *Survival*, Vol. 60, Issue 5 (2018), pp. 227–232. Peter Z. Stockburger, 'Known Unknowns: State Cyber Operations, Cyber Warfare, and the Jus Ad Bellum', *American University International Law Review*, Vol. 31, Issue 4 (2016), pp. 545–590.
237. Anonymous, 'Iran Says Key Natanz Nuclear Facility Hit by 'Sabotage'', https://www.bbc.co.uk/news/world-middle-east-56708778, accessed 30 December 2021.
238. Joe Tidy, 'Predatory Sparrow: Who Are the Hackers Who Say They Started a Fire in Iran?' (July 11 2022), https://www.bbc.co.uk/news/technology-62072480, accessed 11 July 2022.
239. Theohary, 'Iranian Offensive Cyberattack Capabilities'. Marzieh Rahmani, 'Iran Passive Defense Org. Role Vital in Foiling New Threats', https://en.mehrnews.com/news/164314/Iran-Passive-Defense-Org-role-vital-in-foiling-new-threats, accessed 29 December 2021.
240. Anderson and Sadjadpour, 'Iran's Cyber Threat: Espionage, Sabotage, and Revenge' (2018), p. 51.
241. 'Iran Cyber Threat Overview and Advisories', https://us-cert.cisa.gov/iran, Alert (AA20-006A), 'Potential for Iranian Cyber Response to U.S. Military Strike in Baghdad' (January 6 2020), https://us-cert.cisa.gov/ncas/alerts/aa20-006a, both accessed 13 August 2021.
242. Richard J. Campbell, 'Cybersecurity Issues for the Bulk Power System', Congressional Research Service (June 10 2015), p. 9, https://fas.org/sgp/crs/misc/R43989.pdf, accessed 13 August 2019.

243. Kim Min-Seok, 'The State of the North Korean Military' (March 2020), https://carnegieendowment.org/2020/03/18/state-of-north-korean-military-pub-81232, accessed 2 January 2021.
244. 'The Price Is Rights: The Violation of the Right to an Adequate Standard of Living in the Democratic People's Republic of Korea' (May 2019), p. 12. https://www.ohchr.org/Documents/Countries/KP/ThePriceIsRights_EN.pdf. 'Country Reports on Terrorism 2020: Democratic People's Republic of Korea', https://www.state.gov/reports/country-reports-on-terrorism-2020/democratic-peoples-republic-of-korea/, both accessed 2 January 2022.
245. Bruce E. Bechtol Jr., 'North Korea and Support to Terrorism: An Evolving History', *Journal of Strategic Security*, Vol. 3, Issue 2 (Summer 2010), pp. 45–54.
246. Rhoda E. Howard-Hassmann, 'State-Induced Famine and Penal Starvation in North Korea', *Genocide Studies and Prevention*, Vol. 7, Issue 2 (August/December 2012), p. 149.
247. 'The Price Is Rights' (May 2019), p. 11.
248. Howard-Hassmann, 'State-Induced Famine and Penal Starvation in North Korea', p. 150.
249. 'The Price Is Rights' (May 2019), pp. 9–10. See also 'Report of the Panel of Experts Established Pursuant to Resolution 1874' (2019), pp. 4–5, 22–26, https://www.securitycouncilreport.org/atf/cf/%7B65BFCF9B-6D27-4E9C-8CD3-CF6E4FF96FF9%7D/S_2019_691.pdf, accessed 31 December 2021.
250. 'The Price Is Rights' (May 2019), pp. 1–39.
251. Howard-Hassmann, 'State-Induced Famine and Penal Starvation in North Korea', p. 152. Sheena Chestnut, 'Illicit Activity and Proliferation North Korean Smuggling Networks', *International Security*, Vol. 32, Issue 1 (Summer 2007), pp. 80–111. Jenny Jun, Scott LaFoy and Ethan Sohn, 'North Korea's Cyber Operations Strategy and Responses' (December 2015), pp. 52–59, https://www.csis.org/analysis/north-korea%E2%80%99s-cyber-operations, accessed 1 January 2022.
252. Jun, LaFoy and Sohn, 'North Korea's Cyber Operations Strategy and Responses', p. 41.
253. Further details regarding the role played by the army can be found in Jun, LaFoy and Sohn, 'North Korea's Cyber Operations Strategy and Responses', pp. 45–51.

254. Which reportedly evolved from radio and electronic warfare bureaus in the MPAF and GSD.
255. Joseph S. Bermudez Jr., '38 North Special Report: A New Emphasis on Operations Against South Korea', https://38north.org/wp-content/uploads/2010/06/38north_SR_Bermudez.pdf. Another was believed to have been established in January 2021 focused on the theft of Covid-19 vaccines (Bureau 325) made up of two research centers in the DPRK which analyze stolen data from three intelligence units overseas. North Korean Cyber Activity (March 25 2021), https://www.hhs.gov/sites/default/files/dprk-cyber-espionage.pdf, accessed 31 December 2021. Jun, LaFoy and Sohn, 'North Korea's Cyber Operations Strategy and Responses', pp. 39–46.
256. Jun, LaFoy and Sohn, 'North Korea's Cyber Operations Strategy and Responses', p. 40.
257. Jun, LaFoy and Sohn, 'North Korea's Cyber Operations Strategy and Responses', p. 41. 'Gen. Kim Yong Chol', http://www.nkleadershipwatch.org/leadership-biographies/lt-gen-kim-yong-chol/, accessed 1 January 2022.
258. Kong, Lim and Kim, 'The All-Purpose Sword: North Korea's Cyber Operations and Strategies', pp. 4–5. See also Jun, LaFoy and Sohn, 'North Korea's Cyber Operations Strategy and Responses', pp. 41, 45–51.
259. Jun, LaFoy and Sohn, 'North Korea's Cyber Operations Strategy and Responses', pp. 28, 51.
260. Kong, Lim and Kim, 'The All-Purpose Sword: North Korea's Cyber Operations and Strategies', p. 6.
261. Kong, Lim and Kim, 'The All-Purpose Sword: North Korea's Cyber Operations and Strategies', p. 2. Jun, LaFoy and Sohn, 'North Korea's Cyber Operations Strategy and Responses', pp. 31–32.
262. 'Report of the Panel of Experts' (2019), p. 27.
263. Kong, Lim and Kim, 'The All-Purpose Sword: North Korea's Cyber Operations and Strategies', p. 16. North Korean Cyber Activity (March 25 2021).
264. Jun, LaFoy and Sohn, North Korea's Cyber Operations Strategy and Responses, p. 41.
265. Insikt Group, How North Korea Revolutionized the Internet as a Tool for Rogue Regimes (2020), pp. 2, 6, https://go.recordedfuture.com/hubfs/reports/cta-2020-0209.pdf, accessed 31 December 2021.

266. Jun, LaFoy and Sohn, North Korea's Cyber Operations Strategy and Responses, pp. 36, 41. Kong, Lim and Kim, 'The All-Purpose Sword: North Korea's Cyber Operations and Strategies', p. 3.
267. Kong, Lim and Kim, 'The All-Purpose Sword: North Korea's Cyber Operations and Strategies', p. 3. North Korean Cyber Activity (March 25 2021). 'Full Discloser of Andariel, A Subgroup of Lazarus Threat Group' (23 June 2018), https://global.ahnlab.com/global/upload/download/techreport/%5BAhnLab%5DAndariel_a_Subgroup_of_Lazarus%20(3).pdf, accessed 31 December 2021.
268. Alert (AA20-106A), 'Guidance on the North Korean Cyber Threat' (June 23 2020), https://www.cisa.gov/uscert/ncas/alerts/aa20-106a, accessed 1 January 2021.
269. 'All Groups from North Korea' (regularly updated), https://apt.thaicert.or.th/cgi-bin/listgroups.cgi?c=North%2520Korea&v=&s=&m=&x =, accessed 2 January 2021. This site is no longer active and these resources have been moved to Thailand's Electronic Transactions Development Agency (ETDA). 'Threat Group Cards: A Threat Actor Encyclopedia', https://apt.etda.or.th/cgi-bin/aptgroups.cgi, and https://apt.etda.or.th/cgi-bin/listgroups.cgi, both accessed 29 May 2022.
270. Alert (AA20-106A), 'Guidance on the North Korean Cyber Threat' (June 23 2020).
271. 'Report of the Panel of Experts' (2019), p. 4. In one 2018 attack "stolen funds…were transferred through at least 5,000 separate transactions and further routed to multiple countries before eventual conversion to fiat currency, making it highly difficult to track the funds", p. 27. See also pp. 28, 111–112, 138. For more see Daniel Salisbury, 'North Korea's Missile Programme and Supply Side Controls: Lessons for Countering Illicit Procurement', *The RUSI Journal*, Vol. 163, Issue 4 (August/September 2018), pp. 50–61, and Angus Berwick and Tom Wilson, 'How Crypto Giant Binance Became a Hub for Hackers, Fraudsters and Drug Traffickers' (June 6 2022), https://www.reuters.com/investigates/special-report/fintech-crypto-binance-dirtymoney/, accessed 18 June 2022. I am grateful to my PhD student, Kerime Toprak, for pointing me to this source.
272. Jun, LaFoy and Sohn, 'North Korea's Cyber Operations Strategy and Responses', p. 40.
273. Jun, LaFoy and Sohn, 'North Korea's Cyber Operations Strategy and Responses', pp. 56–57.

274. Pierluigi Paganini, 'Cyber Attack on Sony Pictures Is Much More Than a Data Breach—Updated' (December 8 2014), https://resources.infosecinstitute.com/topic/cyber-attack-sony-pictures-much-data-breach/, accessed 30 December 2021.
275. 'Report North Korea Cyber Activity', Recorded Future Insikt Group (June 15 2017), p. 6, https://go.recordedfuture.com/hubfs/reports/north-korea-activity.pdf, accessed 2 January 2022.
276. These have been named by Kong, Lim and Kim as Unit 180, Unit 91, and 413 Liaison Office but no other reference to them can be found. Two others, 128 Liaison Office and 414 Liaison Office, predate the RGB and appear directed at running operatives gathering intelligence in South Korea, including through cyberespionage. Kong, Lim and Kim, 'The All-Purpose Sword: North Korea's Cyber Operations and Strategies', pp. 5–6. See also Jun, LaFoy and Sohn, 'North Korea's Cyber Operations Strategy and Responses', pp. 42–44. Baezner, 'Hotspot Analysis: Iranian Cyber-Activities', p. 9. Sean Gallagher, 'South Korea Claims North Hacked Nuclear Data Hackers Stole Blueprints, Employee Data, and Threatened "Destruction" If Demands Not Met' (March 17 2015), https://arstechnica.com/information-technology/2015/03/south-korea-claims-north-hacked-nuclear-data/, 'Treasury Sanctions North Korean State-Sponsored Malicious Cyber Groups' (September 13 2019), https://home.treasury.gov/news/press-releases/sm774, both accessed 31 December 2021.
277. 'North Korean Cyber Activity' (March 25 2021). 'Full Discloser of Andariel'.
278. 'Treasury Sanctions North Korean State-Sponsored Malicious Cyber Groups' (September 13 2019).
279. Adam Myers, 'Meet CrowdStrike's Adversary of the Month for April: STARDUST CHOLLIMA' (April 6 2018), https://www.crowdstrike.com/blog/meet-crowdstrikes-adversary-of-the-month-for-april-stardust-chollima/, 'Adversary Labyrinth Chollima' (undated), https://adversary.crowdstrike.com/en-US/adversary/labyrinth-chollima/, accessed 3 January 2022.
280. 'APT Group: Kimsuky, Velvet Chollima', https://apt.thaicert.or.th/cgi-bin/showcard.cgi?g=Kimsuky%2C%20Velvet%20Chollima&n=1, accessed 3 January 2022.

281. Kong, Lim and Kim, 'The All-Purpose Sword: North Korea's Cyber Operations and Strategies', p. 6. Gallagher, 'South Korea claims North Hacked Nuclear Data Hackers Stole Blueprints' (March 17 2015).
282. North Korean Cyber Activity (March 25 2021), 'APT Group: Reaper, APT 37, Ricochet Chollima, ScarCruft', https://apt.thaicert.or.th/cgi-bin/showcard.cgi?g=Reaper%2C%20APT%2037%2C%20Ricochet%20Chollima%2C%20ScarCruft&n=1, 'APT37 (Reaper): The Overlooked North Korean Actor' (February 20 2018), https://www2.fireeye.com/rs/848-DID-242/images/rpt_APT37.pdf, both accessed 31 December 2021.
283. FireEye, 'APT38 Un-usual Suspects', p. 2. The extent of their operations is detailed in pp. 6–12, https://content.fireeye.com/apt/rpt-apt38, accessed 31 December 2021.
284. FireEye, 'APT38 Un-usual Suspects', p. 6.
285. 'Report of the Panel of Experts' (2019), p. 27.
286. 'North Korean Cyber Activity' (March 25 2021).
287. 'North Korean Cyber Activity' (March 25 2021).
288. 'Operation Sharpshooter Campaign Targets Global Defense, Critical Infrastructure' (2018), p. 2, https://www.mcafee.com/enterprise/en-us/assets/reports/rp-operation-sharpshooter.pdf, accessed 14 July 2019.
289. Catalin Cimpanu, 'How US Authorities Tracked Down the North Korean Hacker Behind WannaCry: US Authorities Put Together Four Years Worth of Malware Samples, Domain Names, Email and Social Media Accounts to Track Down One of the Lazarus Group Hackers' (September 6 2018), https://www.zdnet.com/article/how-us-authorities-tracked-down-the-north-korean-hacker-behind-wannacry/, accessed 14 July 2019.
290. 'Operation Sharpshooter Campaign' (2018), p. 2.
291. 'Operation Sharpshooter Campaign' (2018), pp. 8–24.
292. 'Three North Korean Military Hackers Indicted in Wide-Ranging Scheme to Commit Cyberattacks and Financial Crimes Across the Globe', https://www.justice.gov/opa/pr/three-north-korean-military-hackers-indicted-wide-ranging-scheme-commit-cyberattacks-and, accessed 10 July 2021.
293. 'APT Group: Lazarus Group, Hidden Cobra, Labyrinth Chollima', https://apt.thaicert.or.th/cgi-bin/showcard.cgi?g=Lazarus%20Group%2C%20Hidden%20Cobra%2C%20Labyrinth%20Chollima&n=1, North Korean Regime-Backed Programmer Charged With Conspiracy to Conduct Multiple Cyber Attacks and Intrusions North Korean

Hacking Team Responsible for Global WannaCry 2.0 Ransomware, Destructive Cyberattack on Sony Pictures, Central Bank Cybertheft in Bangladesh, and Other Malicious Activities, https://www.justice.gov/opa/pr/north-korean-regime-backed-programmer-charged-conspiracy-conduct-multiple-cyber-attacks-and, both accessed 30 December 2021.

294. 'Report of the Panel of Experts' (2019), p. 26. These are detailed in Annex 21, pp. 109–112.
295. Kong, Lim and Kim, 'The All-Purpose Sword: North Korea's Cyber Operations and Strategies', p. 3.
296. Alert (AA20-106A), 'Guidance on the North Korean Cyber Threat' (June 23 2020).
297. Some of these possibilities, including the potential for miscalculation and escalation, are discussed in Jun, LaFoy and Sohn, 'North Korea's Cyber Operations Strategy and Responses', pp. 60–78.
298. Selena Larson, 'Is Russian Social Media Meddling 'Cyberwarfare'?' (November 3 2017), https://money.cnn.com/2017/11/03/technology/business/russian-social-media-info-ops-cyberwar/index.html, accessed 27 September 2018.
299. 'H.R.5220—Cyber Act of War Act of 2016', 114th Congress (2015–2016), https://www.congress.gov/bill/114th-congress/house-bill/5220, accessed 27 September 2018.
300. Selena Larson, 'Is Russian Social Media Meddling 'Cyberwarfare'?' (November 3 2017) and Patrick Beuth, Kai Biermann, Martin Klingst and Holger Stark, 'Merkel and the Fancy Bear' (May 12 2017), https://www.zeit.de/digital/2017-05/cyberattack-bundestag-angela-merkel-fancy-bear-hacker-russia, Carole Cadwalladr, 'Arron Banks, Brexit and the Russia Connection' (June 16 2018), https://www.theguardian.com/uk-news/2018/jun/16/arron-banks-nigel-farage-leave-brexit-russia-connection, all accessed 27 September 2018.
301. Herb Lin, 'What Would Be a Sufficiently Strong Response to Russian Hacking of the U.S. Election?' (December 31 3016), https://www.lawfareblog.com/what-would-be-sufficiently-strong-response-russian-hacking-us-election, accessed 15 August 2019.
302. 'Statement of Admiral Michael S. Rogers Commander United States Cyber Command Before the Senate Committee on Armed Services' (May 9 2017), https://www.armed-services.senate.gov/imo/media/doc/Rogers_05-09-17.pdf, accessed 27 September 2018.

303. Presidential Policy Directive/PPD-20, 'U.S. Cyber Operations Policy', p. 6, https://fas.org/irp/offdocs/ppd/ppd-20.pdf, accessed 13 September 2018.
304. Presidential Policy Directive/PPD-20, 'U.S. Cyber Operations Policy', p. 4.
305. Presidential Policy Directive/PPD-20, 'U.S. Cyber Operations Policy', p. 9.
306. Presidential Policy Directive/PPD-20, 'U.S. Cyber Operations Policy', p. 6.
307. 'Presidential Executive Order on Strengthening the Cybersecurity of Federal Networks and Critical Infrastructure' (May 11 2017), https://www.whitehouse.gov/presidential-actions/presidential-executive-order-strengthening-cybersecurity-federal-networks-critical-infrastructure/, accessed 23 November 2018.
308. 'Federal Cybersecurity Risk Determination Report and Action Plan' (May 2018), https://www.whitehouse.gov/wp-content/uploads/2018/05/Cybersecurity-Risk-Determination-Report-FINAL_May-2018-Release.pdf, accessed 23 November 2018.
309. 'NCSC Mission, Vision, Goals', https://www.dni.gov/index.php/ncsc-who-we-are/ncsc-mission-vision, accessed 13 January 2022.
310. 'National Cyber Strategy of the United States of America' (September 2018), pp. 1–3, https://www.whitehouse.gov/wp-content/uploads/2018/09/National-Cyber-Strategy.pdf, accessed 22 November 2018.
311. Ben Buchanan and Fiona S. Cunningham, 'Preparing the Cyber Battlefield: Assessing a Novel Escalation Risk in a Sino-American Crisis', *Texas National Security Review*, Vol. 3, Issue 4 (Fall 2020), pp. 69–71.
312. 'Industrial Control Systems', https://www.us-cert.gov/ics, accessed 14 July 2019.
313. 'Executive Order on Improving the Nation's Cybersecurity' (May 12 2021), https://www.whitehouse.gov/briefing-room/presidential-actions/2021/05/12/executive-order-on-improving-the-nations-cybersecurity/, accessed 10 July 2021.
314. 'Achieve and Maintain Cyberspace Superiority Command Vision for US Cyber Command', p. 5, https://www.cybercom.mil/Portals/56/Documents/USCYBERCOM%20Vision%20April%202018.pdf?ver=2018-06-14-152556-010, accessed 13 September 2018.
315. 'Achieve and Maintain Cyberspace Superiority Command Vision for US Cyber Command', p. 5.

316. Forward defense in cyberspace and its implications under the LOAC are discussed in Jeff Kosseff, 'The Contours of 'Defend Forward' Under International Law', in Minárik, Alatalu, Biondi, Signoretti, Tolga and Visky (eds.), *2019 11th International Conference on Cyber Conflict: Silent Battle*, pp. 307–320.
317. Anonymous, 'How Does China's First Aircraft Carrier Stack Up?' https://chinapower.csis.org/aircraft-carrier/, Robert Farley, 'Does the US Navy Have 10 or 19 Aircraft Carriers?' (April 17 2014), https://thediplomat.com/2014/04/does-the-us-navy-have-10-or-19-aircraft-carriers/, both accessed 22 November 2018 and Michael E. Haskew, *Aircraft Carriers: The Illustrated History of the World's Most Important Warships* (Minneapolis MA: Zenith Press, 2016).
318. 'Advance Policy Questions for Lieutenant General Paul Nakasone, USA Nominee for Commander, U.S. Cyber Command and Director, National Security Agency/Chief, Central Security Service', p. 24, https://www.armed-services.senate.gov/imo/media/doc/Nakasone_APQs_03-01-18.pdf, accessed 4 December 2018.
319. 'Advance Policy Questions for Lieutenant General Paul Nakasone', p. 8.
320. Committee on Armed Services, 'United States Senate, Nominations' (Thursday March 1 2018), p. 28, https://www.armed-services.senate.gov/imo/media/doc/18-19_03-01-18.pdf, accessed 4 December 2018.
321. Michael Sulmeyer, 'Military Set for Cyber Attacks on Foreign Infrastructure' (April 11 2018), https://www.belfercenter.org/publication/military-set-cyber-attacks-foreign-infrastructure, accessed 1 December 2018.
322. Department of Defense, Defense Science Board, 'Task Force on Cyber Deterrence' (February 2017), p. 12.
323. The FSB report is available from FSB's website but it is not hyperlinked out of caution.
324. Daniil Turovsky, 'Moscow's Cyber-Defense How the Russian Government Plans to Protect the Country from the Coming Cyberwar' (July 19 2018), https://meduza.io/en/feature/2017/07/19/moscow-s-cyber-defense, accessed 7 October 2018. This article names a number of Russian individuals who have been implicated in a variety of both criminal and espionage hacking activities.
325. Turovsky, 'Moscow's Cyber-Defense'.
326. Greg Miller, Ellen Nakashima and Adam Entous, 'Obama's Secret Struggle to Punish Russia for Putin's Election Assault' (June

23 2017), https://www.washingtonpost.com/graphics/2017/world/national-security/obama-putin-election-hacking/, accessed 7 October 2018.
327. Almost certainly the section of the LOAC that concerns self-defense.
328. Miller, Nakashima and Entous, 'Obama's Secret Struggle to Punish Russia for Putin's Election Assault'.
329. Miller, Nakashima and Entous, 'Obama's Secret Struggle to Punish Russia for Putin's Election Assault'.
330. Daniil Turovsky, 'Moscow's Cyber-Defense' and 'Cyberwellness Profile for Russian Federation', https://www.itu.int/en/ITU-D/Cybersecurity/Documents/Country_Profiles/Russia.pdf, accessed 8 October 2018.
331. Article 274.1. of the Russian Criminal Code, http://www.unodc.org/documents/organized-crime/cybercrime/cybercrime-april-2018/SUSHCHIK_Item_3.pdf, accessed 8 October 2018.
332. Anonymous, '70mn Cyberattacks, Mostly Foreign, Targeted Russia's Critical Infrastructure in 2016 – FSB' (January 25 2017), https://www.rt.com/news/374973-cyber-attacks-russian-infrastracture/, accessed 8 October 2018.
333. Rob Sobers, '10 Must-Know Cybersecurity Statistics for 2020' (January 9 2020), https://www.varonis.com/blog/cybersecurity-statistics/, accessed 18 February 2020.
334. Sergey Sukhankin, 'The FSB: A Formidable Player in Russia's Information Security Domain' (March 27 2018), https://jamestown.org/program/fsb-formidable-player-russias-information-security-domain/, and 'Rostec and the Federal Security Service of Russia Signed an Information Security Agreement', https://rostec.ru/en/news/rostec-and-the-federal-security-service-of-russia-signed-an-information-security-agreement/, both accessed 8 October 2018.
335. 'Advance Policy Questions for Lieutenant General Paul Nakasone', p. 24.
336. 'Military Set for Cyber Attacks on Foreign Infrastructure', Belfer Center (April 11 2018).
337. Department of Defense, Defense Science Board, 'Task Force on Cyber Deterrence' (February 2017), p. 14.
338. 'NCSC Mission, Vision, Goals', https://www.dni.gov/index.php/ncsc-who-we-are/ncsc-mission-vision, accessed 8 September 2021.
339. 'Advance Policy Questions for Lieutenant General Paul Nakasone', p. 27.

340. 'Advance Policy Questions for Lieutenant General Paul Nakasone', p. 30.
341. United States Senate, 'Cyber Posture of the Services' (Tuesday March 13 2018), p. 6.
342. Committee on Armed Services, 'United States Senate, Nominations' (Thursday March 1 2018), p. 26.
343. United States Senate, 'Cyber Posture of the Services' (Tuesday March 13 2018), p. 7.
344. United States Senate, 'Cyber Posture of the Services' (Tuesday March 13 2018), p. 33.
345. Nicole Perlroth and David E. Sanger, 'Cyberattacks Put Russian Fingers on the Switch at Power Plants, U.S. Says', *New York Times* (March 15 2018).
346. Michael Sulmeyer, 'How the U.S. Can Play Cyber-Offense', *Foreign Affairs* (March 22 2018). Michèle Flournoy and Michael Sulmeyer, 'Battlefield Internet: A Plan for Securing Cyberspace', *Foreign Affairs*, Vol. 97, Issue 5 (September/October 2018), pp. 40–46.
347. Department of Defense, Defense Science Board, 'Task Force on Cyber Deterrence' (February 2017), p. 10.
348. Robert Kaiser, 'The Birth of Cyberwar', *Political Geography*, Vol. 46 (2015), p. 11.
349. Hughes and Colarik, 'The Hierarchy of Cyber War Definitions', p. 19.
350. David E. Sanger, 'The Age of Cyberwar Is Here. We Can't Keep Citizens out of the Debate', *The Guardian* (July 28 2018).
351. Sanger, 'The Age of Cyberwar Is Here'.

3

Cyberwar: Attacking Critical Infrastructure

Introduction

Developed societies have a high degree of dependency on the computerized control of the sectors that make up critical infrastructure/critical national infrastructure (CI/CNI). That dependency is deepening with the introduction of various Self-Monitoring Analysis and Reporting Technologies (SMART/'smart') in the ever-growing 'Internet of Things' (IoT).[1] In the meantime, legacy systems which can be upwards of twenty years old will still be operating (with minimal security in specific instances). The degree of vulnerability is spread unevenly in all developed and developing nations and has been the subject of active debate for national governments and private industry.[2]

CI encompasses a wide variety of industries. They are protected by regulators, site security, law-enforcement agencies, government security and intelligence services, and ultimately armed forces, but still they have been subject to attacks. This chapter systemically demonstrates the potential for cyberwarfare through attacks on key sectors of CI including public utilities such as electric power generation and distribution; water supplies, water treatment and sanitation; natural gas and oil production; pipelines; shipping and maritime traffic; hydroelectric dams; and traffic lights and train switching systems. These are the backbone of any industrialized state.[3]

For example, in the water utilities sector, local plant systems can include wastewater treatment facilities, whilst 'regional' systems include intake and/or effluent structures, pumping stations, chlorination stations, control valve stations, etc. For electricity generation, computer systems can detect current flow and line voltage, monitor the operation of circuit breakers, or take substations off or onto the national grid. This includes the operation of local, regional, national parts of CI and, if the hubs of the financial services sector such as Wall Street or the Square Mile in London are targeted (both of which are reliant upon electricity and IT systems), can also begin to directly affect the global economy.[4] This could well produce cascade effects like those seen in the 1929 Wall Street Crash and 2008 financial crisis, especially with the volume of non-physical 'virtual' holdings and the dependency on digital trading. Right now, the financial services sector is being targeted mainly for direct monetary gain. This is also the case in the healthcare sector, especially monetizable personal data, and through ransomware.[5] Other sectors have seen other forms of cyberattack, including highly invasive and concerning intrusions and breaches by state-run or state-sponsored cyberespionage actors.

SCADA Systems and Critical Infrastructure

IT and ICT are familiar terms to many people but Operational Technology (OT) less so. This is used to control physical processes and devices across industry and throughout CI. A great deal of CI is controlled through Supervisory Control and Data Acquisition (SCADA) systems. SCADA is an embedded technology in developed and developing states across a wide range of sectors and industries.[6] SCADA is essential for many of the essential services within a developed society. SCADA is a type of Industrial Control System (ICS), dating back to the 1940s and its use in the electric utilities sector with most current SCADA systems dating to the advent of cost-effective 8 and 16-bit minicomputers, and then microcomputers in the 1970s, 1980s, and into the 1990s. The main purpose of SCADA systems is to monitor, physically control, and alarm plant or regional systems from a central location in real time. SCADA allows remote control of often far flung or dispersed sites and the systems at those sites.[7]

There are three main elements to a SCADA system. First, are various Remote Telemetry Units (RTUs) providing communications which relay information. Second, are Human Machine Interfaces (HMIs) which display information in the form of graphics and alphanumeric readouts. The HMI is essentially a PC system running graphic and alarm software programs. As Knapp and Langill describe, "If an attacker is able to successfully compromise the HMI, fully automated systems can be permanently altered through the manipulation of set points. For example, by changing the high-temperature set point to 100 °C, the water in a tank could boil, potentially increasing the pressure enough to rupture the tank".[8] Third, at the facilities themselves are Programmable Logic Controllers (PLCs). These are industrial computer control systems which "continuously monitors the state of input devices and makes decisions based upon a custom program to control the state of output devices".[9] PLCs can be used to control timings and operations such as opening and closing water treatment valves or to manage and determine electric current.

SCADA systems are often customized over time rather than bought off-the-shelf in one go and are component based. SCADA forms part of a wider set of Industrial Control Systems. These include Distributed Control Systems (DCS); Process Control Systems (PCS); Energy Management Systems (EMS); Automation Systems; Safety Instrumented System (SIS); and a number of other automated control systems. In DCS, the data acquisition and control functions are performed near the devices being controlled from which data is being gathered. In years gone by "DCS systems performed most of the technical work and reported back to the SCADA system…SCADA was the optimal technology for monitoring processes and events across a very large geographic region". Now, "the line between SCADA and DCS is becoming more blended as technology advances and bandwidth grows".[10]

There are a significant number of reasons which make attacking ICS and SCADA difficult. First, although there are industry standards, no two SCADA systems are likely to be the same. The vast majority have been built up over time using a variety of bespoke hardware and software. This has generated a degree of 'security by obscurity' (a belief that the uniqueness of bespoke systems makes hacking them very difficult) with a majority of sites operating legacy hardware. At the same time, many now contend that 'security by obscurity' "cannot be considered a valid defensive strategy".[11] It is no longer a sufficient defense because of the advent of the Internet and search engines such as Shodan, which help lower the barriers to entry for would-be hackers, well-resourced and skilled Advanced Persistent Threats (APTs), other espionage actors, and cybercriminals searching for monetary gain. APTs and cybercriminals are very much in evidence.

From a technical standpoint, there are a number of ways malicious actors can exploit vulnerabilities in ICS.[12] This includes through technical standards used by networked systems to communicate with one another. This is known as the Open Systems Interconnection (OSI) model. These contain known vulnerabilities as does the proprietary hardware and software (sold by companies to customers or developed in-house for internal processes). Both are replete in CI and industry as a whole and "in the years following the Stuxnet attack, many researchers have found numerous vulnerabilities with open protocol standards and the systems that utilize them…[alongside] vulnerabilities in the proprietary products". Knapp and Langill note that "These proprietary systems and protocols are at the core of most critical industry, and represent the greatest risk should they be compromised".[13] Defending against industrial vulnerabilities is problematic. As Cherdantseva et al. argue:

> For over forty years confidentiality, integrity and availability—also referred to as the CIA-triad—have been defining the set and priorities of security goals for corporate information systems. In ICS and SCADA systems, the priorities among the goals are different. Among the triad, integrity and availability are highly paramount, while confidentiality is secondary for SCADA systems. In reality, security goals, in what ever

order they appear, are often preceded in SCADA systems by safety, reliability, robustness and maintainability (which are the supreme goal of critical systems).[14]

Attacks on SCADA systems and ICS and devices "can disrupt and damage critical infrastructural operations, cause major economic losses, contaminate [the] ecological environment and...claim human lives".[15] These possibilities are a central aspect of this chapter.

Proof-of-Concept: Aurora and Stuxnet

The most widely known, and most widely reported, attack to date on a SCADA system was Stuxnet. Stuxnet was identified in 2010 and adversely affected the centrifuges in the Natanz nuclear processing plant in Iran—specifically the German designed Siemens S7-417 industrial controllers which helped control the centrifuges—unbeknown to the operators at Natanz. This might have been installed in the plant via a USB device rather than through external infection although later it did 'escape' onto the Internet where it can be further refined beyond the original intent of its programmers (widely suspected to be Israeli and U.S. intelligence agencies).[16]

Stuxnet provided a proof-of-concept demonstration that deliberate alterations to computer code can lead to physical destruction. What is less well-known is an earlier proof-of-concept demonstration codenamed Aurora which was conducted at Idaho National Laboratory (INL) in 2007 by the U.S. government. Perry Pederson, the originator of the demonstration, has said:

> One of the challenges I did envision was finding a way to educate non-technical policy makers about ICS security. In other words, we needed an engineering approach to solve this problem, but we also needed to "sell" the approach to non-technical people and Aurora provided such a vehicle. After briefing the DHS [Department of Homeland Security] Secretary on the proposed test and getting the 'green-light' the DHS and INL crews went into high gear. For some, Aurora was just another test and the outcome was to be determined during actual testing. For those

of us who understood the basic physics involved (and lessons taught in power engineering 101) we knew we were out to destroy a generator. Since that event, there are those who will still deny the validity of what was accomplished on that cold day in Idaho, but the test finally provided empirical evidence that cyber attacks can destroy physical equipment and it captured the event on video.[17]

Aurora was publicized at the time and was the subject of a PBS television documentary in 2015 but retains a low profile.[18] As proof-of-concept demonstrating cyber-physical attacks it predates Stuxnet.[19]

In 2007 CNN reported that "Sources familiar with the experiment said the same attack scenario could be used against huge generators that produce the country's electric power". It continued, "Some experts fear bigger, coordinated attacks could cause widespread damage to electric infrastructure that could take months to fix". The report added, "CNN has honored a request from the Department of Homeland Security not to divulge certain details about the experiment".[20] It is likely that the results of the test were shared with other parts of the U.S. government and there is speculation they were also shared with Israel and this aided the development of Stuxnet.[21]

The Implications of Aurora and Stuxnet

CNN recorded the views of industry insiders who said that Aurora demonstrated that previously unrecognized vulnerabilities can affect large electric systems and do it at scale. Joe Weiss, who has been responsible for engineering and security for public utilities and other industrial companies, said to CNN, that in the past the assumption had been that the worst that could happen was for things to be forced to shut down. However, Weiss also knew that worse was the possibility that valves could be made to open to make them unsafe or hazardous. The main point he made is that very large and very critical pieces of equipment can be controlled and manipulated for hostile intent. This has even wider implications as Weiss and others postulated the possibility of multiple or simultaneous cyberattacks on critical electric sites could take down the power to large geographic areas for a period of months.[22] Weiss repeated these warnings at the DEFCON 25 conference in Las Vegas in 2017.[23]

O. Sami Saydjari, who spent two decades at the National Security Agency (NSA) and three years at the Defense Advanced Research Projects Agency (DARPA) concurred. He told CNN that for as little as $5 million and between 3 and 5 years of development, a nation-state, or a transnational terrorist group could be capable of mounting a strategic attack against the United States.[24] Back in 2007, the economic impact of a national power loss across the U.S. was estimated to be $700 billion which the economist, Scott Borg, placed into context. Borg equated this as the "equivalent to 40 to 50 large hurricanes striking all at once…It's greater economic damage than any modern economy ever suffered…It's greater [sic.] then the Great Depression. It's greater than the damage we did with strategic bombing on Germany in World War II".[25] This was before the Covid-19 pandemic struck.[26]

Pederson stated that "Many U.S. Government agencies were briefed as well as public entities. DHS worked through the North American Electric Reliability Corporation (NERC) and the Nuclear Energy Institute (NEI) in an effort to reach potential targets of Aurora type attacks".[27] Pederson added, "Fast forward to 2014. What have we learned about the protection of critical cyber-physical assets? Based on various open source media reports in just the first half of 2014, we don't seem to be learning how to defend at the same rate as others are learning to breach".[28] Efforts toward remedying this disparity between the defense and offense continue to be made, but are we doing enough?

Real-World Cases

There is a widespread belief that no attack on critical infrastructure has cost lives. This is not true. In 2003, the Blaster worm infected 2 billion computers worldwide. It had "impacts that most technologists had never even considered" and "Within hours of the Blaster worm going viral, the U.S. Northeast suffered a massive power outage. In the following days, U.S. government officials and some private sector business leaders began to tie the power outage to the Blaster event. Over a hundred people died due to that blackout".[29] We cannot afford to be naïve.

Electricity Generation and Distribution

The electric utilities sector is perhaps the most likely sector of critical infrastructure to be targeted because power runs all other sectors.[30] Most sites and facilities have only limited electrical backup. A 2016 U.S. government report by Idaho National Laboratory (INL) highlighted that:

> Threat actors on multiple fronts continue to seek to exploit cyber vulnerabilities in the U.S. electrical grid. Nation-states like Russia, China, and Iran and non-state actors, including foreign terrorist and hacktivist groups, pose varying threats to the power grid. A determined, well-funded, capable threat actor with the appropriate attack vector can succeed to varying levels depending on what defenses are in place.[31]

Mike Assante, when chief security officer of NERC, was similarly unequivocal: "There are absolutely foreign entities that would definitely conduct [cyber] reconnaissance of our power infrastructure…They would be looking to learn, preposition themselves to get a foothold and try to maintain sustained access to computer networks".[32] These remain active concerns and it is a wide attack surface.

The three main functions of the electric grid are: first, electrical generation at power plants; second, transmission from power plants (often over long distances at high voltages); third is distribution. Transmission and distribution systems (including over highly visible overhead pylons) snake countrywide and are connected by electricity substations. At substations, transformers adjust voltage levels down or up, or convert alternating current (AC) to direct current (DC) or vice versa. For most homes and many business customers local to the substation, this is where electricity transmission of current is stepped-down to lower voltages. Depending on where and what they are supplying electricity to, they vary in size and scale and can supply more than one utility. Substations can also operate as switching stations and grid feeders and are a key part of the electrical networks that make up national grids. They will also house fault protection systems (including against lightning strikes, adverse weather, and the local environment), monitoring devices, and communications equipment.

Whilst there are industry standards, there is no one-size-fits-all approach for power plants or grids. Both have developed and evolved over time and reflect the country and locality concerned, including their geography and topography. This also leads to different systems and communications requirements. Also, forms of power generation vary from traditional coal or gas fired power stations and hydroelectric and geothermal systems, through to nuclear plants, wind and solar farms, and through again to next-generation technologies (including nuclear fusion). According to Dragos' study of the Crashoverride/Industroyer malware which hit Ukraine in December 2016:

> This means that the electric grid must be a robust, almost living creature, which moves and balances electricity across large regions...Transmission and distribution owners have their substations in their particular geographical footprint and control centers manage the cross-territory SCADA systems 24/7 by human operators. These control centers often regularly manage the continual demand and response of their customers, respond to faults, and plan and work with neighboring utilities. This simplistic view of grid operations is similar around the world. There are often vendor and network protocol differences between countries but the electrical engineering, and the overall process is largely the same between nations.[33]

There are also contingency plans for power plant or power grid failures known as a 'Black Start'.[34] Within this system of systems, INL showed that three elements are most at risk to cyberattacks. These are the high-voltage substation transformers used in transmission; the transmission towers connecting power lines; and the control centers which handle the delivery of power from power plants. INL assessed that transformers and "other large equipment used to support transformer functions are the most impactful in a cyber-physical event due to replacement time and cost". This is because "The loss of one substation taking on power load from a generator may put too much stress on remaining transformers, creating grid instability" and there are limited spare transformers. Resultantly, "Power surges due to a lack of transmission capacity could lead to cascading failures throughout the grid and long-lasting power outages. Further, while the loss of a transformer is rare, recovery without a spare can take months".[35] Substations could prove an attractive target for cyberattacks (and have also been targeted by drones).[36] There is however comfort in Dragos' assessment that an 'amplifying' attack on the electric grid "would be very difficult to do at scale properly and would require a significant investment on behalf of the adversary".[37]

Electricity Producing Sites Include Nuclear Power Stations

Nuclear power plants have long been considered a prime target for attack or compromise. They are protected proportionately both in terms of onsite safety procedures and protocols. This also includes vetting personnel who have access to sites and restricted access to secure areas including the reactors themselves. These sites, of which there are currently 60 in the United States generate 20% of its electricity. A much larger proportion, 63%, is generated from fossil fuels (coal, oil, natural gas, and other gases) and 17% arrives from renewable energy sources. Wind farms are considered vulnerable because of insecure communications protocols and by physical access. These and other sites are spread across the United States.[38]

For increasingly energy hungry societies an incident at even a single site, such as that at Fukushima following the tsunami which hit eastern Japan in March 2011, can have large-scale and long-term consequences which extend far beyond the effected site, area, sector and country.[39] This was a natural disaster in a country prepared for natural disasters including tsunamis, earthquakes and typhoons but even with preparedness and established procedures in place, Fukushima was a major incident with attendant implications for new and existing nuclear power stations.[40]

"The North American bulk electric system is comprised of more than 200,000 miles of high voltage transmission lines,[41] thousands of generation plants, and millions of digital controls. More than 1,800 entities own and operate portions of the system, with thousands more involved in the operation of distribution networks across North America. These entities range in size from large investor-owned utilities…to small cooperatives. The systems and facilities comprising the larger system have differing configurations, design schemes, and operational concerns".[42] Some of these concerns relate to legacy systems. Moreover, "Much of the U.S. energy system predates the turn of the 21st century. Most electric transmission and distribution lines were constructed in the 1950s and 1960s with a 50-year life expectancy, and the more than 640,000 miles of high-voltage transmission lines in the lower 48 states' power grids are at full capacity".[43]

Canada has 19 nuclear power sites, generating around 15% of its electricity, 19% through fossil fuels, 59% through hydroelectric power, and 7% through other forms of renewable energy. Canada exports 11% of its surplus electricity to the United States across thirty-four major active international transmission lines.[44] As of 2017, the European Union was producing 25.6% of its electricity by nuclear power (spread across 14 EU member states), 26.2% from renewable energy, and 48.3% from fossil fuels.[45] These figures compare with the global average which in 2016 stood at 65.3% through fossil fuels, 10.4% nuclear, and 23.8% through renewables.[46] In 2016, the EU was importing just under 40% of its natural gas from Russia, 31.6% of its crude oil, and 30.2% of its solid fossil fuels.[47] After Russia's invasion of Ukraine, a decision was reached in May 2022 to phase out Russian oil and gas imports by 2030 but with almost a quarter of Europe's energy consumption fueled by Russia, this makes economic sanctions and also any counter cyberattacks on Russian critical infrastructure much more complex and cascadable.[48]

Although energy demands across the EU and the United States are projected to decrease from 2017 through to 2040 there is expected to be a global rise by 30% —"the equivalent of adding another China and India to today's global demand". Indeed, India, China, and South East Asia are projected to be the most energy hungry. These projections are based on "A global economy growing at an average rate of 3.4% per year, a population that expands from 7.4 billion today to more than 9 billion in 2040, and a process of urbanisation that adds a city the size of Shanghai to the world's urban population every four months".[49]

Water Treatment and Sanitation

A different example of the importance of cyberattacking SCADA and critical infrastructure in the utilities sector is the infamous case of Vitek Boden. Boden was a disaffected technician who worked for Hunter Watertech, an Australian firm that installed radio-controlled sewage equipment for SCADA systems for the Maroochy Shire Council in Queensland, Australia. In 2000, Boden crammed his car with stolen radio equipment including a PDS Compact 500 computer and a laptop driving around the area on at least forty-six occasions between 28 February and 23 April 2000.

He used his technical skill and insider knowledge to send radio signals to the SCADA system that he had helped install. This controlled the sewage equipment at 142 pumping stations across a 450^2 mile area servicing 130,000 people. Boden caused 800,000 liters of raw sewage to spill out into local parks, rivers as well as the grounds of the local Hyatt Regency hotel.[50] His car was eventually spotted, and he was pulled over by police who found the equipment he was using. The important element of this case is the 'insider threat'[51] and that he was able to gain access from outside to the Remote Telemetry Units that passed on his malicious instructions to affect operations inside the plants.[52] In February 2013 an external hacker also attempted to poison the water supply of a Florida water treatment facility.[53] The general problem of insider threats and localized hacking onsite will be examined in detail in Chapter 5.

In addition, supposed hacks can generate false positives. In November 2011 it was widely reported that the SCADA system at the Curran Gardner Public Water District outside of Springfield, Illinois had been cyberattacked from an IP address in Russia making a water pump burn out. This led the Illinois Statewide Terrorism & Intelligence Center (STIC) to issue a Daily Intelligence Notes report titled 'Public Water

District Cyber Intrusion'.[54] However, the attacker turned out to be an employee of Navionics Research who had helped install the system and was asked to help with a system problem whilst on vacation in Russia. Then an outside contracted repairman "examined the logs on the SCADA system and saw a Russian IP address connecting to the system in June" with their username appearing in the SCADA system logs next to the IP address.[55] Prior to ICS-CERT's investigation and public clarification, industry experts were suggesting this could have been a "nation-state test" against a sector which provides water treatment for drinking water and sanitation.[56]

Dams and Reservoirs

Groundwater sources such as reservoirs and dams are also used by water treatment plants to provide hydropower production, irrigation, and flood protection. In addition, although electrical plants have access to their own water sources (a reason many are located on rivers, lakes, and coastal sites) national water and energy utilities are "highly interdependent" according to a 2014 U.S. Department of Energy study.[57] Dams are a prime target. This was highlighted in Northern Iraq in 2014 where air strikes from U.S. forces (supported by Kurdish and Iraqi ground forces) cleared an area around the Mosul Dam from invading Islamic State (IS) forces.

Dams regulate the water supply and provide electricity for large geographical areas; in this case along the river Tigris and downstream for over 450 km to Bagdad. These are strategically important assets and IS had already taken control of the dam in Fallujah and attempted to seize the Haditha dam.[58] It was not lost on Islamic State, and is on the radar of other terrorist groups, that dams also represent prime targets in nations they consider to be their enemy. They are also a target for cyber-attacks by states at distance. As was noted in Chapter 2, this includes Iran's attack on the Bowman Dam in Rye, New York in 2013.[59]

Damage, destruction, or disruption to dams could cause drought or flooding, lead to a lack of clean drinking water and sanitation, disrupt navigation and transportation, and effect industrial plants dependent on water supplies (e.g., for cooling) and electricity supplies. It would also place considerable strains on local and national response teams (especially emergency services and military). It would place concomitant pressure on local and central governments and agencies (made worse in crisis situations), create a loss of confidence in the security of these sites, and concern public response or panic. Public concerns over disaster responses are heightened when natural disasters including severe weather and earthquakes strike.[60]

The Oil and Gas Industry: Rigs, Refineries, and Pipelines

Oil and petroleum fields and refineries are vital sectors of critical infrastructure in many countries. Goods and services rely on transport by aircraft, road vehicles, trains, and ships using a myriad of internal combustion engines running on petroleum-based fuels. Rigs and refineries operate at high volume and high capacity in a constant set of interconnected processes. Like many sites, especially public utilities, they run on a 24-hour, 7 day a week, 365 days a year basis. The names and positions of oil rigs are public record as are refineries and easily researched on the Internet. Offshore rigs are found in vast oil and gas fields whilst the vast majority of refineries are, for logistical and economic reasons, located at well-known coastal ports or along estuaries and rivers.

These form part of well-established maritime shipping and trade routes.[61] These critical hubs and trade routes are highly visible. 17% of U.S. crude oil supplies come from the Gulf of Mexico whilst "5% of total U.S. petroleum refining capacity is located along the Gulf coast, as well as 51% of total U.S. natural gas processing plant capacity".[62] In April 2018 it was reported that gas pipelines were targeted in a cyber-attack on a third-party supplier which disrupted supplies but did not cause damage.[63] Responding to this report Bryan Singer, then Director of Security Services at the Seattle-based cybersecurity firm IOActive, warned:

A lot of pipelines have 24-48 hour capacity within the pipelines. If hackers find a way to poison the product, you could have downstream impact for months or more. You could have gas compressors or lift stations where there's a fire or explosion, and where you have to scramble to cap the ends before the fire spreads out. If it's an oil rig, it could certainly be tougher to contain. Hackers can cause some intermediate problems at first, but if they have access long enough, there's a possibility that airports could go down (they often rely on fuel delivered directly) and gas stations could run out of gas. If they're able to maintain an attack for a couple days, there can be very large downstream impact. [In Winter especially] if we don't have power, we're in need of that heat.[64]

In May 2021, Colonial Pipeline paid $5 million (probably in the cryptocurrency Bitcoin) after it was hit by a ransomware attack by the ransomware-as-a-service group 'Darkside', which is believed based in Eastern Europe or Russia. Headquartered in Alpharetta, Georgia, Colonial Pipeline is one of the largest pipeline operators in the U.S. and services around 45% of the eastern United States with gasoline, diesel, home heating oil, and kerosene for aircraft from suppliers across the U.S. During the attack, the company halted their operations as a precaution against further infiltration and potential migration from their corporate IT network and into their OT networks which is where material danger lies. This disrupted supplies for around three days. Quite apart from the ransom being paid (a lucrative incentive to groups such as 'Darkside'), these incidences are growing and carry far more than financial risks.[65]

Four of the world's top ten oil fields are in the United States. Eagle Ford Shale in South Texas, Spraberry/Wolfcamp in West Texas, Prudhoe Bay in Alaska, and the Bakken Field which lies across the west of North Dakota, Eastern Montana, and north into the Canadian provinces of Saskatchewan and Manitoba. The Safaniya Field in the Persian Gulf is the largest offshore oil field in the world. It is believed to contain over 50 billion barrels of oil. It is Saudi Arabia's second largest producing field behind Ghawar, which is the world's largest onshore oil field, and produces 1.5 million barrels per day. Safaniya stretches 50 km by 15 km and is run by the state-owned oil company Saudi Aramco which in 2012 was struck by a major cyberattack known as Shamoon. Shamoon did not seek to damage or destroy but to wipe data from hard drives (wiperware).

It was disruptive and propagated to other oil and gas firms including RasGas.[66] Saudi Aramco is "a key crude oil supplier to six major Asian countries — India, China, Japan, South Korea, Taiwan and the Philippines" and the is world's largest oil company.[67] Saudi Arabia is also home to one of the world's largest chemical plants at Sadara.

Many of the world's largest oil fields are located in the Middle East. These include Zakum in the United Arab Emirates, Burgan in Kuwait, Rumaila and West Qurna-2 in Iraq, and Gachsaran and Ahvaz in Iran. Each of these has seen political turbulence and war and could well be targeted in any future conflict much like Kuwaiti oil fields were in the 1991 Gulf War when they were set on fire by retreating Iraqi forces. Two other giant oilfields are located in Brazil's Lula field in the Santos Basin and the Kashagan field in the Caspian Sea in Russia which is the largest offshore field outside the Middle East. In the North Sea there are over 300 oil and gas fields in the area with the eight largest oil fields divided between Norway and the United Kingdom.

Russia is a peer cyber adversary of the West and to the Russian economy, few things are more central than energy. Energy supplies exert a great degree of influence on its foreign and security policies given its economic importance to Russia.[68] Illustrating this is an analysis from 2014 which left little room for doubt not only of the importance of oil and natural gas to Russia's economy but also the importance of Russia's resources to Europe which is its main export market:

> By an accident of geography and geology Russia contains the largest hydrocarbon resource base in the world and is the largest global producer of oil and gas combined. According to the BP Statistical Review published in 2012 the country contains 88 billion barrels of oil reserves and 45 trillion cubic metres of gas, while producing a combined total of approximately 20 million barrels of oil equivalent (boe) per day. On a graph of global hydrocarbon producers only Saudi Arabia, Venezuela and Iran come close to Russia, and its huge undiscovered resource base marks it out as a country not just of current importance but also of future growth for the oil and gas industry. In a recent survey…the USGS [US Geological Survey] estimated that two thirds of the world's total Arctic resources, amounting to more than 250 billion boe, are located in Russian waters,

and regions such as East and West Siberia, the Black Sea and the Caspian Sea are believed to contain a further 150 billion boe.[69]

Approximately 70% of the oil produced in Russian comes from the West Siberian Basin.[70] With this is mind it is clear that Russian national interests in the energy sector, encompassing oil and gas suppliers, present a clear set of targets for industrial espionage and any economic and political designs which flow out of this. Whether they would be targeted or are being targeted by Western intelligence agencies in response to Russian actions is problematic without reduced European reliance on Russian fossil fuels.

Globally, many more fields are capable of being exploited as technology such as deep-water extraction and fracking makes them economically viable. Within these oil and hydrocarbon fields there are literally thousands of oil and gas platforms with Figs. 3.1 and 3.2 giving an idea of the geographical size, depth, and complexity of the offshore oil and gas industry where multinational corporations operate.[71] In total there are over 65,000 oil and gas fields in the world.

There are considerable logistical elements and small numbers of skilled personnel able to operate the Industrial Control Systems used for computer control of the machinery on these platforms; many of which are only upgradable or fixable in the field. Under these conditions mistakes can and do happen and have to be rolled back.[72] A clear example of the consequences resulting from malicious activity or human error was the *Piper Alpha* oil and gas platform disaster in the North Sea in 1988 which killed 165 of the 220 people on board along with two crewmen from the standby vessel *Sandhaven*.[73] *Piper Alpha* "cost the Lloyd's insurance market more than £1bn, making it the largest insured man-made catastrophe".[74] Although major accidents are rare, the 2010 *Deepwater Horizon* explosion and spill in the Gulf of Mexico which killed 11 people was also attributed to human error. It was also later reported:

> Such accidents, however, are rarely the fault of just a few individuals. Offshore drilling is a complex operation that involves hundreds of people; 126 people were employed on the Horizon drilling the Macondo well on

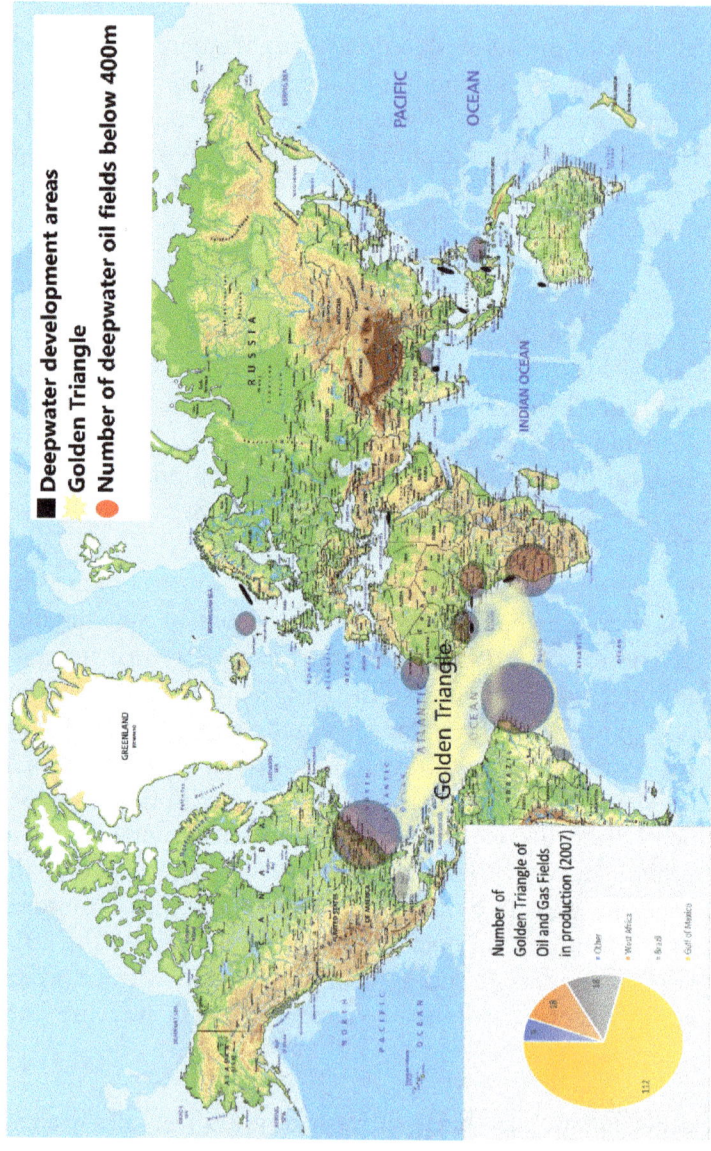

Fig. 3.1 Offshore oil fields (circa 2010) (Author diagram with data from https://www.bbc.co.uk/news/10298342, Infield Systems, and Petroleum Economist)

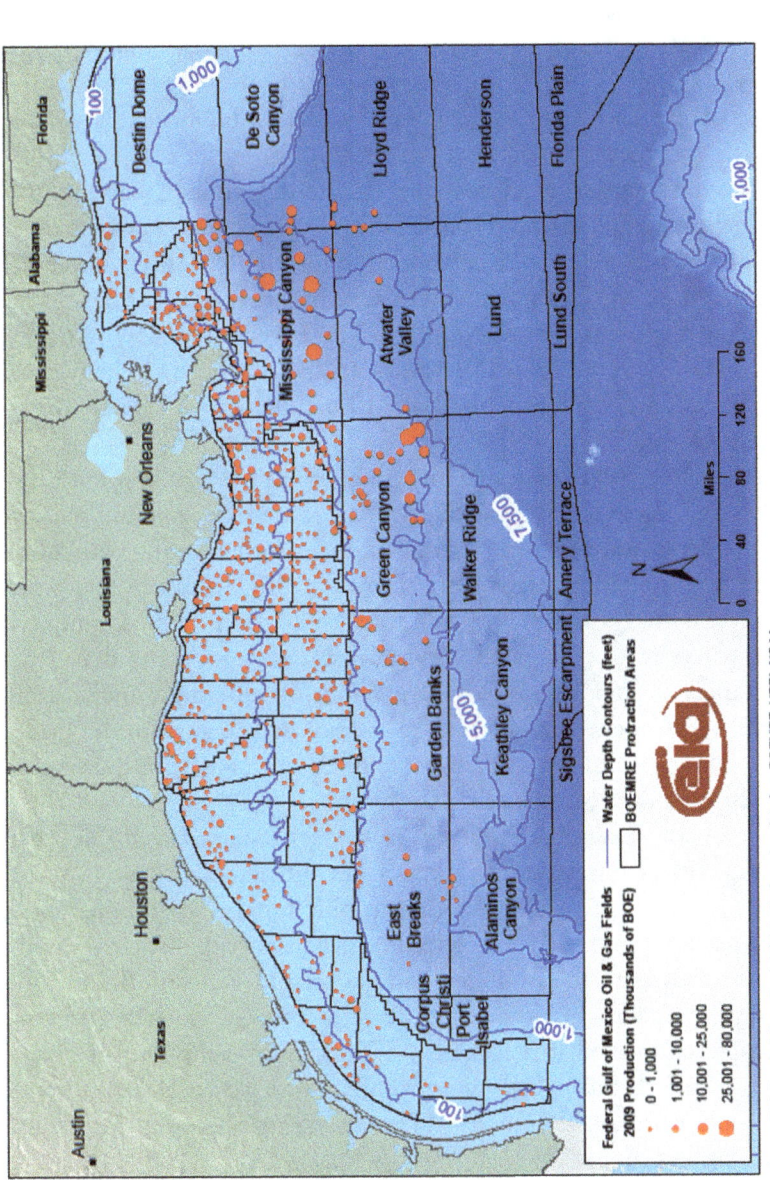

Fig. 3.2 Gulf of Mexico: producing oil and gas fields (By US Energy Information Administration - http://www.eia.gov/pub/oil_gas/natural_gas/data_publications/crude_oil_natural_gas_reserves/current/pdf/gomwaterdepth.pdf, Public Domain, https://commons.wikimedia.org/w/index.php?curid=29929461)

the day of the accident. By then, the project was already running six weeks behind schedule and some $58 million over budget. Everyone was under pressure to finish drilling and get the well into production.[75]

The cost of the *Deepwater Horizon* spill was $1.7 billion to BP, which owned the rig, accompanied by a record $18 billion fine. The environmental cost was arguably even higher. Around 4 million barrels (168 million gallons) of oil spilled into the Gulf of Mexico over a three-month period before the well could be capped. This was deposited across an area stretching from Gulfport, Mississippi to Pensacola, Florida.[76] The disaster was made into a movie in 2016. There were over 1,300 offshore oil and gas rigs operating in 2018 prior to the downturn in extraction consumption resulting from the Covid-19 pandemic.[77] There are many more onshore.

There are also 700 operational oil refineries in existence. For both rigs and refineries there is always a chance that something can and will go wrong at one or more of these at some point—as it has in the past. With each refinery operating high-temperature and high-pressure hydrocarbons, safety is always a prime concern. But "refineries are almost uniquely difficult places to operate safely" and available data indicates that "Any given refinery has about a one in ten chance of suffering a major accident during its operational lifetime".[78] Major accidents at oil refineries number under 100 but the dollar value of these range from the tens of million to the hundreds of millions.

The industry magazine *Offshore Technology* asserted that "a cyber attack on critical infrastructure, such as an oil rig, can result in more than just lost revenue – it can be catastrophic for the environment and have far-reaching impacts".[79] As a result, "cyber security on actual installations is a growing issue in the oil and gas sector, since critical network segments in production sites, which used to be kept isolated, are now increasingly connected to networks".[80] This means "the industry faces many potential threats – including from eco-activists, terrorists, opportunists looking to make money and even countries aiming to disrupt another nation's supply".[81]

These are more than just threats. In late 2017 malware called Triton/Trisis was identified.[82] It infected the hardware of Schneider Electric, one of the largest suppliers of ICS, at a petrochemical plant in Saudi Arabia run by Saudi Aramco.[83] Iran and Russia were separately blamed.[84] In June 2021 the DOJ indicted a member of Russia's Ministry of Defense over the campaign.[85] Triton/Trisis used a zero-day vulnerability and remote access trojan (RAT) in a two-stage attack. The "hackers apparently intended to manipulate the layers of built-in emergency shutdown protocols to keep the system running while they bored deeper and gained more control". This meant it could override safety shutdown features and sabotage the system. However, "the malware accidentally triggered emergency system shutdowns that gave it away".[86]

If the code used by the attackers had been perfected it would have been able to trigger an explosion.[87] Dragos claimed the group behind Triton, "is the only activity group intentionally compromising and disrupting industrial safety instrumented systems, which can lead to scenarios involving loss of life and environmental damage".[88] Brigadier General Danny Bren, the former head of Israel's Cyber Defense Unit, warned "With something like that, you can create great danger to an oil rig, a refinery, a power station. In effect, you have built a bomb".[89] Its source code is now on the Internet.

On 10 June 1999, the *Olympic* pipeline in Bellingham in Washington state ruptured causing 237,000 gallons of gasoline to leak into a nearby creek. An hour and a half after the leak the gasoline ignited along a 1½ mile stretch of the creek leading to eight injuries and three fatalities. The cause was not a single point failure but a multi-point failure which first began during a botched 1994 repair.[90] These kinds of accidents caused by a human error could be replicated by cyberattacks in all of the case studies. Even with safety accorded the highest priority, Murphy's Law, 'Whatever can go wrong, will go wrong', can take effect. This is especially when people are faced with time pressures or have limited choices.[91]

Chemical Plants

Chemical plants are also possible (possibly likely) targets. These handle often toxic chemicals for industry and there are several major sites as well as a number of subsidiary facilities.[92] These are close to, and serviced by, local cities, towns and villages and depending on what they store and process and their integration with CI. Targeting any of these facilities could well have a major impact on health and local life.[93]

In July 2018 Ukraine's intelligence agency prevented a cyberattack at a station providing chlorine for water treatment and sewage plants throughout Ukraine. It was not an attempt to injure but to block plant operations with downstream effects but could have led to an accident. Ukraine's security service stated the attack originated from Russia.[94] Civilian sites are in the line of sight of cyberattackers and targets which can affect civilian populations and private industry are not off-limits. Even failed attacks such as this can be used as proving grounds and to enable pre-positioning for future attacks across critical infrastructure sectors.

A clear example, with tragic consequences, of a major incident at a chemical plant took place at the Union Carbide pesticide plant in Bhopal in India on 3 December 1984. Up to 25,000 people died as a result of the chemical explosion a release of the toxic gas methyl isocyanate with half a million people exposed. Its effects were still being felt 30 years on.[95] The eventual legal settlement in 1989 totaled a meager $470 million, roughly $550 for each victim. It would be wise to assume a far higher figure today which for a country such as the United States, with an established culture of litigation, could be significantly higher still. This is potentially the price of doing too little or nothing in the face of cyber threats to SCADA and ICS. The failed (or miscalculated) Triton attack on a Saudi Aramco petrochemical plant described above is a contemporary example of the dangers being faced.[96]

Ports and Logistics

What have so far been seen at ports are ransomware attacks. In July 2018 the Port of Long Beach in California was infected from the terminal of the China Ocean Shipping Company (COSCO).[97] Two breaches also took place in September 2018 at the Port of Barcelona in Spain and at San Diego Port in the U.S. Both affected only internal IT systems and port operations. In Barcelona land operations were affected, including the loading and unloading of container ships.[98] Barcelona's port authorities cautioned, "While American and Asian ports are keenly aware of this issue and allocate a significant portion of their budgets to protecting against cyberattacks, in Europe the concern with cybersecurity is more recent".[99]

A year previously the Petya/NotPetya attack was designed to make systems, databases, and machines unusable. It initially masked as ransomware designed to look like a variant of the WannaCry ransomware which had struck the month before.[100] It claimed to have encrypted data whilst demanding ransomware payments in Bitcoin. However, it was a wiper—the files could not have been decrypted once NotPetya was installed meaning they were unrecoverable.[101] Petya/NotPetya propagated globally (as had WannaCry). FedEx, the logistics company, said that some of its information was lost forever and it took over a month to restore normal functioning of their information systems and cost them around $300 million.[102] Similarly, shipping conglomerate Maersk reported it affected global shipping, took a month to restore systems, and they also estimated the damage at $300 million.[103] Saint Gobain, Reckitt Beckinser, and several others were also impacted. Petya/NotPetya spread to at least 65 countries in all.[104] WannaCry has been attributed to North Korea and NotPetya to Russia's GRU (specifically Military Unit 74455).[105] Both cases are discussed further in Chapter 6.

Whilst WannaCry and the original Petya were aimed at extracting ransom, not disruption or destruction, NotPetya was dedicated wiperware. However, their effects were largely financial not physically destructive. They were not aimed at ICS. The Industrial Control Systems at ports are essential to operations. The consequences of disruption or destruction of these operations could be catastrophic and could lead to collisions, the release of oil or chemical spillages, toxic gases, or explosions. Emergency or remedial procedures would likely be implemented not only at effected sites but also at related sites, some of which would be dependent on the impact site and could cascade to other sectors and would need to be dealt with both locally and by central government.

Like in other areas of CI, this would place considerable strain on the emergency services. These are contingencies that need careful preplanning and systems of resilience, redundancy, and recovery. This is particularly the case as a plant or port shut down at one site might lead to plant or port shutdowns or lockdowns at other sites considered vulnerable. As these could be speed of light attacks this needs careful preplanning and foresight. The downstream effects could include port energy supplies, communications, water, bridges, dams, or pipelines. This would be combined with an economic cost and short-, medium-, and long-term security concerns and a loss of confidence at these sites. This would ripple out into the energy sector, the country, and into the wider world. A 2017 paper published for NATO's CyCon U.S. conference warned, "cybersecurity of critical maritime and interconnected infrastructure remains largely unregulated with minimal, if any, assessment or mitigation of cybersecurity risks".[106]

Merchant Shipping

Maritime vessels carry 90% of global trade between ports and are also vulnerable. Under the SOLAS convention for international shipping (first adopted in 1914 after the *Titanic* disaster), all merchant vessels over 300 tons are fitted with automatic ship identification systems (AIS) which use transponders.[107] This means shipping is able to be freely tracked through the Internet. This includes many, but not all, military vessels.[108] In authorized penetration tests by the Israeli firm, Navaldome, which specializes in maritime cybersecurity their team:

> hacked into live, in-operation systems used to control a ships' navigation, radar, engines, pumps and machinery. While the test ships and their systems were not in any danger, Naval Dome was able to shift the vessel's reported position and mislead the radar display. Another attack

resulted in machinery being disabled, signals to fuel and ballast pumps being over-ridden and steering gear controls manipulated...[109]

Asaf Shefi, Naval Dome's Chief Technology Officer (CTO), and former Head of the Israeli Naval C4 and Cyber Defense Unit, commented:

> We designed the attack to alter the vessel's position at a critical point during an intended voyage - during night-time passage through a narrow canal. During the attack, the system's display looked normal, but it was deceiving the Officer of the Watch. The actual situation was completely different to the one on screen. If the vessel had been operational, it would have almost certainly run aground...The vessel's crucial parameters - position, heading, depth and speed - were manipulated in a way that the navigation picture made sense and did not arouse suspicion...This type of attack can easily penetrate the antivirus and firewalls typically used in the maritime sector.[110]

That commercial maritime shipping could be a target for external hacking was also demonstrated at DEFCON 29 in Las Vegas in 2021. There small teams of skilled hackers demonstrated a proof-of-concept. They "were able to penetrate different maritime subsystems including navigation, firefighting, and steering systems", as well as "propulsion, steering, and navigation systems". This is accomplished by penetrating the onboard navigation system (systems that are open to compromise) then moving laterally to essential systems including steering and throttle controls.

Demchak and Thomas indicate that "A skilled hacking team typically takes at most 14 hours to penetrate the system safeguards and remotely take control of both steering and throttle controls". This means, "whether conducted for ransom, malicious disruption, piracy, or as part of larger geopolitical conflicts" or terrorism, commercial shipping could be at risk. This could lead to destructive attacks or bring severe disruption to regional and global trade (especially if conducted near maritime chokepoints like the Suez Canal). Moreover, legacy systems continue to be used to varying degrees across the commercial shipping industry. Whilst these demonstrations at DEFCON required wired or "plugged-in" connections to simulate an on-board/air-gapped environment, "remote-access hacking is possible as demonstrated in February 2017, when hackers took control of a German-owned container vessel traveling from Cyprus to Djibouti".[111] Furthermore, and highlighting a wider problem that encompasses all CI sectors, Navaldome also pointed out that, "*manufacturers themselves can be targeted* [emphasis added], when they take control of onboard computers to carry out diagnostics or perform software upgrades, they can inadvertently open the gate to a cyber attack and infect other PC-based systems on board the ship".[112]

With 30 million barrels of oil alone a day shipped around the world, there is also increasing reliance on automation and computer-controlled shipping. The possibility that a malicious actor could remotely hijack a vessel at sea is currently a very real possibility, especially in a situation of military conflict by major developed powers.[113] Autonomous shipping is regulated by the International Maritime Organization (IMO) which defines Maritime Autonomous Surface Ships (MASS) and is working on a framework for their operation.[114]

There are considerable benefits to autonomous shipping, given the growth in global trade. Additionally, naval traffic is monitored under maritime law by both national and international security services and by large insurance companies (such as Lloyds Register) as well as the maritime transport and logistics industry.[115] The United Nations estimated in 2015 the "total seaborne trade volume surpassed 10 billion tons for the first time — roughly a four-fold increase since 1970".[116] It is the main method of international transport and a key enabler of globalization and development but without a measured approach to current and future cybersecurity vulnerabilities there could be icebergs ahead.

Road and Rail

Road and rail also use ICS and SCADA including sensors and wireless repeaters which relay real-time information back to control points. Many of these are vulnerable to hacking. Cesar Cerrudo, a former hacker who now runs a cybersecurity company, told the DEFCON 22 conference in 2014 that these sensors and repeaters are distributed worldwide and can be disabled or made unusable. They can also be spoofed into sending false signals misreporting traffic flows. This can be done in the field using relatively inexpensive commercial-off-the-shelf-technologies (COTS).[117]

There are also wireless vehicle detectors operating without security features which can be locally hacked from a distance of around 1,500ft. Traffic controllers can also be spoofed with false Global Positioning Signals (GPS). External attacks through the networks of traffic controllers as well as through physical access to Traffic Signal Controller cabinets on roadsides or power loss can each affect operations. These are all increasingly networked to allow remote maintenance and are increasingly accessible via Bluetooth or Wi-Fi. These can control timing signals for various times of the day and the operation of traffic lights with sensors used to detect vehicles and pedestrians. Most cabinets are not connected to anything but each other in the relay, usually through microwave links or GPS synchronization, but some will have an RTU which is used for remote access. Communications can be over broadband Internet, fiber optic cable or wireless signal. Street and road lights work in a similar way. Resultantly, cybersecurity researchers have found "the range of vulnerabilities for traffic signal controllers is large…[but] the vehicle volumes may be low enough that compromised sensors or timing offsets may be tolerable when compared with the expense of securing the systems".[118] This is concerning and both commercial and military GPS has seen disruption for unreported reasons.[119]

The firmware hardwired onto these commercially sold devices is also hackable and this renders the attack landscape wider still as an unknown number of "sensors and repeaters can be accessed and manipulated over the air by anyone, including firmware updates". Although this does not apply to all COTS products, a number are neither encrypted or digitally signed (meaning they have not been authenticated).[120] In short, if exploited by cybercriminals or APTs, this can cause gridlock and accidents. Cerrudo concluded:

> Any third world guy can easily get devices used by US critical infrastructure, hack them and then attack the US. Anyone can build a $100 worth device to cause traffic problems on most important cities on US (some other world cities too). Smart cities are not so smart when data that

feeds them is blindly trusted and can be easily manipulated. Cyberwar is cheap.[121]

GCN, which assesses technology for the public sector out of its offices in Virginia, said in 2018 "Transportation systems are ripe targets for cybercriminals...and many state and local government officials are only now waking up to the threat and realizing they need to beef up their defences".[122] It is not only cybercriminals who could disrupt or harm transportation but they are the most likely, especially through ransomware. With transportation privatized in many nations and with pressures on owner-operators to run low-cost efficient services attacking them would place strain on already under-resourced sector of CI.

Attacking road and rail might be focused on cities—with the metro systems of San Francisco hit in November 2016 and Colorado in February and April 2018 by separate ransomware attacks—but these could spread to effect wider areas with concomitant effects on national transportation, the wider economy, and social cohesion in affected areas.[123] Sen. Mark R. Warner (D-Va.) said in response to the San Francisco attack that although attackers might currently be opportunistic, metropolitan transport "represent a particularly enticing target for more advanced threats".[124]

For major capitals, which rely on public transport to ferry people from suburban areas into cities, disruptive cyberattacks on metro lines can bring about transport chaos if services are severely disrupted or forced to stop for days or weeks. Unless security-by-design is built into the architecture, IoT enabled SMART systems will suffer from the same problems.[125] Warner's response also highlighted several common concerns. San Francisco's ransomware targeted their legacy systems. The status of their backup systems was uncertain. In the case of a full system outage could continuity of operations be assured? What systems were Internet-facing and what were 'mission critical'? What preventive measures were there to prevent attackers moving laterally and burrowing deeper into systems? What plans were in place to cope and was there liaison with local, regional, and national agencies? Warner was concerned that Washington DC could be next.

In 2016–2017 the Washington Metropolitan Area Transit Authority (WMATA), saw scheduled IT maintenance cause a failure at the Rail Operations Control Center. This saw them unable to remotely operate switches used by trains to change tracks and operate and ventilation fans during the morning rush hour.[126] The worst effects were remedied in minutes "but problems persisted throughout the day as some devices required reboots even after connections were restored".[127] Most, if not all, major metro systems are subject to daily interruptions and delays but systemwide slowdowns or failures lead to ripple effects of further delays and disruption with knock-on effects as people use alternative forms of transport.[128]

The concerns of technical cybersecurity researchers who regularly report these and other vulnerabilities are shared with many on the front line. Peter Rahn, the Department of Transportation Secretary for Maryland, warned industry officials that if hackers penetrated the networks of transport operators "they can play with our trains, traffic signals, variable message boards. We've never had to think about these things before".[129] To place this in further context, "cybersecurity funding remains a serious challenge for state governments. In 2016–2017 WMATA was operating with a $290 million budget deficit and a 2016 survey of top IT security officers from 48 states found that in most states, spending on cybersecurity was a fraction of the overall IT budget, ranging from zero to 2 percent".[130]

Civil Aviation

For both ground controllers and pilots, data integrity is of paramount importance. With safety and security priority concerns for civil aviation operators worldwide, it is another heavily regulated industry. Also, airports are "a gateway to the world for travelers and business", and aid development and economic growth. As they carry international passengers this does not make them a natural target in terms of cyberwarfare. This cannot be ruled out but most of what has been reported is cybercrime related. In either scenario, part of the underlying reasons for breaches (in common with CI attacks more widely and part of their defense) "is a large variation in the way airports implement measures to protect networked infrastructures and design cybersecurity solutions" and because they employ a wide variety of ICT systems.[131] The growth of the civil aviation sector has seen cybersecurity, and physical security checks, improve. This includes Air Traffic Control/Management systems (ATC/ATM). Operators also share data and exchange information. This could be (perhaps is) another risk factor, and there are also security gaps common to other industries, including gaps in cybersecurity awareness among their employees. As in all other industries, "the use of ICT in civil aviation has increased exponentially over recent years, from the development and construction of aircraft to communications and navigation instruments, along with all the thousands of connections that link the various parts of an airport".[132]

The same dangers that are apparent in other sectors are felt in civil aviation and airports themselves are vulnerable to cyberattacks. The Internet is used in daily operations, including for information and messaging. At airports, SCADA controls include HVAC (heating, ventilation, air conditioning systems), building management systems, baggage handling, and power distribution. However, "SCADA systems are in their infancy stage overall in airports" and currently implemented only in around 20%. At many bring-your-own-device controls for staff (and passengers) are limited but so are instructions to change default credentials and knowledge of application security and encouragement of security-by-design procedures from design to installation and operation. Airports fair better on implementing firewalls and network segmentation, software and hardware updates, disaster recovery plans, strong user authentication protocols, data encryption, and Intrusion Detection Systems.[133] They are however vulnerable to physical attack as well as infection through USB drives, SD cards, alongside the increasing use of SMART devices by employees and travelers. They are also susceptible to drones, surveillance through digital cameras, or compromise of their CCTV systems.[134]

Lykou, Anagnostopoulou, and Gritzalis point out that "Any smart airport system with an available attack surface, where [sic.] security fix has not applied and system is not running with all the latest security patches is a likely target of malicious software attack". In addition,

3 Cyberwar: Attacking Critical Infrastructure 177

SMART devices operating in aviation facilities, including airports, can be tampered with. This could lead to the alteration of central reservation systems, IT systems, and stored sensor data. Current risks include data deletion or corruption such as those seen in ransomware and wiperware attacks. Attacks could affect automated check-in machines, passport control gates, and building management systems. The same set of problem areas that impact all other areas of CI (and beyond) are also in play, including phishing and social engineering, outsider and insider threats, credential theft, and privilege escalation.[135] There are known cases we can draw lessons from. These include:

- 2006, July: a cyberattack forces America's Federal Aviation Administration (FAA) to shut down several ATC systems in Alaska.
- 2008, August: At the now defunct Spanair, their central computer systems at Spain's Madrid-Bajaras airport (which was used to monitor technical problems on aircraft) was infected with trojan malware which prevented the reception and activation of an alarm message from a DC-9 aircraft. It crashed after takeoff. 154 passengers died on Flight 5022. Whether the Trojan was a direct or indirect cause of the crash is disputed.[136]
- 2009, February: 48,000 personnel files were hacked from FAA computers.
- 2011, June: Three engineers were accused of "interruption of computer services" that resulted in check-in congestion and flight delays.
- 2013, July: At Istanbul Ataturk and Sabiha Gökçen airports, a cyberattack hit passport control systems resulting in numerous flight delays.
- 2013: Prolonged spear phishing campaigns were aimed at approximately 75 American airports.
- 2014, March: Malaysia Airlines flight MH370 disappears from radar. Its disappearance remains unexplained. It has been suggested that the aircraft, a Boeing 777-200ER, was commandeered through a cellphone and/or a USB drive. This has been roundly rejected by Boeing and no supporting evidence of cyberattack has come to light. This has not prevented speculation.[137]

- 2014, December: Iranian hackers were accused of a coordinated cyber-attack on over 16 countries. This included airports and airlines in Pakistan, Saudi Arabia, South Korea, and the United States.
- 2015, February: The FAA discovered various forms of malware in personnel e-mail accounts.
- 2015, April: The Irish airline Ryanair suffered losses of €4.6 million through a fraudulent electronic transfer made via a Chinese bank.
- 2015, June: 1,400 passengers were grounded at Warsaw's Chopin airport when the flight plan system failed following a DDoS attack.[138]
- 2015, June: LOT, Poland's national airline, saw ten flights grounded and a further ten delayed from a cyberattack on their flight plan system.[139]
- 2018, May: A dissident hacktivist group hacked into Mashhad airport in Iran and displayed protest messages on airport monitors.[140]
- 2019, Christmas: Albany airport in upstate New York was hit by a ransomware attack. A six-figure ransom was paid to restore services.[141]
- 2020, January–May. The UK low-cost carrier EasyJet saw the travel information of 9 million customers compromised when its IT systems were breached.[142]
- 2021, March–June. The European Air Traffic Management Computer Emergency Response Team (EATM-CERT) reported an increase of 530% in reported cyberattacks on the aviation industry. None of those reported "were directly against safety–critical aircraft systems or passenger mobile devices connected to in-flight internet". 95% were financially motivated. The other 5% included Intellectual Property (IP) theft. This included ransomware attacks on VT San Antonio Aerospace in June 2020 and on the U.S. carrier Spirit Airlines in March 2021.[143]
- 2022, June. Russia's Federal Air Transport Agency (Rosaviatsiya) was hit by a coordinated wiperware attack. This led to temporary restrictions on flights to and from 11 airports in central and southern Russia. This included the airports of Anapa, Belgorod, Bryansk, Voronezh, Gelendzhik, Krasnodar, Kursk, Lipetsk, Rostov-on-Don, Simferopol, and Elista.[144]

'Only' two of these cases (Spanair and MH370) resulted in downed aircraft. Neither are believed to be caused by cyberattack. However further safety and security measures have been introduced since MH370 went missing. Introduced in 2017, the Automatic Dependent Surveillance–Broadcast (ADS–B) system provides a valuable contingency to GPS failures (and those of similar systems like Russia's GLONASS, China's BeiDou, and the EU's Galileo) but it transmits an unauthenticated and unencrypted signal. This means "the system is susceptible to hacking" and eavesdropping. Additionally, "hundreds of [sic.] in-flight aircrafts are accessible and vulnerable to message jamming, replaying of injection and other active attacks", because the system broadcasts aircraft identifiers and flight paths through satellite communications. Depending on the aircraft type, malware brought onboard could also be used gain access to airplane systems and airport systems could be similarly vulnerable—and could cascade to other airports. Relatedly, the Common Use Passenger Processing System (CUPPS), might also be vulnerable to hacking and could lead to unidentified passengers being allowed to board.[145]

So far cyberattacks on civil aviation have resulted in little more than temporary check-in and flight delays which can be a major inconvenience (at best). This includes attacks for the exfiltration of customer records which were almost certainly for financial gain but also carry security implications. Airports have also experienced ransomware attacks which in one case forced a UK airport to close down its flight information system.[146] In the United States, the Government Accountability Office (GAO) found in 2015 that despite improvements at the FAA:

> significant security control weaknesses remain, threatening the agency's ability to ensure the safe and uninterrupted operation of the national airspace system (NAS). These include weaknesses in controls intended to prevent, limit, and detect unauthorized access to computer resources, such as controls for protecting system boundaries, identifying and authenticating users, authorizing users to access systems, encrypting sensitive data, and auditing and monitoring activity on FAA's systems.[147]

The GAO also found the "FAA also did not fully implement its agency-wide information security program. As required by the Federal Information Security Management Act of 2002".[148]

The situation differed little in Europe. The European Cockpit Association (ECA) which represents over 38,000 European pilots warned in 2017 that "aircraft, ground facilities and other critical infrastructures are vulnerable to cyber-attacks and therefore are at significant risk of unsafe situations that may, ultimately, even cause loss of life", adding, "This risk increases with connectivity".[149] The ECA highlighted common problems including threats to data integrity; vulnerabilities in legacy systems and procedures associated with them; insecure communications; low-barriers to entry for would-be attackers; and problems of collateral damage from cyberattacks beyond civil aviation. The ECA went even further, fearing that data theft could be used "to manipulate or erase information and/or to control or destroy systems or services that could even result in uncontrollability of aircraft".[150]

The ECA recorded that protective measures cannot be conducted in isolation given the interdependencies within the commercial aviation industry. Instead, the European Union (EU) should regulate and ensure compliance across Europe. Currently the European Organisation for the Safety of Air Navigation (Eurocontrol), headquartered in Brussels, directs flights into and out of EU airspace under the European Commission's Single European Sky initiative with a dedicated cybersecurity hub—the European Centre for Cybersecurity in Aviation (ECCSA). ECCSA offers a secure platform to exchange information on vulnerabilities and incidents allowing airlines and owner-operators to assess risks, but this is as far as its mandate goes as it is a voluntary arrangement (including information sharing[151]) with organizations having to apply for membership.[152]

The ECA was critical of this process of voluntary information sharing (often referred to as 'responsible disclosure') saying the EU needs to adopt mandatory reporting. With the General Data Protection Regulation (GDPR) now mandating organizations in the EU to disclose breaches, sharing would be a preventative measure with part of the design of the GDPR to raise cybersecurity standards. The ECA also argued, echoing the 2015 GAO report, "Such regulation should not only provide technical requirements. Proper training of all personnel is a significant factor in the mitigation of cyber risks. Contingency planning is a key aspect as well, and therefore all personnel that use safety critical systems, including flight deck crew, should also be adequately trained to detect actual cyber-attacks and act accordingly".[153]

In July 2018 the EU enacted Regulation (EU) 2018/1139 which gives the European Aviation Safety Agency (EASA) a coordinating role for cybersecurity and for international cooperation, but the bulk of the regulation was directed to concerns over commercial drones.[154] As a result, whether the directive will satisfy the concerns of the ECA and the extent to which airlines and airports will be legally required to train and plan for cybersecurity risks remains to be seen. The system they operate called the Enhanced Tactical Flow Management System (ETFMS) went down in April 2018 leaving half of all European flights delayed. This gives an idea of what a non-lethal cyberattack could do.[155]

Aviation safety is a paramount concern and up until the Covid-19 pandemic hit, aviation was a booming industry carrying ever-increasing numbers. These should be risks that are still addressed as a high priority. The potential dangers are too high to do otherwise. This might not be a worst-case scenario that sees aircraft fall from the sky but even this cannot be ruled out. It has been technically demonstrated at a conference organized by EASA that an app on a smartphone can be used to override aircraft controls and crash it. Flight navigation systems can be compromised through satellite provided Wi-Fi or through physically connecting through the on-board entertainment system. Air Traffic Control and Air Traffic Management systems are also a potential target as is Ground Control.[156] Similar attack vectors are used across all of the sectors of critical infrastructure detailed in the above case studies. These sector-by-sector case studies should be used to advance our understanding of cyberwar(fare) by attacking critical infrastructure. They can also be used to highlight why considered and holistic thought needs to be given over to the issues of ICS and SCADA architecture, their vulnerabilities, and supply chains who support these industries with ICT products and services.

The Good News

There is good news. Awareness of cybersecurity threats is increasing, and industry and governments have got better at coordinating and sharing threat intelligence. This is accompanied by increasing numbers of defensive products, services, and techniques. No sector is safe, but some are better protected than others although we continue to see major breaches of the cybersecurity fence. Voluntary cyber-standards have raised the bar across critical infrastructure and, in the U.S. at least, cybersecurity is a priority in the electric power, banking and finance, sectors.[157] This is largely the case across the developed world, although Asia–Pacific is subject to more breaches than anywhere else, but it is also true that "industrial control systems are of interest to malicious actors, and that systems are both accessible and vulnerable".[158] Moreover, as Robert M. Lee of Dragos explained, "We're not to a point of cascading grid failure … it's actually much harder than people realize".[159]

There are protective barriers.[160] Defense in depth models incorporating firewalls, segmentation, and access privileges are widely in place. There are also a wide variety of bespoke and customized products and services available to prevent intrusions. If an intrusion does occur Intrusion Detection Systems (IDS) are also widely in use and cyber kill-chain models are used to try and neutralize threats. There are many bolt-on solutions available to enhance industrial security alongside IDS, but security-by-design is now the preferred solution alongside cryptographic technologies like Blockchain. Patches are also issued for known threats, but SCADA systems are rarely patched as they are usually in constant operation or because they might well crash the system.

For companies that cannot afford customized IDS, SNORT is a valuable option. SNORT "is an open-source, free and lightweight network intrusion detection system (NIDS) software for Linux and Windows to detect emerging threats".[161] SNORT is owned and run by Cisco, the U.S. networking multinational, which bought out the original developer, Sourcefire, in 2013. SNORT is leveraged by Cisco for its Next Generation Intrusion Prevention System (IPS) and Next Generation Firewall.[162] GitHub, an open-source project allowing developers to cooperate on code, is also widely used in a wide variety of cybersecurity projects.

In developed nations, intelligence services and policing and law enforcement organizations are relatively well-resourced and do proactively prevent attacks from occurring. National and supranational Computer Emergency Response Teams (CERTs) are proliferating and increasingly sharing threat intelligence with the industry. There are also a number of government-run cybersecurity centers like the United Kingdom's National Cyber Security Centre (NCSC). The U.S.-based National Institute of Standards and Technology (NIST) is an extremely valuable center promoting good and best practice in the field of cybersecurity which includes detailed advice on ICS security.[163] The UK-based International Cyber Security Protection Alliance (ICSPA) is a nongovernmental organization that seeks to provide a private sector-financed hub that highlights these kinds of dangers.[164] Another is the U.S.-based Internet Security Alliance, which offers guidelines for corporate cybersecurity and public–private engagement with input from a number of large multisector corporations.[165] The Repository of Industrial Security Incidents (RISI) also hosts a publicly accessible database for "Incidents of a cyber security nature that directly affect industrial Supervisory Control and Data Acquisition (SCADA) and process control systems, accidental cyber-related incidents, as well deliberate events such as external hacks, Denial of Service (DoS) attacks, and virus/worm infiltrations".[166] The UN's International Civil Aviation Organization (ICAO) has also initiated several working groups to raise standards in aviation cybersecurity which works alongside national and supranational regulatory authorities.[167] Protective organizations are further detailed in Chapter 5.

Universities are also enjoying a period of boom in computer science as companies look to plug the skills gaps that currently exist with consortiums like CANVAS.[168] High-level conferences also take place regularly with good and best practices shared. Web-based resources such as Metasploit "verify vulnerabilities, manage security assessments, and improve security awareness".[169] The now defunct Open Sourced Vulnerability Database (OSVDB), which ran from 2004 to 2016 "catalogued more than 100,000 flaws affecting a large number of products" and was "free for non-commercial use".[170]

Several pieces of legislation are also worthy of note. In January 2018 the Cyber Vulnerability Disclosure Reporting Act was passed by the U.S. Senate. This allows policymakers to annually scrutinize the coordination of cyber vulnerability disclosures through the DHS and "the degree to which such information was acted upon by industry and other stakeholders".[171] This means "Under the Homeland Security Act of 2002 and the Cybersecurity Information Sharing Act of 2015 ("CISA"), DHS is responsible for working with industry to develop DHS policies and procedures for coordinating the disclosure of cyber vulnerabilities".[172] Through CISA, "Utilities expect more qualitative, timely threat intelligence from existing federal information sharing programs…[and] clarity about the conditions of information sharing programs based on new national cyber security policy".[173]

This includes reconnaissance, including communications designed to gather technical information related to a cybersecurity threat or security vulnerability; any means to defeat security or exploit security vulnerabilities; attempts at phishing or 'man-in-the-middle attacks'[174]; identifying malicious cyber command and control; identify any actual or potential harm caused by an incident, including descriptions of any information exfiltrated; and to attribute cyberattacks where possible.[175] There is also a proviso that DHS must also report how it works with "critical infrastructure owners and operators to prevent, detect, and mitigate cyber vulnerabilities".[176]

In April 2018 NIST also revised part of its Framework for Improving Critical Infrastructure Cybersecurity to improve reporting, analysis, and responses to disclosed vulnerabilities.[177] Relatedly the Data Security and Breach Notification Act (2018) seeks to punish and standardize reporting of data breaches which have hit the headlines with increased frequency and high financial, political, and security costs.[178] Although this repeats similar state laws in 48 of the 50 states this federal law is to be welcomed. The GDPR adopted by the European Union in May 2018 has similar goals but placed its emphasis on data privacy.[179] The 2016 Directive on security of network and information systems (NIS Directive) aims to raise cybersecurity standards across the EU, improve cooperation and information exchanges on incidents and risks, and embed "a culture of security" across critical infrastructure.[180] The NIS Directive is explicitly designed to enhance risk management and cyber resiliency within CI and digital services providers.[181]

However, national security falls largely outside of the purview of EU law, notwithstanding cooperation on transnational cybercrime and terrorism. Therefore, the NIS Directive and EU law more broadly "differentiates between physical and digital critical infrastructures as regards incident information sharing and common approaches".[182] This being said, issues like the disruption of international logistics and trade, such as that resulting from WannaCry, threatens internal security (one of the EU's pillars). Moreover, "content-related challenges such as hate speech and fake news, and the use of encryption by criminals or terrorists" also affect the EU's single market which enables the free flow of goods, services, capital, and people. This means "Dealing with cyber-incidents in the single market is also of ever greater importance in EU sectoral policies for industries such as finance, transport and energy. For example, end-to-end continuity of energy supply ('energy security') requires reliable interconnection of electricity lines across Europe and the joint ability to withstand cyberattacks".[183]

This was one of the reasons the EU Cybersecurity Act was enacted in June 2019 which strengthened the mandate of the European Network and Information Security Agency (ENISA)[184] and established "the European cybersecurity certification framework".[185] Alongside the NIS Directive and GDPR these mark a more interventionist policy for the EU in the arena of cybersecurity. This also "included a coordinated response (a 'Blueprint') to large-scale cybersecurity incidents and crises, addressing the growing concerns about a major 'black swan' incident" [an unlikely but major event like 9/11]…drawing on existing Integrated Political Crisis Response".[186]

Whilst national security remains within the purview of national governments and NATO, from both a national political and external perspective there is a great deal that binds EU states through common interests and concerns. However, "Fragmentation of the internal market means that EU countries impose different approaches…such as for ICT security requirements and standards. As a consequence EU-internal trade and cross-border business is not as smooth as it could be. This manifests itself in hesitation to buy from another country in the EU, or overlap in standards, duplication in certification schemes or uncertainty about how to deal with cross-border cyber incidents".[187] This security–economic–political nexus has seen "Digital policy and cybersecurity policy become highly politicised when viewed through the lens of sovereignty, even more so when the balance between national and European sovereignty is added to the agenda" but it is a top EU priority.[188]

The Bad News

The bad news is that we still see a multitude of serious cyberattacks. The most severe are from nations who are Advanced Persistent Threats but there are also a multiplicity of other actors who can do harm to critical infrastructure. The most dangerous of these are cybercriminals and nations like Russia are known to leverage them and hide behind them. Thankfully, terrorists have yet to grasp the disruption and damage that could be caused or have been stopped before they could do harm. This paradigm might not hold, and they too could be leveraged by a nation-state.

In May 2017 Admiral Mike Rogers, whilst Director of the NSA and U.S. Cybercommand (USCYBERCOM), testified before the Senate Armed Services Committee that "Power and basic infrastructure is something that always concerns me because the potential impact on the nation is very significant… for the utilities…for dams…for nuclear power plants. Clearly, in those areas, if someone with intent could get into an operating system, they could do a significant amount of damage. Perhaps bodily injury as well".[189] This chapter has demonstrated this is not only a clear and present danger but depending on the target, it could be a mass casualty event or series of events. Hackers could also breach emergency alert systems, which includes warnings of nuclear attack.[190]

In October 2017, the DHS and FBI issued a joint warning of an ongoing APT campaign targeting government agencies and companies in energy, nuclear, water, aviation, and critical manufacturing sectors. This was "a multi-stage intrusion campaign by threat actors targeting low security and small networks to gain access and move laterally to networks of major, high value asset owners within the energy sector".[191]

This campaign, dubbed 'Dragonfly 2.0', was attributed to Russia with the cybersecurity firm Symantec assessing that "The Dragonfly group appears to be interested in both learning how energy facilities operate and also gaining access to operational systems themselves, to the extent that the group now potentially has the ability to sabotage or gain control of these systems should it decide to do so".[192]

There is also a wider problem that will only grow with SMART and IoT devices. Mike Assante, who was heavily involved in setting cybersecurity standards for the North American electric power industry, has claimed that power generators and distributors are also now struggling with the size and complexity of the networks they manage. Moreover, highly trained ICT specialists are at a premium, and there are simply not enough to cope with demand. APT threats in particular are difficult to defend against.[193] This was highlighted by INL's 2016 report into cyber threats and vulnerabilities in the electric sector:

> Utilities often lack full scope perspective of their cyber security posture. Total awareness of all vulnerabilities and threats at all times is improbable, but without enough cyber security staff and/or resources utilities often lack the capabilities to identify cyber assets and fully comprehend system and network architectures necessary for conducting cyber security assessments, monitoring, and upgrades.[194]

If these problems are to be taken seriously, more investment is needed to meet this demand. This means companies need to recognize the scale of the problems they are facing as well as firefighting them when they appear (then taking remedial action). These practices are inconsistent across industrial sectors nationally and globally.[195]

Even before the Covid-19 pandemic, for cost and logistical reasons companies saw benefits in remotely operating their systems, but this also opens attack vectors. Many ICT specialists are already well aware of the types of attack vectors that even unsophisticated attackers are able to exploit and which websites and search engines can generate. Alastair O'Neill of the Insecurety computer security research collective and others discovered from the Internet "public control interfaces for heating systems, geo-thermal energy plants, building control systems and manufacturing plants...treatment systems, power plants and traffic control systems."[196]

Part of the reason we continue to see cyberattacks large and small is because the attack surface is so wide. This attack surface is increasing. By 2020 the IoT incorporated 30 billion devices and this number will only grow.[197] The IoT represents "one aspect of a Future Internet", for which "many application areas have been postulated – not only in industrial domains such as manufacturing, logistics, retail, service management, energy, public security, and insurance, but also for the life of every citizen – where IoT can bring significant improvements, and leading to new business models".[198]

It is estimated that 40% of the IoT will be healthcare related, making it the biggest user.[199] These trends are collectively known as 'Industry 4.0' which includes manufacturing and the 'factory of the future' alongside further developments in the use of robotics.[200] However, without commensurate 'security by design' measures in place throughout global supply chains, developing nations are facing the same cybersecurity problems that have affected the developed world. This includes insecurity in legacy ICT systems and, by rushing to bring SMART IoT devices to market, and not ensuring that 'security-by-design' is built into products.

However, security-by-design cannot be applied to legacy systems; only products that are new to market. In a highly competitive global market, we are risking repeating the same or similar cybersecurity problems. These will be felt across the developed world and by developing nations. This will be throughout industry, in new infrastructure projects, and in urban planning. The IoT is also designing people out of these processes in favor of increasing automation and in the race to bring products and services to market security is still not accorded as high a priority as it needs.

A joint McAfee/CSIS report in 2011 assessed "that 24 hours of down time from a major attack would cost their own organization U.S. $6.3 million. Costs were highest in the oil/gas sector, where the average estimate was U.S. $8.4 million a day. They were lowest in the government and water/sewage sectors".[201] Who should pay for losses remains a difficult subject. Half of McAfee's respondents "expected insurance to cover

3 Cyberwar: Attacking Critical Infrastructure

the cost, while nearly one in five said the cost would fall on rate-payers or customers and just over a quarter expecting a government bail-out. Expectations that insurance would defray the cost were highest in Italy, Spain and Germany and lowest in India and Saudi Arabia".[202]

One of the important issues that business has to address is how risk is identified and measured beyond financial liabilities.[203] One of these areas of risk remains the "insider threat." The insider threat remains a problem for all businesses, as it does for government organizations despite security vetting procedures.[204] Improving resilience beyond 'defense in depth' also requires serious study and robust physical or reroutable power backups for extended periods should be a high priority.[205] These are active sets of issues. They also have implications for cyberwarfare.

Ukraine and Russia's 2022 Invasion

An hour before Russia invaded Ukraine on February 24, 2022, a cyberattack was felt Europe-wide. It targeted Viasat, a Californian-based provider of satellite communications for commercial and military users. It was quickly mitigated by Viasat in coordination with Mandiant "as well as law enforcement and U.S. and international government agencies".[206] According to UK and U.S. intelligence, the primary target was believed to be Ukraine's military and the successful attack saw tens of thousands of terminals become inoperable.[207] *Wired* reported it "caused a significant disruption to Ukraine's military communications".[208] It also had wider effects and "caused outages for several thousand Ukrainian customers, and impacted windfarms and internet users in central Europe". The attack was attributed by the United Kingdom's National Cyber Security Centre (NCSC) to the GRU.[209] They had used new wiper malware called 'Whispergate' (first detected by Microsoft in mid-January).[210]

For industry insiders the Viasat cyberattack highlighted that it was possible high-end actors like Russia could get into space systems and ground-based uplinks. Uplinks could be infiltrated, jammed, or subjected to ransomware demands. With ever growing numbers of satellites with increasing commercial satellite constellations in low-earth orbit (like Elon Musk's Starlink), some using shared inter-satellite links, the potential for intrusions and breaches is yet another significant cybersecurity issue. Potential insecurity in satellite supply chains also provides a problematic vector for compromise or attack.[211] All but the largest satellites are likely to burn up on re-entry but if one or more satellites were made to collide in space this would produce a debris field that poses a risk to national space powers and commercial operators alike. Alongside anti-satellite (ASAT) weapons, this poses a further threat.[212]

In early February, oil and port storage facilities across Europe had also been hit by ransomware gangs dubbed BlackCat and Conti. Thought to be cybercrime rather than state-sponsored, these came at a point when there were rising tensions and (well founded) fears over disruption of energy supplies and wholesale price rises. The Kremlin has long used Western dependence on gas and oil imports from Russia as a bargaining chip and energy security as leverage against Western Europe.[213] It has also been known to leverage cybercriminal gangs as proxies and privateers. In addition, Russian patriotic hacking groups were also active against Ukrainian and Western targets.[214] These were part of a series of cyberattacks aimed at Ukraine. In parallel, DDoS attacks had been conducted with another new wiper (dubbed HermeticWiper) discovered on 23 February. Its primary targets were organizations in finance, defense, aviation, and IT services (all sectors of CI). Although primarily directed against Ukraine, HermeticWiper also spread to Lithuania, a NATO member.[215]

NATO's long-standing Secretary General Jens Stoltenberg had told the Atlantic Council in late January 2022 that Russian cyberattacks could be both a first step or "precursor to a kinetic military attack" on Ukraine. This could be combined with "combat troops, heavy armor, planes, missiles, and all the stuff they have lined up along the borders of Ukraine". Stoltenberg declared that Russia's use of cyber could also be part of an attempted coup to topple the government in Kyiv aided by Russian intelligence officers operating in Ukraine.[216] Seeing Ukraine fall would destabilize European security and bring grave concern to NATO's former Soviet states. For these reasons, Ukraine has received significant alliance (and wider international) support short of direct intervention.

NATO has long recognized that direct confrontation with Russia has escalation through to potential nuclear use baked-in.

With these stakes in mind, since 2017 under the Trump administration the United States began working closely with the Ukrainian government to improve Ukraine's cyber resilience. This included $40 million in "cyber capacity development".[217] The Biden administration was well aware in the lead up to the invasion that the use of cyber offensive tools by Russia was an ever-present probability. It has since been made public that pre-emptive U.S. support for Ukraine's cyber defense was coordinated across the government. This included "hands-on support" to "essential service providers within the Ukrainian government including government ministries and critical infrastructure operators to identify malware and restore systems after an incident has occurred". In addition, "6,750 emergency communications devices, including satellite phones and data terminals, to essential service providers, government officials, and critical infrastructure operators in key sectors such as energy and telecommunications" were provided.[218] This included a satellite phone to Ukraine's inspirational President, Volodymyr Zelensky.[219]

Enhanced cybersecurity measures for Ukraine's electrical grid were also fostered alongside integration with the European Network of Transmission System Operators for Electricity (ENTSO-E). The Cybersecurity and Infrastructure Security Agency (CISA) was used as the vehicle to exchange technical information to Ukraine, especially relating to Russian cybersecurity threats. The Department of Energy (DOE) was also involved as were the national laboratories who coordinated with Ukrainian utility providers to enhance resilience. The Treasury Department also worked with the National Bank of Ukraine (NBU), via the Software Engineering Institute (SEI) and Ukraine's CSIRT to bolster Ukraine's financial sector. From December 2021 to February 2022, U.S. Cyber Command also worked with Ukraine's State Services for Special Communication and Information Protection (SSSCIP)—sometimes referred to as Ukraine's Cyber Command—to bolster resilience in critical networks. This saw cybersecurity experts from both nations sit side-by-side analyzing adversary activities and identifying vulnerabilities. Remote analytic and advisory support was also provided, specifically directed to protecting critical networks. Encouraging connections between CI operators and Ukraine's civilian private sector was also nurtured.[220] It is also believed that the NSA and CIA also helped support Ukraine.[221]

The UK government has similarly supported Ukraine's cyber defenses.[222] So too did the European Union's newly created Cyber Rapid Response Team (CRRT). Like USCYBERCOM, CRRT aided Ukraine remotely and onsite whilst volunteers to Ukraine's new civilian 'IT Army' were domestically coordinated via Telegram.[223] How many other nations might have supported Ukraine's cyber defense and the nature of their involvement is uncertain but financial and humanitarian aid alongside military equipment and intelligence has flooded into Ukraine from nations worldwide. It is also uncertain how many (or how) private cybersecurity providers have assisted Ukraine or offered their support. However, a number have gone public.[224] These include Microsoft and Cisco Systems.[225] Separately Elon Musk supplied Starlink terminals in Ukraine to bolster Internet access.[226] Microsoft's assistance includes targeting the Internet domains of 'Strontium' (aka the GRU's 'Fancy Bear'/APT28). Tom Burt of Microsoft disclosed on April 7, 2022:

> Before the Russian invasion, our teams began working around the clock to help organizations in Ukraine, including government agencies, defend against an onslaught of cyberwarfare that has escalated since the invasion began and has continued relentlessly. Since then, we have observed nearly all of Russia's nation-state actors engaged in the ongoing full-scale offensive against Ukraine's government and critical infrastructure, and we continue to work closely with government and organizations of all kinds in Ukraine to help them defend against this onslaught.[227]

Specifically, Microsoft detected new forms of offensive and destructive malware (including a trojan they dubbed 'FoxBlade') as part of renewed cyberattacks against Ukraine. Microsoft immediately notified the Ukrainian government and officials in Europe and America. They also provided "threat intelligence and defensive suggestions" to "a range of targets, including Ukrainian military institutions and manufacturers and several other Ukrainian government agencies". Russia's attacks were precisely targeted and included "the financial sector, agriculture sector, emergency response services, humanitarian aid efforts, and energy sector organizations and enterprises". Microsoft's president, Brad Smith, commented that as civilian targets they "raise serious concerns under the Geneva Convention, and we have shared information with the Ukrainian government about each of them".[228]

Ukraine's 400,000-strong ICT sector also mobilized (into Ukraine's 'IT Army'). They were on high alert alongside the rest of Ukraine and took defensive action encouraged by Ukraine's SSSCIP. They also took offensive action against Russia's military systems (a potential breach of the LOAC) and provided intelligence.[229] This is further evidence of the blurring of lines between state-controlled/military combatants and sub-state/state-sponsored actors but when Russia finally launched its expected invasion, they were manning the barriers.[230] This also helped blunt and limit the potential damage and opportunities.[231] What was seen were DDoS attacks and website defacements prior to and during the invasion.[232] An uptick in Russian cyberespionage was also seen as was an uptick in cybercrime activities. In both cases, this included new malware variants.

In April 2022, a joint cybersecurity advisory (CSA) notice was issued by the 'Five Eyes' (the United States, Canada, the United Kingdom, Australia, and New Zealand) which had input from industry partners through the Joint Cyber Defense Collaborative (JCDC) body. It warned that Russia was seeking revenge for the imposition of economic sanctions rolled out following the invasion. Organizations across the 'Five Eyes' and beyond were warned to guard CI against possible attacks by patching systems, enforcing multi-factor authentication, securing Remote Desktop Protocols (RDPs), and educating end users regarding renewed threats from Russia. This included threats from patriotic cybercrime groups from within Russia and Russian-aligned groups operating outside of Russia against nations supplying material aid to Ukraine. Eight separate groups were highlighted. Each had different specialties or tactics, including hacking and leaking, botnets, DDoS attacks, ransomware, worms, trojans, or malware-as-a-service offerings.[233]

The CSA notice urged "critical infrastructure network defenders to prepare for and mitigate potential cyber threats—including destructive malware, ransomware, DDoS attacks, and cyber espionage—by hardening their cyber defenses and performing due diligence in identifying indicators of malicious activity". Of particular concern, given past cases like BlackEnergy and NotPetya, were IT and OT networks (including ICS). The FSB's Center 16 (Military Unit 71330) and Center 18 (Unit 64829), the SVR (*Sluzhba Vneshnei Razvedki*)—Russia's Foreign Intelligence Service—and General Staff Main Intelligence Directorate (GRU), 85th Main Special Service Center (GTsSS) were also flagged. Specifically, this included the GRU's Main Center for Special Technologies (GTsST/ Unit 74455), and Russia's Ministry of Defense, Central Scientific Institute of Chemistry and Mechanics (TsNIIKhM). Each has been responsible for, or were implicated in, global campaigns targeting CI and other sectors.[234]

Between March and April 2022, the GRU tried again to attack CI by targeting high-voltage electrical substations in Ukraine. 'Sandworm' (Unit 74455) deployed a new upgraded version of Industroyer/Crash Override malware directed against ICS and used alongside wiperware.[235] Industroyer 2.0 sent commands to substation devices that control power flows. It was first identified by the Slovakian cybersecurity firm ESET which has been working with Ukraine's CERT (CERT-UA). This led to an advisory notice from CERT-UA and at one utility company, nine electrical substations were temporarily switched off as a precaution. Over two million people lived in the affected area.

The utility company had been penetrated prior to or during the February invasion and Industroyer 2.0 had lain undetected until the order was given to activate it. According to ESET and SSSCIP, "the new version of Industroyer had the ability to send commands to circuit breakers to trigger a blackout, just as the original did. ESET found, too, that the malware had the ability to send commands to protective relays".[236] It was not clear how the compromise had occurred, but it spread from the IT network to the OT (ICS) network. ESET describes it as "a fully-modular platform with payloads for multiple ICS protocols" and "highly configurable" meaning it is portable to other victims and environments. Part of its design is to cover its tracks and "slow down the recovery process and prevent operators of the energy company from regaining control of the ICS consoles".[237] According to Viktor Zhora, a senior official at SSSCIP, "We were very lucky".[238]

David Sanger and Julian Barnes cautioned, "The Ukrainian grid was built in the days of the Soviet Union, connected to Russia's. It has been upgraded with Russian parts. The software is as familiar to the attackers as to its operators". The BlackEnergy and Industroyer attacks of 2015 and 2016 had exposed flaws, and whilst remedial work was undertaken, patching and upgrades take significant time and resources. Added to this, "Russian hackers understand every linkage in the design — and most likely have insiders who can help them" according to Sean Plankey, a former DOE cybersecurity specialist.[239]

They were also 'lucky' or prepared when Russia also went after Ukraine's railroad network. 'Wiperware' was found on its servers. This was discovered *before* it was activated but it had penetrated the rail networks cyber defenses. This could have had dire consequences. Not only were they a vital supply line inward for weapons and humanitarian aid but they were a lifeline for Ukrainian refugees fleeing the fighting. In the first ten days alone, 1 million Ukrainian civilians fled to safety on the rail network.[240] Upwards of 5 million have fled in total, many by rail. The importance of CI was already clear to NATO, and their annual cyber exercise, Locked Shields, wargamed CI attacks and defense in 2022.[241]

Cellular networks were left largely free from disruption. Russia might have initially had faith that a military victory would be swift as they sought to encircle Kiev and attacked the Donbas from the east and from Crimea to the south. In prospective anticipation of this, Politico reported that "Even before the invasion, Russian surveillance of Ukrainian telephone networks was pervasive". This was because "most of Ukraine's telecommunications providers were either owned by Russians or Russian-Ukrainian businesspeople" and they could "lean on the private sector for help infiltrating networks".[242] Another reason why telecommunications were spared might lie in interception technologies collectively known as the System of Operative-Investigative Measures (SORM). SORM allows intelligence agencies and police to lawfully intercept and record landline and cell phone telecommunications within Russia. Ukraine adopted it in 2010 (along with several other former Soviet states).[243]

SORM could still come into play during the war or in any occupation. This would aid Russian intelligence gathering and the military offensive. It could provide access to top level politico-military communications and detailed knowledge of ground-based operations. However, for a variety of unrelated reasons it is the communications of Russian forces that have appeared more susceptible to interception. Electronic jamming, a lack of sufficient number and quality of secure military communications equipment, and the rank-and-file disorder of Russian troops have all been contributory factors. This situation has seen stolen Ukrainian cell phones as well as their own phones used to communicate between Russian forces within Ukraine and back home to Russia and senior generals deployed to combat zones where many have been killed in ambushes or sniper attacks.[244] These insecure communications on commercial networks have been intercepted and widely leaked on the Internet.[245] They provide valuable and actionable real-time intelligence including real-time geolocation and other metadata for Ukrainian forces.

President Biden's 2022 State of the Union Address acknowledged the value of accurate intelligence vis-à-vis Russian designs on Ukraine. Biden was unambiguous: "we knew [what] Putin was planning and precisely how he would try to falsely and justify his aggression".[246] Much of this intelligence has been shared with Ukraine and some made public. The proliferation of Open Source Intelligence (OSINT) has been a feature of the conflict (and a growing feature of conflict, investigations, and accredited and citizen journalism more widely). It is related to the parallel information war and a war over 'truth' (a feature of modern hybrid warfare) with mis/dis and malinformation embedded into public political narratives and discourse through mainstream media, and social media-driven ecosystems and echo chambers.[247]

Chris Krebbs, the former director of CISA, has also highlighted other potential factors for the direction of the war. Inside Russia, decision-making was kept to a very small circle—headed by an autocratic leader that has gradually exercised micro-management over decision-making (including down to the tactical military level). This means those responsible for cyber operations in the FSB, SVR, and GRU had to scramble as decision-making unfolded and orders issued.[248] Even senior level government and corporate officials were sidelined and blindsided by the scale of Vladimir Putin's plans.²⁴⁹ Over confidence that Ukraine (and the West) would buckle under Russia's blitzkrieg and threats (including thinly veiled nuclear threats) may also have contributed to indecision and preparedness to go fully after CI.²⁵⁰

It was only in Fall 2022 that the Russian military, following a change in leadership, consistently began to target CI with military force through missile strikes. Another potential issue is the exodus of key people from Russia after the invasion. This includes tens of thousands of ICT professionals. Whether this contains state cybersecurity professionals or cybercriminals from Russia's underworld is uncertain, but it has negatively impacted their ICT sector.²⁵¹ A Belorussian anti-war dissident hacktivist group calling themselves 'Cyber Partisans' also conducted attacks against the network of Belorussia's state railway system (using ransomware) ahead of the invasion. They also conducted physical attacks modeled on the sabotage of Nazi railway lines during World War II in addition to hacking and leaking information from Russian government databases.²⁵²

Broader factors were also at work. Strategic miscalculations were made by Vladimir Putin and the Kremlin leadership. President Biden, in his 2022 State of the Union Address, said "he badly miscalculated. He thought he could roll into Ukraine and the world would roll over. Instead, he met with a wall of strength he never anticipated or imagined. He met the Ukrainian people".²⁵³ Despite these explanations, more will undoubtedly emerge. For example, in June 2022 Microsoft released a new intelligence report which outlined that Russia had targeted Ukraine's governmental data center in an early cruise missile attack, and that other onsite servers were similarly vulnerable to attacks by conventional weapons. However, Ukraine had pre-emptively moved key services to cloud servers hosted by data centers throughout Europe.

48 distinct cyberattacks against Ukrainian agencies and enterprises have since been seen. Their modus operandi is almost always the same. First, they try to penetrate the network and then try to inject malware (wiperware in these cases). 128 organizations in 42 countries outside Ukraine have also been targeted in this way, with the United States as Russia's number one target. 29% of these attacks have proved successful. A quarter of these have seen data exfiltrated. The targets have included

foreign ministries of NATO and EU members as well as humanitarian organizations, think tanks, IT companies, and energy and other critical infrastructure suppliers. Because these are low visibility operations, these are difficult to track by journalists "and even many military analysts". Within Ukraine they are being used alongside conventional munitions to target ICT infrastructure and they have adapted their targets according to Russia's mutating war aims. Preparedness, cloud-based services, and active Cyber Threat Intelligence (including through Artificial Intelligence) has meant that so far this cyber defense has "proven stronger than offensive cyber capabilities". But sustaining this "defensive advantage" requires ongoing vigilance and innovation. This is especially in respect to Russia's APTs which have consistently demonstrated their abilities to slip through defenses. Prior to the invasion this had been demonstrated (again) by hacking the U.S. company SolarWinds, a key vendor of network monitoring tools used by major businesses and U.S. government agencies in late 2019–early 2020.

At the same time, Russian agencies have been conducting renewed influence operations focusing on the war in Ukraine "with the goal of undermining Western unity and deflecting criticism of Russian military war crimes. And they are starting to target populations in nonaligned countries, potentially in part to sustain their support at the United Nations and in other venues". Brad Smith makes an interesting observation in this respect. His view is that just as Russia's APTs work within Russia's intelligence services so too do these Advance Persistent Manipulator (APM) teams. Smith's analysis is that as "the Russian Government does not pursue them as separate efforts…we should not put them in separate analytical silos".[254] This has merit, and the Russo-Ukraine War remains active at the time of writing and the tables can turn toward Russia and/or fulfill Putin's objectives. The fog of war also applies, and available information is incomplete (and might remain so) regarding the scale or attempts at cyberwarfare in Ukraine. How long, and in what forms, the war continues or how peace will manifest remain to be seen. Despite the future uncertainty of whether this is a relatively short war, a protracted war, or negotiated 'long peace' settlement, it will be wise to continue to harden defenses and raise barriers against the threats Russia poses under Putin.

Conclusion

Writing in 2011, researchers from Berkley warned that "Cyber attacks on SCADA system can take routes through Internet connections, business or enterprise network connections and or connections to other networks, to the layer of control networks then down the level of field devices".[255] A decade on too little has changed. Cyberattacks on ICS and SCADA could cause electricity blackouts, water shortages, or be used to disrupt financial services. There are cases of directed attacks and publicly spoken fears of a 'cyber Pearl Harbor', a 'cyber 9/11', or even a state wide 'Cybergeddon' attack aimed at crippling or seriously damaging a nation and which could cascade to other states.[256]

As Chapter 2 showed, the use of cyber as weapons or tools of disruption is redefining what we think of as conflict. It is challenging conventional military doctrine founded upon territorial defense and the abilities of governments to defend territory through the use of force. Cyberattacks on these and other systems can pass through borders at the speed of light and current models of deterrence are proving ineffective or redundant against this form of asymmetric attack.[257] Instead, technical defense is prioritized. Currently cyberattacks are being used as undeclared war-fighting tools containing code that can be weaponized for destruction. Aimed at CI they can disrupt, degrade, damage, or even cripple a state and its ability to defend before a declaration of war has been made. Prevailing ideas of perimeter defenses and defense in depth models are capable of being bypassed throughout industry.

Admiral Mike Rogers was clear in his assessment to the Senate in 2017: "The worst case scenario in my mind…[includes] outright destructive activity focused on some aspects of critical infrastructure".[258] Across the Atlantic, then head of Britain's NCSC, Ciaran Martin, was perhaps even more unequivocal. In a public address in January 2018, he foresaw that a major cyberattack on UK infrastructure is a case of "when, not if". Martin added, "we will be fortunate to come to the end of the decade without having to trigger a category one attack".[259] This would be a sustained disruption of essential services or affecting national security "leading to severe economic or social consequences or to loss of life".[260]

These concerns are not unique and with Ukraine having experienced two direct cyberattacks from Russia on its electrical grid in 2015 and 2016, with others following in 2022, the proof-of-concept exists but currently it would be very difficult to accomplish at scale. This being said, it was reported in *The New York Times* in March 2018, that U.S. officials and the wider cybersecurity community was warning of the "ominous sign of what the Russian cyberstrikes may portend in the United States and Europe in the event of escalating hostilities".[261] The report continued, "by December 2015…Russian hacks had taken an aggressive turn…[and] no longer aimed at intelligence gathering, but at potentially sabotaging or shutting down [power] plant operations".[262] This perhaps points to a further reason why the U.S. and its allies have been drawn to Ukraine's cyber defense.

Attacks upon critical infrastructure were recognized by *Tallinn 2.0* as an area of growing concern for states and qualify as grave and imminent peril.[263] *Tallinn 2.0* also recognized that the temporal nature of these threats. They might not be immediate but over time can constitute 'grave peril' with immanency only when the 'window of opportunity' for preventive measures is about to close.[264] Should there be the kind of major cyberattacks on critical infrastructure that Admiral Rogers and Ciaran Martin forecast this 'window of opportunity' could close. Diplomacy and deterrence will have failed and this could be accompanied by, or help pave the way for, economic or political action or the use of military force. To do this, attackers first need to gain access. This is the subject of Chapter 4.

Notes

1. Saleh Soltan, Prateek Mittal, and H. Vincent Poor, 'BlackIoT: IoT Botnet of High Wattage Devices Can Disrupt the Power Grid', *Proceedings of the 27th USENIX Security Symposium* (August 15–17 2018, Baltimore, MD, USA), pp. 15-32, https://www.usenix.org/conference/usenixsecurity18/presentation/soltan. I am grateful to Professor Ariel Pinto for this source.

2. The two most polarised examples of these debates are to be found in Richard A. Clarke and Robert K. Knake, *Cyber War: The Next Threat to National Security and What to Do About It* (London: ECCO Press, 2010) and Thomas Rid, *Cyber War Will Not Take Place* (London: Hurst, 2013).
3. Pierluigi Paganini, 'Improving SCADA System Security' (December 6 2013), http://resources.infosecinstitute.com/improving-scada-system-security/, accessed 26 July 2015.
4. Oliver Wyman, 'Large-Scale Cyber-Attacks on the Financial Services System A Case for Better Coordinated Response and Recovery Strategies A White Paper to the Industry' (March 28 2018), https://www.oliverwyman.com/content/dam/oliver-wyman/v2/publications/2018/march/Large-Scale-Cyber-Attacks-DTCC-2018.pdf, Rob Neave, 'Cyber Attacks: Protecting the Financial Service Sector's Data Centres' (September 9 2019), https://thefintechtimes.com/cyber-attacks-data-centres/, Emily Hardy, 'Failure to Launch: Stock Market Open Delayed by Nearly Two Hours as Technical Issue at LSE Stalls FTSE Share Trading' (August 16 2019), https://www.thisismoney.co.uk/money/markets/article-7363411/Failure-launch-UK-share-trading-delayed-nearly-2-hours-technical-issue-LSE.html, all accessed 30 September 2019.
5. Verizon, 2021 Data Breach Investigations Report, http://verizon.com/dbir/, pp. 75–76.
6. 'What Is SCADA?', http://www.dpstele.com/dpsnews/techinfo/what_is_scada.php, accessed 22 July 2014.
7. 'SCADA Systems', http://www.engineersgarage.com/articles/scada-systems, accessed 22 July 2014.
8. Eric D. Knapp, Joel Thomas Langill, *Industrial Network Security: Securing Critical Infrastructure Networks for Smart Grid, SCADA, and Other Industrial Control Systems*, Second Edition (New York: Elsevier, 2015), p. 77.
9. 'What Is a Programmable Logic Controller', http://www.amci.com/tutorials/tutorials-what-is-programmable-logic-controller.asp, accessed 22 July 2014.
10. 'What Is the Difference Between SCADA and DCS Systems?' (March 21 2018), https://www.maderelectricinc.com/blog/what-is-the-difference-between-scada-and-dcs-systems, accessed 6 November 2018.
11. Knapp and Langill, *Industrial Network Security*, p. 77.
12. https://scadahacker.com/, accessed 8 October 2018.
13. Knapp and Langill, *Industrial Network Security*, p. 18.

14. Yulia Cherdantseva, Peter Burnap, Andrew Blyth, Peter Eden, Kevin Jones, Hugh Soulsby, and Kristan Stoddart, 'A Review of Cyber Security Risk Assessment Methods for SCADA Systems', *Computers & Security*, Vol. 56 (February 2016), p. 5.
15. Bonnie Zhu, Anthony Joseph and Shankar Sastry, 'A Taxonomy of Cyber Attacks on SCADA Systems', 2011 IEEE International Conferences on Internet of Things, and Cyber, Physical and Social Computing, p. 380, https://ieeexplore.ieee.org/stamp/stamp.jsp?tp=&arnumber=6142258, accessed 21 July 2019.
16. For an excellent synopsis of Stuxnet see Ralph Langer, 'Stuxnet's Secret Twin', *Foreign Policy*, 19 November 2013. See also Stuart Winer, "Dutch Mole' Planted Stuxnet Virus in Iran Nuclear Site on Behalf of CIA, Mossad' (September 3 2019),https://www.timesofisrael.com/dutch-mole-planted-infamous-stuxnet-virus-in-iran-nuclear-site-report/, accessed 1 October 2019.
17. Perry Pederson, 'Aurora Revisited—By Its Original Project Lead' (July 9 2014), https://www.langner.com/2014/07/aurora-revisited-by-its-original-project-lead/, accessed 12 September 2018. Aaron Turner, who worked on the project along with Mike Assante describes it as "one of the most important cybersecurity projects in our lifetimes". Turner also recounted, "I'll never forget one meeting in which we presented some of the initial findings to a group of infrastructure regulators who actually shouted us out of the room, accusing us of being fear-mongering psychopaths who were suffering from technology-induced delusions". Aaron Turner, 'Mike Assante's Lasting Impact on Critical Infrastructure Security (and Me)' (June 18 2019), https://www.csoonline.com/article/3403656/mike-assantes-lasting-impact-on-critical-infrastructure-security-and-me.html, accessed 26 July 2019. See also Curtis Waltman, 'Aurora: Homeland Security's Secret Project to Change How We Think About Cybersecurity' (November 14 2016), https://www.muckrock.com/news/archives/2016/nov/14/aurora-generator-test-homeland-security/, accessed 25 June 2022. I am grateful to Perry Pederson for pointing me to this source which contains declassified documentation on the Aurora test.

18. Jeanne Meserve, 'Sources: Staged Cyber Attack Reveals Vulnerability in Power Grid' (September 26 2007), http://edition.cnn.com/2007/US/09/26/power.at.risk/. PBS-Nova, 'CyberWar Threat' http://www.pbs.org/wgbh/nova/military/cyberwar-threat.html. 'PBS - NOVA (CyberWar Threat)', https://www.youtube.com/watch?v=DAl7Cre3BeI, all accessed 12 September 2018.
19. Although Symantec postulates that Stuxnet's development began in 2005. Symantec, 'Stuxnet 0.5: How It Evolved' (February 26 2013), https://www.symantec.com/connect/blogs/stuxnet-05-how-it-evolved, accessed 7 November 2018.
20. Meserve, 'Sources: Staged Cyber Attack Reveals' (September 26 2007).
21. Kim Zetter, 'Did a U.S. Government Lab Help Israel Develop Stuxnet?' (January 17 2011), https://www.wired.com/2011/01/inland-stuxnet/, accessed 6 November 2018.
22. Meserve, 'Sources: Staged Cyber Attack' (September 26 2007).
23. Aside from those present, as of 2022 it has only received a little over 4,000 views. 'DEF CON 25 ICS Village—Joe Weiss—Cyber Security Issues with Level 0 Through 1 Devices' (October 16 2017), https://www.youtube.com/watch?v=UgvVaniZhsk, accessed 12 September 2018.
24. Meserve, 'Sources: Staged Cyber Attack' (September 26 2007).
25. Meserve, 'Sources: Staged Cyber Attack' (September 26 2007).
26. Alvin Powell, 'What Might COVID Cost the U.S.? Try $16 Trillion' (November 10 2020), https://news.harvard.edu/gazette/story/2020/11/what-might-covid-cost-the-u-s-experts-eye-16-trillion/, accessed 10 July 2021).
27. Pederson, 'Aurora Revisited—by Its Original Project Lead' (July 9 2014).
28. Pederson, 'Aurora Revisited—By Its Original project Lead' (July 9 2014).
29. Turner, 'Mike Assante's Lasting Impact on Critical Infrastructure Security (and Me)' (June 18 2019).
30. Benjamin Monarch, 'Black Start: The Risk of Grid Failure from a Cyber Attack and the Policies Needed to Prepare for It', *Journal of Energy and Natural Resources Law*, Vol. 38, Issue 2 (April 2020), pp. 131–160.
31. Cyber Threat and Vulnerability Analysis of the U.S. Electric Sector Prepared by: Mission Support Center Idaho National Laboratory

August 2016, p. ii, https://www.energy.gov/sites/prod/files/2017/01/f34/Cyber%20Threat%20and%20Vulnerability%20Analysis%20of%20the%20U.S.%20Electric%20Sector.pdf, accessed 12 August 2019.
32. McAfee/CSIS, 'In the Crossfire: Critical Infrastructure in the Age of Cyber War', p. 5, https://www.govexec.com/pdfs/012810j1.pdf, accessed 26 July 2019.
33. 'Crashoverride Analysis of the Threat to Electric Grid Operations', p. 6, https://dragos.com/blog/crashoverride/CrashOverride-01.pdf, accessed 7 October 2018. See also https://www.osha.gov/etools/electric-power/illustrated-glossary/sub-station, 'U.S. Researchers Simulate Compact Fusion Power Plant Concept' (September 21 2021), https://www.energy.gov/science/fes/articles/us-researchers-simulate-compact-fusion-power-plant-concept, accessed 4 June 2022.
34. Jose R. Gracia, Patrick W. O'Connor, Lawrence C. Markel, Rui Shan, D. Thomas Rizy, and Alfonso Tarditi, 'Hydropower Plants as Black Start Resources' (May 2019), https://www.energy.gov/sites/prod/files/2019/05/f62/Hydro-Black-Start_May2019.pdf, accessed 4 June 2022.
35. However, if an attacker can physically access a Distribution SCADA Master via a substation this poses a greater risk. Cyber Threat and Vulnerability Analysis of the U.S. Electric Sector, pp. 10, 12.
36. Alex Marquardt, 'New Report Reveals Apparent Plot Against Electrical Grid' (November 4 2021), https://edition.cnn.com/videos/politics/2021/11/04/drone-threat-power-grid-new-details-marquardt-lead-pkg-vpx.cnn, accessed 20 May 2022.
37. 'Crashoverride Analysis of the Threat to Electric Grid Operations', p. 25. INL came to similar conclusions. Cyber Threat and Vulnerability Analysis of the U.S. Electric Sector, pp. 9–10.
38. Blake Sobczak, 'Hackers Warn of 'Tipping Point' for Critical Infrastructure' (July 27 2017), https://www.eenews.net/stories/1060057993, accessed 14 July 2019.
39. Caroline Baylon with Roger Brunt and David Livingstone, 'Cyber Security at Civil Nuclear Facilities Understanding the Risks' (September 2015), https://www.chathamhouse.org/sites/default/files/field/field_document/20151005CyberSecurityNuclearBaylonBruntLivingstoneUpdate.pdf, accessed 17 September 2018.
40. Jeff Kingston, 'Mismanaging Risk and the Fukushima Nuclear Crisis' in Paul Bacon and Christopher Hobson (eds.), *Responding to the 2011 Earthquake, Tsunami and Fukushima Nuclear Crisis* (Abingdon: Routledge, 2014), pp. 39–58.

41. Outlined in another study by Jennifer Weeks, 'U.S. Electrical Grid Undergoes Massive Transition to Connect to Renewables' (April 28 2010), https://www.scientificamerican.com/article/what-is-the-smart-grid/, accessed 19 February 2020.
42. Cyber Threat and Vulnerability Analysis of the U.S. Electric Sector, pp. 8–9.
43. 'Overview', https://www.infrastructurereportcard.org/cat-item/energy/ (2017). See also 'U.S. Power Grid' (undated), http://www.earthlyissues.com/uspower.htm, both accessed 19 February 2020.
44. Depending on the source these numbers vary. See 'How Many Nuclear Power Plants Are in the United States, and Where Are They Located?', https://www.eia.gov/tools/faqs/faq.php?id=207&t=3, 'What Is U.S. Electricity Generation by Energy Source?', https://www.eia.gov/tools/faqs/faq.php?id=427&t=3, 'Uranium and Nuclear Power Facts', https://www.nrcan.gc.ca/energy/facts/uranium/20070, 'What Is Electricity', https://www.nrcan.gc.ca/energy/facts/electricity/20068, all accessed 24 October 2018. See also 'North American Cooperation on Energy Information (NACEI) North American Infrastructure Map' (actively updated), http://nacei.org/#!/maps, accessed 19 February 2020.
45. 'Nuclear Energy', https://ec.europa.eu/energy/en/topics/nuclear-energy, and 'Electricity Production, Consumption and Market Overview', https://ec.europa.eu/eurostat/statistics-explained/index.php/Electricity_production,_consumption_and_market_overview#Electricity_generation, both accessed 19 February 2020.
46. EU energy in figures Statistical pocketbook 2018, pp. 16, 26. Available from https://publications.europa.eu/en/publication-detail/-/publication/99fc30eb-c06d-11e8-9893-01aa75ed71a1/language-en, accessed 24 October 2018.
47. EU energy in figures Statistical pocketbook 2018, pp. 16, 26, 'Electricity Generation Statistics—First Results', https://ec.europa.eu/eurostat/statistics-explained/index.php?title=Electricity_generation_statistics_%E2%80%93_first_results, and https://ec.europa.eu/eurostat/cache/infographs/energy/bloc-2c.html, both accessed 24 October 2018.
48. Douglas Broom, 'What Is the EU Doing to End Its Reliance on Russian Energy?' (April 26 2022), https://www.weforum.org/agenda/2022/04/europe-russia-energy-alternatives/, accessed 4 June 2022.

49. 'Global Shifts in the Energy System', https://www.iea.org/weo2017/ (November 14 2017), accessed 24 October 2018.
50. Marshall Abrams and Joe Weiss, 'Malicious Control System Cyber Security Attack Case Study–Maroochy Water Services, Australia', http://csrc.nist.gov/groups/SMA/fisma/ics/documents/Maroochy-Water-Services-Case-Study_report.pdf, accessed 21 September 2014.
51. Boden had been sacked—hence his grudge against the company.
52. Nabil Sayfayn and Stuart Madnick, 'Cybersafety Analysis of the Maroochy Shire Sewage Spill', Working Paper CISL# 2017–09 May 2017, http://web.mit.edu/smadnick/www/wp/2017-09.pdf, accessed 5 November 2018.
53. WASD continues to monitor cyber-threat assessments, while ensuring safe delivery of drinking water (February 9 2021), https://www.miamidade.gov/releases/2021-02-09-WASD-cyber-security-finalone.asp, accessed 10 July 2021.
54. 'ICSB-11-327-01—Illinois Water Pump Failure Report', https://www.us-cert.gov/ics/tips/ICSB-11-327-01, accessed 23 July 2019.
55. Kim Zetter, 'Exclusive: Comedy of Errors Led to False 'Water-Pump' Hack Report' (November 30 2011), https://www.wired.com/2011/11/water-pump-hack-mystery-solved/, accessed 23 July 2019.
56. Richard J. Brennan, 'Cyber Attack on Small Illinois Water Treatment Plant Has Serious Implications: Security Expert' (November 21 2011), https://www.thestar.com/news/world/2011/11/21/cyber_attack_on_small_illinois_water_treatment_plant_has_serious_implications_security_expert.html, accessed 23 July 2019.
57. U.S. Department of Energy the Water-Energy Nexus: Challenges and Opportunities June 2014, p. 6, https://www.energy.gov/sites/prod/files/2014/07/f17/Water%20Energy%20Nexus%20Full%20Report%20July%202014.pdf, accessed 23 July 2019.
58. Alex Milner, 'Mosul Dam: Why the Battle for Water Matters in Iraq' (August 18 2014), http://www.bbc.co.uk/news/world-middle-east-28772478, accessed 22 September 2014.
59. Mark Thomson, 'Iranian Cyber Attack on New York Dam Shows Future of War' (March 24 2016), https://time.com/4270728/iran-cyber-attack-dam-fbi/, accessed 30 September 2019.
60. The Recent Storms and Floods in the UK (February 2014), https://www.ceh.ac.uk/sites/default/files/Recent%20Storms%20Briefing.pdf, accessed 14 April 2020.

61. This is widely recognised and a great deal of very interesting work on these routes and ports has been undertaken. See for example Jean-Paul Rodrigue, *The Geography of Transport Systems* (New York: Routledge, 2013).
62. 'Gulf of Mexico Fact Sheet', https://www.eia.gov/special/gulf_of_mexico/, accessed 2 November 2018.
63. Phil Muncaster, 'US Gas Pipelines Hit by Cyber-Attack' (April 4 2019), https://www.infosecurity-magazine.com/news/us-gas-pipelines-hit-by-cyberattack/, accessed 14 July 2019.
64. Eduard Kovacs, 'Multiple U.S. Gas Pipeline Firms Affected by Cyberattack' (April 4 2018), https://www.icscybersecurityconference.com/multiple-u-s-gas-pipeline-firms-affected-by-cyberattack/, accessed July 14 2019.
65. House Homeland Security Committee Hearing on the Colonial Pipeline Cyber Attack (June 9 2021), https://www.c-span.org/video/?512332-1/colonial-pipeline-ceo-joseph-blount-testifies-house-homeland-security-committee, 'Shining a Light on DARKSIDE Ransomware Operations' (May 11 2021), https://www.fireeye.com/blog/threat-research/2021/05/shining-a-light-on-darkside-ransomware-operations.html and Mike Hoffman and Tom Winston, Recommendations Following the Colonial Pipeline Cyber Attack (May 11 2021), https://www.dragos.com/blog/industry-news/recommendations-following-the-colonial-pipeline-cyber-attack/, all accessed 10 July 2021.
66. Chris Bronk and Eneken Tikk-Ringas, 'The Cyber Attack on Saudi Aramco', *Survival*, Vol. 55, Issue 2 (April–May 2013), pp. 81–96. Drones have also been used for low-cost attacks on Saudi Arabian oil production from Yemen during their civil war. Ben Hubbard, Palko Karasz and Stanley Reed, 'Two Major Saudi Oil Installations Hit by Drone Strike, and U.S. Blames Iran' (September 14 2019), https://www.nytimes.com/2019/09/14/world/middleeast/saudi-arabia-refineries-drone-attack.html, accessed 19 February 2020.
67. 'Who We Are', https://www.saudiaramco.com/en/who-we-are/overview/global-presence, accessed 5 November 2018.
68. Nikolai Kuznetzov, 'Russia's Energy Sector Set to Thrive in 2017' (January 11 2017), https://www.forbes.com/sites/nikolaikuznetsov/2017/01/11/russias-energy-sector-could-thrive-in-2017/#7b62ef7b9595, accessed 7 September 2017 and Ernest Wyciszkiewicz (ed.), *Geopolitics of Pipelines Energy Interdependence and Inter-State*

Relations in the Post-Soviet Area (Warsaw: Polski Instytut Spraw Miedzynarodowych, 2009).
69. James Henderson and Alastair Ferguson, *International Partnership in Russia Conclusions from the Oil and Gas Industry* (London: Palgrave Macmillan, 2014), p. 2.
70. 'West Siberian Oil Basin', http://petroneft.com/operations/west-siberian-oil-basin/, accessed 5 November 2018.
71. See also Terry Macalister, 'Who Would Get the Oil Revenues if Scotland Became Independent?', *The Guardian*, 2 March 2012.
72. Eric Byres, keynote address to the 2nd International Symposium for ICS & SCADA Cyber Security attended by author. See also Zuoning Yin, Ding Yuan, Yuanyuan Zhou, Shankar Pasupathy, Lakshmi Bairavasundaram, 'How do Fixes Become Bugs?', Proceedings of the 19th ACM SIGSOFT symposium and the 13th European conference on Foundations of software engineering (September 5–9 2011), pp. 26–36, http://opera.ucsd.edu/paper/fse11.pdf, accessed 29 March 2020.
73. 'Piper Alpha Platform, North Sea', http://www.offshore-technology.com/projects/piper-alpha-platform-north-sea/, accessed 27 September 2014. See also the 1990 Cullen Report into the disaster.
74. Terry Macalister, 'Piper Alpha Disaster: How 167 Oil Rig Workers Died' (July 4 2013), https://www.theguardian.com/business/2013/jul/04/piper-alpha-disaster-167-oil-rig, accessed 4 November 2018.
75. 'Blowout' (October 2016), https://www.texasmonthly.com/articles/deepwater-horizon-prosecution/, accessed 4 November 2018.
76. 'Deepwater Horizon—BP Gulf of Mexico Oil Spill', https://www.epa.gov/enforcement/deepwater-horizon-bp-gulf-mexico-oil-spill, accessed 4 June 2022.
77. 'Number of Offshore Oil Rigs Worldwide as of January 2018 by Region', https://www.statista.com/statistics/279100/number-of-offshore-rigs-worldwide-by-region/, accessed 4 November 2018.
78. Jim Thompson, 'Refineries and Associated Plant: Three Accident Case Studies' (2013), http://www.safetyinengineering.com/FileUploads/Refineries%20and%203%20accident%20case%20studies%20v2_1370770924_2.pdf, accessed 21 September 2014.
79. Heidi Vella, 'Fighting Cyber Crime in the Offshore Oil and Gas Industry' (December 13 2016), https://www.offshore-technology.com/digital-disruption/cybersecurity/featurefighting-cyber-crime-in-the-offshore-oil-and-gas-industry-5692000/, accessed 7 November 2018.

80. Vella, 'Fighting Cyber Crime in the Offshore Oil and Gas Industry' (December 13 2016).
81. Vella, 'Fighting Cyber Crime in the Offshore Oil and Gas Industry' (December 13 2016).
82. Blake Johnson, Dan Caban, Marina Krotofil, Dan Scali, Nathan Brubaker, Christopher Glyer, 'Attackers Deploy New ICS Attack Framework 'Triton' and Cause Operational Disruption to Critical Infrastructure' (December 14 2018), https://www.fireeye.com/blog/threat-research/2017/12/attackers-deploy-new-ics-attack-framework-triton.html, accessed 7 November 2018.
83. This was denied by Saudi Aramco.
84. Elias Groll, 'Cyberattack Targets Safety System at Saudi Aramco' (December 21 2017), https://foreignpolicy.com/2017/12/21/cyber-attack-targets-safety-system-at-saudi-aramco/, accessed 7 November 2018.
85. 'Four Russian Government Employees Charged in Two Historical Hacking Campaigns Targeting Critical Infrastructure Worldwide' (March 24 2022), https://www.justice.gov/opa/pr/four-russian-government-employees-charged-two-historical-hacking-campaigns-targeting-critical, accessed 4 June 2022.
86. Lily Hay Newman, 'Menacing Malware Shows the Dangers of Industrial Systems Sabotage', https://www.wired.com/story/triton-malware-dangers-industrial-system-sabotage/, accessed 7 November 2018.
87. Nicola Perlroth and Clifford Krauss, 'A Cyberattack in Saudi Arabia Had a Deadly Goal. Experts Fear Another Try' (March 15 2018), https://www.nytimes.com/2018/03/15/technology/saudi-arabia-hacks-cyberattacks.html, accessed 7 November 2018.
88. 'Xenotime', https://dragos.com/blog/20180524Xenotime.html (undated), accessed 7 November 2018.
89. Thomas Mackie, 'Revealed: New Era of State Sponsored Hacking Can Turn Oil Rigs into 'Bomb That Can Kill'' (February 18 2018), https://www.express.co.uk/news/world/920437/computer-hacker-cyber-hack-saudi-arabia-cyber-criminals-oil-rigs, accessed 7 November 2018.
90. Thompson, 'Refineries and Associated Plant: Three Accident Case Studies' (2013). For more information on this incident see Joseph Weiss, *Protecting Industrial Control Systems from Electronic Threats* (New York: Momentum Press, 2010), pp. 123–128.
91. See for example James Reason, *Human Error* (Cambridge: Cambridge University Press, 1990), James Reason, 'Human Error: Models and

Management', *British Medical Journal*, Vol. 320 (March 2000), pp. 768–770.
92. Information on these sites can be found at (for example) http://www.exxonmobil.co.uk/UK-English/about_what_chemicals.aspx, http://www.warwickchem.com/, http://www.synthite.co.uk/, http://www.euticals.com/, http://chemexfranchises.co.uk/, http://www.chemsol.co.uk/, all accessed 25 September 2014.
93. Further detail on chemical plants can be found in Ronald L. Krutz, *Securing SCADA Systems* (Indianapolis, IN: Wiley, 2006), pp. 38–41.
94. Anonymous, 'SBU Thwarts Cyber Attack from Russia Against Chlorine Station in Dnipropetrovsk Region' (July 11 2018), https://en.interfax.com.ua/news/general/517337.html, accessed 8 November 2018.
95. Alex Masi, Sanjay Verma, Maddie Oatman, 'Photos: Living in the Shadow of the Bhopal Chemical Disaster' (June 2 2014), http://www.motherjones.com/environment/2014/06/photos-bhopal-india-union-carbide-sanjay-verma-pesticides-explosion, accessed 26 September 2014.
96. Dustin Volz, 'Researchers Link Cyberattack on Saudi Petrochemical Plant to Russia' (October 23 2018), https://www.wsj.com/articles/u-s-researchers-link-cyberattack-on-saudi-petrochemical-plant-to-russia-1540322439, accessed 8 November 2018.
97. One of the world's largest, which threatened to spread into COSCO's corporate network.
98. Catalin Cimpanu, 'Port of San Diego Suffers Cyber-Attack, Second Port in a Week After Barcelona' (September 27 2018), https://www.zdnet.com/article/port-of-san-diego-suffers-cyber-attack-second-port-in-a-week-after-barcelona/, accessed 8 November 2018.
99. Anonymous, 'Are Ports Prepared to Deal with Threats from Hackers?' (April 6 2018), https://piernext.portdebarcelona.cat/en/technology/are-ports-prepared-to-deal-with-threats-from-hackers/, accessed 8 November 2018.
100. Anonymous/Gordon Corera, 'Cyber-Attack: US and UK Blame North Korea for WannaCry' (December 19 2017), http://www.bbc.co.uk/news/world-us-canada-42407488, accessed January 20 2018.
101. 'New Petya / NotPetya / ExPetr Ransomware Outbreak' (June 28 2017), https://www.kaspersky.com/blog/new-ransomware-epidemics/17314/, accessed January 20 2018.

102. 'FedEx Corp. Reports First Quarter Earnings' (September 19 2017), http://about.van.fedex.com/newsroom/fedex-corp-reports-first-quarter-earnings-2/, accessed January 20 2018.
103. Jonathan Saul, 'Global Shipping Feels Fallout from Maersk Cyber Attack' (June 29 2017), https://www.reuters.com/article/us-cyber-attack-maersk/global-shipping-feels-fallout-from-maersk-cyber-attack-idUSKBN19K2LE, accessed January 20 2018.
104. Brett Molina, Jon Swartz and Rachel Sandler, 'Petya Cyberattack Spreads to 65 Countries' (June 28 2017), https://www.usatoday.com/story/tech/talkingtech/2017/06/28/petya-cyberattack-spreads-65-countries/435016001/, accessed January 20 2018.
105. https://www.justice.gov/opa/press-release/file/1328521/download, accessed 4 June 2022.
106. Daniel Trimble, Jonathon Monken and Alex F. L. Sand, 'A Framework for Cybersecurity Assessments of Critical Port Infrastructure', 2017 International Conference on Cyber Conflict (CyCon U.S.), 2017, p. 1, https://ieeexplore.ieee.org/document/8167506/authors#authors, accessed 4 June 2022.
107. Under SOLAS, they use the "Global Maritime Distress and Safety System (GMDSS)". This means "All passenger ships and all cargo ships of 300 gross tonnage and upwards on international voyages are required to carry equipment designed to improve the chances of rescue following an accident, including satellite emergency position indicating radio beacons (EPIRBs) and search and rescue transponders (SARTs) for the location of the ship or survival craft". International Convention for the Safety of Life at Sea (SOLAS), 1974, http://www.imo.org/en/About/conventions/listofconventions/pages/international-convention-for-the-safety-of-life-at-sea-(solas),-1974.aspx, See also 'AIS Transponders', https://www.steamshipmutual.com/publications/Articles/Articles/04_AIS_Transporters.asp, both accessed 2 November 2018.
108. Tom Batchelor, 'Tracking Apps That Reveal Location of British Warships Spark Security Fears' (February 5 2018), https://www.independent.co.uk/news/uk/home-news/royal-navy-tracking-app-warship-nato-russia-china-military-security-a8191896.html, accessed 4 November 2018.

109. Anonymous, 'Tests Show Ease of Hacking ECDIS, Radar and Machinery' (December 21 2017), https://www.maritime-executive.com/article/tests-show-ease-of-hacking-ecdis-radar-and-machinery, accessed 4 November 2018.
110. Anonymous, 'Tests Show Ease of Hacking ECDIS, Radar and Machinery'.
111. Chris C. Demchak and Michael L. Thomas, 'Can't Sail Away from Cyber Attacks: 'Sea-Hacking' from Land' (October 15 2021), https://warontherocks.com/2021/10/cant-sail-away-from-cyber-attacks-sea-hacking-from-land/, accessed 15 November 2021.
112. Anonymous, 'Tests Show Ease of Hacking ECDIS, Radar and Machinery'.
113. Rolls Royce, 'Autonomous Ships the Next Step' (2016), https://www.rolls-royce.com/~/media/Files/R/Rolls-Royce/documents/customers/marine/ship-intel/aawa-whitepaper-210616.pdf, accessed 4 November 2018.
114. Anonymous, 'IMO Moves Forward to Address Autonomous Ships' (May 27 2018), https://worldmaritimenews.com/archives/253639/imo-moves-forward-to-address-autonomous-ships/, accessed 4 November 2018. 'Autonomous Ships: Regulatory Scoping Exercise Completed' (May 25 2021), https://www.imo.org/en/MediaCentre/PressBriefings/pages/MASSRSE2021.aspx, accessed 4 June 2022.
115. See for example http://www.exactearth.com/, https://www.marinetraffic.com/ and http://shipfinder.co/, all accessed 24 September 2014.
116. Jon Walker, 'Autonomous Ships Timeline—Comparing Rolls-Royce, Kongsberg, Yara and More' (May 29 2018), https://www.techemergence.com/autonomous-ships-timeline/, accessed 4 November 2018.
117. Cesar Cerrudo, 'Hacking US Traffic Control Systems' (2014), https://www.defcon.org/images/defcon-22/dc-22-presentations/Cerrudo/DEFCON-22-Cesar-Cerrudo-Hacking-Traffic-Control-Systems-UPDATED.pdf, accessed 12 November 2018.
118. Joseph M. Ernst and Alan J. Michaels, 'Framework for Evaluating the Severity of Cybervulnerability of a Traffic Cabinet', *Transportation Research Record: Journal of the Transportation Research Board*, Vol. 2619, Issue 1 (January 2017), pp. 55–63 and p. 62.
119. '2019-005-Eastern Mediterranean and Red Seas-GPS Interference', https://www.maritime.dot.gov/content/2019-005-eastern-mediterranean-and-red-seas-gps-interference, accessed 4 October 2019.

120. Barış Egemen Özkan and Serol Bulkan, 'Hidden Risks to Cyberspace Security from Obsolete COTS Software', in Tomáš Minárik, Siim Alatalu, Stefano Biondi, Massimiliano Signoretti, Ihsan Tolga and Gábor Visky (eds.), *2019 11th International Conference on Cyber Conflict: Silent Battle* (Tallinn: CCD COE Publications, 2019), pp. 61–80.
121. Cerrudo, 'Hacking US Traffic Control Systems'.
122. Jenni Bergal, 'How Hackers Could Cause Chaos on America's Roads and Railways' (April 25 2018), https://gcn.com/articles/2018/04/25/hacks-transportation-systems.aspx, accessed 12 November 2018.
123. Faiz Siddiqui, 'Cyberattack on San Francisco Transit Agency Prompts Senate Questions for Metro' (January 9 2017), https://www.washingtonpost.com/news/dr-gridlock/wp/2017/01/09/cyberattack-on-san-francisco-transit-agency-prompts-senate-questions-for-metro/, accessed 12 November 2018 and Tamara Chuang, 'Cyber Attack on CDOT Computers Estimated to Cost up to $1.5 Million so Far', https://www.denverpost.com/2018/04/05/samsam-ransomware-cdot-cost/, 5 April 2018.
124. Siddiqui, 'Cyberattack on San Francisco Transit Agency' (January 9 2017).
125. Gurcan Comert, Jacquan Pollard, David M. Nicol, Kartik Palani, and Babu Vignesh, 'Modeling Cyber Attacks at Intelligent Traffic Signals', *Transportation Research Record: Journal of the Transportation Research Board*, Vol. 2672, Issue 1 (December 2018), pp. 76–89.
126. Siddiqui, 'Cyberattack on San Francisco Transit Agency' (January 9 2017).
127. Faiz Siddiqui, 'System Glitch Leaves Metro's Nerve Center Unable to Control Its Tracks, Wrecking Morning Commute', *The Washington Post*, 5 January 2017.
128. Yiheng Feng, Shihong Huang, Qi Alfred Chen, Henry X. Liu1, and Z. Morley Mao, 'Vulnerability of Traffic Control System Under Cyberattacks with Falsified Data', *Transportation Research: Record Journal of the Transportation Research Board*, Vol. 2672, Issue 1 (December 2018), pp. 1–11.
129. Bergal, 'How Hackers Could Cause Chaos on America's Roads and Railways' (April 25 2018).
130. Bergal, 'How Hackers Could Cause Chaos on America's Roads and Railways' (April 25 2018).

131. Georgia Lykou, Argiro Anagnostopoulou and Dimitris Gritzalis, 'Smart Airport Cybersecurity: Threat Mitigation and Cyber Resilience Controls', *Sensors*, Vol. 19, Issue 19 (January 2019), pp. 1, 21.
132. Tommaso De Zan, Fabrizio d'Amore and Federica Di Camillo, 'The Defence of Civilian Air Traffic Systems from Cyber Threats' (December 2015), p. 9, http://www.iai.it/sites/default/files/iai1523e.pdf, accessed 13 November 2018.
133. Lykou, Anagnostopoulou and Gritzalis, 'Smart Airport Cybersecurity', pp. 5–6.
134. De Zan, d'Amore and Di Camillo, 'The Defence of Civilian Air Traffic Systems from Cyber Threats'. See also Martin Strohmeier, Matthias Schäfer, Matt Smith, Vincent Lenders and Ivan Martinovic, 'Assessing the Impact of Aviation Security on Cyber Power', in Nikolaos Pissanidis, Henry Rõigas, Matthijs Veenendaal (eds.), *2016 8th International Conference on Cyber Conflict Cyber Power* (Tallinn: CCD COE Publications, 2016), pp. 223–242.
135. Lykou, Anagnostopoulou and Gritzalis, 'Smart Airport Cybersecurity', p. 8.
136. Ed Bott, 'Fact Check: Malware Did Not Bring Down a Passenger Jet' (August 24 2010), https://www.zdnet.com/article/fact-check-malware-did-not-bring-down-a-passenger-jet/, accessed 5 June 2022.
137. Pierluigi Paganini, 'Cyber Threats Against the Aviation Industry' (April 8 2014), https://resources.infosecinstitute.com/topic/cyber-threats-aviation-industry/, accessed 6 June 2022.
138. Wiktor Szary, Eric Auchard, 'Polish Airline, Hit by Cyber Attack, Says All Carriers Are at Risk' (June 22 2015), https://www.reuters.com/article/us-poland-lot-cybercrime-idUKKBN0P21DC20150622, accessed 5 June 2022.
139. De Zan, d'Amore and Di Camillo, 'The Defence of Civilian Air Traffic Systems from Cyber Threats'.
140. Toi Staff, 'Screens at Iran Airport Said Hacked with Anti-regime Messages' (May 25 2018), https://www.timesofisrael.com/screens-at-iran-airport-said-hacked-with-anti-regime-messages/, accessed 5 June 2022.
141. Anonymous, 'Christmas Ransomware Attack Hit New York Airport Servers' (January 10 2020), https://apnews.com/article/fbefe0ccdfac9279df8c817068482b1b, accessed 5 June 2022.

142. Anonymous, 'Top 5 Cyber Attacks in the Aviation Industry' (April 16 2021), https://cnsight.io/2021/04/16/top-5-cyber-attacks-in-the-aviation-industry/, accessed 6 June 2022.
143. Woodrow Bellamy III, 'New Eurocontrol Data Shows Airlines Increasingly Becoming Targets for Cyber Attacks' (July 12 2021), https://www.aviationtoday.com/2021/07/12/new-eurocontrol-data-shows-airlines-increasingly-becoming-targets-cyber-attacks/, accessed 6 June 2022.
144. Anonymous, 'Rosaviatsiya Extends Temporary Closure of 11 Airports in Southern Russia' (June 4 2022), https://azeritimes.com/2022/06/04/rosaviatsiya-extends-temporary-closure-of-11-airports-in-southern-russia-2/, accessed 6 June 2022.
145. Lykou, Anagnostopoulou and Gritzalis, 'Smart Airport Cybersecurity', pp. 10–13.
146. Anonymous, 'Blank Screens at Bristol Airport After Cyber Attack' (September 16 2018), https://www.itv.com/news/westcountry/2018-09-16/blank-screens-at-bristol-airport-after-cyber-attack/, accessed 1 October 2019.
147. US Government Accountability Office (GAO), Report to Congressional Requesters, FAA Needs to Address Weaknesses in Air Traffic Control Systems, January 2015, https://www.gao.gov/assets/670/668169.pdf, accessed 13 November 2013.
148. GAO Report, FAA Needs to Address Weaknesses in Air Traffic Control Systems.
149. Anonymous, Cyber threat to civil aviation position paper 2 May 2017, https://www.eurocockpit.be/positions-publications/cyber-threat-civil-aviation, accessed 13 November 2018.
150. Cyber threat to civil aviation position paper 2 May 2017.
151. 'EASA Engaged in a "Responsible Disclosure" of Cybersecurity Issues in Coordination with Operators and Authorities', https://www.easa.europa.eu/newsroom-and-events/news/easa-engaged-%E2%80%9Cresponsible-disclosure%E2%80%9D-cybersecurity-issues-coordination, accessed 13 November 2018.
152. https://www.easa.europa.eu/eccsa, accessed 13 November 2018.
153. Cyber threat to civil aviation position paper 2 May 2017.
154. https://eur-lex.europa.eu/legal-content/EN/TXT/?uri=celex:32018R1139, accessed 13 November 2018.

155. Alix Culbertson, 'Half of European Flights Face Delay After Computer Failure' (April 4 2018), https://news.sky.com/story/half-of-european-flights-face-delay-after-computer-failure-11315397, accessed 28 September 2019.
156. De Zan, d'Amore and Di Camillo, 'The Defence of Civilian Air Traffic Systems from Cyber Threats'.
157. Meserve, 'Sources: Staged Cyber Attack' (September 26 2007).
158. Knapp and Langill, *Industrial Network Security*, p. 50. See also 2021 Data Breach Investigations Report, pp. 93–99.
159. Sobczak, 'Hackers Warn of 'Tipping Point' for Critical Infrastructure' (July 27 2017).
160. This includes Lockheed-Martin's Cyber Kill Chain model which is widely used to analyze, discuss, and dissect serious cyberattacks. This analyzes the following campaign elements: reconnaissance, weaponization, delivery, exploitation, installation, command and control. https://www.lockheedmartin.com/en-us/capabilities/cyber/cyber-kill-chain.html, accessed 10 January 2022.
161. https://www.snort.org/, accessed 9 November 2018.
162. 'Does Cisco Sell Snort?', https://www.snort.org/faq/does-cisco-sell-snort, accessed 9 November 2018.
163. Keith Stouffer, Victoria Pillitteri, Suzanne Lightman, Marshall Abrams, Adam Hahn, NIST Special Publication 800-82 Revision 2 Guide to Industrial Control Systems (ICS) Security, https://nvlpubs.nist.gov/nistpubs/specialpublications/nist.sp.800-82r2.pdf, accessed 1 October 2019.
164. International Cyber Security Protection Alliance, https://www.icspa.org/, 15 July 2014.
165. Internet Security Alliance, http://www.isalliance.org/isa-publications/, 19 July 2014.
166. 'RISI Online Incident Database', https://www.risidata.com/, accessed 9 November 2018.
167. https://www.icao.int/cybersecurity/Pages/Working-Groups.aspx, accessed 6 June 2022.
168. 'What Is CANVAS?', https://canvas-project.eu/canvas/what-is-canvas/, accessed 9 November 2018.
169. 'Metasploit the World's Most Used Penetration Testing Framework', https://www.metasploit.com/, accessed 9 November 2018.

170. Eduard Kovacs, 'OSVDB Shut Down Permanently' (April 7 2016), https://www.securityweek.com/osvdb-shut-down-permanently, accessed 9 November 2018.
171. 'H.R.3202—Cyber Vulnerability Disclosure Reporting Act', https://www.congress.gov/bill/115th-congress/house-bill/3202/text, accessed 9 November 2018.
172. Calvin Cohen, 'House Passes Cyber Vulnerability Disclosure Reporting Act' (January 12 2018), https://www.insideprivacy.com/united-states/congress/house-passes-cyber-vulnerability-disclosure-reporting-act/, accessed 9 November 2018.
173. Cyber Threat and Vulnerability Analysis of the U.S. Electric Sector, p. iii.
174. See Chapter 4.
175. Cohen, 'House Passes Cyber Vulnerability Disclosure Reporting Act' (January 12 2018). See also Cyber Threat and Vulnerability Analysis of the U.S. Electric Sector, pp. 30–31.
176. Cohen, 'House Passes Cyber Vulnerability Disclosure Reporting Act' (January 12 2018). See also Cyber Threat and Vulnerability Analysis of the U.S. Electric Sector, pp. 23–26.
177. Casey Ellis, 'NIST: Vulnerability Disclosure as a Requirement for Every Organization' (January 18 2018), https://www.bugcrowd.com/nist-vulnerability-disclosure-as-a-requirement-for-every-organization/, accessed 9 November 2018.
178. Jesse Rifkin, 'Data Security and Breach Notification Act Would Create the First-Ever Federal Standard for Penalizing Hacks of Consumer Information' (December 22 2018), https://govtrackinsider.com/data-security-and-breach-notification-act-would-create-the-first-ever-federal-standard-for-9842596a27ba, and Dan Swinhoe, 'The 17 Biggest Data Breaches of the 21st Century' (January 26 2018), https://www.csoonline.com/article/2130877/data-breach/the-biggest-data-breaches-of-the-21st-century.html, both accessed 9 November 2018.
179. https://eugdpr.org/, accessed 9 November 2018.
180. The Directive on security of network and information systems (NIS Directive), https://ec.europa.eu/digital-single-market/en/network-and-information-security-nis-directive, accessed 9 November 2018.
181. Paul Timmers, 'The European Union's Cybersecurity Industrial Policy', *Journal of Cyber Policy*, Vol. 3, Issue 3 (December 2018), pp. 371–372.

182. Timmers, 'The European Union's Cybersecurity Industrial Policy', p. 365. What effect Brexit will have remains unclear for Britain and the EU.
183. Timmers, 'The European Union's Cybersecurity Industrial Policy', p. 364.
184. Now the European Union Agency for Cybersecurity.
185. 'The EU Cybersecurity Act Brings a Strong Agency for Cybersecurity and EU-Wide Rules on Cybersecurity Certification' (June 26 2019), https://ec.europa.eu/digital-single-market/en/news/eu-cybersecurity-act-brings-strong-agency-cybersecurity-and-eu-wide-rules-cybersecurity, accessed 15 September 2019.
186. Timmers, 'The European Union's Cybersecurity Industrial policy', pp. 368–370.
187. Timmers, 'The European Union's Cybersecurity Industrial Policy', p. 366.
188. Timmers, 'The European Union's Cybersecurity Industrial Policy', p. 367.
189. 'Cybersecurity Threats and Defense Strategy' (May 9 2017), https://www.c-span.org/video/?428023-1/nsa-director-rogers-russia-poses-threat-congressional-elections&start=1166, accessed 27 September 2018.
190. Anonymous, 'Spoofing Presidential Alerts', https://systems.cs.colorado.edu/headlines/cmas.html, accessed 20 January 2020.
191. 'Alert (TA17-293A) Advanced Persistent Threat Activity Targeting Energy and Other Critical Infrastructure Sectors' (October 20 2017, updated March 15 2018), https://www.us-cert.gov/ncas/alerts/TA17-293A, accessed 6 December 2018.
192. Bill Gertz, 'DHS, FBI Warn Companies of Ongoing Cyber Attacks on Critical Infrastructure Russia Seen as Behind Cyber Targeting of Electric Grid, Other Public Infrastructures' (October 24 2017), https://freebeacon.com/national-security/dhs-fbi-warn-companies-ongoing-cyber-attacks-critical-infrastructure/, accessed 6 December 2017.
193. Bryan Krekel, Patton Adams, and George Bakos, *Occupying the Information High Ground: Chinese Capabilities for Computer Network Operations and Cyber Espionage* Prepared for the U.S.-China Economic and Security Review Commission by Northrop Grumman Corp (McLean, VA: Northrop Grumman Corporation, March 7 2012), pp. 11–12, https://info.publicintelligence.net/USCC-ChinaCyberEspionage.pdf, accessed 16 April 2019.

194. Cyber Threat and Vulnerability Analysis of the U.S. Electric Sector, p. ii.
195. Andy Bochman, 'Michael Assante Holds Forth on Cybersecurity Leadership', *Smart Grid Security Blog*, 1 August 2012, http://smartgridsecurity.blogspot.co.uk/2012/08/michael-assante-holds-forth-on.html, accessed 15 July 2014. See also Cyber Threat and Vulnerability Analysis of the U.S. Electric Sector, p. iii.
196. Mark Ward, 'How to Hack a Nation's Infrastructure' (May 20 2013), http://www.bbc.co.uk/news/technology-22524274, accessed 15 July 2014.
197. Tim Stack, 'Internet of Things (IoT) Data Continues to Explode Exponentially. Who Is Using That Data and How?' (February 5 2018), https://blogs.cisco.com/datacenter/internet-of-things-iot-data-continues-to-explode-exponentially-who-is-using-that-data-and-how, accessed 27 October 2018.
198. Stephan Haller, Carsten Magerkurth, 'The Real-time Enterprise: IoT-enabled Business Processes', IETF IAB Workshop on Interconnecting Smart Objects with the Internet, March 2011, p. 1. For in-depth discussion see Alec Ross, *The Industries of the Future* (New York: Simon & Shuster, 2016).
199. Dimiter V. Dimitrov, 'Medical Internet of Things and Big Data in Healthcare', *Healthcare Informatics Research*, Vol. 22, Issue 3 (July 2016), pp. 156–163.
200. See for example, Klaus-Dieter Thoben, Stefan Wiesner and Thorsten Wuest, "'Instrustrie 4.0" and Smart Manufacturing—A Review of Research Issues and Application Examples', *International Journal of Automation Technology*, Vol. 11, Isuue 1 (2017), pp. 4–16.
201. McAfee/CSIS, 'In the Crossfire Critical: Infrastructure in the Age of Cyber War', p. 10.
202. McAfee/CSIS, 'In the Crossfire Critical: Infrastructure in the Age of Cyber War', p. 10.
203. 'Cyber Security 2014', http://www.corporatelivewire.com/round-tables.html?id=cyber-security-2014, accessed 20 July 2014.
204. Tom Groenfeldt, 'Insiders Pose a Serious Threat to Corporate Information' (May 8 2014), http://www.forbes.com/sites/tomgroenfeldt/2014/05/08/insiders-pose-a-serious-threat-to-corporate-information/, accessed 20 July 2014.
205. Louise Comfort, *Designing Resilience: Preparing for Extreme Events* (Pittsburgh: University of Pittsburgh Press, 2010).

206. 'KA-SAT Network Cyber Attack Overview' (March 30 2022), https://www.viasat.com/about/newsroom/blog/ka-sat-network-cyber-attack-overview/#, accessed 22 May 2022.
207. 'Russia Behind Cyber-Attack with Europe-Wide Impact an Hour Before Ukraine Invasion' (May 10 2022), https://www.gov.uk/government/news/russia-behind-cyber-attack-with-europe-wide-impact-an-hour-before-ukraine-invasion, accessed 19 May 2022.
208. Andy Greenberg, 'Russia's Sandworm Hackers Attempted a Third Blackout in Ukraine' (April 12 2022), https://www.wired.com/story/sandworm-russia-ukraine-blackout-gru/, accessed 20 May 2022.
209. Russia behind cyber-attack with Europe-wide impact an hour before Ukraine invasion (May 10 2022).
210. Alert (AA22-057A) Update: Destructive Malware Targeting Organizations in Ukraine (April 28 2022), https://www.cisa.gov/uscert/ncas/alerts/aa22-057a, accessed 20 May 2022.
211. Mark Holmes, 'Experts Say Viasat Cyber Attack Exposed Ground Terminal, Satellite Supply Chain Vulnerabilities' (April 11 2022), https://www.satellitetoday.com/cybersecurity/2022/04/11/experts-say-viasat-cyber-attacks-exposed-ground-terminal-satellite-supply-chain-vulnerabilities/, accessed 22 May 2022.
212. Gil Baram and Omree Wechsler, 'Cyber Threats to Space Systems Current Risks and the Role of NATO' (June 2020), https://www.japcc.org/read-aheads/joint-air-space-power-conference-2020-read-ahead/, accessed 22 May 2022.
213. 'Destructive Malware Targeting Ukrainian Organizations', (January 15 2022), https://www.microsoft.com/security/blog/2022/01/15/destructive-malware-targeting-ukrainian-organizations/, Joe Tidy, 'European Oil Facilities Hit by Cyber-Attacks' (February 3 2022), https://www.bbc.co.uk/news/technology-60250956, Anna Ribeiro, 'Cyberattacks Continue to Extend Across Europe, BlackCat Ransomware May Be Involved' (February 4 2022), https://industrialcyber.co/threats-attacks/cyberattacks-continue-to-extend-across-europe-blackcat-ransomware-may-be-involved/, both accessed 20 May 2022. See also Mikael Wigell and Antto Vihma, 'Geopolitics Versus Geoeconomics: The Case of Russia's Geostrategy and Its Effects on the EU', *International Affairs*, Vol. 92, Issue 3 (May 2016), pp. 605–627.
214. 'Pro-Russian Hacking Group Hits Back at Anonymous—RT' (February 2022), https://www.rt.com/russia/551080-killnet-hackers-anonymous-retaliation-russia/, accessed 20 May 2022.

215. 'Ukraine: Disk-Wiping Attacks Precede Russian Invasion' (February 24 2022), https://symantec-enterprise-blogs.security.com/blogs/threat-intelligence/ukraine-wiper-malware-russia, accessed 23 May 2022.
216. https://www.atlanticcouncil.org/news/transcripts/transcript-nato-head-jens-stoltenberg-on-russian-aggression-ukraines-capabilities-and-expanding-the-alliance/ (January 28 2022), accessed May 22 2022.
217. 'AA22-110A U.S. Support for Connectivity and Cybersecurity in Ukraine' (May 10 2022), https://www.state.gov/u-s-support-for-connectivity-and-cybersecurity-in-ukraine/, accessed 19 May 2022.
218. 'AA22-110A U.S. Support for Connectivity and Cybersecurity in Ukraine' (May 10 2022).
219. Kylie Atwood and Zachary Cohen, 'US in Contact with Zelensky Through Secure Satellite Phone' (March 1 2022), https://edition.cnn.com/europe/live-news/ukraine-russia-putin-news-03-02-22/h_6b5c8062541ddb6c36dd43ca70391608, accessed 23 May 2022.
220. 'AA22-110A U.S. Support for Connectivity and Cybersecurity in Ukraine' (May 10 2022).
221. Greg Myre, 'How Does Ukraine Keep Intercepting Russian Military Communications?' (April 26 2022), https://www.npr.org/2022/04/26/1094656395/how-does-ukraine-keep-intercepting-russian-military-communications, accessed 21 May 2022.
222. 'Director GCHQ's Speech on Global Security Amid War in Ukraine' (March 31 2022), https://www.gchq.gov.uk/speech/director-gchq-global-security-amid-russia-invasion-of-ukraine, accessed 21 May 2022.
223. https://crrts.eu/index.html, Joe Tidy, 'Ukraine: EU Deploys Cyber Rapid-Response Team (February 22 2022), https://www.bbc.co.uk/news/technology-60484979, Pascale Davies, 'Cyber Espionage Is Key to Russia's Invasion of Ukraine. The International Community Is Fighting Back' (March 9 2022), https://www.euronews.com/next/2022/03/09/cyberespionage-is-key-to-russia-s-invasion-of-ukraine-the-international-community-is-fight, all accessed 21 May 2022.
224. Michael Hill, 'How Security Vendors Are Aiding Ukraine' (March 2 2022), https://www.csoonline.com/article/3651685/how-security-vendors-are-aiding-ukraine.html, accessed 21 May 2022.
225. Nick Biasini, Michael Chen, Alex Karkins, Azim Khodjibaev, Chris Neal and Matt Olney, with contributions from Dmytro Korzhevin, 'Ukraine Campaign Delivers Defacement and Wipers, in Continued

Escalation' (January 21 2022), https://blog.talosintelligence.com/2022/01/ukraine-campaign-delivers-defacement.html, accessed 21 May 2022.
226. Marina Koren, 'The War on Ukraine Is Testing the Myth of Elon Musk', https://www.theatlantic.com/science/archive/2022/02/elon-musk-ukraine-starlink-satellites/622954/, accessed 21 May 2022.
227. 'Disrupting Cyberattacks Targeting Ukraine' (April 7 2022), https://blogs.microsoft.com/on-the-issues/2022/04/07/cyberattacks-ukraine-strontium-russia/, accessed 19 May 2022.
228. 'Digital Technology and the War in Ukraine' (February 28 2022), https://blogs.microsoft.com/on-the-issues/2022/02/28/ukraine-russia-digital-war-cyberattacks/, accessed 19 May 2022.
229. Sam Schechner, 'Ukraine's 'IT Army' Has Hundreds of Thousands of Hackers, Kyiv Says' (March 4 2022), https://www.wsj.com/livecoverage/russia-ukraine-latest-news-2022-03-04/card/ukraine-s-it-army-has-hundreds-of-thousands-of-hackers-kyiv-says-RfpGa5zmLtavrot27OWX, accessed 20 May 2022.
230. Kate Conger and Adam Satariano, 'Volunteer Hackers Converge on Ukraine Conflict with No One in Charge' (March 4 2022), https://www.nytimes.com/2022/03/04/technology/ukraine-russia-hackers.html, accessed 20 May 2022.
231. Christopher Krebbs and Robert Chesney, 'Gray Zone, Twilight Zone, or Danger Zone? Russian Cyber and Information Operations in Ukraine' (March 18 2022), https://warontherocks.com/2022/03/gray-zone-twilight-zone-or-danger-zone-russian-cyber-and-information-operations-in-ukraine/, accessed 21 May 2022.
232. Deborah Haynes, 'Ukraine: 'Massive Cyber Attack' Shuts Down Government Websites' (January 14 2022), https://news.sky.com/story/ukraine-says-massive-cyber-attack-has-shut-down-government-websites-12515487, accessed 19 May 2022.
233. 'Joint Cybersecurity Advisory Russian State-Sponsored and Criminal Cyber Threats to Critical Infrastructure' (April 20 2022), https://www.cisa.gov/uscert/sites/default/files/publications/AA22-110A_Joint_CSA_Russian_State-Sponsored_and_Criminal_Cyber_Threats_to_Critical_Infrastructure_4_20_22_Final.pdf. See also Richard Hichman, 'Conti Ransomware Gang: An Overview', https://unit42.paloaltonetworks.com/conti-ransomware-gang/, accessed 19 May 2022.
234. 'Joint Cybersecurity Advisory' (April 20 2022).

235. Covered in detail in my book Russia's Cyber Offensive against the West.
236. Greenberg, 'Russia's Sandworm Hackers Attempted a Third Blackout in Ukraine' (April 12 2022).
237. 'Industroyer2: Industroyer Reloaded' (April 12 2022), https://www.welivesecurity.com/2022/04/12/industroyer2-industroyer-reloaded/, accessed 20 May 2022.
238. Greenberg, 'Russia's Sandworm Hackers Attempted a Third Blackout in Ukraine' (April 12 2022).
239. David E. Sanger and Julian E. Barnes, 'U.S. and Britain Help Ukraine Prepare for Potential Russian Cyberassault' (December 20 2021), https://www.nytimes.com/2021/12/20/us/politics/russia-ukraine-cyberattacks.html, accessed 23 May 2022.
240. Chris Krebbs, 'The Cyber Warfare Predicted in Ukraine May Be Yet to Come', https://www.ft.com/content/2938a3cd-1825-4013-8219-4ee6342e20ca, accessed 21 May 2022.
241. 'World's Largest International Live-Fire Cyber Exercise Launches in Tallinn' (undated 2022), https://ccdcoe.org/news/2022/locked-shields-2022-exercise-to-be-launched-next-week/, accessed 21 May 2022.
242. Sam Sabin and Laurens Cerulus, '3 Reasons Moscow Isn't Taking Down Ukraine's Cell Networks' (March 7 2022), https://www.politico.com/news/2022/03/07/ukraine-phones-internet-still-work-00014487, accessed 21 May 2022.
243. Andrei Soldatov and Irina Borogan, 'Russia's Surveillance State', *World Policy Journal*, Vol. 30, Issue 23 (Fall 2013), pp. 23–30.
244. Daniel Ong, 'Russian General Brutally Dies in Ukraine: 'Collected His Guts…Back In His Belly" (April 27 2022), https://www.ibtimes.com/russian-general-brutally-dies-ukraine-collected-his-gutsback-his-belly-3488076, accessed 23 May 2022.
245. Including on Ukraine's SBU channel on YouTube.
246. https://www.whitehouse.gov/state-of-the-union-2022/, accessed 21 May 2022.
247. John Grady, 'Intel Sharing Between U.S. and Ukraine 'Revolutionary' Says DIA Director' (March 18 2022), https://news.usni.org/2022/03/18/intel-sharing-between-u-s-and-ukraine-revolutionary-says-dia-director, Neveen Shaaban Abdalla, Philip H. J. Davies, Kristian Gustafson, Dan Lomas, and Steven Wagner, 'Intelligence and the War

in Ukraine', https://warontherocks.com/2022/05/intelligence-and-the-war-in-ukraine-part-1/, https://warontherocks.com/2022/05/intelligence-and-the-war-in-ukraine-part-2/, all accessed 22 May 2022.
248. Krebbs, 'The Cyber Warfare Predicted in Ukraine May Be Yet to Come'.
249. Andrey Pertsev, 'Blindsided Russia's Top Officials Were Caught off Guard by Putin's War in Ukraine. Many of Them Want to resign—But Can't' (March 9 2022), https://meduza.io/en/feature/2022/03/09/blindsided, accessed 22 May 2022.
250. Krebbs, 'The Cyber Warfare Predicted in Ukraine May Be Yet to Come'.
251. Gian M. Volpicelli, 'Russia Is Facing a Tech Worker Exodus', https://www.wired.com/story/russian-techies-exodus-ukraine/, accessed 22 May 2022.
252. Mehul Srivastava, 'Russia Hammered by Pro-Ukrainian Hackers Following Invasion' (May 6 2022), https://arstechnica.com/information-technology/2022/05/russia-hammered-by-pro-ukrainian-hackers-following-invasion/2/, accessed 22 May 2022.
253. https://www.whitehouse.gov/state-of-the-union-2022/.
254. Brad Smith, 'Defending Ukraine: Early Lessons from the Cyber War' (June 22 2022), https://blogs.microsoft.com/on-the-issues/2022/06/22/defending-ukraine-early-lessons-from-the-cyber-war/, accessed 23 June 2022.
255. Zhu, Anthony and Sastry, 'A Taxonomy of Cyber Attacks on SCADA Systems', p. 384. See also Liam Tung, Cisco Warning: These Routers Running IOS Have 9.9/10-Severity Security Flaw' (September 26 2019), https://www.zdnet.com/article/cisco-warning-these-routers-running-ios-have-9-910-severity-security-flaw/, Lily Hay Newman, 'A Cisco Router Bug Has Massive Global Implications' (May 13 2019), https://www.wired.com/story/cisco-router-bug-secure-boot-trust-anchor/, both accessed 4 October 2019.
256. See for example Clarke and Knake, *Cyber War*, Elisabeth B. Miller and Thom Shanker, 'Panetta Warns of Dire Threat of Cyberattack on U.S.' (October 11 2012), http://www.nytimes.com/2012/10/12/world/panetta-warns-of-dire-threat-of-cyberattack.html?pagewanted=all&_r=0, accessed 23 July 2014. David Aaro, 'Massive Electrical Failure Cuts Power to Argentina and Uruguay' (June 16 2019), https://www.foxnews.com/world/massive-electrical-failure-cuts-power-to-argentina-and-uruguay, accessed 4 October 2019.

257. Department of Defense Defense Science Board Task Force on Cyber Deterrence February 2017, https://www.armed-services.senate.gov/imo/media/doc/DSB%20CD%20Report%202017-02-27-17_v18_Final-Cleared%20Security%20Review.pdf, accessed 1 December 2018.
258. 'Cybersecurity Threats and Defense Strategy' (C-Span May 9 2017).
259. Anonymous, 'Major Cyberattack on UK Infrastructure Is 'When, Not if'' (January 23 2018), https://news.sky.com/story/major-cyberattack-on-uk-infrastructure-is-when-not-if-11219026, accessed 27 September 2018.
260. 'New Cyber Attack Categorisation System to Improve UK Response to Incidents' (April 12 2018), https://www.ncsc.gov.uk/news/new-cyber-attack-categorisation-system-improve-uk-response-incidents, accessed 27 September 2018.
261. Nicole Perlroth and David E. Sanger, 'Cyberattacks Put Russian Fingers on the Switch at Power Plants, U.S. Says', *New York Times*, 15 March 2018.
262. Perlroth and Sanger, 'Cyberattacks Put Russian Fingers on the Switch at Power Plants, U.S. Says'.
263. Michael N. Schmitt (ed.), *Tallinn Manual 2.0 on the International Law Applicable to Cyber Operations* (New York: Cambridge University Press, 2017), pp. 25–26, 37–38, 135–141.
264. Schmitt (ed.), *Tallinn 2.0*, pp. 135–142.

4

Gaining Access: Attack and Defense Methods and Legacy Systems

Introduction

The last chapter detailed the threats that exist across critical infrastructure and ended with an assessment of cyberwarfare in Ukraine. The introduction had previously outlined a belief within the cybersecurity industry that the first steps to intrusions and breaches begin with reconnaissance. Against critical infrastructure, this is not the first stage of an attack or campaign. For nation-state actors representing Advanced Persistent Threats (APTs), the first steps are to strategize and target. The second stage begins with reconnaissance. The reconnaissance phase helps to identify key individuals and parts of the business. Technical reconnaissance is used to aid the hacker/s gain access to external portals and internal systems and begin the process of infiltration, allied to social engineering (SE) which is detailed in Chapter 5. This is increasingly used as an inroad into an organization to help bypass cybersecurity defenses like firewalls and can include hacking personal e-mail and home office equipment.

Technical reconnaissance generally begins with scans of Internet Protocol addresses (IPs) and open ports, then searching for specific services, then testing for specific Common Vulnerabilities and Exposures (CVEs). Lastly, they attempt Remote Code Execution (RCE) to gain access.[1] After reconnaissance there follows initial intrusions into a network eventually leading to the compromise of that network. This can be through an individual user, computer, or hardware that can provide network access. There follows the establishment of a backdoor and foothold into the network (sometimes called a beachhead as if one were landing troops for an invasion). User credentials such as passwords or authenticators can, and have been, exploited for the initial intrusion or breach and to escalate privileges and move laterally and vertically into critical infrastructure (and other) networks.[2]

This process is often accompanied by the installation of malicious software (malware). Widely termed computer viruses, "Malware has threatened computers, networks, and infrastructures since the eighties". The majority of organizations "rely almost exclusively on just one approach, the decade's old *signature-based* methodology. A more advanced method of detecting malware via *behavior analysis* is gaining rapid traction, but is still largely unfamiliar".[3] There are also YARA rules which help triage through pattern matching; one of a series of methods to search for malware and Indicators of Compromise.[4] Malware often includes Remote Access Trojans (RATs). RATs are one of the main methods for maintaining covert persistent access to individual computers and the networks they are connected to. This provides further reconnaissance openings and means to act on their designs or instructions including locating further targets of opportunity.[5] On the 'Dark Net' "Malware and software bugs can sell for tens or hundreds of thousands of dollars".[6]

Common Technical Attack Methods

To get various forms of malware into systems and to be able to exploit access, a growing variety of means lay at the disposal of the attackers. This is aided by a variety of open-source tools and by shared knowledge among the hacking community and for those looking to gain advanced skills, sharing knowledge is based on hierarchy and demonstrated capabilities. Tools include DMitry (Deepmagic Information Gathering Tool) which "is able to gather possible subdomains, email addresses, uptime information, tcp [Transmission Control Protocol] port scan, whois lookups, and more".[7] Shodan helps locate Internet-connected devices.[8] Meanwhile, long-standing communications protocols used by Industrial Control Systems (ICS) including Modbus and DNP3 "have little or no security measures: lacking authentication capabilities, [meaning] messages may be intercepted, spoofed, or altered, potentially causing a dangerous event in an operations environment".[9] In addition, much of the data is carried by ethernet cables, and PLCs and RTU are frequently connected through old serial ports which are still used in [electricity] substations. These too are a less secure "blind spot".[10]

Drive-by Downloads

Drive-by download attacks are a common way of dispersing malware. This is where hackers search for insecure websites and then install a short piece of malicious code called a script through the Hypertext Transfer Protocol (HTTP)—the 'language' used by Internet browsers (to communicate with websites and the prefix for web addresses) or PHP (the computer language used to code web pages) to change the code on pages. This can be used to install malware onto the computers of unsuspecting victims visiting a website. It can also redirect the victim to a site controlled by the hackers.

Drive-by downloads can occur when visiting a website or reading an e-mail message or through a pop-up window. This is a passive form of cyberattack, as nothing has been done to actively enable the attack. Nothing was clicked on to enable the download of a malicious application (app)/program and a malicious e-mail attachment did not need to be clicked. Drive-by downloads can exploit an app, operating system (OS), or Internet/web browser containing security flaws.[11] Internet browsers like Google Chrome or Firefox regularly automatically update or patch any discovered flaws/bugs but drive-by downloads can also occur because of failed updates or a failure to update.[12]

Fox Business reported that through no fault of their own anyone visiting legitimate and sometimes well-known websites can fall victim. In particular, banking trojans and ransomware are closely associated with drive-by downloads and "These two particularly virulent strains of malware can cause enormous disruption to companies and may even put SMBs out of business".[13] Small to medium sized businesses (SMBs) and enterprises (SMEs) are increasingly targeted because they are part of supply chains for larger companies (the bigger fish) and seen as easier to pick lower hanging fruit.[14]

Watering Hole Attacks

Watering hole attacks are very similar to drive-by downloads. These too hijack legitimate websites to spread malware. What differentiates them is that Watering hole attacks are highly targeted and uncovering them difficult. Attackers are increasingly proactive, reactive, and opportunistic. They also play with human tendencies to trust known sites. Watering hole attacks can take patience and are another passive threat. They either compromise an existing trusted site or create a site appearing to be legitimate, for example, by subtly misspelling addresses (URLs) like amricanexpress.com. These are deliberately designed to catch people out. In addition, insecure or less secure company websites (for example those of SMEs in supply chains) can be attacked.

For example, a Java script, which enables functionality on websites, can be falsely inserted (an example of a Cross-Site Scripting (XSS) attack).[15] These can be used to redirect unsuspecting users to a fake site. These can contain malware and exploitation tools. This can put businesses and reputations at risk. As AT&T explains, "With watering holes, attackers target anywhere from a single company or government agency to larger communities of interest – such as industries or groups of companies".[16] This can include a Structured Query Language (SQL) injection attack. SQL is a computer language used to maintain most databases, and cross-site scripting can also be used to attack through websites.[17]

Man-in-the-Middle/Session HIJACKING

E-mail can be used for something called a man-in-the-middle attack (MITM). An MITM attack consists of three 'players': a "victim, the entity with which the victim is trying to communicate, and the "man in the middle," who's intercepting the victim's communications. Critical to the scenario is that the victim isn't aware of the man in the middle".[18] The ease by which fictitious e-mail accounts and even fake profiles and companies can be created adds to the problem.[19]

MITM, allied to spear phishing (targeted e-mails) and watering hole attacks, are a means to deliver a payload through existing defenses and into sites that run critical infrastructure (and many other sectors). In 2016, Idaho National Laboratory (INL) cautioned: "phishing attacks have increased against production engineers and those on the plant floor area, as well as watering hole attacks against sites with information for ICS engineers. Attackers are starting to trojanize ICS files and components that are available for updating firmware and finding ways to replace them in the supply chain in order to get malware over the firewall and into production environments".[20]

Zero-Days

The U.S. cybersecurity provider Norton outlines that "In the world of cyber security, vulnerabilities are unintended flaws found in software programs or operating systems. Vulnerabilities can be the result of improper computer or security configurations and programming errors. If left unaddressed, vulnerabilities create security holes that cybercriminals can exploit".[21] It is not just cybercriminals that exploit these flaws (bugs) and when companies discover or are informed that vulnerabilities exist (including by professional bug hunters who get 'bug bounties' to identify vulnerabilities) they issue new code known as patches to repair them.

Joss Meakins points out that consumer software typically has between 0.2 and 2.0 bugs for every 1,000 lines of code (LOC). Microsoft averages 0.5 per every 1,000 leading to an estimated 25,000 bugs from the 50 million LOC in Microsoft Windows Operating Systems.[22] Until discovered, these are zero-days which according to Norton security "refers to the fact that the developers have "zero days" to fix the problem that has just been exposed — and perhaps already exploited by hackers".[23] Another major cybersecurity provider, FireEye, adds "These attacks are rarely discovered right away. In fact, it often takes not just days but months and sometimes years before a developer learns of the vulnerability that led to an attack".[24]

Additionally, Meakins rightly flags that currently it is extremely difficult to accurately assess how many zero-days there are, and the ongoing march of software development continues apace. In addition, software code is becoming more complex. In addition, the ongoing evolution of the Internet of Things means an increasing number of devices are connecting to the Internet. This means zero-days will remain valuable and potent and although many bugs get reported and ironed out over time, all new software produces fresh potential exploits. Whilst 'bug bounties' of up to $250,000 are quite common and can go up to $1 million or more (making it lucrative for companies to operate in this market), a mindset of security in tandem with functionality and operability could reduce bugs and with it improve security.[25]

On the 'Dark Net', zero-days are bought and sold on sites like 0DAY.today, TheRealDeal Market, Rutor, and Dream Market.[26] Whilst 'bug bounties' are offered to 'White Hat' hackers by a number of vendors, they are competing with this underworld.[27] There is also a 'grey market' where zero-days are proactively located and sold to buyers for exploitation. These buyers are largely governments or suppliers to governments and include intelligence and law enforcement agencies, militaries, defense contractors, and private surveillance companies. 'Grey market' prices are historically 10–100 times more than on the 'white market'.[28]

For example, to reveal individuals accessing child pornography via the Tor browser the FBI employed a zero-day. Zero-day exploits were also used for Stuxnet, in Operation Aurora, as well as the OPM breach, WannaCry, and NotPetya.[29] The value of zero-days is threefold. First, they are, and need to remain, secret. Once exposed or used they can be patched. Second, products are patched regularly so they have a shelf life. This has been variously estimated to be anywhere from almost seven years to between one and three. Third, the value of the zero-day also depends on the value of the software. An exploitable zero-day in Microsoft Windows or Apple's iOS will have far greater value in the upper price range than one in a little used app. The limited shelf life of zero-days also means a dilemma exists.

For governments and their intelligence agencies and law enforcement, in particular, vulnerabilities are extremely useful. Disclosing them to the vendor gives them the opportunity to patch or remediate and close off the vulnerability/zero-day. This makes systems more secure but means the chance for exploitation decreases significantly. Depending on how widespread the software (or hardware) is and who the users are, that balance between disclosure and secrecy needs to be evaluated. Whether Russian and Chinese operational zero-days intersect with those of Western intelligence, especially the U.S. Intelligence Community, is open to question. However, non-disclosure or non-use of exploits still means they can be found and exploited by an adversary and be used against you. These can include fundamental and widely exploitable zero-days. This makes it a zero-sum game and states as varied as the United States, United Kingdom, France, Israel, Russia, China, India, Brazil, Iran, Malaysia, Singapore, North Korea, and countries across the Middle East have all bought zero-days from the grey market.[30]

Rootkits

Rootkits is a "term loosely applied to a subset of malware tools that are designed specifically to stay hidden on infected computers and enable the attacker to remotely control" a PC or other hardware.[31] They have been used since the 1990s and rootkits can burrow deep into an operating system making detection difficult by anti-virus or anti-malware software and hardware. They routinely operate as a backdoor allowing an attacker to gain remote access at will. Rootkits can contain sections of program code known as modules containing "malicious tools, such as a keystroke logger, a password stealer, a module for stealing credit card or online banking information, a bot for DDoS attacks or functionality that can disable security software".[32] Rootkits allow privileged access to the inner workings of operating systems providing widespread capabilities and target both known and unknown (zero-day) vulnerabilities in an OS.

Remote Access Trojans (RATs)

There are common tactics used which can vary in sophistication depending on the actor/s involved. Understanding these tactics, techniques, and procedures (TTPs) help forensic analysis of the actor or actors involved. The first stage of reconnaissance is where a choice of target/s is made. This is usually followed by social engineering individuals in the target organization and a ruse built through social engineering. If it involves a nation-state/APT against a high-value (and often hard) target these ruses can be highly sophisticated. One of their goals will be to install a Remote Access Trojan.

An example of what a RAT can do can be demonstrated by their open-source variants, Remote Access Tools. One such tool is TeamViewer. TeamViewer is a free piece of software offering remote access "to your work computer, from anywhere, anytime. So you can continue working with desktop applications remotely. And access desktop files, as if you were sitting in front of your workstation. Quickly. Easily. Securely".[33] Unfortunately, TeamViewer is also used by scammers who call unsuspecting people at home over Voice over Internet Protocol (VoIP) claiming to be from 'Microsoft technical support', the Internal Revenue Service, or some other seemingly legitimate organization.[34]

TeamViewer is one of several remote access tools that provide external client access to a remote (host) computer providing them with unfettered access and generally needs software installed on the host to enable remote access.[35] Examples of how remote access works and how the scammers trick users and the malicious acts they perform (and how they can have these same remote access tools turned against them) abound on YouTube. For various reasons, in the critical infrastructure/Operational Technology environment "Strong passwords, authentication, and data encryption may seem to be obvious measures to employ when remote access to ICS networks or devices is necessary, but are often overlooked or ignored". Moreover, "some prominent vendors are slow or resistant to acknowledge vulnerabilities in their own software, even simply refusing to address vulnerabilities that exist, or inadvertently serving as the distributor of malware concealed in vendor-provided web updates as demonstrated in the Havex campaign" (a Russian APT campaign in critical infrastructure operating from at least 2013).[36] An advisory from the cybersecurity provider McAfee states that a backdoor:

> gives an attacker unauthorized access to a system by bypassing normal security mechanisms. This threat works in the background, hiding itself from the user, and it's very difficult to detect and remove.

This is designed to give actors from cybercriminals through to APT's backdoors through widely available malware giving them remote access to a system. This can be done to any networked computer or mobile device. Once remote access is enabled, backdoor access "can potentially modify files, steal personal information, install unwanted software, and even take control of the entire computer". This can include sensitive or proprietary files, including personal identifiers and passwords.[37] This can include installing keyloggers which record every keystroke.

Microsoft's Remote Desktop Protocol (RPD) is also being exploited. RDP was designed by Microsoft to help system administrators access remote computers (similar to TeamViewer) but is now being exploited by organized cybercriminals. Access via RDP to "multiple government systems" is "being sold worldwide, including those linked to the United States, and dozens of connections linked to health care institutions, from hospitals and nursing homes to suppliers of medical equipment".[38] Of at least equal concern, McAfee has found "access linked to security and building automation systems of a major international airport could be bought for only US$10" and being sold on a Russian 'Dark Net' underground shop called Ultimate Anonymity Service (UAS). They also discovered that this provided access to the "airport's automated transit system, the passenger transport system that connects terminals…openly accessible from the Internet". Exploiting RDP requires "no zero-day exploit, elaborate phishing campaign, or watering hole attack".[39]

The Use of Mobile/Cellular Devices and Remote Access

The increasing use, power, and sophistication of mobile devices (many of which are likely to be used on critical infrastructure sites) including smart phones and tablet computers widen still the attack surface and add another layer for defenders. An eightfold increase in mobile data is being predicted and although customer data are the 'crown jewels' for cell phone/mobile providers this does not offer a guarantee of security. Major telecom suppliers are also part of the critical infrastructure of every state they operate in and a target in themselves.[40] The e-mail read on cell phones is also a growing problem with fake hyperlinks or attachments able to be clicked on and opened.[41] 18 percent of phishing e-mails that are clicked on are from a mobile device. Many also receive cold calls and SMS texts.[42]

The use of wireless technology as an attack vector is not limited only to smart phones. Arden Bement, when Director of the U.S. National Institute of Standards & Technology (NIST), discussed a case he encountered back in 2003 when addressing a Workshop on Critical Infrastructure Protection and the need for a 'wakeup call'. Bement recounted that a large southwestern power utility in the United States servicing around four million customers had hired a team of 'White Hat' security professionals to assess SCADA vulnerabilities and they had driven out to a substation. Whilst still in their vehicle they spotted a wireless network antenna. Using their netbook computers, in under five minutes they had connected to the internal systems. Ten minutes after this "they had mapped every piece of equipment in the facility. Within 15 min, they mapped every piece of equipment in the operational control network. Within 20 min, they were talking to the business network and had pulled off several business reports. They never even left the vehicle". Bement warned his audience:

> These incidents and many others demonstrate the vulnerability of far too many of our critical infrastructure control systems. And as we all know too well since September 11, 2001, there are people in the world who would be only too happy to use those vulnerabilities in the most harmful

ways possible. So what's to be done? There is, dare I say it, a vital role here for measurements and standards. The root of the problem in critical infrastructure control systems is a large legacy of hardware and software that was designed with two thoughts in mind: performance and reliability. Security, by and large, was not on the list. But now it must be.[43]

It would be all too tempting to think that this could not still happen almost two decades later, but it could.[44] It was said in the introduction that both Operational Technology (OT) networks and corporate IT (enterprise) networks use accepted international standards like Transmission Control Protocol/Internet Protocol (TCP/IP) and Ethernet protocols to communicate. The OT and IT networks are converging making previously hermetically sealed Local Area Networks (LAN), including home networks, as well as Wide Area Networks (WAN) increasingly accessible from remote offices or from homes—including via 'dial-in' Virtual Private Networks (VPNs) which are designed to be a secure portal between two computers. Still more widely, Twitter has been used to activate malware and the source code sharing platform GitHub has been targeted. It contains code used in everything from 3D modeling to the XML computer language. This code is used by governments, companies, and individuals across the world for tools and applications in everyday use.[45]

Script Kiddies or Nation-States?

Analysis of the actor or actors involved is vitally important, especially for governments and the companies and sectors that make up critical infrastructure. Attackers could potentially be anyone from the stereotypical image of a teenage hacker in a hoodie known as a 'script kiddie', to terrorists, patriotic hackers, or hacktivists (an individual or group who hack for political or social reasons) through to criminals with links to organized crime gangs. Each might operate as individual lone wolf or, more likely, as part of a collective.[46] To cover their tracks they will hide behind VPNs which help mask their Internet Protocol (IP) address (which can reveal your location, Internet Service Provider, Operating System, and Web

Browser) and use multi-hop proxies to route their connections through several servers in various countries.[47] The Onion Router (TOR), a freely available encrypted Internet browser, is also widely used alongside other means of encryption to conceal their tracks.[48]

Given the vital importance of computer-controlled critical infrastructure any of these individuals or groups might be leveraged or under the control of state intelligence agencies or even in the pay of rival companies—who in turn might be acting on behalf of governments or seen to be acting in the national interest. Against critical infrastructure, attackers are more likely to be intelligence agencies or militaries operated by nation-states. These are known as Advanced Persistent Threats (APTs) and are the most virulent of the actors moving against the companies and sectors that make up critical infrastructure.

These actors make the cybersecurity threat landscape diverse and complex but there are guides which help map and understand the nature of the threats in cyberspace. One such guide is the annual Data Breach Investigation Reports (DBIR) produced since 2008 by the American telecommunications giant Verizon. It combines "data from public and private organizations around the world, including law enforcement agencies, national incident-reporting entities, research institutions, private security firms and Verizon" itself.[49] As their dataset has expanded, Verizon's DBIR has become a very useful bellwether for measuring and mapping cybersecurity incidents and breaches.[50] There is cold comfort in terms of CI for 2021 DBIR stated, "As in past years, financially motivated attacks continue to be the most common (Fig. 4.1), likewise, actors categorized as Organized crime continue to be number one (Fig. 4.2)".[51]

It is interesting to note that the 2019 DBIR found that cyberespionage had increased after dipping in 2014 and the first half of 2017. These coincide with negotiations on the Obama–Xi agreement on cyberespionage between the U.S. and China, signed in September 2015, and Russian interference in the U.S. election cycle during 2016, culminating in the November 2016 election of President Trump.[52] During 2017, breaches continued to rise as did 'state-affiliated' activities. Against the public sector, cyberespionage was described as 'rampant'.[53]

The use and usefulness of stolen credentials like passwords, including from commonly used cloud-based e-mail servers, was at the top of the list

4 Gaining Access: Attack and Defense Methods ...

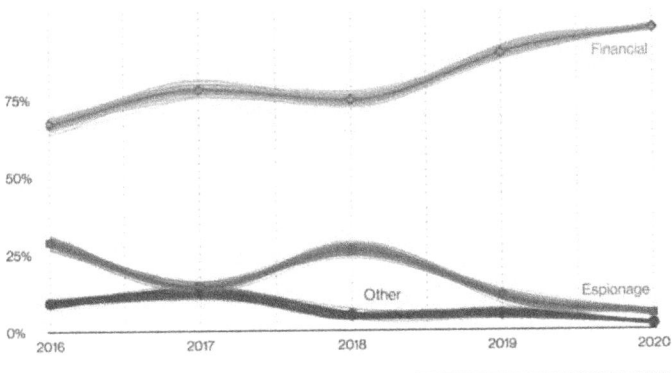

Fig. 4.1 Top threat actor motive over time in breaches (*Source* 2021 Data Breach Investigations Report, http://verizon.com/dbir/, p. 12)

of reasons for breaches in 2019. Only 6 percent was through exploiting vulnerabilities in hardware and software. 94 percent of incidents were found to have originated in e-mail distributed malware.[54] As in years gone the 2019 DBIR found that "when breaches are successful, the time to compromise is typically quite short", usually minutes, but with "no way of knowing how many resources were expended in activities such as intelligence gathering and other preparations".[55] Discovery can take months and depend on the type of attack or campaign. Verizon analogized:

> a golfer navigating a golf course is a lot like an adversary attacking your network. The course creator builds sand traps and water hazards along the way to make life difficult. Additional steps, such as the length of grass in the rough and even the pin placement on the green can raise the stroke average for a given hole. In our world, you've put defenses and mitigations in place to deter, detect, and defend. And just like on the golf course, the attackers reach into their bag, pull out their iron, in the form of a threat action, and do everything they can to land on the attribute they want in the soft grass of the fairway…[and with] attack paths…much more likely to be short than long…why hit from the tees unless you absolutely have to? Just place your ball right there on the green and tap it in for a birdie or a double eagle…[56]

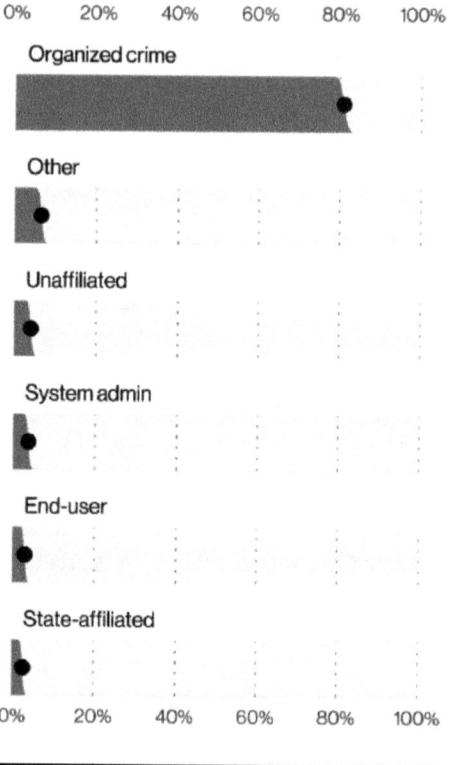

Fig. 4.2 Top threat actor varieties in breaches (n = 2,277)

Of those 'threat actors' engaged in cyberespionage, 96 percent were attributed to nation-states or 'state affiliated' groups. These are the professional players on the tour. Phishing was used in 78 percent of cyberespionage incidents along with the use of backdoors and malware for the command and control of victim machines.[57] In addition, Internet-based 'timestomping' tools that modify files to change create, modify, and access dates of files are also used in an attempt to cover tracks.[58]

Common TTPs

The identifiers used by the cybersecurity community to classify attackers are called Tactics, Techniques, and Procedures (TTPs). The tactics of a cyber threat actor encompass the tools used "while techniques give a more detailed description of behavior in the context of a tactic, and procedures an even lower-level, highly detailed description in the context of a technique".[59] The TTPs help identify the intent and possible actors behind these intrusions and breaches. One of the smallest cybersecurity providers, the Denver-based Optiv, is one of a number who has examined the maturation of the TTPs of threat actors. These include:

- Rapid triage and contextualization of an event or incident by correlating it to TTPs of known actors or groups potentially related to an attack.
- Supporting the investigative process by providing probable paths for research and focus, based upon former TTPs used in a campaign or attack.
- Supporting identification of possible sources or vectors of attack.
- Supporting the incident response and threat identification and mitigation processes by helping identify which systems are likely to be compromised.

This means "When an incident does take place, TTPs related to that incident help to establish potential attribution and an attack framework".[60] Verizon's 2016 DBIR noted that basic but well-executed practices "continue to be more important than complex systems". It recommended:

- Know what attack patterns are most common for your industry.
- Utilize two-factor authentication for your systems and other applications, such as popular social networking sites.
- Patch promptly.[61]
- Monitor all inputs: Review all logs to help identify malicious activity.
- Encrypt your data: If stolen devices are encrypted, it's much harder for attackers to access the data.

- Train your staff: Developing security awareness within your organization is critical especially with the rise in phishing attacks.
- "Know your data and protect it accordingly. Also limit who has access to it".[62] This is also part of the 'zero trust' model outlined below.

Counters and Defenses

The front line of IT departments and the wider cybersecurity industry backed by academia and national governments have not been standing still under this barrage. Whilst long-standing defensive methods continue to be employed, with existing hardware and software known vulnerabilities are disclosed and shared, education of the threats being faced continues to teach us a hard lesson. In future the next generation of systems, including the SMART system should (in theory) have security baked in through security-by-design. The extent to which security, which adds cost, complexity, and delays to bringing products to market in a highly contested marketplace is prioritized, is open to question.[63] To varying degrees upgrades will continue to be layered over older technologies, some of which form the backbone of companies, industries, and sectors. One of the first lines of existing defenses is firewalls.

Firewalls

Firewalls first came into widespread use during the 1990s. A firewall is a software or hardware device, or both at the same time, which monitors both incoming and outgoing information known as network traffic. These include Internet and e-mail filters on the network and act as a fence between an internal system or network and the outside world. They are based on rules or permissions of what to allow in and what to block. They are now evolving into 'next-generation firewalls' (NGFW) which integrate improvements in intrusion prevention and application controls. They include updates based on known threats and methods to address developing security threats. NGFWs can also help companies identify which assets are most at risk and to aid the detection of evasive or suspicious network activity.[64] Firewalls are an important first line of defense for SCADA and ICS which control critical infrastructure. Hardware firewalls can be expensive and can require companies to enter into costly maintenance contracts and "vendor-specific configurations".[65]

Demilitarized Zones (DMZs)

Demilitarized Zones are a form of network segmentation and work with firewalls to prevent breaches. A DMZ will consist of an external firewall, such as protecting a web server, and an internal firewall to protect internal systems or intranets. DMZs can house Intrusion Detection or Intrusion Prevention systems (IDS/IPS), honeypots, and proxy servers. This is part of a layered defense known as a 'defense in-depth' model—another military term deployed in cybersecurity—as part of the Confidentiality, Integrity, and Availability (CIA) triad. DMZs are vulnerable to packet sniffing which looks for pathways into networks and devices such as routers which are used to transmit and receive data packets.[66]

Intrusion Detection Systems (IDS): HIDS/SIDS/HIPS

An IDS is designed to detect reconnaissance into systems and networks and "In many of today's IDS solutions, a hybrid approach is used, with both signatures and anomaly detection used in tandem to provide the best possible intrusion detection".[67] IDS can be used to detect attacks against many different devices. They can incorporate anomaly detection which "capture traffic over a period of time and use this as a reference for what is valid. These systems then compare new traffic to what is considered to be 'normal' and look for anomalies".[68]

However, this can generate many false positives because network traffic will alter over time as businesses change or network component pieces are added or removed. Signature-based IDS (SIDS) are useful in flagging known threats and generate less false positives, but newer threats might well be missed, and they also require regular updates. There are variants which include a Network-based Intrusion Detection System (NIDS) and a Host Intrusion Detection System (HIDS). Both are available to corporate clients.[69] There are also free providers of HIDS.[70] Separately, a number of major providers offer protection against e-mail threats at the mail gateway which offers protection for inbound and outbound data.[71] Cybersecurity product vendors are also now developing proactive variants in the form of Network-based Intrusion Prevention Systems (NIPS).[72] The Semantic Analytics Stack (SANSA), leveraging Machine Learning allied to libraries of known attacks, is also a valuable and active development.[73]

HIDS are designed only to protect a PC or file server but not a network. Servers store increasingly large amounts of data. With data usage globally continuing to grow apace, this has led to the creation and expansion of cloud service providers who host data remotely. This is also being fueled by the growth of SMART devices and the IoT. These cloud servers, and IoT devices, are a growing attack vector, especially for ransomware attacks, but the extent to which cloud services are being used by critical infrastructure owner-operators is difficult to assess.[74]

Honeypots and Honeytraps

In traditional intelligence gathering, honeypots and honeytraps have been real-life people who seek to entrap persons of interest (often through sexual promise), which have been used from antiquity through to the present. In cybersecurity, they have been in existence for decades and are generally thought of as a computer or computer system designed to simulate or imitate targets of cyberattacks. They are used to discover and identify attacks or draw off attackers from their genuine targets. They are useful to gather information and intelligence on how attackers are operating. As Norton describes, "If you, for instance, were in charge of IT security for a bank, you might set up a honeypot system that, to outsiders, looks like the bank's network". The same logic applies to other sectors, especially where IT or OT systems connect to internal networks and the Internet. Honeypots help understand who is attacking, from where, and what is being targeted (whether cybercriminals or APTs) but

"More importantly, you can determine which security measures you have in place are working — and which ones may need improvement".[75]

Web analytics which analyze website visits (hits) can also help forensic analysis and can (without the use of tools such as VPNs and proxy connections) help identify the geolocation of visitors.[76] There is also an offensive character to honeypots. They can be an agent of social engineering and precursor for gateway entry or exploitation "In addition to phishing and cracking attacks" or enticement through social media. Today's online honeypots can appear as Facebook or LinkedIn friends or professional contacts and can:

> include a component of sexual appeal or attraction, but they just as often appear to be people who share a target's political views, obscure personal hobbies, or issues related to family history. Through direct messaging or email conversations, honeypots seek to engage the target in conversations seemingly unrelated to national security or political influence. These honeypots often appear as friends on social media sites, sending direct messages to their targets to lower their defenses through social engineering. After winning trust, honeypots have been observed taking part in a range of behaviors, including sharing content from white and gray active measures websites, attempting to compromise the target with sexual exchanges, and most perilously, inducing targets to click on malicious links or download attachments infected with malware.[77]

Signature and Behavior-Based Malware Detection

There are a variety of methods to detect and to try to neutralize malware. The baseline approach is signature-based detection. John Cloonan, a 25-year veteran of the cybersecurity industry, explains that "In computing, all objects have attributes that can be used to create a unique signature". 'Objects' in computer science refer to a resource that exists in memory, on, disk, in transit, or in the cloud and algorithms have been developed which rapidly and effectively establish these digital signatures. Cybersecurity vendors can then list these 'objects' (which have values that are distinguishable) as malicious and their signatures are then added to a database of known malware and publicized to the cybersecurity community at large. These databases can contain hundreds of millions of signatures that identify malware. Cloonan remarks that:

> This method of identifying malicious objects has been the primary technique used by malware products and remains the base approach used by the latest firewalls, email and network gateways. Signature-based malware detection technology has a number of strengths, the main being simply that it is well known and understood – the very first anti-virus programs used this approach. It is also speedy, simple to run, and widely available. Above all else, it provides good protection from the many millions of older, but still active threats.[78]

Unfortunately, new types of malware are generated daily and existing malware is evolved rapidly and can have an immediate impact and devastating effects (like WannaCry) because of hyper connectivity and "Verifying that a new file is malicious can be complex and time consuming". Moreover, "today's advanced malware can alter its signature to avoid detection".[79]

This is known as polymorphic malware and has led governments and the cybersecurity industry to look less at the underlying code and more on the effects they generate—the behavior of the malware.[80] This can help detect suspicious activities. Heuristic scanning is also used. This examines signatures and looks for commands which help flag malicious intent. Most anti-virus providers use both signature and heuristic scans against malware looking to escape detection and heuristic scanning does not reveal their discovery to malware authors. However, it has proved to be limited in application. Malware can elude detection and heuristic scanning can generate false positives.[81] Further technical evaluation of "malicious behavior as it executes is called **dynamic analysis.** Threat potential or malicious intent can also be assessed by **static analysis,** which looks for dangerous capabilities within the object's code and structure". And "While no solution is completely foolproof, behavior-based detection still leads technology today to uncover new and unknown threats in near real-time".[82] Sandboxing can be used alongside these methods.

Sandboxing

A sandbox is a computer system sealed off from an external connection to the Internet designed as a testing environment. Because the system and any components are isolated this allows code to be tested, including malware. In their analysis of sandboxing capabilities, Cisco points out that whilst established signature-based detection mechanisms work well against known malware, new malware and malware variants can be missed. One of the main reasons is because of polymorphic malware. These polymorphic threats (and the actors behind them) are dynamic and can be more agile than the organizations defending against them. This is "making signature-based detection a futile effort" because "new signatures are created for each victim…repackaging threats until they evade detection by all the major vendors". Moreover, "Attackers test their malware on copies of the vendor's software and do not release their malware until it bypasses the exact defenses they are up against".[83] Because of this sandboxing technology is very useful, including when aligned with honeytraps as this executes dangerous or unknown files in a safe environment to record its actions. This said, there are limitations to sandboxing because "If malware determines it's running in a sandbox, it'll attempt to avoid detection by curtailing malicious activities. It's critical that a sandbox remains undetectable, and most are not".[84]

Packet Sniffers

Packet sniffers like Wireshark[85] "are commonly used by network administrators to diagnose network-related problems such as bottlenecks or system-wide slow-downs. Hackers are also able to use packet sniffers to spy on network traffic or gather or assemble passwords These can intercept and log wired and wireless data packets and traffic".[86]

Application Whitelisting

Application whitelisting is a system which stops users from running unauthorized software. With the growth of malware distributed by e-mail its use has increased. However, to be effective whitelisting needs to be combined with other controls. It has limits as a mitigation measure and if a whitelisted (authorized) application has a vulnerability, whitelisting will not help against an attacker. Naturally, this depends on an attacker knowing what software is installed but this is part of the reconnaissance phase of an attack. Whitelisting is linked to signature or publisher keys and anti-virus software and related products. Built in to modern operating systems like Windows 10, they will ask if you wish to install software and inform you of the publisher's certificate advising you if it's legitimate or potentially harmful.

Installation can require administrator rights designed to limit what can be installed and access to system files and root directories which effect system operations. However, despite having benefits, application whitelisting can be bypassed meaning even known 'blacklisted' malware or programs can be installed.[87] 'Dropper's, trojan malware designed to bypass firewalls and anti-virus products, can delete their files leaving no trace behind and have been used in a number of cases.[88] 'Dropper's are similar to malware downloaders, but have malware pre-packaged with them with 'malware downloaders' needing an Internet connection. Both can infect Internet-connected endpoints including desktop computers, laptops, smart phones, and tablets.

Other Internet-connected devices including printers or specialized hardware such point-of-sale (POS) terminals and SMART meters are also vulnerable and have been subject to compromise. Each of these is a potential entry point for attackers.[89] Blacklisting sites and applications is an available option whilst Internet filtering and the use of proxy servers to block access to certain sites is also common practice for many businesses. Data Execution Prevention (DEP) is also employed, including within Microsoft Windows to attempt malware from executing.[90] Software restriction policies (whitelisting) has been available since Windows XP and Windows Server 2003.[91] Modern iterations include dedicated support for whitelisting software such as AppLocker.[92]

Security Information and Event Management

A Security Information and Event Management (SIEM) system observes and reacts to dynamic security events by gathering event data and analyzing it. This includes new network connections to an internal system and changes made to SCADA system files. As Knapp explains, "The SIEM, if properly configured, will also detect a new IP address within the control system".[93] SIEMs offer an additional layer of protection within critical infrastructure. Moreover, "The latest SIEM technology also provides the means to deal with hundreds of different technologies logging data and creating audit trails — something that is particularly relevant for ICS and SCADA". However, "Implementing fit for purpose security technology on ageing ICS and SCADA systems isn't without its challenges…".[94]

SIEMs work with Microsoft's Windows Management Instrumentation (WMI) software and allow third parties, such as industrial IT departments, to write their own "scripts or applications to automate administrative tasks on remote computers".[95] WMI also provisions management data to other parts of the operating system and its functions mean it is "most useful in enterprise applications" and for system administrators to code their own scripts.[96] Unfortunately, WMI has also been used in conjunction with Windows' PowerShell language which "control and automate the administration of the Windows operating system and applications that run on Windows" and have aided Russian APT campaigns through a backdoor technique (demonstrated at the Black Hat Asia conference in 2018).[97]

Blockchain

Blockchain is a foundational innovation in computer science that first appeared in 2008 in a whitepaper, 'Bitcoin: A Peer-to-Peer Electronic Cash System', published under the pseudonym Satoshi Nakamoto. Its code was released as open source the following year. Over the intervening years, Blockchain has been developed and evolved. It has been used to validate and support cryptocurrencies, but from around 2014 was seen to have wider applications.[98] It continues to underpin cryptocurrencies and cybersecurity applications.

It is a distributed ledger that records transactions "in a verifiable and permanent way" through an open and distributed system.[99] It has security-by-design built in through its distributed database, combined with peer-to-peer sharing and transparency with pseudonymity. This makes for unalterable records through the use of computer logic and algorithms. These protect whatever is stored on the ledger/database from deletion, interference, or unauthorized alteration. Its system of encryption and digital ledger has found uses across critical infrastructure sectors and into the service economy from banking and financial services to travel and transport, healthcare, and the mining industry.[100] Indeed, 'mining' the sophisticated mathematical problems that underpin Blockchain technology is one of the keys to its success (frequently conflated with mining Bitcoin and mining other cryptocurrencies).[101] Through it, "Businesses are able to authenticate devices and users without the need for a password...".[102]

Although it holds much promise for businesses and transactions, and the global economy, in creating increased security: "technological innovation tells us that if there's to be a blockchain revolution, many barriers—technological, governance, organizational, and even societal—will have to fall. It would be a mistake to rush headlong into blockchain innovation without understanding how it is likely to take hold".[103] Relatedly, cryptocurrencies also face similar questions—especially through their use by cybercriminals for ransom.

Pressing the Reset

Should the worst happen companies retain the option to reboot and restore systems, but this is highly disruptive and time consuming and across critical infrastructure always-on availability is paramount. Although there are likely to be backups of files, settings, and configurations, a reset of systems or of an entire network (including attempts to restore factory settings as the only way to restore operations and process integrity) is going to lead to downtime of critical services.

4 Gaining Access: Attack and Defense Methods … 251

Private sector owner-operators of critical infrastructure, and many wider organizations also house data and process historians as backup. These include vital data such as temperature, flow, pressure, water levels, and so on. These are vitally important redundancy systems. Not only do they contain possibly highly confidential process information, but "many control algorithms rely on past process data to make correct decisions". Historical process data is not only helpful (or essential) for safety reasons "but also for business purposes, such as electricity pricing".[104]

There are three main problems. First, "Although we've assumed the algorithms of these [sic.] softwares are trustworthy, there are still vulnerabilities associated with their implementations".[105] Second, the off-site 'cold storage' of data historians and air-gapping might not be a safeguard as highly inventive ways have been found to bridge air-gaps and access secure sites.[106] Third, data historians, as well as industrial plant workstations and servers, are also vulnerable to something called a buffer overflow attack. A buffer overflow is when "part of the physical memory storage that is temporarily used to store data…[and] a program or process tries to write or read more data from a buffer than the buffer can hold".[107] Buffer overflows "are more problematic in SCADA systems than in traditional IT" because SCADA and ICS contain embedded systems that run years without rebooting".[108] These are one of a number of attack vectors against SCADA and ICS and the critical infrastructure they run.[109]

The Zero Trust Security Model

A more recent ground up and holistic approach has been a 'zero trust model'. This assumes that breaches are inevitable and instead "embeds comprehensive security monitoring; granular risk-based access controls; and system security automation in a coordinated manner throughout all aspects of the infrastructure in order to focus on protecting critical assets (data) in real-time within a dynamic threat environment".[110] It shares elements from earlier work on dynamic security policies, where uncertainty is an attribute, and systems are founded on a need-to-know/need-to-access basis.[111]

Legacy Systems: In-Built Vulnerabilities in Critical Infrastructure

Across critical infrastructure, business needs increase demand for connectivity which in turn produces increased vulnerabilities—vulnerabilities which are attenuated by legacy systems. Legacy systems are a problem in terms of protecting critical infrastructure because of SCADA systems. SCADA systems are an embedded technology in Industrial Control Systems and integral to the vulnerabilities that exist in CI.[112] They were put in place when the Internet was in its infancy and computer security was not a national security issue back in the 1980s and 1990s (and even further back).

Legacy systems are made up of a variety of hardware and software configurations embedded on a technology base which is now more than a generation old and like everything will falter and fail at some point. The failure rate of SCADA systems is represented as a 'bathtub curve' (due to its shape). When systems are developed and installed the failure rate is high, this decreases as operational experience matures, and rises again as they reach the end. Across this timespan, the longevity of ICT professionals with expertise of legacy systems decreases over the 15–20 + year cybersecurity lifecycle. By the end, even if documentation is available for their successors to maintain the systems, maintenance, and spares become problems to solve.

Legacy devices struggle to adapt to new capabilities including encryption or added security features. In addition, old and obsolete communication protocols are still being used which makes them vulnerable to snooping and interception.[113] They also have backdoors through links to third-party maintenance staff and contractors.[114] Moreover, many of the computers, hardware products, and software used in industrial networks were put in place when the system was first installed and have not been updated with newer software, operating systems, or updated with patches.[115]

In most cases, computers within SCADA networks can be made more secure by ensuring they are running a current operating system with the latest security patches and security software. However, in some cases, they are running SCADA specific software that is only supported by an older and unsupported version of the operating system (or they run on a bespoke or obscure operating system). This prevents them from being upgraded and creates known security issues. Wholesale replacement is expensive, labor intensive (especially with skilled technicians in short supply), and will lead to down times. Resultantly, legacy devices will continue to be used as next-generation systems are gradually rolled out; including for SMART devices leading to large-scale 'SMART grids' which monitor and deliver electric power aided by SMART meters.[116]

There is a need to upgrade. However, "The addition of modern digital technology to legacy equipment never designed to be digitally connected or the replacement of analog equipment with digitally connected devices creates cyber vulnerabilities to systems that were previously immune".[117] This all means legacy systems add complexity to the security calculus. Legacy systems are a problem. They are also physically robust and hard to remove and frequently responsible for core applications. Legacy systems are to be found everywhere, from major corporations to local governments. SCADA is also used in military systems such as ship propulsion which makes this a wider problem than just Industrial Control Systems.[118] This includes even U.S. military systems. The U.S. military, like their civilian counterparts, faces the problem of unpatched and some un-patchable systems. This "can be particularly problematic when physically inaccessible systems (such as those deployed to remote sites) require replacement or upgrade".[119]

This makes them increasingly vulnerable to attack. Most embedded computers and systems in these networks were designed before cybersecurity became a major concern and therefore can contain few, if any, security measures. Given the growing scale and complexity of the threats now faced this is a major concern. This has been brought to light by several publicly reported cases including Russia's Dragonfly campaign. These business-critical processes are especially prevalent in the industry and help run critical infrastructure, operating on the basis of Confidentiality, Integrity, and Availability (known as the CIA triad). For critical infrastructure, always-on availability is prioritized because the systems are critical for running essential processes. For other industries like financial services, confidentiality, and integrity are paramount.

With government policies in the developed world reflecting the need for increased energy efficiency, reduced carbon emissions, and stringent safety provisions, as well as savings through a reduced emphasis on personnel where possible, greater care, and targeted investment is needed.[120] This is also when installing new systems and especially if third-party subcontractors are allowed to connect to a public network to enable remote monitoring, control, and maintenance with simple authentication procedures—sometimes unbeknownst to the corporate/enterprise IT department.[121]

Kyle Wilhoit, a threat researcher from Trend Micro who focuses on locating Internet-facing systems and helping companies improve their cybersecurity practices, points out that elements of CI are "in far-flung places and it was much cheaper to keep an eye on them via the internet than to send an engineer out".[122] This will be especially true in geographically dispersed sites where it is much more difficult, impractical, and less cost-effective to micro-manage.

Increased use of ICT technologies and remote access also means that middle managers can now have greater awareness of plant floor issues; especially issues that can improve or downgrade productivity. This also means greater visibility in reporting any issues to senior plant managers through intranets and the Internet. In this context, modernization can be a route to improved performance through hybrid systems where state-of-the-art systems are linked to legacy systems.[123] However, increasing connectivity to legacy systems without a wholescale upgrade can open attack paths. Moreover, SCADA systems across CI sectors use a combination of PCs and tailored embedded systems which run a variety of obscure operating systems including VxWorks, INTEGRITY, and MQX. Across industry and the defense-industrial base, these "operating systems and applications running on SCADA systems, which are unconventional to typical IT personnel, may not operate correctly with commercial off-the-shelf IT cyber security solutions".[124]

Legacy Systems of the U.S. Government

Legacy systems operated by the U.S. military include those of Strategic Automated Command and Control System (SACCS) which, according to Exelis (a cleared subcontractor with the contract to develop modern data storage and retrieval technology for SACCS) "provides orders and pass codes authentication for Air Force nuclear weapons platforms, bombers, and support units".[125] SACCS was initiated after the 1962 Cuban Missile Crisis when John F. Kennedy was U.S. president. It "runs on 1970s-era IBM computer systems and uses 8-inch floppy disks. Each disk holds 80 kilobytes of data – meaning it would take more than 3.2 million floppy disks to equal the storage power of a 'single modern flash drive'".[126] SACCS is part of the U.S. Department of Defense's nuclear command, control, and communications system (NC3). The Department of Defense is modernizing NC3. NC3:

> includes terrestrial and space-based sensors that monitor the globe for threats, and a communications architecture that connects the nation's decision makers to nuclear forces under any conditions. It is a classified system designed in the 1960s and 1970s during the height of the Cold War and is projected to stay in service for years to come.[127]

Elizabeth Durham-Ruiz, STRATCOM's director of command, control, communications, and computer systems, said in March 2019 that it "will be a long process. There are more than 100 programs that make up the NC3 system today".[128] Whether nuclear weapons themselves are vulnerable to cyberattack has been speculated.[129] It is the subject of detailed analysis in books by Andrew Futter and Herbert Lin.[130]

Critical systems such as those found in the nuclear power industry are disconnected from networks entirely. However, legacy Windows operating systems, which Microsoft had stopped supporting, dating from the 1980s and 1990s were still being used. Ten of the most critical systems were essential to government operations including emergency management, health care, and wartime readiness. The Federal Emergency Management Agency (FEMA), part of the DHS, had 168 high risk or critical vulnerabilities.[131] A June 2019 study by the U.S. Government

Accountability Office (GAO) initiated by President Trump's Executive Order 13800, *Strengthening the Cybersecurity of Federal Networks and Critical Infrastructure*, found that:

> the Department of Education runs on Common Business Oriented Language (COBOL)—a programming language that has a dwindling number of people available with the skills needed to support it. In addition, the Department of the Interior's system contains obsolete hardware that is not supported by the manufacturers. Regarding cybersecurity, the Department of Homeland Security's system had a large number of reported vulnerabilities, of which 168 were considered high or critical risk to the network as of September 2018.[132]

COBOL is still being used in a number of agencies. This included the U.S. Air Force (USAF). Seven out of 10 government agencies were planning to modernize these systems and of the $90 billion expected to be spent on IT in 2019, 80 percent was directed on legacy systems which are increasingly costly to maintain and "more exposed to cybersecurity risks".[133]

At the Department of the Interior's Bureau of Reclamation, their SCADA system running dams and power plants date from the turn of the millennium and houses obsolete hardware no longer supported by the original manufacturers with long-term vendor support not built into the original contract. This hardware "may have had long-term exposure to security and performance weaknesses". Relatedly, the Director of National Intelligence testified in January 2014 that "ICS and SCADA systems used in electrical power distribution provided an enticing target to malicious actors and that, although newer architectures provide flexibility, functionality, and resilience, large segments of the systems remain vulnerable to attack, potentially causing significant economic or human impact".[134]

A system operated by the Federal Aviation Administration (FAA), part of the Department of Transportation, housing data on aircraft and pilots and "also provides information to other government agencies, including those responsible for homeland security and investigations of aviation accidents" is DOS-based with some of its core components mainframe applications that have operated since 1984.[135] The FAA also runs unsupported software. More widely, the GAO was categorical:

> federal legacy systems are becoming increasingly obsolete. In May 2016, we reported that many of the government's IT investments used outdated software languages and hardware parts that were unsupported. We also reported instances where agencies were using systems that had components that were at least 50 years old or the vendors were no longer providing support for hardware or software…The consequences of not updating legacy systems has contributed to, among other things, security risks, unmet mission needs, staffing issues, and increased costs…[and] may operate with known security vulnerabilities that are either technically difficult or prohibitively expensive to address…[and] may not be able to reliably meet mission needs because they are outdated or obsolete.[136]

As in industries the world over, "Agencies have had difficulty finding employees with such knowledge and may have to pay a premium to hire specialized staff or contractors". Moreover, because "agencies are not required to identify, evaluate, and prioritize existing IT investments to determine whether they should be kept as-is, modernized, replaced, or retired" mean they are not mandated to decide on the retirement of legacy systems.[137]

Industry and the Costs of 'Keeping the Lights On'

Legacy systems can also incorporate a large number of changes made over many years. Many different people have been involved in making these changes and it is unusual for any one person to have a complete understanding of any one system. Moreover, people frequently move between businesses and sectors and although businesses regularly replace their equipment, they too incorporate legacy hardware and software including SCADA and ICS. Terence Milholland, the Chief Technology Officer of the Internal Revenue Service, in testimony to Congress saw this situation as "analogous to operating a 1960s automobile with the original chassis, suspension and drivetrain, but with a more modern engine, satellite radio and a GPS navigation system".[138]

Bipin Patel, a veteran Chief Information Officer, outlined two cases illustrating problems that legacy systems can have on front line operations, reputation, and profitability (let alone from problems that could ensue from malicious action). The first stemmed from a power loss at Delta Airlines' Atlanta technology center in August 2016 where restoration was delayed after backup systems failed. Even when power was restored, for several hours afterwards the reservation system had difficulties connecting to the check-in and boarding system and this saw the cancelation of around 2,300 flights. The second was at Deutsche Bank in 2015, where poor trading and risk management were blamed on their outdated IT systems which their chief executive labeled a 'Horlicks' needing total replacement.

For Patel, these legacy systems are a potentially existential threat to some businesses. Instead of being seen as core business and efficiency assets that need considered attention and investment (as well as posing a potential security risk), they are instead seen as 'IT problems'. Across many medium- to large-scale businesses, these systems reflect the world prior to massive connectivity and the reliance upon ICT as well as past business needs and practices. Behind the computer and display screens sit legacy code and data flows. Some of these are long forgotten and where the original designers, installers, and maintenance technicians have left the business or retired, and new code has been layered over old code as business needs to be changed. Much like an onion, these layers enable, inhibit, or house security risks as well as the functioning of the business concerned. Because these systems work it means ignoring or storing up problems, including against cybercriminal ransomware and potential threats from APTs. These issues are becoming clearer to executives, especially as more and more cases come to light.[139]

Large businesses are unlikely to upgrade their systems in one go, however. This means parts of the business operating on legacy systems, whilst some parts will have newer systems. Hardware and software critical to business needs are particularly sensitive to changes and require careful upgrades. Moreover, duplication of assets and systems, as well as personnel, can result from company mergers and acquisitions. This can include financial information and accounts, customer files, and manufacturing systems or process controls. Each of these companies could well have different practices and technologies in operation and this too adds to the complications.[140] Much like a house move, things can go wrong or missing, and things take longer than anticipated.

For companies the hardware backbones of mainframes, servers, and connected computer architecture will frequently come from multiple vendors. In some cases, the hardware has not been changed since the first installation, including default passwords. To meet business needs, especially for staff not trained to operate old legacy systems, these passwords are available on the Internet. This is valuable from a practical perspective as some of the original vendors might have gone out of business or been merged into other companies, but it also carries significant security risks. In addition, it is less likely as the years pass that the original designers or installers are still around. Hiring former staff can be expensive but to 'keep the lights on' ongoing maintenance is vital. Running legacy systems is also time consuming, can result in operational instability, difficulties creating new products or services, and make responding to regulatory changes and cyberattacks more challenging and costly. Increasingly, connectivity between different parts of the business might well help streamline operations, but these are also what help cyberattackers move laterally and vertically within and between companies.

For these reasons, Patel recommended that board level understanding needs to improve. This can help recognize and initiate changes where needed before problems or attacks necessitate an intervention. This includes a more complete boardroom understanding of what ICT and OT systems and assets are currently operating and what needs remedial action. Many of these could represent decades of investment and be extremely complex and interconnected. Retiring or combining legacy code also imposes operational changes and migrating (often large-scale) data from old and diverse legacy systems is also time consuming. Finding, hiring, and retaining personnel with the expertise and experience of old code and new systems and migrating from one to the other carries further costs. This means that (at scale) in most cases upgrading needs to be done piecemeal. Choices of new systems and platforms also need preplanning, costing and budget, considerations of future proofing, as well as long-term maintenance and support. This includes hardware, software, and third-party cloud services.[141]

Patching

If systems are running proprietary software such as Windows (especially older versions of Windows such as Windows NT or XP—both of which are no longer being supported or patched by Microsoft[142]) then their security flaws are already well known and understood both in the hacking community, by Information Communications Technology (ICT) specialists, and by states such as Russia and China (which themselves have this problem).[143] Although XP and even earlier mainstream operating systems (OS) are retired they are still running across critical infrastructure. There is no public data on the numbers of industries, but they are well established. In any event, patching might not be possible in ICS and SCADA environments. Here, safety and uninterrupted availability outrank cybersecurity concerns.

According to Kelly Jackson Higgins, this is cultural within the ICS/SCADA world where experience has taught that patching can create problems. In 2014, it was believed that no more than 10–20 percent of organizations were installing patches released by their own bespoke vendors, let alone those from Microsoft. It was more likely that workstations and servers would be patched as they have shorter lifespans and are less likely to effect critical plant floor operations. However, there is a bigger issue as the human–machine interface (HMI) and other applications that run on OSs like XP are more vulnerable still. Billy Rios, the director of threat intelligence at Qualys, part of whose job is to test for security flaws in ICS/SCADA environments, cautions that "They really don't patch, anyway…And even if they did update, it's the software that's on top that's most vulnerable. The HMI software to run power plants and oil refineries is so riddled with bugs… it doesn't matter what OS it's running…When you have a backdoor password in the HMI…Someone can log in, regardless. You could upgrade to Windows 8 and still have problems".[144]

Russia's 'Dragonfly' campaign targeted Windows XP. Other operating systems used in Industrial Control Systems include Unix and the more widely used Unix-like Linux OS. The disclosure of the Bash (the Bourne Again Shell) vulnerability in September 2014 affected these Linux-based systems with Bash widespread in ICS and SCADA systems and embedded devices.¹⁴⁵ Industry standards aim to raise cybersecurity including through Ethernet/IP, the Open DeviceNet Vendors Association (ODVA), and RSLogix 5000 (a software package that designs and programs SCADA systems and operations). However, these remain issues which are also felt widely including in the large, complex, and interconnected supply chains which help make, enable, and support critical infrastructure.

Targeting Supply Chains

Rather than target sites directly it is also viable to target upstream suppliers and the supply chains which produce and supply the hardware and software and, in a number of cases, supply the necessary expertise to run and maintain components or systems. Companies supply the hardware, software, and technical experts who install SCADA and Industrial Control Systems for industry although large companies will have in-house technicians and proprietary software. Supply chains were hit particularly hard in 2021–22, including vendors, partners, and third parties.¹⁴⁶

Companies such as Rockwell, Cisco Systems, General Electric (GE), ABB, and Siemens all supply hardware to a variety of industries at a global level. These companies also source components for their hardware, including rare earth elements¹⁴⁷ from a number of states many of whom are developing, are politically or socially unstable, or in conflict zones.¹⁴⁸ This can and will affect supplies. The source code to operate hardware devices and their software applications are easily accessible through the Internet.¹⁴⁹ The most popular code sharing website on the Internet, GitHub, contains much of value to the cybersecurity community and would-be hackers. This includes a hardcoded password list.¹⁵⁰ The 'Dark Net' houses illicit trading forums such as infamous Silk Road/Silk Road 2.0 (best known for selling illegal drugs), Hansa, AlphaBay, Dream Market, Wall Street Market, and Olympus Market. Many get taken down by police and international law enforcement, including the Silk Road and Hansa, but are replaced and continue to provide active marketplaces for trading malware, exploits, ideas, and services.[151]

In addition, owner-operators and companies across the industry could either be unaware of intrusions or breaches or not reporting incidents for fear of reputational loss and adverse effects on their business.[152] This was a prime reason for the introduction of the EU's General Data Protection Regulation (GDPR) in May 2018.[153] Equally disturbing is the issue of complacency. The UK-based Information Assurance Advisory Council (IAAC) reported in June 2014:

> There are currently very few drivers for small companies to work in a secure way for their own ends. Although small companies would like to be secure, they do not see the business case - even issues such as IP [Intellectual Property] protection are only narrowly understood. Many smaller firms work on a month-by-month or - at best - quarterly basis. They face many more urgent short term threats to their business when compared to security breaches. Even larger companies have been seen to take an attitude that 'it does not matter if no one knows'.[154]

This is troubling, with implications for SCADA/Industrial Control Systems architecture and the supply chains they are part of. This was part of the reason for the adoption of the GDPR and the 2016 EU Directive on security of Network and Information Systems (the NIS Directive). There are also several central government and self-help initiatives (including those outlined above) but there is already some concern from Small to Medium-sized Enterprises (SMEs) that "Some companies might focus on compliance with the scheme rather than on security per se".[155] This applies mainly to security in supply chains but could also play a part in boardroom decision-making and an issue that extends globally. McAfee warned in 2011 that "Cyber security investment is made often at the CIO/CISO[156] level as a technology conversation for the technology budget vs. where it really needs to be—at the CEO/CFO[157] level where business risk is assessed. Cyber security is a business risk—if the lights go out, everyone loses money".[158] The economic impact is unlikely to be the only, or the most major, effect.

Conclusion

John Cloonan suggests that "Too many security officers are misled by vendors promoting 'next-generation' firewalls and other 'state-of-the-art' security tools. They don't realize that these 'latest' products are relying exclusively on the decades old signature-based approach to malware detection that will miss evasive malware and zero-day attacks".[159] This means that "No organization with sensitive data or critical operations to protect should be without behavior-based malware detection to augment the capabilities of existing security tools".[160] Whilst "cybercriminals are eager to weaponize both new and old vulnerabilities" the same is true for state intelligence run APTs and cybercriminals. Cybercriminals and corporate or industrial espionage actors can be used as proxies by states seeking access, or to plant false flags which try to mislead and misdirect security professionals, law enforcement, and intelligence agencies like the National Security Agency and Central Intelligence Agency.[161]

Companies, government departments, and other organizations will seek to strengthen network security policies, but this depends on recognition of the threats that exist and cost–benefit analysis. Preventive protection is best practice but in the absence of a known threat or attack it is a difficult judgment to make (potentially heavy) investments in cybersecurity products and services. Lighter protection, especially for start-ups or for SMBs/SMEs in the supply chains of large companies, is looking increasingly vulnerable. Added to this, vulnerability disclosure policies are being driven not by good or best practices but by economic and reputational concerns. This is an area where governments can legislate (as the EU has through the GDPR and NIS Directive).[162]

Protection from highly organized and well-resourced actors such as Russia and China and their highly orchestrated intelligence apparatus remains problematic. State-sponsored attacks combine off-the-shelf tools, custom tools, and 'living-on-the-land' techniques and are patiently carried out. No criminal organization does this, only states. These represent known Advanced Persistent Threats in cyberspace. This is not only apparent at the state versus state level and is part of a much broader series of activities against companies, organizations, and individuals. Arden Bement, from 2004–2010 the Director of the U.S. National Science

Foundation (NSF), cautioned in his 2003 keynote address at a workshop on Critical Infrastructure Protection for SCADA & IT:

> The notion that SCADA systems are highly customized, highly technical, and therefore the guys in the black hats won't be able to figure them out is something Eric Byres of the British Columbia Institute of Technology (BCIT) calls the 'Myth of Obscurity.' Forget it. SCADA documents have been recovered from al Qaeda safe houses in Afghanistan...There is no security through obscurity.[163]

These recovered files included detailed schematics of a dam using professional structural analysis software which could be used to "simulate catastrophic failure". A separate investigation by the FBI in San Francisco and Lawrence Livermore National Laboratory discovered much wider plans to hit critical infrastructure. These included "emergency telephone systems, electrical generation and transmission, water storage and distribution, nuclear power plants and gas facilities".[164] Byres, a leading industry expert on cybersecurity, warned that Stuxnet has already provided "so many advanced attack and malware techniques to the black hat world – basically it gave the technology of malware design a major leap forward. Perhaps more serious, it showed exactly how to build and then use weaponized software against critical control systems".[165] That a major cyberterrorist attack has not emerged is due to successful counterterrorism activities, the defensive barriers in place, and the complexity of attacking. It may also be down to luck because the skills and knowledge are available and the attack surface is vast. Lucky or prudent, counterterrorism needs to be successful all the time, terrorists need only be successful once to perpetrate a high-impact/low-probability 'black swan' event potentially as damaging as 9/11.

Ronald Krutz in his 2005 book *Securing SCADA Systems* detailed a large number of protective measures for SCADA systems and this book is not attempting to reinvent the wheel.[166] Rather it is drawing attention to a wider set of issues which highlight not only issues faced nationally, but globally. In this regard, it is also worth stressing Krutz's valid contention that:

A paradigm shift in thinking is required to address issues such as attacks through the organization's enterprise network, attacks through the Internet, the types of threats, SCADA system vulnerabilities, incident response, application of SCADA security standards, firewall management, disaster recovery, and risk management. In order to raise security awareness, employees have to be informed and educated about various security topics.[167]

To protect critical infrastructure, this paradigm shift should not be restricted to those directly concerned with SCADA and Industrial Control Systems but should be educationally driven (both formally and informally); combine a dual-approach from the top-down (corporate), in partnership with local and central government, and bottom-up and be inclusive of the end user. This requires a partnered response between industry and government, individual awareness, and support at all rungs of the educational ladder and throughout the critical infrastructure. Attack vectors are not only limited to fixed hardware and software. People are widely considered the weakest link in the cybersecurity fence and although this book focuses on critical infrastructure, the attack paths and vulnerabilities also apply to all other areas of cybersecurity; including the defense-industrial base. How and why people as 'end users' are vulnerable to these threats is the subject of Chapter 5, 'Hacking the Human'.

Notes

1. Verizon Data Breach Investigations Report 2022, p. 31, https://www.verizon.com/business/resources/reports/2022/dbir/2022-data-breach-investigations-report-dbir.pdf, accessed 11 June 2022.
2. Cyber Threat and Vulnerability Analysis of the U.S. Electric Sector Prepared by: Mission Support Center.
    Idaho National Laboratory August 2016, p. 4, https://www.energy.gov/sites/prod/files/2017/01/f34/Cyber%20Threat%20and%20Vulnerability%20Analysis%20of%20the%20U.S.%20Electric%20Sector.pdf, accessed 12 August 2019.

3. John Cloonan, 'Advanced Malware Detection—Signatures vs. Behavior Analysis' (11 April 2017), https://www.infosecurity-magazine.com/opinions/malware-detection-signatures/, accessed 19 July 2019. Behavior analysis for the cybersecurity community refers to the behavior of the malware, not the attackers.
4. Nitin Naik, Paul Jenkins, Nick Savage, Longzhi Yang, Kshirasagar Naik, Jingping Song, Tossapon Boongoen and Natthakan Iam-On, 'Fuzzy Hashing Aided Enhanced YARA Rules for Malware Triaging', *2020 IEEE Symposium Series on Computational Intelligence* (SSCI) (Red Hood, NY: Curran Associates/IEEE: 2021), pp. 1138–1145.
5. 'Mandiant APT1 Exposing One of China's Cyber Espionage Units' (2013), pp. 63–65, https://www.fireeye.com/content/dam/fireeye-www/services/pdfs/mandiant-apt1-report.pdf, accessed 17 January 2019.
6. Dan Patterson, 'Gallery: The Top Zero Day Dark Web Markets' (January 9 2017), https://www.techrepublic.com/pictures/gallery-the-top-zero-day-dark-web-markets/, accessed 27 July 2019.
7. 'DMitry', https://tools.kali.org/information-gathering/dmitry, accessed 2 August 2019.
8. 'Shodan is the World's First Search Engine for Internet-Connected Devices', https://www.shodan.io/, accessed 6 September 2018.
9. Cyber Threat and Vulnerability Analysis of the U.S. Electric Sector, p. 13.
10. Cyber Threat and Vulnerability Analysis of the U.S. Electric Sector, p. 13.
11. 2021 Data Breach Investigations Report, http://verizon.com/dbir/, pp. 58–61.
12. Jeff Melnick, 'Top 10 Most Common Types of Cyber Attacks' (May 15 2018), https://blog.netwrix.com/2018/05/15/top-10-most-common-types-of-cyber-attacks/, accessed 18 July 2019.
13. Jason Glassberg, 'What You Need to Know About 'Drive-By' Cyber Attacks' (February 4 2015), https://www.foxbusiness.com/features/what-you-need-to-know-about-drive-by-cyber-attacks, accessed 19 February 2019.
14. For helpful advice see Glen Jackson, 'How to Actively Protect Your Website from Cyberattacks' (22 October 2018), https://www.siliconrepublic.com/enterprise/website-protection-cyberattacks, accessed 19 September 2019.

15. Satyam Singh, 'Practical Scenarios for XSS Attacks' (October 4 2018), https://pentest-tools.com/blog/xss-attacks-practical-scenarios/, accessed 19 July 2019.
16. Kate Brew, 'Watering Hole Attacks: Detecting End-User Compromise before the Damage is Done' (April 26 2016), https://www.alienvault.com/blogs/security-essentials/watering-hole-attacks-detecting-end-user-compromise-before-the-damage-is-done, accessed 18 July 2019.
17. Melnick, 'Top 10 Most Common Types of Cyber Attacks' (May 15 2018).
18. 'What is a Man-in-the-Middle Attack?' (undated), https://us.norton.com/internetsecurity-wifi-what-is-a-man-in-the-middle-attack.html, accessed 15 July 2019. For technical analysis and preventive solutions see Peter Maynard, Kieran McLaughlin and Berthold Haberler, 'Towards Understanding Man-In-The-Middle Attacks on IEC 60,870-5-104 SCADA Networks'. Paper presented at International Symposium for ICS & SCADA Cyber Security Research (ICS-CSR), St. Polten, Austria, (2014), http://www.qub.ac.uk/sites/CSIT/ACEpublications/2016Papers/Filetoupload,734096,en.pdf, accessed 15 July 2019.
19. See for example Austin Hummert, 'Reviewing X Sender Headers: How to Prevent Email Spoofing From Fake Senders' (February 20 2019), https://www.alienvault.com/blogs/security-essentials/how-hackers-manipulate-email-to-defraud-you-and-your-customers, accessed 16 July 2019.
20. Cyber Threat and Vulnerability Analysis of the U.S. Electric Sector, p. 20.
21. 'Zero-Day Vulnerability: What It Is, and How It Works' (undated), https://us.norton.com/internetsecurity-emerging-threats-how-do-zero-day-vulnerabilities-work-30sectech.html, accessed 19 July 2019.
22. Joss Meakins, 'A Zero-Sum Game: The Zero-Day Market in 2018', *Journal of Cyber Policy*, Vol. 4, No. 1 (January 2019), p. 60.
23. 'Zero-Day Vulnerability: What it is, and How it Works'.
24. 'What is a Zero-Day Exploit? Zero-Day Exploit: An Advanced Cyber Attack Defined' (undated), https://www.fireeye.com/current-threats/what-is-a-zero-day-exploit.html, accessed 19 July 2019.
25. Meakins, 'A Zero-Sum Game: The Zero-Day Market in 2018', pp. 61, 65.

26. Dan Patterson, 'Gallery: The Top Zero Day Dark Web Markets' (January 9 2017), https://www.techrepublic.com/pictures/gallery-the-top-zero-day-dark-web-markets/, accessed 27 July 2019. See also Robert Koch, 'Hidden in the Shadow: The Dark Web—A Growing Risk for Military Operations?', in Tomáš Minárik, Siim Alatalu, Stefano Biondi, Massimiliano Signoretti, Ihsan Tolga and Gábor Visky (eds.), *2019 11th International Conference on Cyber Conflict: Silent Battle* (Tallinn: CCD COE Publications, 2019), pp. 267–290.
27. Marleen Weulen Kranenbarg, Thomas J. Holt and Jeroen van der Ham, 'Don't Shoot the Messenger! A Criminological and Computer Science Perspective on Coordinated Vulnerability Disclosure', *Crime Science*, Vol. 7, Issue 16 (December 2018), pp. 1–9.
28. Meakins, 'A Zero-Sum Game: The Zero-Day Market in 2018', p. 62.
29. Meakins, 'A Zero-Sum Game: The Zero-Day Market in 2018', p. 61.
30. Meakins, 'A Zero-Sum Game: The Zero-Day Market in 2018', *Journal of Cyber Policy*, pp. 62–64.
31. Serge Malenkovich, 'What is a Rootkit and How to Remove It' (March 28 2013), https://www.kaspersky.co.uk/blog/rootkit/1508/, accessed 19 July 2019.
32. Malenkovich, 'What is a Rootkit and How to Remove It'.
33. https://www.teamviewer.com/en/solutions/remote-access/#gref, accessed 16 July 2019.
34. 'TeamViewer and Scamming', https://community.teamviewer.com/t5/Knowledge-Base/TeamViewer-and-scamming/ta-p/4715, Olivia Solon, 'What happens if you play along with a Microsoft 'tech support' scam?', https://www.wired.co.uk/article/malwarebytes, both accessed 16 July 2019.
35. Tim Fisher, '14 Free Remote Access Software Tools' (July 12 2019), https://www.lifewire.com/free-remote-access-software-tools-2625161, accessed 16 July 2019.
36. Cyber Threat and Vulnerability Analysis of the U.S. Electric Sector, pp. 13–14, 15. Reference Number: JAR-16-20296A December 29, 2016 Grizzly Steppe—Russian Malicious Cyber Activity, https://www.us-cert.gov/sites/default/files/publications/JAR_16-20296A_GRIZZLY%20STEPPE-2016-1229.pdf, accessed 8 October 2018.
37. Robert Siciliano, 'What is a Backdoor Threat?', https://securingtomorrow.mcafee.com/consumer/identity-protection/backdoor-threat/, accessed 15 July 2019.

38. John Fokker, 'Organizations Leave Backdoors Open to Cheap Remote Desktop Protocol Attacks' (July 11 2018), https://securingtomorrow.mcafee.com/other-blogs/mcafee-labs/organizations-leave-backdoors-open-to-cheap-remote-desktop-protocol-attacks/, accessed 25 July 2019.
39. Fokker, 'Organizations Leave Backdoors Open to Cheap Remote Desktop Protocol Attacks'.
40. Remarks given at Chatham House Conference: Cyber Security Building Resilience Reducing Risk 19 May 2014 attended by author.
41. Verizon 2019 Data Breach Investigation Report, p. 14, https://enterprise.verizon.com/resources/reports/2019-data-breach-investigations-report.pdf, accessed 13 July 2019.
42. Verizon Data Breach Investigations Report 2022, pp. 45–46.
43. 'Keynote Address by Dr. Arden Bement Director, National Institute of Standards & Technology At the NSF Workshop on Critical Infrastructure Protection for SCADA & IT (As Prepared) October 20, 2003', http://www.nist.gov/director/speeches/bement_102003.cfm, accessed 1 August 2015.
44. Further, more detailed, attack pathways can be found in Bonnie Zhu, Anthony Joseph and Shankar Sastry, 'A Taxonomy of Cyber Attacks on SCADA Systems', 2011 IEEE International Conferences on Internet of Things, and Cyber, Physical and Social Computing, pp. 380–388, https://ieeexplore.ieee.org/stamp/stamp.jsp?tp=&arnumber=6142258, accessed 21 July 2019.
45. Ashley Carman, 'Hammertoss Malware Represents Culmination of 'Best Practices' for Cyber Attackers' (July 29 2015), https://www.scmagazine.com/home/security-news/hammertoss-malware-represents-culmination-of-best-practices-for-cyber-attackers/, Naveen Goud, 'Malware Attack Via Twitter' (undated), https://www.cybersecurity-insiders.com/malware-attack-via-twitter/, Kim Crawley, 'Insecure Key Collection on GitHub Is a Dream Come True for Cyber Attackers' (April 1 2019), https://dzone.com/articles/insecure-key-collection-on-github-is-a-dream-come and 'Browse the top used topics on GitHub', https://github.com/topics, all accessed 21 July 2019.
46. Anonymous, 'Daniel Kelley: The Teen Behind the Cybercrime Screen' (10 June 2019), https://www.bbc.co.uk/news/uk-wales-48120428, accessed 17 July 2019.

47. 'What's My IP Address?', https://www.privateinternetaccess.com/pages/whats-my-ip/, and 'Multi-Hop Proxy', https://attack.mitre.org/techniques/T1188/, accessed 3 August 2019.
48. https://www.torproject.org/, accessed 17 July 2021 and Rhyme Upadhyaya and Aruna Jain, 'Cyber Ethics and Cyber Crime: A Deep Delved Study into Legality, Ransomware, Underground Web and Bitcoin Wallet', *2016 International Conference on Computing, Communication and Automation (ICCCA)*, 2016, https://ieeexplore.ieee.org/document/7813706/similar#similar, pp. 143–148.
49. Margaret Rouse, 'Verizon Data Breach Investigations Report (DBIR)' (undated), https://searchsecurity.techtarget.com/definition/Verizon-Data-Breach-Investigations-Report-DBIR, accessed 13 July 2019.
50. Other useful industry reporters include Norton, Cisco, and FireEye.
51. Verizon 2021 Data Breach Investigations Report, pp. 12, 26.
52. Detailed in my books *China and its Embrace of Offensive Cyberespionage* and *Russia's Cyber offensive Against the West*.
53. 2019 Data Breach Investigation Report, p. 55.
54. 2019 Data Breach Investigation Report, pp. 6–7, 10, 13.
55. 2019 Data Breach Investigation Report, p. 19.
56. 2019 Data Breach Investigation Report, p. 20.
57. 2019 Data Breach Investigation Report, p. 25.
58. 'TimeStomp' (undated), https://www.offensive-security.com/metasploit-unleashed/timestomp/, accessed 21 July 2019.
59. 'Tactics, Techniques and Procedures (TTPs)', https://csrc.nist.gov/glossary/term/Tactics-Techniques-and-Procedures, accessed 9 July 2019.
60. 'Tactics, Techniques and Procedures (TTPs) Within Cyber Threat Intelligence' (January 19 2017), https://www.optiv.com/blog/tactics-techniques-and-procedures-ttps-within-cyber-threat-intelligence, accessed 11 July 2019.
61. For SCADA systems, especially legacy systems, this might not be an option for a variety of reasons.
62. 'Verizon's 2016 Data Breach Investigations Report finds cybercriminals are exploiting human nature'.
63. Uchenna P. Daniel Ani, Hongmei (Mary) He and Ashutosh Tiwari, 'Review of Cybersecurity Issues in Industrial Critical Infrastructure: Manufacturing in Perspective', *Journal of Cyber Security Technology*, Vol. 1, Issue 1 (December 2016), pp. 32–74.

64. 'What Is a Firewall?' (updated, https://www.cisco.com/c/en_uk/products/security/firewalls/what-is-a-firewall.html, accessed 21 July 2019.
65. Dhaval Satasiya, Raviya D. Rupal, 'Analysis of Software Defined Network firewall (SDF)', 2016 International Conference on Wireless Communications, Signal Processing and Networking (WiSPNET), https://ieeexplore.ieee.org/abstract/document/7566125/authors#authors, accessed 21 July 2019.
66. Susan Young and Dave Aitel, *The Hacker's Handbook: The Strategy Behind Breaking into and Defending Networks* (Boca Raton: Auerbach, 2003), pp. 246–250.
67. Vidthya Redya, K. Shahu Chatrapati and V. N. Kamalesh, 'Paper on Types of Firewall and Design Principles', *International Journal of Science and Research*, Vol. 6, No. 14 (2015), p. 1587.
68. Redya, Chatrapati and Kamalesh, 'Paper on Types of Firewall and Design Principles', p. 1587.
69. Vivek Saxena, 'Description of the Difference Between HIDs & NIDs' (undated), https://www.techwalla.com/articles/description-of-the-difference-between-hids-nids, accessed 21 July 2019.
70. These include SNORT and OSSEC. 'Open Source IDS Tools: Comparing Suricata, Snort, Bro (Zeek), Linux' (October 26 2018), https://www.alienvault.com/blogs/security-essentials/open-source-intrusion-detection-tools-a-quick-overview, and 'The World's Most Widely Used Host-based Intrusion Detection System Used by tens of thousands of organizations around the world', https://www.ossec.net/, both accessed 6 August 2019.
71. For example, Trend Micro, Cisco, Kaspersky, and McAfee all have competing offerings.
72. Karen Scarfone and Peter Mell, Guide to Intrusion Detection and Prevention Systems (IDPS)(Draft) Recommendations of the National Institute of Standards and Technology, Special Publication 800–94 Revision 1 (Draft) (July 2012), https://csrc.nist.gov/csrc/media/publications/sp/800-94/rev-1/draft/documents/draft_sp800-94-rev1.pdf, accessed 21 July 2019. Ansam Khraisat, Iqbal Gondal, Peter Vamplew & Joarder Kamruzzaman, Survey of intrusion detection systems: techniques, datasets and challenges, *Cybersecurity*, Vol. 2, No. 20 (July 2019), pp. 1–22. Ansam Khrasat and Ammar Alazab, 'A critical review of intrusion detection systems in the internet of things: techniques, deployment strategy, validation strategy, attacks, public

datasets and challenges', *Cybersecurity*, Vol. 4, No. 18 (March 2021), pp. 1–27.
73. https://sansa-stack.net/, https://github.com/SANSA-Stack/SANSA-Stack, both accessed 30 December 2021. Jens Lehmann, Gezim Sejdiu, Lorenz Bühmann, Patrick Westphal, Claus Stadler, Ivan Ermilov, Simon Bin, Nilesh Chakraborty, Muhammad Saleem, Axel-Cyrille Ngonga Ngomo, Hajira Jabeen, 'Distributed Semantic Analytics Using the SANSA Stack', in Claudia d'Amato, Miriam Fernandez, Valentina Tamma, Freddy Lecue, Philippe Cudré-Mauroux, Juan Sequeda, Christoph Lange, and Jeff Heflin (eds.), *The Semantic Web—ISWC 2017 16th International Semantic Web Conference, Vienna, Austria, October 21–25, 2017, Proceedings, Part II* (Cham, Switzerland: Springer, 2017), pp. 147–155.
74. Aaron Zimba, Zhaoshun Wang and Hongsong Chen, 'Multi-Stage Crypto Ransomware Attacks: A New Emerging Cyber Threat to Critical Infrastructure and Industrial Control Systems', *ICT Express*, Vol. 4, Issue 1 (March 2018), pp. 14–18.
75. 'What is a Honeypot? How It Can Lure Cyberattackers' (undated), https://us.norton.com/internetsecurity-iot-what-is-a-honeypot.html. See also 'What is a Honey Pot?' (September 15 2018), https://resources.infosecinstitute.com/what-is-a-honey-pot/#gref, both accessed 11 July 2019.
76. Ahmed Abbasi, Weifeng Li, Victor Benjamin, Shiyu Hu and Hsinchun Chen, 'Descriptive Analytics: Examining Expert Hackers in Web Forums', 2014 IEEE Joint Intelligence and Security Informatics Conference, pp. 56–63, https://ieeexplore.ieee.org/stamp/stamp.jsp?arnumber=6975554, accessed 21 July 2019.
77. Andrew Weisburd, Clint Watts, and J. M. Berger, 'Trolling for Trump: How Russia Is Trying to Destroy Our Democracy' (November 6 2016), https://warontherocks.com/2016/11/trolling-for-trump-how-russia-is-trying-to-destroy-our-democracy/, accessed 26 October 2019.
78. Cloonan, 'Advanced Malware Detection—Signatures vs. Behavior Analysis.'
79. Cloonan, 'Advanced Malware Detection—Signatures vs. Behavior Analysis.'
80. Nate Lord, 'What is Polymorphic Malware? A Definition and Best Practices for Defending Against Polymorphic Malware' (September 11

2018), https://digitalguardian.com/blog/what-polymorphic-malware-definition-and-best-practices-defending-against-polymorphic-malware, accessed 19 July 2019.
81. Yiyi Miao, 'Understanding Heuristic-based Scanning vs. Sandboxing' (July 13 2015), https://www.opswat.com/blog/understanding-heuristic-based-scanning-vs-sandboxing, accessed 19 July 2019.
82. Cloonan, 'Advanced Malware Detection—Signatures vs. Behavior Analysis.'
83. Cisco White Paper, Cisco Advanced Malware Protection Sandboxing Capabilities, p. 3, https://www.cisco.com/c/en/us/products/collateral/security/whitepaper_c78-733277.pdf, accessed 19 July 2019.
84. Cloonan, 'Advanced Malware Detection—Signatures vs. Behavior Analysis.'
85. https://www.wireshark.org/, accessed 16 September 2016.
86. 'What is a Packet Sniffer?', http://netsecurity.about.com/od/informationresources/a/What-Is-A-Packet-Sniffer.htm, accessed 16 September 2016.
87. 'Implementing Application Whitelisting', https://www.cert.govt.nz/it-specialists/critical-controls/application-whitelisting/implementing-application-whitelisting/, accessed 17 July 2019.
88. This is illustrated in Symantec's Internet Security Threat Report of July 2017, Living off the land Defining fileless attack methods, https://www.symantec.com/content/dam/symantec/docs/security-center/white-papers/istr-living-off-the-land-and-fileless-attack-techniques-en.pdf, accessed 17 July 2019.
89. Diana Rose Brandon, 'Getting Serious About Security Breaches with Endpoint Protection' (25 October 2016), https://www.insight.com/en_US/learn/content/2016/10252016-getting-serious-about-endpoint-security.html, accessed 19 July 2019.
90. 'A detailed description of the Data Execution Prevention (DEP) feature in Windows XP Service Pack 2, Windows XP Tablet PC Edition 2005, and Windows Server 2003' (11 July 2017), https://support.microsoft.com/en-gb/help/875352/a-detailed-description-of-the-data-execution-prevention-dep-feature-in, accessed 21 July 2019.
91. 'How To use Software Restriction Policies in Windows Server 2003' (17 April 2018), https://support.microsoft.com/en-gb/help/324036/how-to-use-software-restriction-policies-in-windows-server-2003, accessed 21 July 2019.

92. 'AppLocker' (16 October 2017), https://docs.microsoft.com/en-us/windows/security/threat-protection/windows-defender-application-control/applocker/applocker-overview, accessed 21 July 2019.
93. Eric Knapp, 'SCADA Mischief Episode 2: Context and Correlation' (March 6 2012), https://www.securityweek.com/scada-mischief-episode-2-context-and-correlation, accessed 21 July 2019.
94. Tim Ferguson, 'SCADA and ICS: Combating the Security Risk' (17 January 2018), https://threatmanagement.info/scada_and_ics_combating_the_security_risk/, accessed 21 July 2019.
95. 'Windows Management Instrumentation' (31 May 2018), https://docs.microsoft.com/en-us/windows/win32/wmisdk/wmi-start-page, accessed 21 July 2019.
96. 'Windows Management Instrumentation'.
97. 'Windows PowerShell' (July 8 2013), https://msdn.microsoft.com/library/dd835506.aspx, Mathew Dunwoody, 'Dissecting One of APT29's Fileless WMI and PowerShell Backdoors (POSHSPY)' (April 3 2017), https://www.fireeye.com/blog/threat-research/2017/03/dissecting_one_ofap.html and Ionut Arghire, 'Researchers Dissect PowerShell Scripts Used by Russia-Linked Hackers' (May 31 2019), https://www.securityweek.com/researchers-dissect-powershell-scripts-used-russia-linked-hackers, all accessed 21 July 2019.
98. Bernard Marr, 'A Very Brief History Of Blockchain Technology Everyone Should Read' (February 16 2018), https://www.forbes.com/sites/bernardmarr/2018/02/16/a-very-brief-history-of-blockchain-technology-everyone-should-read/#17ef6b77bc47, accessed 27 July 2019.
99. Marco Iansiti and Karim R. Lakhani, 'The Truth About Blockchain', *Harvard Business Review*, January–February 2017, p. 4, https://hbr.org/2017/01/the-truth-about-blockchain, accessed 27 July 2017.
100. 'Rewire Your Industry with IBM Blockchain', https://www.ibm.com/uk-en/blockchain/industries and Leon Cosgrove, 'How will Blockchain technology change the mining industry' (undated), https://www.wipro.com/natural-resources/how-will-blockchain-technology-change-the-mining-industry/, both accessed 27 July 2019.
101. Jordan Tuwiner, 'What is Bitcoin Mining and How Does it Work?' (February 8 2019), https://www.buybitcoinworldwide.com/mining/, accessed 27 July 2019.
102. Savaram Ravindra, 'The Role of Blockchain in Cybersecurity' (8 January 2018), https://www.infosecurity-magazine.com/next-gen-infosec/blockchain-cybersecurity/, accessed 27 July 2019.

103. Iansiti and Lakhani, 'The Truth About Blockchain', p. 4.
104. Zhu, Joseph and Sastry, 'A Taxonomy of Cyber Attacks on SCADA Systems', p. 384.
105. Zhu, Joseph and Sastry, 'A Taxonomy of Cyber Attacks on SCADA Systems', p. 384.
106. Andy Greenberg, 'Mind the Gap: This Researcher Steals Data With Noise, Light, and Magnets' (February 7 2018), https://www.wired.com/story/air-gap-researcher-mordechai-guri/, accessed 8 June 2022.
107. 'Buffer Overflow', https://www.enisa.europa.eu/topics/csirts-in-europe/glossary/buffer-overflow, accessed 21 July 2019.
108. Zhu, Joseph and Sastry, 'A Taxonomy of Cyber Attacks on SCADA Systems', p. 381.
109. Zhu, Joseph and Sastry, 'A Taxonomy of Cyber Attacks on SCADA Systems', pp. 380–388.
110. National Security Agency, 'Embracing a Zero Trust Security Model' (February 2021), https://media.defense.gov/2021/Feb/25/2002588479/-1/-1/0/CSI_EMBRACING_ZT_SECURITY_MODEL_UOO115131-21.PDF, p. 1.
111. Helge Janicke, Antonio Cau, François Siewe, Hussein Zedan, and Kevin Jones, 'A Compositional Event & Time-Based Policy Model', *Seventh IEEE International Workshop on Policies for Distributed Systems and Networks (POLICY'06)*, 2006, pp. 173–182, Helge Janicke, François Siewe, Kevin Jones, Antonio Cau, and Hussein Zedan, 'Analysis and Run-Time Verification of Dynamic Security Policies', in Simon G. Thompson and Robert Ghanea-Hercock (eds.), *Defence Applications of Multi-Agent Systems* (DAMAS 2005) (Berlin: Springer, 2006), pp. 92–103.
112. A. Nicholson, S. Webber, S. Dyer, T. Patel and H. Janicke, 'SCADA Security in the Light of Cyber-Warfare', *Computers & Security*, Volume 31, Issue 4 (June 2012), pp. 418–436.
113. Ronald L. Krutz, *Securing SCADA Systems* (Indianapolis, IN: Wiley, 2006), pp. 43–72. See also Shaw, *Cybersecurity for SCADA Systems*, p. 6.
114. Roadmap to Secure Control Systems in the Water Sector (March 2008) Developed by March 2008 Water Sector Coordinating Council Cyber Security Working Group Sponsored by American Water Works Association/Department for Homeland Security, pp. 13–15, https://www.n-dimension.com/wp-content/uploads/NDSI-WATER-CybersecurityRoadmap08-1.pdf, accessed 29 March 2020.

115. On the problem of patching SCADA systems see 'Window of Exposure… A Real Problem for SCADA Systems? Recommendations for Europe on SCADA patching', ENISA, December 2013, https://www.enisa.europa.eu/publications/window-of-exposure-a-real-problem-for-scada-systems, accessed 29 March 2020.
116. Annarita Giani, Eilyan Bitar, Manuel Garcia, Miles McQueen, Pramod Khargonekar and Kameshwar Poolla, 'Smart Grid Data Integrity Attacks: Characterizations and Countermeasures', Second International Conference On Smart Grid Communications (October 2011). Available from http://www5vip.inl.gov/technicalpublications/Documents/5250261.pdf, accessed 20 August 2015.
117. Cyber Threat and Vulnerability Analysis of the U.S. Electric Sector, p. ii.
118. P.W. Singer and Allan Friedman, *Cybersecurity and Cyberwar What Everyone Needs to Know* (Oxford: Oxford University Press, 2014), pp. 130–131.
119. Joint Publication 3–12 Cyberspace Operations Joint Chiefs of Staff (8 June 2018), p. II-13, https://www.jcs.mil/Portals/36/Documents/Doctrine/pubs/jp3_12.pdf?ver=2018-07-16-134954-150, accessed 8 March 2020.
120. A good analysis of these issues from the point of view of U.S. municipal authorities is provided by Ginger Armbruster, Barbara Endicott-Popovsky and Jan Whittington, 'Threats to Municipal Information Systems Posed by Aging Infrastructure', *International Journal of Critical Infrastructure Protection*, Vol. 6, Issues 3–4 (December 2013), pp. 123–131.
121. Eric Luiijf, 'Why Are We So Unconsciously Insecure?', *International Journal of Critical Infrastructure Protection*, Vol. 6, Issues 3–4 (December 2013), pp. 179–180.
122. Mark Ward, 'How to Hack a Nation's Infrastructure' (May 20 2013), https://www.bbc.co.uk/news/technology-22524274, accessed 15 July 2014.
123. Adam A. Creery and Eric J. Byres, 'Industrial Cybersecurity for Power System and SCADA Networks', Proceedings of the IEEE Petroleum and Chemical Industries Conference, Institute of Electrical and Electronics Engineers, Denver, September 2007, p. 3, https://www.tofinosecurity.com/professional/industrial-cybersecurity-power-system-and-scada-networks, accessed 29 March 2020..

124. Zhu, Joseph and Sastry, 'A Taxonomy of Cyber Attacks on SCADA Systems', p. 381.
125. 'Exelis awarded US Air Force contract to upgrade Strategic Automated Command Control System digital memory technology' (December 3 2014), https://www.harris.com/press-releases/2014/12/exelis-awarded-us-air-force-contract-to-upgrade-strategic-automated-command, accessed 24 July 2019.
126. Jack Moore, 'The application programming languages undergirding some federal IT systems were new when "The Andy Griffith Show" premiered' (May 25 2016), https://www.nextgov.com/cio-briefing/2016/05/10-oldest-it-systems-federal-government/128599/, accessed 4 September 2018.
127. Sandra Erwin, 'STRATCOM to Design Blueprint for Nuclear Command, Control and Communications' (March 29 2019), https://spacenews.com/stratcom-to-design-blueprint-for-nuclear-command-control-and-communications/, accessed 24 July 2019.
128. Erwin, 'STRATCOM to Design Blueprint for Nuclear Command, Control and Communications'.
129. Page O. Stoutland and Samantha Pitts-Kiefer (Foreword by Ernest J. Moniz, Sam Nunn, and Des Browne), 'Nuclear Weapons in the New Cyber Age Report of the Cyber Nuclear Weapons Study Group' (September 2018), https://media.nti.org/documents/Cyber_report_finalsmall.pdf, Sico van der Meer, 'Cyber Warfare and Nuclear Weapons: Game-changing Consequences?', in Oliver Meier and Elisabeth Suh (eds.), Reviving nuclear disarmament. Paths towards a joint enterprise, SWP Working Paper, FG03 No. 6, (December 2016), https://www.swp-berlin.org/en/publication/reviving-nuclear-disarmament/, pp. 37–38.
130. Andrew Futter, *Hacking the Bomb: Cyber Threats and Nuclear Weapons* (Washington DC: Georgetown University Press, 2018). Herbert Lin, *Cyber Threats and Nuclear Weapons* (Stanford: Stanford University Press, 2021). See also 'Cyber Threats and Nuclear Weapons: A Book Talk with Professor Herb Lin' (6 December 2021), https://www.youtube.com/watch?v=TvZvh6IbWWY, accessed 6 December 2021. See also Eva Cohn, Caitlin Listek, Hannah Schuller, and Celeste Travis, 'Reforming the U.S. Nuclear Enterprise to Account for Emerging Cyber Threats', Report prepared for the Nuclear Threat Initiative, May 2022.

131. GAO-19–471 Report to Congressional Requesters Information Technology Agencies Need to Develop Modernization Plans for Critical Legacy Systems, June 2019, pp. 7–8, 13–14, https://www.gao.gov/assets/700/699616.pdf, accessed 24 July 2019.
132. GAO-19–471 Report to Congressional Requesters, p. i.
133. GAO-19–471 Report to Congressional Requesters, pp. i, 40.
134. GAO-19–471 Report to Congressional Requesters p. 48.
135. GAO-19–471 Report to Congressional Requesters p. 52.
136. GAO-19–471 Report to Congressional Requesters, pp. 1, 4–5.
137. GAO-19–471 Report to Congressional Requesters, pp. 5, 19.
138. Moore, "The Andy Griffith Show".
139. Bipin Patel, 'Legacy Systems are Problems for Boardrooms Not Computer Geeks' (February 1 2017), https://www.ft.com/content/5bf9de84-d665-11e6-944b-e7eb37a6aa8e, accessed 6 September 2018.
140. Patel, 'Legacy Systems are Problems for Boardrooms not Computer Geeks'.
141. Patel, 'Legacy Systems are Problems for Boardrooms Not Computer Geeks'.
142. http://windows.microsoft.com/en-gb/windows/end-support-help, accessed 27 July 2014.
143. Windows XP has been widely pirated in China with 70% of these never patched and this led to high-level discussions with the Chinese authorities over continued support for the operating system. See for example, Michael Kan, 'Windows XP Will Continue Receiving Security Support in China' (March 3 2014), http://www.pcworld.com/article/2103680/chinas-windows-xp-users-to-still-get-security-support.html and Mark Ward, 'XP—The Operating System That Will Not Die' (5 March 2014), http://www.bbc.co.uk/news/technology-26432473 both accessed 27 July 2014 and 'Operating System Replacing Windows Given Go-ahead in Russia' (May 29 2019), https://russiabusinesstoday.com/featured/operating-system-replacing-windows-given-go-ahead-in-russia/, accessed 26 July 2019.
144. Kelly Jackson Higgins, 'Windows XP Alive & Well in ICS/SCADA Networks' (April 10 2014), http://www.darkreading.com/informationweek-home/windows-xp-alive-and-well-in-ics-scada-networks/d/d-id/1204385, accessed 27 July 2014.
145. Michael Mimoso, 'Patching Bash Vulnerability a Challenge for ICS, SCADA' (September 25 2014),

http://threatpost.com/patching-bash-vulnerability-a-challenge-for-ics-scada, accessed 26 September 2014.
146. Verizon Data Breach Investigations Report 2022, pp. 29–30.
147. 'What are 'Rare Earths' Used For?' (March 13 2012), http://www.bbc.co.uk/news/world-17357863, accessed 29 September 2014.
148. Joe Miller, 'Intel Vows to Stop Using 'Conflict Minerals' in New Chips', http://www.bbc.co.uk/news/technology-25636001, accessed 29 September 2014.
149. Sandra Keen, 'Open Source Software Exposes ICS Device Vulnerabilities to Hackers' (August 14 2018), https://www.nozominetworks.com/blog/open-source-software-exposes-ics-device-vulnerabilities-to-hackers/ and Warwick Ashford, 'Cyber Attack Warnings Highlight Need to Be Prepared' (14 June 2018), https://www.computerweekly.com/news/252443085/Cyber-attack-warnings-highlight-need-to-be-prepared, both accessed 26 July 2019.
150. https://github.com/hslatman/awesome-industrial-control-system-security, https://github.com/rmusser01/Infosec_Reference/blob/master/Draft/SCADA.md, both accessed 26 January 2019.
151. McAfee Labs Threats Report (December 2018), pp. 1–9, https://www.mcafee.com/enterprise/en-us/assets/reports/rp-quarterly-threats-dec-2018.pdf, and Andy Greenberg, 'Operation Bayonet: Inside the Sting That Hijacked an Entire Dark Web Drug Market Dutch police detail for the first time how they secretly hijacked Hansa, Europe's Most Popular Dark Web Market' (August 3 2018), https://www.wired.com/story/hansa-dutch-police-sting-operation/?verso=true, both accessed 25 July 2019.
152. Views expressed at 2nd International Symposium for ICS & SCADA Cyber Security, 11th-12th September 2014 at University of Applied Sciences St. Pölten, Austria, attended by author.
153. https://eugdpr.org/, accessed 9 November 2018.
154. Information Assurance Advisory Council, 'SMEs and the supply chain White Paper, June 2014' (author copy).
155. The Directive on security of network and information systems (NIS Directive), https://ec.europa.eu/digital-single-market/en/network-and-information-security-nis-directive, accessed 9 November 2018.
156. Chief information officer/chief information security officer.
157. Chief Executive Officer/Chief Financial Officer.

158. 'In the Dark: Crucial Industries Confront Cyberattacks' (April 19 2011), https://www.mcafee.com/blogs/enterprise/in-the-dark-crucial-industries-confront-cyberattacks/, accessed 26 July 2019.
159. Cloonan, 'Advanced Malware Detection—Signatures vs. Behavior Analysis.'
160. Cloonan, 'Advanced Malware Detection—Signatures vs. Behavior Analysis.'
161. McAfee Labs Threats Report (December 2018), p. 5.
162. Meakins, 'A Zero-Sum Game: The Zero-Day Market in 2018', pp. 67–68.
163. Quoted in Krutz, *Securing SCADA Systems*, p. 75.
164. Barton Gellman, 'Cyber-Attacks by Al Qaeda Feared' (June 27 2002), https://www.washingtonpost.com/archive/politics/2002/06/27/cyber-attacks-by-al-qaeda-feared/5d9d6b05-fe79-432f-8245-7c8e9bb45813/, accessed 19 February 2020.
165. Bill Lydon, 'Cyber Security Threats: Expert Interview with Eric Byres, Part 1', http://www.automation.com/automation-news/article/cyber-security-threats-expert-interview-with-eric-byres-part-1, accessed 28 September 2014.
166. Krutz, *Securing SCADA Systems*, pp. 89–108.
167. Krutz, *Securing SCADA Systems*, p. 106.

5

Hacking the Human

Introduction

This chapter will follow a three-step process which will specifically be geared in the direction of SCADA-ICS and threats to critical infrastructure (CI) but has wider applicability. The first step will analyze social engineering (SE) methods and practices. These have been embraced by criminals and criminal gangs and lone wolf hackers/crackers to gain access to information or systems and who can be used by nation states as deniable 'privateers'. The second step will engage with the world of professional nation-state espionage and their intelligence agencies in order to set a wider context to social engineering practices and applications. The third will analyze insider threats, physical security, and mitigation methods bearing in mind the global skills gap in ICT with people widely recognized as the weakest link in the cyber security fence—a link that might never be fully remedied or reconciled.[1]

Social Engineering

Social engineering is an established series of techniques that exploit human psychology, rather than technical hacking, to gain access to individual computers, internal systems, and organizations, the data they hold, and is also used to access physical sites and buildings. Most commonly, it is used remotely. For example, instead of finding or using a technical vulnerability, a social engineer might pose as an IT support technician via e-mail (a technique called phishing) and try to trick someone into divulging their password. Telephone scams are also a common way for organized crime gangs to extort money and social engineering is the primary techniques used by scam call centers. These frequently prey on scare tactics and a lack of technical knowledge. This sees "Americans lose around $1.5bn to tech-support scams every year" with 86% of these from India.[2] One of the most widely used ruses begins with "I am calling from Microsoft technical support".[3]

People are exploited by simple tactics such as this and even wary and technically informed professionals are duped by highly skilled activities of state intelligence agencies and industrial/corporate espionage to gain access into systems, organizations, and facilities. For all the technical measures in place, the leading cyber security firm FireEye-Mandiant has remarked that "All cyber attacks involve a human adversary". The most organized of these are often "groups of people under the same organizational umbrella, with multiple teams of people assigned to specific tasks as part of a common mission. Because attackers are living, breathing people — not pieces of mindless code — they are motivated, organized, and unpredictable".[4]

People are often the first line of attack and "when phishing attacks slip through network perimeters, people become the last line of defense".[5] According to Verizon's 2021 Data Breach Investigations Report (DBIR), phishing was the origin of the vast majority of breaches with cloud-based e-mail servers in particular a "target of choice". Business e-mail compromises (BECs) were "the second most common form of Social Engineering".[6] These are where an attacker gains access to a corporate e-mail account and uses that account to defraud a company or target its employees, customers, or partners. Also referred to as 'man-in-the email' attacks, BECs are mainly used for monetary gain.[7] More widely, Verizon recorded "a jump in Social Engineering breaches as a pattern from last year with an overall upward trend since 2017".[8] According to the 2022 DBIR, "roughly 4 in 5 breaches can be attributed to organized crime" 82% were down to people (again due largely to phishing, stolen credentials, misuse, and human error). 80% of breaches were external and aimed at monetary gain whilst espionage was second on the list at around 15%.[9]

Although only 2.9% of employees click on phishing e-mails (a percentage that is relatively stable), "that is still more than enough for criminals to continue to use it".[10] More broadly, people remain susceptible to cybercriminal hackers and by the professional activities of state intelligence agencies and through industrial/corporate espionage actors who engage in SE. At the state-on-state level, this was highlighted by Russia's 'Dragonfly' campaign, where "Authorized internal personnel initiated actions in each phase of the attack, using components that were obtained from trusted sources and "assumed" authentic. These "insiders" could have been staff at the target companies or third-parties and subcontractors providing maintenance under service level agreements (SLA)".[11]

SE uses Open Source Intelligence (OSINT) gathering which is used to profile individuals and organizations and tailor an attack to them. Against private sector owner-operators of CI, Google and company websites are a useful starting point as is LinkedIn and social media accounts. OSINT and SE can also be used to help understand and produce an organizational chart of a business if one is not publicly available. This helps understand the structure of an organization and the relationships and relative ranks of its parts and positions. OSINT and SE can also help find business connections within and between companies not listed on organizational charts.

SE is also increasingly used to target individuals to extort money from them or, more commonly, the organizations they work for as part of a criminal enterprise/organized criminal gangs. This usually involves 'scareware' and (increasingly) 'ransomware'. It has long been realized that rather than attack organizations directly it might well be easier, more cost-effective, and more efficient to instead co-opt unwitting insiders to gain access to internal networks. These practices encompass all major industries including individuals and organizations in the SCADA-ICS-CI ecosystem.

SE is a successful methodology to exploit individuals and a method to gain access to systems. This has also been honed over time by state intelligence agencies and in corporate espionage activities which use mature and well-resourced methods and often highly trained human assets to access targets. SE is a practice used by states, state-based corporate espionage, by insiders such as moles (an agent of one organization sent to penetrate another organization, including intelligence agencies, by gaining employment—where counterintelligence tries to catch and expose them), and external hackers. Each of these could be acting on behalf of a foreign government.[12] It remains a very effective method which does not necessarily require high-level knowledge of code or computer systems. As a SANS Institute paper simply puts it, "Social engineering exploits a number of human traits and tendencies; with the goal of inducing the target to provide information or access that otherwise is not available".[13]

The widespread use of 'social engineering' by the hacking and cracking communities has been widely studied in the discipline of computer science and by the wider cybersecurity community. It is a sub-field of professional Human Intelligence (HUMINT) and the tradecraft of espionage through recruiting spies, running sources, and related espionage activities. HUMINT plays a vital role in the intelligence cycle—a cycle of collection and analysis that has matured over considerable time. HUMINT and espionage are not synonymous, however. HUMINT has now developed:

> into a complex discipline, richly documented by collective experience, taken partly from social and behavioural studies (regarding issues of understanding and anticipation of human behaviour, and even as part of agent's recruiting and running methodologies, investigation, etc.) and partially from exact sciences (in relation to intelligence technologies based on encryption, and later on the SIGINT intelligence etc.). Countries such as Great Britain, the United States and the Soviet Union carried out intelligence research, documenting issues such as recruitment, running

methodologies, intelligence analysis, as well as teaching the subject in intelligence schools and academies.[14]

Adjunctly, the term CYBERINT has now entered the lexicon as cyberspace has become a domain for intelligence agencies and the fifth domain of warfare. However, for all the lines of code and technology it is a socially constructed domain where humans interact. Bryan Sartin of Verizon recorded in the foreword to their 2016 DBIR, "Despite advances in information security research and cyber detection solutions and tools, we continue to see many of the same errors we've known about for more than a decade now. How do you reconcile that?"[15] A significant part of the explanation is human nature and social behavior.

Examples of Social Engineering

There is a growing literature on SE. Much of this describes the various techniques being employed. These offer case examples and mitigation measures against social engineering attacks.[16] These include Internet-based examples drawn from industry professionals and hackers alongside contributions from computer scientists in mainstream academia. SE encompasses human, cognitive, and social psychology. The way we are wired (human nature) and associate as social actors with others means successful and well-orchestrated social engineering can be extremely difficult if not impossible to defeat all the time. Our psychosocial and psychological tendencies and triggers will not always ring alarm bells or make us act on red flags, especially when we are acting alone. Awareness building and training are positive first steps, but SE-based attacks continue to be successfully used. The most common of these are:

- **Pretexting**: Pretexting is an example of a 'confidence trick'. At a base level it is simply someone pretending to be someone they are not or inventing a scenario designed to give access rights such as passwords or to volunteer confidential details.

- **Shoulder surfing**: This is little more than looking over the shoulder of an unsuspecting target to observe or record their computer interaction and can include observing passwords.
- **Dumpster diving**: This involves rooting for hardware or information that has been discarded but contains valuable data such as names, addresses, telephone numbers, bank and credit card statements etc. This requires an attacker or affiliate to be in physical proximity to the target.
- **Forensic recovery** through the analysis of non-securely disposed devices including storage media such as hard drives. Again, this requires physical proximity.
- **Baiting**: This is a practice of placing an infected storage device, such as a USB stick, in an area where it is likely to be picked up and plugged into a computer. This is purported to be one of the methods used to get the Stuxnet worm into the Natanz nuclear facility in Iran.
- **Tailgating** involves following someone in order to attempt to access or assess secured locations.
- **Diversion theft** is used to redirect couriers to deliver to locations chosen by the attacker.
- **Quid pro quo**: A 'quid pro quo' attack is the offer of something in exchange for information—for example, an unsolicited contact to fix an issue (i.e., someone purporting to be from 'technical support' enticing the target to provide login details to a device to fix a system-related problem). The attacker will seek to extract identifying information such as a username or password to initiate the attack.
- **Impersonation**: Much like quid pro quo and tailgating, this is a widespread technique used to deceive people. Widely used by confidence tricksters and fraudsters, the most common means against critical infrastructure are posing as an employee of the same organization. Posing as a colleague from an external organization can also be used. In a physical environment it might appear harmless to tell someone who appears to be lost where the computer room is located or to let someone into the building who 'forgot' their security credentials or appear to be visitors doing legitimate business. This might be principally successful in companies used to high volume visitors or with high staff turnover.

- **Phishing**: Directing an end user via e-mail to click on an attachment or link hosting malware designed to extract information.
- **Spear Phishing**: is an e-mail or electronic communications scam tailored towards a specific individual and/or organization with the aim of defrauding the target, extracting information for malicious purpose, to install malware, and/or to provide a gateway for deeper activity towards individuals and their associates or organizations. SE is being increasingly used in spear phishing.
- **Vishing**, is where a fraudster requires Personal Identification Numbers to be entered into an automated system or (in other cases) provided to a human operator/fraudster after first soliciting the contact through e-mail or telephone calls.
- **Smishing** uses similar techniques but does so via SMS text message.[17] With an increased blending of work, home, and social life and the ubiquity of SMART devices and bring-your-own-device policies (where people are permitted to bring their own devices to work and connect to networks) smishing is a growing concern and target vector.[18]

Exploiting Cognitive and Behavioral Psychology

Many of the above activities can be exploited or enhanced by confidence tricks and techniques used by both 'amateur' social engineers and intelligence professionals which exploit cognitive or behavioral psychology. These include 'foot-in-the-door' approaches. These involve asking a series of ostensibly innocuous requests and, if granted, moving on from there. This is what the script kiddie group, 'Crackas With Attitude', did when gaining access to the personal e-mail accounts of the Director of the CIA and the Director of National Intelligence (detailed in Chapter 6). Cognitive and behavioral psychology plays an important part in these exchanges and can be exploited. This includes posing as an authority figure or trusted insider.

This approach is perhaps best demonstrated by the famous experiment conducted by the social psychologist Stanley Milgram at Yale University in 1961 (an experiment repeated many times around the world with similar results). Milgram's experiment found that people have an innate disposition to obey authority figures either out of fear or to be cooperative—even when this contradicts with their principles or judgment. During his career Milgram came to believe that "Obedience is...[a] basic...element in the structure of social life".[19] Milgram's experiment:

> consists of ordering a naive S[ubject] to administer increasingly more severe punishment to a victim in the context of a learning experiment. Punishment is administered by means of a shock generator with 30 graded switches ranging from Slight Shock to Danger: Severe Shock. The victim is a confederate of the E[xperiment]. The primary dependent variable is the maximum shock the S[ubject] is willing to administer before he refuses to continue further. 26S[ubject]s obeyed the experimental commands fully, and administered the highest shock on the generator. 14Ss broke off the experiment at some point after the victim protested and refused to provide further answers.[20]

What these subjects heard ranged from audible discomfort, to pained cries, and finally silence. Related to obedience is the social psychological phenomenon of the 'diffusion of responsibility' whereby decisions that involve risks and payoffs by individuals belonging to a group mean people feel less responsible for actions that occur individually outside of the group context. The most well-known example being Philip Zimbardo's Stanford Prison Experiment which saw students take roles as prisoners or guards and immersing themselves in these roles to such a degree that the experiment was brought to a premature conclusion.[21]

Ingratiation and trust building also can be used as psychological and behavioral triggers for 'social engineers'. Being seen to be helpful or unhelpful and associations with moral guilt and ethics also play a part in SE. The same is true for impressions of competency or incompetency and tendencies towards 'benefit of the doubt' thinking. This plays into scenarios of cost–benefit-opportunity/opportunism/advancement and job security/insecurity. This bears on psychological and behavioral issues of acceptance, compliance, values, and norms. This also involves thinking about organizational desirables like bonding and team building.

Exploitation of these cognitive and behavioral psychological characteristics is also apparent in the work of performers, magicians, and illusionists. For example the British TV hypnotist and conjuror Derren Brown, a self-styled practitioner of "the art of psychological manipulation", has "convinced middle-managers to commit an armed robbery in the street, led the [British] nation in a seance, stuck viewers to their sofas, successfully predicted the National Lottery, motivated a shy man to land a packed passenger plane at 30,000 feet, exposed psychic and faith-healing charlatans, and hypnotised a man to assassinate Stephen Fry".[22]

Brown has written four books on psychological manipulation: *Absolute Magic*, *Pure Effect*, *Tricks of the Mind*, and *Confessions of a Conjuror*, as well as appearing in a number of television series and stage shows.[23] Brown employs techniques derived from hypnosis, such as the ideomotor principle which triggers neurologically driven motor reflexes that can be interpreted or 'read' and in so doing indicating and influencing thoughts and actions in people.[24] These are techniques that are used and studied by others. This includes Ian Mann who explicitly refers to Brown in his 2008 book *Hacking the Human: Social Engineering Techniques and Security Countermeasures*.[25]

Brown and many others also employ so-called Neuro-Linguistic Programming (NLP). NLP originated in the work of Richard Bandler and John Grinder in the United States in the 1970s and its use in psychotherapy to cure phobias and other psychological issues.[26] It has been described by many people in both mainstream comment and academia as a pseudo-science lacking in empirical evidence. Nevertheless, it has taken root and "is a popular form of inter-personal skill and communication training…NLP is big business with large numbers of training courses, personal development programmes, therapeutic and educational interventions purporting to be based on the principles of NLP".[27] Although NLP is difficult to pin down Paul Tosey and Jane Mathison describe it as follows:

NLP makes a number of claims about the ways in which communication can influence people. The meta-model is essentially a model of the relationship between language, and how people have constructed information about events at a level of which they may not be aware. In selling, NLP is used to influence people's views about products and services. Ericksonian language patterns [the use of storytelling and vocal tonality],[28] which are an integral part of the armoury of NLP practitioners, are claimed to enable the user to bypass the listener's conscious, analytical mind, and exert an unconscious influence….[and] there are worrying reports about how NLP is used…for example it has been linked to teaching people seduction techniques, and to promises of gaining power, earning a fortune, or easily changing people's lives.[29]

Each of these techniques plays to ideas of suggestion and suggestibility.

Hacking the Human

Further highlighting some of the methodologies being used is the case of Ian Hurst, a former member of the British Army Intelligence Corps and Force Research Unit (FRU) who between 1980 and 1991 recruited and ran agents within Irish republican terrorist groups. It was revealed to him by the BBC that his computer had been hacked by an individual paid by the now defunct British Sunday tabloid *The News of the World*. This "individual [was] known to me, having served with him in the intelligence community in Northern Ireland for over 3 years whilst he was attached to the FRU".[30] Hurst went on to detail the intelligence gathering methods used against him.

> [H]e had placed a computer Trojan on my hard drive (by sending me an email from a bogus address which I then opened) and had, over a three month period, obtained all email traffic which was sent and received by me. I now cannot recall whether I opened an email from a bogus address in 2006. However, I can say that it would have been much more likely to have opened an email that had come from a trusted contact rather than an unfamiliar or unknown name. I do know that an email was sent to me by a trusted media contact within Times Newspapers Ltd

around the time the Trojan became active and was collecting information from my computer. I have recently seen evidence that this individual was in email contact with X [later identified as Philip Campbell Smith—a former member of the FRU who became a private investigator] regarding my activities during these months, something which had previously been denied.[31]

Hurst believed the newspaper wanted to obtain information on an Irish Republican Army informer and alleged that they had cultivated links with individuals in London's Metropolitan Police to further their ends. Ranged against the 'black hats' who targeted Hurst (depending on your point of view) are 'white hats' or 'ethical hackers'.

A whole micro-industry of 'white hat' social engineers has grown up from within the cybersecurity community and former industry personnel. These include U.S.-based start-ups like Social-Engineer, Inc. founded by Christopher Hadnagy,[32] Digital Guardian established by Stu Sjouwerman and Kevin Mitnick,[33] and UK-based Antisocial Engineer Limited.[34] YouTube is littered with examples of how to mount social engineering attacks and 'human hacking' and well as some of the safeguards.[35] The literature on social engineering and 'human hacking' is also growing. This includes books such as Ian Mann's *Hacking the Human 2* and Christopher Hadnagy's *Social Engineering: The Art of Human Hacking*.[36]

Firms such as Exabeam and CrowdStrike run webinars, paid training events, and courses on behavioral analytics in comprehending incident responses by security analysts.[37] Security Operations Centers (SOCs) and Incident Response teams are overworked and highly trained or experienced cyber security personnel in short supply and expensive to recruit and retain (due to a widely acknowledged global skills gap[38]). This means demand for the services of firms such as Exabeam and CrowdStrike are likely to continue to grow as will the marketplace they operate in.

At the same time whether the right balance between human factors such as behavioral modeling and computer interactions (including machine learning and statistical analysis) is being struck is a valid question to ask companies, the ICT industry, and national governments.

These activities are part of the broader espionage and counterintelligence picture conducted by national governments which, given the resources and capacity at their disposal, is as likely to fuel insecurity and the global cyber arms race in trying to further state-based foreign and security policies. Rogue insiders acting for personal gain and corporate espionage are also at work as are criminals and terrorists. The most widely used, and most successful, means of breaching a system and gaining access to the network/s it connects to is through spear phishing.

Spear Phishing

Phishing e-mails (or for that matter telephone or texting scams) often offer 'too good to be true' offers, enticing people to part with personal details, information, or money, or to click on links or open attachments (often accompanied by giving a sense of urgency) are a worrying and growing trend. Some can be convincing or target the vulnerable. When they are crafted against individuals this is known as spear phishing. This is a growing attack vector within cybercrime. These and other phishing attacks still make it through corporate firewalls and anti-virus software despite ongoing attempts to spot and stop them.

Phishing and spear phishing have been around since the 1990s and is "still one of the most widespread and pernicious" gateways into systems "with phishing messages and techniques becoming increasingly sophisticated".[39] There are a number of 'phishing kits' available. Sites such as Phishtank and OpenPhish offer an open-source repository of kits on the 'Dark Net'. These kits enable anyone, even with limited technical skills, to conduct phishing campaigns.

Even basic phishing attempts can attempt to disguise their origins by tricking recipients into believing that they are coming from real e-mail addresses and a trusted source (known as a 'man-in-the-middle' attack). This includes setting up bogus websites appearing to be trusted sites (called a 'watering hole' attack). These can utilize international character sets to hide web addresses (known as URLs). Phishing's two main purposes are to get targets to hand over sensitive information and/or to entice them to click links or files in e-mails containing or hosting malware. Spear phishing accounts for some of the most popular attack injections at all levels—from 'mostly harmless' to dangerous.

Phishing continues to increase according to Proofpoint's 2019 State of the Phish survey. They also found that there is a greater awareness of phishing and ransomware among those aged 22 and over than there is from millennials who are seen as 'digital natives'. Proofpoint found that despite millennials having "been raised on smart devices and applications" this "doesn't necessarily lead to a clear understanding of cybersecurity".[40] They added that 'baby boomers' born after World War 2 and 'Generation X', born from the mid-1960s to early-1980s, had a greater awareness of cybersecurity. This they believed was due to greater exposure. For millennials, "organizations should not assume that younger workers have an innate understanding of cybersecurity threats and/or best practices simply because they are more cyber-savvy…security awareness training is needed across all age groups".[41]

Phishing leads directly to financial losses, including fraudulent online transfers; compliance issues; fines and legal fees; an increased burden on IT teams; reputational damage within information security (infosec) teams and companies themselves; and investments in new technology. This includes multi-factor authentication (MFA) and annoyance or legal liability from customers and employees following data breaches.[42]

In addition, "Whale phishing, or whaling, is a form of spear phishing aimed at the very big fish —CEOs or other high-value targets. Many of these scams target company board members, who are considered particularly vulnerable".[43] It is the case that "Gathering enough information to trick a really high-value target might take time, but it can have a surprisingly high payoff".[44] Many of these simple tactics not only continue to be successfully and widely deployed but might need only to work once to pay dividends. Illustrating this is a case from 2008 where "cybercriminals targeted corporate CEOs with e-mails that claimed to have FBI subpoenas attached. In fact, they downloaded keyloggers onto the executives' computers—and the scammers' success rate was 10%, snagging almost 2,000 victims".[45] These can be gateways for a business e-mail compromise.

Although good and best cyber hygiene practices tell us to be cautious, carefully crafted spear phishing can catch out even the wariest. Noa

Bar-Yosef described a decade ago that "Kits exist for just about everything. Automation is what made a hacker community into a hacker enterprise".[46] She added, "They are growing bigger, faster and stronger by the day. It does not help to simply recognize the problem".[47] This remains the case a decade later. Verizon's 2019 DBIR noted: "The typical utilization of phishing attacks to convince users to install remote access tools that establish footholds and begin the journey towards stealing important competitive information from victims remains the same".[48] The 2021 DBIR noted "They were the second most common form of Social attacks".[49] The 2022 DBIR recorded a similar picture.[50] Theft of competitive information is one of the targets.

Critical infrastructure owner-operators in particular need heightened awareness given the possibilities a malicious actor could unleash on their systems and processes. This also increasingly applies to small-to-medium sized businesses and enterprises (SMBs/SMEs) in supply chains. Proofpoint's 2019 State of the Phish survey (a valuable guide) found the most common types which lured end users were: 'Toll Violation Notification; [EXTERNAL]': 'Your Unclaimed Property'; 'Updated Building Evacuation Plan'; 'Invoice Payment Required'; 'Updated Organization Chart'; and 'Urgent Attention required' (leading to requests for a password change).[51] Front-facing departments including commercial, purchasing, communications, and sales topped the list but there was an even spread between these and other parts of the business including maintenance, security, executive level, legal services, marketing, and procurement. Of greater concern for critical infrastructure owner-operators is that the highest failure rates, and also the most targeted areas, were in production and operations.[52]

Mitigating Spear Phishing

Establishing systematic training mechanisms or formal programs to inform and educate end users on good e-mail and Internet usage practices, including common indicators of phishing to be aware of, is helpful and valuable. Spear phishing is more difficult to address. This is especially the case in time-pressured business and where there can be a need to trust and not skeptically pause before replying, clicking a link, or downloading an attachment.[53] Defensive technology from cybersecurity vendors and monitoring from IT departments can help in this respect.

Allying this to employee training, although burdensome and time consuming, against general and specific threats is also beneficial. Although formal cybersecurity education programs are becoming more systematic and driving increasing wariness of phishing and spear phishing within companies, changing end user behavior requires both a more general appreciation of the cybersecurity environment and the threats and threat actors driven by profit or malicious/mischievous intent.[54] For others who see themselves as whistleblowers or patriots, conscience and morality are also factors.

Some of these attack vectors can be blunted relatively simply. For example, minimizing the possibility of 'shoulder surfing' by looking around and having situational awareness; 'dumpster diving' by paper shredding; or preventing forensic recovery by physical or magnetic destruction of devices such as hard drives. Similarly, 'tailgating', quid pro quo, and impersonation can be vetted or minimized by challenge protocols (see below) or healthy skepticism. Nevertheless, 'social engineers' can be highly convincing or be able to catch people unawares or at opportune moments.[55] The growing use of malware 'toolkits' is also an aid to attackers. McAfee reported in February 2011:

> In 2010, we entered a new decade in the world of cybersecurity. The prior decade was stained with immaturity, reactive technical solutions, and a lack of security sophistication that promoted critical outbreaks, such as Code Red, Nimda, Blaster, Sasser, SQL Slammer, Conficker, and myDoom—to name a few. The security community has evolved and grown smarter about security, safe computing, and system hardening but so have our adversaries. This decade is setting up to be the exponential jumping off point. The adversaries are rapidly leveraging productized malware toolkits that let them develop more malware than in all prior years combined, and they have matured from the prior decade to release the most insidious and persistent cyberthreats ever known.[56]

Many case studies bear McAfee's concerns out. In addition to those already mentioned other options include conducting a regular review of employee privileges and access rights and changing administrative passwords to servers and networks following changes in IT personnel. Also expire passwords so that employees have to change passwords regularly because "in many instances, default passwords are provided by IT staff and are never changed".[57] There have also been cases of passwords shared by senior staff to give junior staff access.[58] This can give them access rights not only to the system it is being provided for but also to other parts of the network. It is good practice not to have the same login and password for multiple platforms, servers, or networks—but many do.

Further safeguards, such as informing third-party service companies providing ICT services or customer support, that an employee has left an organization is particularly important when an employee has been dismissed, made redundant, or has taken a role with a competitor. Restricting Internet access on corporate computers to cloud storage (another attack vector) and not allowing employees to download and run unauthorized remote login applications on corporate computers can also be considered good practice. Mandating daily backups of critical information held on servers across all computer networks can also be used to restore systems after breaches but these backups, a standard IT practice, are also a possible attack vector for advanced cybercriminals and APTs. Another standard practice is to terminate all accounts associated with an employee or contractor immediately upon dismissal or redundancy although in the latter there can be a business case to continue to grant access.[59]

Spear Phishing Attack Tools and Websites

There are a large number of attack tools and websites to guide both ethical 'White Hat' hackers, malicious 'Black Hats', and even 'Gray Hats' who fall somewhere in between like Marcus Hutchins. Hutchins, who like many other hackers is self-taught, was a malware author who is also credited with preventing the further spread of the WannaCry ransomware in May 2017.[60] He had little formal education in computer science and instead used the Internet to develop his skills.

Websites like 'Wonderhowto' offer do-it-yourself guides on 'How to Use Persuasion to Compromise a Human Target', 'The Ultimate Social Engineering Hack', and 'How to Spear Phish with the Social Engineering Toolkit (SET)' whilst also offering guidance on escaping 'unlawful' stops and police searches.[61] Whilst branded as training for ethical 'White Hat' hackers, sites like these can clearly be used by Black and Gray hats alike. The techniques are the same as are the tools.

The Social Engineering Toolkit for example "has quickly became a standard tool in a penetration testers arsenal" with "The attacks built into the toolkit…designed to be targeted and focused attacks against a person or organization used during a penetration test".[62] It utilizes Metasploit, a software tool which is used to create a 'payload' consisting of malware, bugs to be exploited, and web browser exploits. Metasploit, a collaboration between the open-source community and the cybersecurity provider Rapid7, automates methods of information gathering, gaining access, maintaining persistence, and evading detection.[63]

Metasploit now contains 1,500 exploits and 500 payload options and "Pretty much every reconnaissance tool you can think of integrates with Metasploit, making it possible to find the chink in the armor you're looking for".[64] Moreover, "Once you've identified a weakness, hunt through Metasploit's large and extensible database for the exploit that will crack open that chink and get you in. For instance, the NSA's EternalBlue exploit, released by the hacking group the Shadow Brokers in 2017, has been packaged for Metasploit and is a reliable go-to when dealing with unpatched legacy Windows systems".[65] In addition, "Once on a target machine, Metasploit's quiver contains a full suite of post-exploitation tools, including privilege escalation, pass the hash,[66] packet sniffing, screen capture, keyloggers, and pivoting tools. You can also set up a persistent backdoor in case the machine in question gets rebooted. More and more features are being added to Metasploit every year…"[67] It is packaged with Kali Linux; a computer-based toolbox designed for ethical penetration testing which companies can pay to test their defenses. However, it can also be used by malicious hackers. It saw increased popularity after the TV show *Mr. Robot* was released in 2015. *Mr. Robot* demonstrated to a wide audience what can be accomplished by hacking.[68]

As with most open-source projects like Kali Linux, tutorials are widely available. Other hacking tools include Sn1per, Owasp Zed, Wireshark, and Nikto (all vulnerability scanners of different kinds), Aircrack-NG, a password and Wi-Fi cracking tool, and NMAP which will scan for ports and is used to map networks.[69] Maltego, which analyzes "real-world relationships" between publicly accessible information on the Internet…can be used to determine the footprint of Internet-facing infrastructure within an organization and "gathering information about the people and organisation who own it".[70]

This "can be used to determine the relationships between…People, Names, Email addresses, Aliases, Groups of people (social networks), Companies, Organizations, Web sites" and Internet infrastructure including "Domains, DNS names, Netblocks, IP addresses, Affiliations" and "Documents and files".[71] This helps with network enumeration which helps gather user names, machine names, and network information with still more specific tools also available.[72] These connections are generated using OSINT "by querying sources" and Maltego can display these interconnections through a number of wide-ranging graphical representations. Up to 10,000 separate entities can be analyzed and clustered in this way and "this makes it possible to see hidden connections even if they are three or four degrees of separation apart".[73]

This idea of separation is built on the hypothesis that we inhabit a 'small world' and the concept widely attributed to Stanley Milgram's work in the 1960s that everyone is separated by at most six degrees of separation.[74] The idea of six degrees of separation and the small world phenomenon (also evident in ideas of the global village) is superficially simple. Such as: "what is the probability that any two people, selected arbitrarily from a large population, such as that of the United States, will know each other? A more interesting formulation, however, takes account of the fact that, whilst persons a and z may not know each other directly, they may share one or more mutual acquaintances". These "acquaintance chains to extend an individual's contacts to a geographically and socially remote target".[75]

Around the millennium there was "An explosion of interest…in the small world problem because mathematicians…developed computer models of how the small world phenomenon could logically work".

However, "mathematical modeling is not a substitute for empirical evidence. At the core of the small world problem are fascinating psychological mysteries", especially in a world continuing to display and be divided by "social barriers such as race and class".[76] The development of the Internet and social media has shrunk or collapsed many of the inhibitors to the 'small world' idea.[77]

In addition, password hacking tools like John the Ripper, THC Hydra, and Cain and Abel are also available. Each of these can be used to exploit publicly known unpatched vulnerabilities and zero-days and aid man-in-the-middle attacks. Website copying tools are also available which can assist with 'website mirroring' which hosts a duplicate website in another part of the world (e.g. one in Europe and a duplicate in Asia) and is often used by multinational companies. Website copying can be used to clone websites.[78] Tools such as Burp Suite can also be used to find and exploit web applications that enable functionality like hyperlinks, shopping carts, and embedded video on websites.[79]

If this does not work then hackers who target local organizations familiar to them, or those willing to travel to physically go after high-value or hard targets can attempt local monitoring and physical contacts and identify or closely target employees. This is more likely to be attempted by APT groups who will be provided with supporting field agents as was the case with the attempted hack by Russia's *Glavnoye Razvedovatel'noye Upravlenie* (GRU) on the Organisation for the Prohibition of Chemical Weapons (OPCW) in the Netherlands in April 2018. It is also possible to capture the information stored on cards and devices that use Radio Frequency Identification (RFID); a technology common to many modern swipe cards (including credit cards). These can be cloned to grant access to unauthorized visitors.[80]

A number of other methods under the wider term 'semantic attacks' are also used. These include 'file masquerading' (hiding malicious file types by looking like Microsoft Office documents or PDFs); web pop-ups (opening new windows when visiting websites) including 'malvertisements'[81]; and 'friend injection' through social network sites.[82] A combination of one or more of these tactics can be used by actors ranging from fraudsters and organized criminal gangs, script kiddies and hacktivists, and also by state intelligence agencies and in corporate espionage.

These attacks are commonplace and are becoming more sophisticated as awareness grows and social engineering utilized.[83]

According to Verizon's 2021 DBIR, once inside a target organization one of the main aims is to steal credentials (seen as a form of hacking). However, phishing e-mails might also have 'dropped' malware—including trojans or backdoors. It noted: "As with past years, Social actions are predominantly Phishing, though Pretexting, normally associated with the BEC, also makes a strong showing". The 2021 DBIR also recorded that the majority of social engineered incidents were discovered externally and "This means that when employees are falling for the bait, they don't realize they've been hooked. Either that, or they don't have an easy way to raise a red flag and let someone know they might have become a victim".[84] The 2022 DBIR showed that the main method of delivering malware continued to be e-mail. Once in, attackers try to establish a backdoor or form of command and control leading to other types of malware to be inserted through a malware downloader.[85]

State Intelligence: HUMINT Beyond Social Engineering

One of the themes of this book is that states are both targets and attackers and what applies to them as offensive actors also apply to their defensive posture and capabilities. In terms of HUMINT, espionage and social engineering, Nigel Inkster, a former Deputy Director of SIS (and Director of operations), argues that the days when intelligence officers could credibly embrace numerous identities and personas are ending. Technologies including biometrics (including facial and voice recognition), CCTV, forensics, and DNA, make operating in hostile regions or foreign countries much more difficult.[86] Inkster describes this as a digital footprint or exhaust leaving traces that can be tracked.[87] As Russia found out in the botched hack of the OPCW, because of facial recognition technologies, smart phones, CCTV, and big data analytics, the creation and maintenance of cover for a field agent and their handler (a case officer responsible for handling agents in operations) is much more difficult, if not impossible.

State intelligence agencies can also take SE to degrees way beyond cybercriminals. They are skilled at producing intelligence on individuals, groups, organizations, and states, which are not publicly available. Through OSINT and covert intelligence and analysis, this helps form a 'pattern of life' including "potential refuge locations, work locations, travel patterns, known vehicles, and social activities" of persons of interest.[88] Accurate and timely knowledge of a target, target organization, the culture of that organization, and their *modus operandi* and culture of their host country or national affiliation is used to enhance this intelligence and the ability to act. This same methodology applies to a target's social, political, religious, or financial status as well as a host of related factors. This is a form of biographical accumulation and analysis which can also be used for the purposes of entrapment.

Entrapment can take several forms but the most common are financial or sexual but this form of 'coercive recruitment' is far from a guarantee of success, much less so willingness. Resentment of blackmail, coercion, or exploitation, can lead people to turn themselves in. These then can provide information and intelligence on those 'handling' them. Willingness is found in 'walk-ins' who volunteer to spy for a foreign intelligence service and intelligence can be used as a stick and carrot to co-opt the willing, unwilling, and unwitting (including so-called 'useful idiots'). Joel Brenner, a former NSA Inspector General argues: "Ordinarily, it takes months, even years to compile this kind of profile on a potential target. With sites like Facebook you can do it in ten minutes. Again, this is a blessing when you're recruiting agents and a nightmare when you're in counterintelligence…These sites are part of the world now, and they're not going away".[89] In this world of SE, intelligence, counterintelligence, and espionage, social media plays an increasingly important role—especially Facebook which was used by Russia to target key voters and influence both the 2016 U.S. presidential elections and 2016 Brexit referendum.[90]

FBI Counter Intelligence Assistant Director, Frank Figliuzzi, issued a reminder in the wake of the discovery of deep cover Russian spies (called Illegals) and their spy ring in 2010: that: "The public needs to know this threat continues…Spying has been with us since the Old Testament; spying is with us now".[91] The social and psychological techniques they used also included pretexts and backstories (known as 'legends') to deceive and exploit socio-psychological tendencies to trust. These are a core component of the training and tradecraft of spying and espionage and are built on establishing trust and/or to enable coercion. Since antiquity, intelligence agencies and agents the world over have long used these and other methodologies. Alex Younger, the Chief of Britain's Secret Intelligence Service (SIS/MI6) stated in a rare public speech in December 2018:

> if you strip away the mystique that envelops our organisation…we provide human intelligence. Our task is to create human relationships that bridge forbidding cultural and linguistic boundaries, in some of the most challenging environments on earth and online. We do this for a specific reason: in order to obtain information and take actions required by the British government to keep this country safe. Our skill lies in our ability to create relationships of trust between our officers and people inside the organisations we need to understand.[92]

The 'Birds Eye' Macro View and the Micro Level of HUMINT

Younger's comments point to a central part of espionage tradecraft which is often allied with other intelligence collection and analytical methods through Signals or Electronic Intelligence (SIGINT and ELINT). Combined with satellite reconnaissance and intelligence gathering (SATINT) this can provide a 'birds eye' or macro view of threats and threat actors and enable secure one-to-one communications (including through encrypted devices). At the micro interpersonal level, social and psychological tools used in mainstream social engineering have wide applications against targets—including in cyberspace. This embraces the use and utilization of social media through Social Media Intelligence (SOCMINT) which is also used by non- or sub-state actors.[93] For example, SOCMINT and OSINT were used to identify individual Russian soldiers and their units who crossed into Ukraine in 2014 and were used by the private investigative organization Bellingcat to identify the Russians responsible for poisoning Sergei and Yulia Skripal with the chemical agent Novichok in Salisbury, England in March 2018.[94]

Cyber may already be the prime avenue intelligence agencies now take due to its relatively low cost and high returns. Remote computer-based cyberespionage does not require the heavy investment in equipment that traditional land-air-sea-space-based intelligence requires through ground stations, surveillance aircraft, ships, submarines, and satellites, and whilst it can have its limits it is highly cost-effective. Moreover, modern militaries depend on ICT for navigation and positioning via satellite to electronic reconnaissance through sensors on land, air, sea, and space. Realtime data sharing can provide commanders and deployed forces with shared awareness of the battlespace. Logistics and resupply and refueling are also affected alongside further developments in aircraft control, landing, instrumentation, and propulsion. Radar and sonar are similarly computer dependent. Militaries have long had units tasked with gathering and interpreting signals intelligence and for offensive and defensive electronic warfare. This mission has now evolved to include cyber penetration and disruption.[95]

Nevertheless, field agents have been used against people and places that require something beyond cyber where there is a need for someone on the ground be it in Moscow, Beijing, or Kabul. These field agents utilize technology and cyberespionage in their tradecraft. Secure encrypted communications are also an aid when running human sources or conducting covert operations. Despite the birds-eye view SIGINT, ELINT, Measurement and Signature Intelligence (MASINT) and Geospatial Intelligence (GEOINT) bring—much of which is enabled through satellites and through Internet monitoring—field agents continue to be used.[96] In addition, state intelligence agencies see that CYBERINT has grown in importance alongside SIGINT, MASINT, GEOINT, and HUMINT.[97]

ELINT, CYBERINT, and technological development also continue to play a role in the tradecraft of espionage. Avi-Tal and David Siman-Tov, point out that not only have small, portable consumer devices like SD cards and USB sticks made it easier to store and exfiltrate intelligence but they also argue that "The cyber era makes it possible to vanish into the vast sea of information and under assumed identities and roles, thereby substantially ensuring safe communications with agents…[and] reduced the need for face-to-face meetings" which has diminished the risk of discovery.[98] However, as Nigel Inkster alluded to earlier, this sea of information also makes it more difficult to remain concealed, especially for field agents. The use of facial recognition, biometrics, and OSINT has seen the identification of individuals accused of hacking and led to deportations, indictments, and prosecutions.

Human Sources and Human Agency

HUMINT requires a great deal of time and resources in creating human 'assets', running them, analyzing the intelligence/information they produce and integrating this into the intelligence cycle. This cycle includes intelligence collection/gathering, including from HUMINT assets, and evaluation and assessment of the intelligence gathered (intelligence analysis[99]). This forms the basis for the production of intelligence reports and the dissemination and consumption of the intelligence product by governments.[100]

HUMINT and espionage can be high risk and requires a great deal of complex effort. Intelligence officers are frequently required to possess or learn foreign languages, asset detection techniques, surveillance and surveillance avoidance abilities, recruitment skills, as well as self-defense and weaponry skills. The use of the latter two belongs more in the fictional worlds of James Bond and Jason Bourne than real life but this said field agents are potentially a last line of defense and a first line of attack.[101] Younger told his audience: "the people who agree to work in secret for MI6—to do extraordinary things and run great risks. And I will not hide from you that some have paid the ultimate price".[102]

A former field intelligence officer of SIS/MI6 noted that in days gone by field agents would 'work a room' including embassy cocktail parties and high-profile events and gather useful intelligence and the weak spots of persons of interest. Now, identities and affiliations (and much else besides) are posted on social media. This makes the job of trying to identify and recruit potential assets much easier. Combined with data analysis, this makes targeting and potential recruitment simpler and more accurate. Real identities are also harder to hide. Previously this would have taken months. Allied to technological developments, this can make recruiting and running sources much easier than during the Cold War and immediately after 9/11.

In these eras one of the most significant problems, especially with deeply embedded sources, was how to maintain contact without compromising them. Through encrypted communications and devices, messages can now be passed with relative confidence they will remain secret. For sources with the right access, including the 'crown jewels' of intelligence (documents and other primary sources) digitization also makes these easier to exfiltrate. To them, the case of Edward Snowden also showed that even relatively low-level sources can have extensive access.[103]

Spies, moles, and insiders can also be used to infect devices and computer systems as well as gathering and exfiltrating intelligence.[104] Robert Wallace, the former head of the CIA's equipment lab (an equivalent to James Bond's Quartermaster or 'Q'), in the Office of Technical Services (OTS) points out that miniaturized, audio and visual surveillance equipment that were "once the exclusive providence of a handful of intelligence services, is [now] available over the Internet or from a local "spy shop"".[105] From high resolution camera's and end-to-end encrypted communications on smart phones to the kinds of devices used by penetration testers and malicious hackers, more and more hardware and software devices for hacking are available commercially. There is no need not go to a 'spy shop'. Many of these devices are available on Amazon and eBay.

During the Cold War era OSINT made up a significant part of intelligence assessments. Now its ubiquity means it is an invaluable part of the intelligence analysis profession in its own right. In the security environment of the twenty-first century it helps put meat on the bones of sensitive covert reporting and adds evidential weight. This includes on complex subjects including political and societal shifts, such as riots and revolutions, alongside data analytics combined with cultural and area-specific knowledge. This aids meaningful analysis and helps provide context, insights, and to see connections that would be much harder to spot.[106] The same can be said of counterterrorism operations.

SOCMINT and OSINT can also be used for Social Network Analysis (SNA). This can be used to better understand 'persons of interest' through their familial ties and wider social groups and map continuity and changes in people and their affiliations over time. SNA methodology, and the sophistication able to be employed, can provide shortcuts or pathways to target individuals, groups, and organizations. This aids the increasing ways states, and perhaps well-resourced non-state actors, can 'spear phish' targets.[107]

Cyber Defense and Offense

With these constant developments in mind, Joel Brenner, warned in early 2015:

> We have been walking backward on cyber defense whilst ignoring the real issues. First, we adopted a moat-and-drawbridge approach. This didn't work for two reasons. We had barbarians inside the gates, and the gates themselves, which we fancied as "firewalls," were merely flimsy filters…All defense strategies are variants on these models, and all of them are variants of Whac-A-Mole. We are playing a losing game.[108]

Not all are so pessimistic—at least in public. Alex Younger, also known as 'C' sees that:

> We are in the early stages of a fourth industrial revolution that will further blur the lines between the physical, the digital and biological realms. Lawfully used, technology such as bulk data, modern analytics and machine learning is a golden opportunity for society at large, including for MI6 as an organisation. But I have also witnessed the damage new technologies can do in the hands of a skilled opponent unrestrained by any notion of law or morality, as well as the potentially existential challenge the data age poses to the traditional operating methods of a secret intelligence agency. We and our allies face a battle to make sure technology works to our advantage, not to that of our opponents.[109]

This encompasses both offensive cyber operations (OCO) and defensive cyber operations (DCO).[110] Under these conditions Aaron F.

Brantly of West Point Military Academy, also points to the benefits and indistinguishable nature of cybersecurity in asymmetric contexts which sees states defending multiple attack surfaces against a well-motivated and advanced attacker (and/or simultaneous attacks from others).[111] This is especially true in the context of CI where successful attacks can (and have) be used as beachheads for longer term campaigns.

Defending Insider Threats

Espionage practices, including social engineering, are also used for covert action.[112] Operation Olympic Games, which successfully infiltrated the Iranian nuclear facility at Natanz with Stuxnet, was a covert action with plausible deniability. This could well have utilized an insider and SE to help penetrate the site.[113] The threat posed by insiders is difficult, if not impossible, to defend against 100% of the time. There are many examples of the insider threat but few that, so far, have been known to effect critical infrastructure sites. One was Vitek Boden in 2000 (detailed in Chapter 3). Another occurred in May 2014 when "Ricky Joe Mitchell was sentenced to four years in federal prison and ordered to pay $428,000 in restitution for responding to the news that he was going to be fired from his position as a network engineer with the oil and gas company EnerVest by resetting all network servers to factory settings, disconnecting critical pieces of network equipment, and disabling the equipment's cooling systems".[114]

The National Cybersecurity and Communications Integration Center (NCCIC), which works under CISA, through the Department of Homeland Security (DHS), states that insider threats "include sabotage, theft, espionage, fraud, and competitive advantage". These "are often carried out through abusing access rights, theft of materials, and mishandling physical devices." DHS cautioned that "Threats can also result from employee carelessness or policy violations that allow system access to malicious outsiders. These activities typically persist over time, and occur in all types of work environments, ranging from private companies to government agencies".[115] The DHS lists a wide number of organizations, portals, and advice through which companies can seek to mitigate insider threats. These are:

- The DHS National Cybersecurity and Communications Integration Center's 'Combating the Insider Threat'. This designates "the importance of distinguishing between normal and risky behavior to detect and deter insider threats".
- The DHS Science and Technology Insider Threat Cybersecurity Program. This "seeks advanced R&D solutions to provide needed capabilities to address six areas: Collect & Analyze, Detect, Deter, Protect, Predict, and React".
- The DHS National Intellectual Property Rights Coordination Center which "connects 23 federal and international partners to defend against global intellectual property theft and enforcement of international trade laws".
- The DHS US-CERT Assessments: Cyber Resilience Review—"a no-cost, non-technical, self-assessment to evaluate an organization's operational resilience and cybersecurity practices".
- The DHS US-CERT Critical Infrastructure Cyber Community Voluntary Program (C^3VP). This is designed to improve "the resiliency of critical infrastructure's cybersecurity systems by supporting and promoting the use of the National Institute of Standards and Technology's (NIST) Cybersecurity Framework".
- The CERT Insider Threat Center. Based at Carnegie Mellon University's Software Engineering Institute (SEI), its goal is to "help identify potential and realized insider threats in an organization, institute ways to prevent them, and establish processes to deal with them if they do happen".
- The Department of Justice 'Reporting Intellectual Property Crime: A Guide for Victims of Copyright Infringement, Trademark Counterfeiting, and Trade Secret Theft' provides government liaison following intellectual property theft.
- The FBI Internet Crime Complaint Center (IC3): a portal through which "internet-related crimes can be reported and public awareness materials can be published".

In addition, several fact sheets are available. These include the FBI's Intellectual Property Protection Fact Sheet; the FBI Checklist for Reporting an Economic Espionage or Theft of Trade Secrets Offense and an "Economic Espionage brochure for information on protecting trade secrets". Also, the "FBI Insider Threat: An Introduction to Detecting and Deterring an Insider Spy is an introduction for managers and security personnel on behavioral indicators, warning signs and ways to more effectively detect and deter insiders from compromising organizational trade secrets and sensitive data".[116] These are a recognition that the United States is the prime target of intellectual property theft and cyberattacks more generally. Other nations have their own reporting systems.

Major leaks, such as that perpetrated by Edward Snowden, have shown that even a single trusted insider has the capacity to do great harm to an organization. If there are multiple trusted insiders who 'go rogue' in one or more organizations, industries, or governments then the risk and potential harms they could do are multiplied. If the NSA, perhaps the world's most well-resourced and security conscious organization, had an undetected insider threat in Edward Snowden then any organization is vulnerable. The most common reasons rogue insiders offer are: financial considerations/greed; revenge, grudges, or dissatisfaction; blackmail or entrapment; national pride or corporate competitiveness; emotional involvement with someone hostile to the target organization or country; false flag approaches/exercises; naiveté; ideology; and whistleblowing.

Snowden and many other hackers are self-taught but if hackers cannot breach remotely there remains the option of hacking key sites themselves. In addition, when former insiders become 'outsiders' (e.g. by being made redundant in a business) this can open up new possibilities for recruitment or exploitation. Both targets and 'walk-ins' can carry risks. They might, for example, report an approach, be what the CIA calls 'a dangle'—an enemy-controlled agent or defector, masquerading as a traitor, often used to 'out' intelligence officers in an embassy, or they may simply have superfluous or outdated information-intelligence. Identifying what and who is of value are important markers.

Relatively low-level personnel can be in a position to know a great deal about the people and processes within an organization or can have access to sensitive areas. These can include security guards and cleaners as in the 1987 case of Marine Sgt. Clayton Lonetree, a guard at the U.S. embassy in Moscow. Lonetree was convicted of espionage after being seduced by a KGB 'swallow' (a seductress engaged in 'sexpionage').[117] They can also have daily access to people higher up in an organization or smaller subcontractors with access to sensitive data, systems, or personnel in the supply chain. Both private and public sector organizations can contain and process sensitive or valuable information. These 'low hanging fruit' might be able to access, or be able to gain access, to secure locations or data. These people might be able to exfiltrate data knowingly or unwittingly which can be used to gather valuable information-intelligence or to discover gateways or vulnerabilities.

Mitigation and the Insider Threat

There are a number of private sector security and penetration testing companies who specialize in reducing these risks by highlighting corporate security weaknesses. This includes gaining physical access to buildings or terminals armed with enough knowledge to exfiltrate data or to conduct malicious attacks. One such 'white hat' professional social engineer is Colin Greenlees who has worked as an IT consultant for Eurostar and Siemens. He is frequently hired by companies to test their security procedures and what he finds is alarming but also commonplace. According to Greenlees:

> It is all about confidence. I walked into the building [of a FTSE-listed firm] having an imaginary conversation on my mobile and the swipe-card operated lift was held open for me by what turned out to be the managing director…I remained there for five days working from a third floor meeting room.[118]

After gaining access he discovered information from the human resources department, and learned of deals, mergers, and acquisitions, and gained the cell phone numbers of all senior managers. Another Siemens consultant he brought in to work with him, using the same methods to gain access, used pretexting to access multiple usernames and passwords and infiltrate their internal network through his own laptop. The BBC reported in 2009, "Mr Greenlees is convinced that anyone intent on industrial espionage would be able to find a social engineer capable of wreaking havoc".[119] Greenlees warned, "There are people in the criminal underworld who know people who know people. They don't advertise but they are out there".[120]

Although this story is now more than a decade old, what it reported remains as pertinent today as the vast majority of data breaches are the result of insiders giving access (largely unwittingly) to outsiders. This is added to by the ever-increasing use of mobile technology including laptops, smartphones, and memory sticks. This "means more and more information is leaving buildings, some of it never to return". Additionally, "There have been high profile cases of laptops left on trains and, according to a recent survey from data security firm Credant Technologies, 9,000 USB sticks have been found at laundrettes in the past year; left in pockets when clothes are taken to be dry-cleaned. It illustrates just how difficult it is for firms to secure their data. The next big breach could be just around the corner".[121]

In 2015, Greenlees recounted the continued prevalence of these issues. From his perspective, this includes surveillance from hanging around coffee shops or restaurants frequented by staff. From these vantage points, much information can be gleaned (including targeting individuals and other companies through ID cards on lanyards). Through this covert surveillance a fake ID can be gained (either customized or bought online). This can be followed up with research into company events (including through LinkedIn). This can then be used alongside manipulation techniques and people's inability to keep up with staff turnover and internal changes, business practices, and dealings—all of which are accelerated by technology trends. He argued why use brute force when softer methods and social engineering (which encompasses well-honed intelligence activities) can be used?

He also described the kinds of psychological triggers that can be used as part of the practice of social engineering. This included that 'acting like you belong' in an organization, where confidence and body posture are important to yield results. Also, by talking to the security guard/s, paying attention to them, and taking an interest in their work, enough information can be gained to compromise internal security and with it information security. These techniques have seen Greenlees explore everywhere in 'secure' buildings, including server rooms, and he could have gone much further on several occasions. On one occasion, he took this information to the company concerned and was asked to stop. He did a retest on the organization three years later and he could not get into the building, but he persisted and did manage on one occasion but was then challenged and denied further access.

In this vein Greenlees asked when was the last time companies risk assessed their people? In the hierarchy of risk, he restated that the weakest link in the cybersecurity fence remains people with ICT access. In addition, there can be a lack of awareness of the value of the information individuals hold within organizations. If a tailored and structured awareness program is fostered as part of a secure and resilient culture, then these security failures can be mitigated—but the residual risk will remain. The security teams, whether the front of house or in ICT departments, also must be aware of business needs. It is also worth considering that security has to be benchmarked against the cost to that business. This is a balance that needs to be struck but at critical infrastructure sites in particular, security challenges are a better practice than a more trusting 'Can I help you?' attitude.[122] There is a good reason for this.

Physical Security I

On YouTube alone, there are a multiplicity of videos which demonstrates how to hack—including how to hack critical infrastructure specifically. Some embrace breaching physical security to hack key sites. To take but one example. A White Hat team of ethical penetration testers, RedTeam Security, was hired by a power company in the U.S. Midwest to test its defenses over a three-day period in a video published in May 2016. They began by reconnoitering the physical location and its physical security and, using widely used social engineering techniques, they looked for points of access into buildings.

This included driving into the main parking lot "like you belong" there and dressing up as technicians with a GoPro in a satchel for on-site recording (note that cell phones can be taken off you at secure sites along with other forms of electronic equipment). They had gathered names of individuals who worked at the site already from public websites. They were immediately issued with visitor badges for technicians without ID checks. Calls by service providers are not unusual. They played on the receptionist's willingness to want to help and created within her a sense of inconvenience which aided their entrance. Given they were first-time visitors to the site an employee even offered to show them where they had asked to go. However, they were challenged by a supervisor on this visit and asked to leave. Although failing on their first point of access, under different conditions they could have been escorted in (or alternatively seen site security or the local police called).

After failing in their first attempt, they then attempted to physically break-in. To accomplish this, two teams were organized. The first entered through a wooded area of the site. The second team came through the employee entrance again acting like they belonged there. They used a 'Shove-It' tool, a widely available device used by locksmiths to gain access against physical locks, which they used to gain access to locked parts of the site.[123] Another tool was used to gain access from beneath locked doors "but sometimes the doors aren't locked". In other cases, they could pick locks or disassemble the padlocks around sites. They then freely wandered, including the area of the front desk where they had previously been denied access. On their travels they found company laptops, iPads, and credit cards with Personal Identification Numbers (PINs) to hand.

They then looked for a port where they could install a 'PlugBot'. This is a piece of hardware resembling a power adapter which uses a single board, low-cost disposable computer such as a Raspberry Pi with their own software. The PlugBot is "designed to be a proof-of-concept / experimental foray into the development of software that could potentially support the concept of a hardware botnet...Alternatively, PlugBot can be used as a vulnerability assessment or penetration testing device (bot) for covert use during physical penetration tests".[124] Once installed the PlugBot gave them persistent access to the network after they left and

RedTeam Security themselves recognize that "it can be used for nefarious purposes".[125] The software is available as open source on GitHub ready to be installed on a Raspberry Pi or similar single board computer as these have a small physical footprint and can remain hidden out of sight.[126]

Next, they broke into the server room and in a single visit had free reign of the offices and gained administrative privileges in the network. They then left after being told that the coast was clear from their surveillance team outside. The following day they broke into a power substation. This had two motion detectors and a CCTV camera. They flew a small drone over the site from a safe distance. One detector used microwaves and the other infrared. They identified a blind spot towards the rear of the substation. The CCTV camera had a 280° view which meant it could not see behind the pole it was mounted on and they used a physical heat shield in the cold conditions to mask themselves from the infrared detector.[127] There are a multitude of other ways to spoof or shield from an infrared detector.[128] At the substation they wore protective clothing to safeguard themselves from electrical arcs. At the substation they installed a device and used a tablet to control it which gave them network access.

Separately, they installed spyware and malware via USB drives which they had left on desks and in workspaces (in highly secure environments these ports should be physically blocked). This enabled them to turn on webcams and microphones and create screen captures. They also found open doors and unlocked computers. RedTeam Security stated in their video that confidence was very important and came from "solidifying the pretext" based on research into the company. Bearing in mind the methods used to distribute electric power detailed in Chapter 3:

> controllers and other devices increasingly used in substation automation are often sources of the thousand-plus ICS vulnerabilities discovered since 2010 that can serve as entry points to networks. Once inside the digital operations of a substation, an attacker with the necessary skills and tools could disrupt, desynchronize, or impact data communications necessary for communications and controls causing load instability. Substation

networks without detection capabilities to identify intrusions and malicious data injection could allow an attacker to manipulate multiple substations over time without discovery. In these networks, the risk of a coordinated cyber attack powerful enough to disrupt a portion of the grid is greater. ICS experts also note that if a threat actor can physically access a substation there is virtually no limit to potential damage: malware can be directly introduced to computers and devices, protective relays manipulated, and equipment physically destroyed.[129]

RedTeam used similar techniques to break into eight locations. This included pretending to be students at one facility where they cloned ID badges using a concealed RFID reader. These are commercially available devices, which are also used for credit card fraud. These read the data contained in the radio signals emitted by the chip on their ID cards. These are used widely on a variety of products besides and they used the captured data to encode their own cards.[130] These systems use a reader/writer attached to a handheld computer to clone the card.[131]

Since it was posted in 2016, by 2020 this one video by RedTeam was viewed 7.26 million times. In mid-2022 9.07 million times.[132] It would be prudent to assume that not all viewers are ethical hackers. What the relationship is between these White Hat penetration testers/security teams and law enforcement and security agencies (including Homeland Security, the CIA, FBI, DoD, and NSA) is unclear as is who can hire them. How many of these penetrating testing teams are founded or staffed by former government officials or former military is also unclear as is their background, composition, and hiring practices and whether they are security cleared by governments or agencies.[133]

George Wrenn, who was vice president of cybersecurity for Schneider Electric before joining MIT, is one of many who have warned of vulnerabilities in physical security and cyber-physical systems. For industry experts like Wrenn, reliance on commercial-off-the-shelf (COTS) technologies for the IoT, alongside problematic supply chain security, new technologies being rolled as a result of market pressures and the high clock speed of technological development, alongside continuing use of legacy systems is a recipe for a 'perfect storm'. This too is helping breed David defeating Goliath scenarios and the potential for cyberwarfare.

This applies both in terms of both physical destruction but also severe disruption and dislocation that falls beneath the threshold of armed attack (as argued in Chapter 2).[134] Another disturbing case from 2021 involved the use of drones against substations.[135]

Physical Security II: The CIA Triad and 'Full Disclosure'

Red Teaming can complement non-physical penetration testing, which normally occurs only from the outside in, not usually by trying to physically get inside premises. Red Teaming "operates much more like a true attacker by leveraging any and all means possible (within safety and legal limits, of course) to achieve the objective".[136] This helps test for vulnerabilities but insecure products are still being rolled out with vendors allowing them to be tested freely which, for software such as smartphone apps, allows them to discover or iron out bugs, but which in CI provides a target (as well as an opportunity to patch, remove, or remedy). Meanwhile Bitcoin and other cryptocurrencies have the potential to be used for hiding or hiring malicious actors and activities with a high degree of anonymity. Markets for malware continue to grow via the 'Dark Net' with organized criminal gangs funded through cryptocurrencies.[137]

The policy of 'full disclosure' requires that full details of a security vulnerability are publicly disclosed, including details of the vulnerability and how to detect and exploit it. As Idaho National Laboratory (INL) reported in 2016 "Discovery, publication, and mitigation of cyber threats are often the work of cyber researchers and cyber security teams, acting either independently or as surveyors on behalf of a commissioning body".[138] The theory behind full disclosure is that releasing vulnerability information immediately results in improved security. Fixes are produced faster because vendors and application authors are forced to respond in order to protect their hardware and software from potential attacks as well as to protect their reputation. Security is improved because the 'window of exposure', the amount of time the vulnerability is open to attack, is reduced.

The Operational Technology (OT) networks which RedTeam successfully breached follow the Confidentiality, Integrity, and Availability (CIA) triad. This is firmly established within the Information Technology (IT) environment in this order. However, in the OT environment, with systems that are designed for constant operation over periods of years or decades (hence the problem with legacy systems) that triad is reversed into Availability, Integrity, and Confidentiality. When members of APT 28 (the GRU's 'Fancy Bear') were indicted by the U.S. Department of Justice (DOJ) in October 2018 it was detailed that they used many of the widely known hacking techniques used by RedTeam Security. The DOJ indictment indicated:

> When the conspirators' remote hacking efforts failed to capture log-in credentials, or if the accounts that were successfully compromised did not have the necessary access privileges for the sought-after information, teams of GRU technical intelligence officers, including [Aleksei] Morenets, [Evgenii] Serebriakov, [Oleg] Sotnikov, and [Alexey] Minin, traveled to locations around the world where targets were physically located. Using specialized equipment, and with the remote support of conspirators in Russia, including [Ivan] Yermakov, these close access teams hacked computer networks used by victim organizations or their personnel through Wi-Fi connections, including hotel Wi-Fi networks. After a successful hacking operation, the close access team transferred such access to conspirators in Russia for exploitation.[139]

These 'on-site' or 'drive-by' attacks on Wi-Fi and other forms of network access (such as that conducted by Vitek Boden in the Maroochi Shire case and by RedTeam Security) are difficult to outright prevent because of the hardwired protocols used.[140] In centrally controlled, distributed systems like SCADA and ICS, downtimes through the loss of availability or the need for system reboots and restorations from backup (outlined in Chapter 4) mean the cure might kill the patient. The fragility of the systems and the need for always-on availability also means that many systems cannot be penetration tested.

System restorations take time, and large geographical or temporal power outages will have cascade effects to other parts of critical infrastructure. This was apparent in a widespread power loss in England in

August 2019.[141] The attack might not be restricted to electrical generation and distribution although this is most likely to have the largest effect. Water treatment and supply would also have a demonstrable impact to life and health. As Chapter 3 detailed, a series of orchestrated attacks at a regional or countrywide scale could affect all public utilities in affected areas and produce cascade effects to all other elements of critical infrastructure. Whether from an 'insider' or outsider's point of view, cyber security threats are growing and they are real with a capacity to do untold economic, social, political, and physical harm. It is not safe to assume 'this couldn't happen to me'.[142]

The Cybersecurity Workforce Deficit

Despite high levels of investment in technical cyber security products and services, currently cyber security threats continue to grow.[143] The global shortfall in personnel involved in cybersecurity is not helping combat this situation. 209,000 cybersecurity vacancies went unfilled in the United States alone in 2015.[144] By 2019 this had grown to nearly 314,000. Globally, that number was projected by the Center for Strategic and International Studies (CSIS) study to be in excess of 1.8 million by 2022.[145] This might be un underestimate, even prior to the Covid-19 pandemic which increased remote working. The annual (ISC)² Cybersecurity Workforce Study in 2021 estimated this gap to be 2.72 million (a decrease for the second year in a row from 3.12 million). Those in the 'available pool' of cybersecurity professionals was estimated at 4.19 million worldwide.[146] Although this is being addressed, this is partly a result of a shortage of people trained in Science, Technology, Engineering, and Mathematics (STEM) subjects and years of relative underinvestment across industry and from governments. A 2016 joint report by Intel and CSIS, 'Hacking the Skills Shortage: A study of the international shortage in cybersecurity skills', argued there was a direct correlation. James A Lewis, the director of the Strategic Technologies Program at CSIS noted that the report's findings pointed that:

A shortage of people with cybersecurity skills results in direct damage to companies, including the loss of proprietary data and IP…This is a global problem; a majority of respondents in all countries surveyed could link their workforce shortage to damage to their organization…there are no signs of this workforce shortage abating in the near-term…The demand for cybersecurity professionals is outpacing the supply of qualified workers, with highly technical skills the most in need across all countries surveyed.[147]

Moreover, "ICS security is a specialized field, generally with fewer practitioners than the IT security field" and staff shortages can directly affect the vital utilities that we have come to rely and depend on.[148] This specialized shortfall increases the drive towards increasing automation and remote accessibility and with it the attendant risks that work-from-home or work-from-anywhere business practices bring.[149]

Although more companies are employing Chief Information Security Officers (CISOs), "there is not a great wealth of talent to fill those positions however high paying" and "Cyber-security has long been plagued by a yawning skills gap, and a drought of real world experience in the field".[150] Moreover, Chief Information Officers (CIOs) and CISOs have traditionally reported to company boards, not sat on them.[151] This might be changing but hitherto cybersecurity has been treated as a costly but "necessary burden…adding little to the bottom line".[152] In addition, "Data breaches can now cost a chief executive his or her job and for a company, remediation can come with a large price tag. Ransomware, an increasingly popular weapon, can cost a company millions in lost work hours and revenue".[153] It can also cost those tasked with protecting their assets their job or force SMBs/SMEs to downsize, go out of business, or into mergers.

ISACA, a long-standing computer industry body, found (pre-Covid) in the 'State of Cybersecurity 2019' that "enterprises continue to struggle with cybersecurity staffing" because "Technically proficient cybersecurity professionals continue to be in short supply and difficult to find".[154] Retention was also an issue and 58% of posts went unfilled. This has led to an "increased use of contract employees and outside consultants" and a reliance on artificial intelligence.[155] This increased use of contract employees and outside consultants too is opening up attack vectors for corporate espionage and especially for state-based APTs based in Russia and China.

Respondents were not asked about "the effectiveness of these artificial intelligence tools as replacements for trained and experienced human beings" but did find that "Although these software solutions and networked algorithms are becoming more prevalent in all fields, they have yet to prove themselves as capable of complete workforce replacement".[156] 70% believed they were understaffed, and 20% significantly understaffed. This is likely not only to add pressure, increase retention problems, but also lead to shortcuts and mistakes. This is leading to "an overall perception of under-preparedness..."[157] Cybersecurity also remains a male-dominated profession.[158]

ISACA's 2019 global study also found that "although some academic institutions are implementing successful technical programs, most are still perceived as training cybersecurity in abstraction, rather than training it as a technical, hands-on field, which, by its very nature, requires some business intelligence".[159] This means that the:

> cybersecurity workforce gap is becoming more pronounced as talent becomes more difficult to find. These strong headwinds are compounded by a suppliers' market for the talent that exists, creating a highly competitive environment in which traditional retention strategies—such as education and certification incentives—lose out to incentives like greater pay and career advancement that entice individuals to seek employment at a different enterprise...[and] expectations of lower budgets will continue to feed the reality of underfunded security programs and insufficient staff.[160]

As a result of Covid-19, ISACA's 2021 report cautioned "Because 2020 was anything but typical, readers are cautioned against interpreting any sizable shifts in workforce estimates during this period. Location and government mandates highly influenced which work was permissible and how that work was to be done. Government responses to the pandemic varied by country, region and locality".[161]

Before the Covid-19 pandemic struck, 3.5 million cybersecurity jobs were anticipated to go unfilled by 2021. In the United States in 2017,

780,000 were employed in cybersecurity but by 2021, 500,000 more were already needed. Demand in countries such as India is even greater; mostly to combat cybercrime. The cost to business globally from cybercrime was set to double from $3 trillion in 2015 to $6 trillion by 2021.[162] At least as importantly "The likelihood for cyber attacks against utilities is increasing in frequency and severity of attacks" coupled with a "lack of detection and monitoring capabilities" making it problematic to analyze the methods used; including APT intrusions.[163] This shortfall is also borne out by the Cybersecurity Workforce Study conducted by (ISC)^2, the world's largest nonprofit association of certified cybersecurity professionals.[164] More widely, the Organization for Economic Cooperation and Development (OECD) indicated that 40% of new jobs created between 2005 and 2016 demanded ICT skills, evenly spread across all 38 OECD member states.[165] A 2018 OECD study also found that shortages in ICT are the top requirement for the labor market in OECD and EU countries.[166]

A 2018 report by the World Economic Forum also highlighted how increasing movement towards new technologies and increasing automation in the 'Fourth Industrial Revolution' is set to change employment practices across all sectors.[167] The Covid-19 pandemic might have accelerated these pre-existing trends as more people were compelled to work from home. During the pandemic ISACA's 2021 "reporting suggests that the cybersecurity profession— albeit understaffed and overworked— rose to the occasion, enabling enterprises across the globe to pivot very quickly to a wholly or mostly remote workforce".[168]

These efforts notwithstanding, cybercrime continued to be a growth area. Interpol cautioned that as a result of Covid-19, "the growing number of people relying on online tools overburdens the security measures put in place prior to the virus outbreak, [as] offenders search for more chances of exposure to steal data, make a profit or cause disruption".[169] In terms of the future, ISACA's 2021 survey postulated that: "Although not every industry or occupation is conducive to remote work, the pandemic is proving that a great deal of work can be performed outside the traditional office—often with little impact on business. Some enterprises are pleasantly discovering an increase in productivity whilst employees work remotely during the pandemic, which may forever sunset business-as-usual mindsets that have long bolstered exorbitant travel budgets and expansive capital expenditures".[170]

Computer Emergency Response Teams

Whether or not increased remote working is set to remain, a number of organizations provide detailed guidelines and helpful advice on good and best practice. Perhaps the most important are national Computer Emergency Readiness/Response Teams (CERTs). As far back as 1997, during the Clinton administration, the vulnerability of U.S. critical infrastructure to cyberattack and catastrophic failure was recognized in a report by the President's Commission on Critical Infrastructure Protection.[171] In 2003 President George W. Bush established the first national government CERT under the DHS (US-CERT). In late 2018 the Cybersecurity and Infrastructure Security Agency Act was signed by President Trump designed for "leading cybersecurity and critical infrastructure security programs, operations, and associated policy".[172]

With around eighty percent of critical infrastructure owned and operated by the private sector, the act "elevated the mission of the former National Protection and Programs Directorate (NPPD) within the Department of Homeland Security (DHS) and established CISA [the Cybersecurity and Infrastructure Security Agency], which includes the National Cybersecurity and Communications Integration Center (NCCIC)".[173] Established in 2009, the NCCIC is the "flagship cyber defense, incident response, and operational integration center" which since 2017 begun integrating "functions previously performed independently by the U.S. Computer Emergency Readiness Team (US-CERT) and the Industrial Control Systems Cyber Emergency Response Team (ICS-CERT)".[174]

CISA also established the National Risk Management Center (NRMC) as a planning, analysis, and collaboration center for critical infrastructure. Additionally, "Like NPPD before it, CISA also oversees within DHS the Federal Protective Service (FPS), the Office of Cyber and Infrastructure Analysis (OCIA), the Office of Cybersecurity & Communications (OC&C) and the Office of Infrastructure

Protection (OIP)". This placed CISA on "par with the Secret Service or Federal Emergency Management Agency (FEMA)". It aims are to tackle long-standing problems as "Neither government nor the private sector alone has the knowledge, authority, or resources to do it. Public-private partnerships are the foundation for effective critical infrastructure security and resilience strategies, and timely, trusted information sharing among stakeholders...essential to the security of the nation's critical infrastructure".[175]

This 'whole-of-nation' approach is hampered by political division in the United States but also because, as in many other neo-liberal economies, private industry is responsible for self-protection. It is not a direct government responsibility although there are industry regulators. Whilst both government and private industry have shared goals and do cooperate and share intelligence-information in defined areas who has clearance is often limited to a small number. Improved cooperation and intelligence sharing is paramount in defending CI and a reason Information Sharing and Analysis Centers (ISACs) were established. These partnered with US-CERT alongside the Protected Critical Infrastructure Information (PCII) Program run by the DHS.[176]

In Europe, the EU's European Network and Information Security Agency (ENISA) states that national Computer Emergency Response Teams (CERTs)/Computer Security Incident Response Teams (CSIRTs) have different roles across the EU and beyond. One common denominator is that "every country that is concerned about protecting its digital assets, starting from sensitive government information to its citizens and their information".[177] To varying degrees CERTs and CSIRTs also protect critical infrastructure and to these ends work with law enforcement and other agencies, including intelligence agencies, to protect information and assets and to practice crisis management procedures and exercises. They are an important part of the cybersecurity fence but there are different levels of 'maturity' of national CERTs/CSIRTs for financial, logistical, and other reasons.[178]

ENISA, in a December 2015 document, recognized that "Some countries begin with defining a national cyber security strategy, following later with appropriate legislation and the establishment of a CSIRT.

Other countries start with the establishment of a CSIRT and the legislative basis and strategies follow later". Furthermore, "Historically, many national and governmental CSIRTs have developed from very informal, sometimes ad hoc groups of highly skilled and motivated people" but now there is growing recognition this needs greater organization and governance and an evolution of capabilities.[179] The need for better coordination and a 'one stop shop' for cybersecurity incident reporting and response was a prime motivator in establishing the UK's National Cyber Security Centre (NCSC) in 2016.[180]

Various national CERTs have however been criticized for ad hoc decision-making, for taking credit for reducing incidents of cybercrime when it is widely believed most incidents go unreported, and for governments being slow to respond with multiple (and overlapping) layers of responsibility.[181] Moreover, sharing threat intelligence can be a sticking point between national governments who seek not only to guard their security arrangements against espionage but how, where, and when to share that intelligence with a friendly or allied government. Sharing intelligence can mean that it is open to compromise by adversaries and by corrupt insiders or moles.[182]

As Chapter 3 and this chapter have described, sharing good and best practices is encouraged by and facilitated by governments, supranational organizations like the EU, and industry bodies as well as through a series of public and private bodies and initiatives. Whilst the EU has the European Programme for Critical Infrastructure Protection, "a package of measures aimed at improving the protection of critical infrastructure in Europe" and the European Programme for Critical Infrastructure Protection (EPCIP), the EU and European Economic Area (EEA) remain a patchwork quilt of national regulation, regulators, organizations, and action plans that vary widely in their scope and maturity. This patchwork quilt is also reflected in intelligence sharing arrangements; a significant portion of which are founded on self-help groups and contacts between private industry and between private industry and governments.[183]

Cyber Threat Intelligence and the Cybersecurity Community

The cybersecurity industry actively tries to find workable and cost-effective solutions to cyber threats. This includes in the analysis, development, and uses of malware (which are often grouped together as malware families as they are developed or repurposed) and the tactics, techniques, and procedures (TTPs) of the actors using them.[184] This is part of the growing field of Cyber Threat Intelligence (CTI). Increasingly cybersecurity firms, several of which work with governments, offer CTI as part of their portfolio of offerings.

Although large companies, including multinationals, are likely to have their own in-house teams, small to mid-size businesses are likely to require bought-in solutions and even governments and multinationals employ external cybersecurity providers from the private sector. These private sector providers include Cisco, Symantec, FireEye-Mandiant, McAfee, and Palo Alto Networks of California's famed Silicon Valley. Other notable providers include Proofpoint of Sunnyvale, California, and San Francisco's Splunk whilst the Atlanta-based SecureWorks is one of the providers of Security Operations Centers (SOCs) and services. One of the first ICT industry giants, IBM, headquartered in New York, also provides a range of cybersecurity offerings including in Security Information and Event Management (SIEM), providing real-time analysis of security alerts produced by applications and networked hardware. Microsoft, best known for its operating systems, also provides cybersecurity products. Forcepoint, from Austin, Texas specializes in Cloud and network security. Massachusetts-based RSA (owned by Dell Technologies) include SIEM and endpoint security and Dell EMC cloud and data protection. In addition, Florida-based KnowBe4 is a leading provider of cybersecurity training, especially against phishing.

Outside of America, Tel Aviv is home to the Israeli firm Check Point and British Telecom, based in London and Sophos of Oxford are two of the UK's leading providers whilst Tokyo is home to Trend Micro. Other providers include Carbon Black, Darktrace, and Moscow-based Kaspersky Labs. FireEye is a key provider of CTI as is Kaspersky (although its Russian origins have made many wary of its products and services).[185] A number of these companies feature heavily in this book.

Finally, anyone who has been to a cybersecurity expo, industry conference, or event will find a bewildering array of cybersecurity products and services, several of whom offer similar or competing offerings and new start-ups can quickly become major players or can be bought out by larger, established providers. Finding the right products and services can be mystifying and is often left to CIOs, CISOs, middle managers, or small IT teams to decide. Board awareness of ICT issues only tends to be prioritized after an incident or breach.

Industry and Government Backed Self-Help Groups

There are good and best practices they try and follow. Information security polices flow not only from companies but from industry self-help groups, the wider cybersecurity community (which include universities who actively work with industry and governments), and government bodies such as CERTs. The implementation of best practices is trickier and depends on available resources and personnel. Creating a business continuity plan should things go wrong is helpful if not already in place.

Should the worst happen having disaster recovery plans, which includes a prior audit through the use of network discovery tools which will list and locate ICT assets, will help. Automated network discovery tools are increasingly important as "Utilities may lack comprehensive awareness of all control system assets online at any given time and/or a party knowledgeable of all network interconnections. This also diminishes utilities' detection capabilities for cyber intrusions, as well as unauthorized devices or connections to networks".[186] Network traffic scanning will help to decide if threats are ongoing and if immediate isolation of critical networks is needed. Pre, post, and active defenses should be able to isolate affected systems. If standard data storage policies have been adhered to, restoration from backup files (such as data historians) will be rapidly required.

One of the industry standards, especially in the field of critical infrastructure is the advice provided by the U.S. National Institute of Standards and Technology (NIST) based in Gaithersburg, Maryland.[187] The aforementioned ISACA promotes the NIST framework.[188] The closely named Information Sharing and Analysis Centers (ISACs) introduced in the U.S. 1999 by the Clinton administration are designed around cooperation between national government and private sector owner-operators of critical infrastructure with a National Council of ISACs (NCI) formed in 2003. There are currently twenty-six ISACs. Each oversees their own distinct sector.

The NCI acts as a hub for the ISACs and "organizes its own drills and exercises and participates in national exercises". As well as this, a member of the NCI council is housed within the "National Cybersecurity and Communications Integration Center (NCCIC) watch floor" and "representatives can embed with National Infrastructure Coordinating Center (NICC) during significant national incidents".[189] ISACs cover a broad range of sectors and are designed to "help critical infrastructure owners and operators protect their facilities, personnel and customers from cyber and physical security threats and other hazards. ISACs collect, analyze and disseminate actionable threat information to their members and provide members with tools to mitigate risks and enhance resiliency".[190]

A decade after their establishment in the U.S., ISACs were created in the European Union by ENISA (now the European Union Agency for Cybersecurity) which helps support the NIS Directive.[191] India also began their own ISAC in the early 2000s.[192] In addition, the U.S.-based Internet Security Alliance (ISA) offers guidelines for corporate cyber security and public–private engagement with input from a large number of multi-sector corporations.[193] The Center for Internet Security based in New York is another industry self-help group founded by ISACA, so is ISC^2 as is the SANS Institute (a widely respected cybersecurity training provider). The Pittsburgh-based National Cyber-Forensics and Training Alliance (NCFTA) is another nonprofit public–private partnership against cybercrime.[194] Meanwhile the Council on Foreign Relations has run since 2005 a very useful open-source Cyber Operations Tracker of known state-sponsored incidents.[195] Outside of the U.S., the UK-based International Cyber Security Protection Alliance (ICSPA) seeks to provide a private sector financed hub to aid law enforcement and until it was disbanded, the Information Assurance Advisory Council (IAAC) advised UK government and industry on the use of secure hardware and software.[196] This has now given way to the UK Cyber Security Council.[197] These efforts and similar bodies can be found in a number of other nations and regions. This is not to say improvements cannot still be made.[198]

Conclusion

The BBC reported a decade ago that:

> Social engineering has become the confidence trick of the 21st century. The term will be familiar to anyone in the online security world as hackers and cyber criminals increasingly use social engineering methods to manipulate people into handing over information or giving them access to a computer. In response to the growing phenomenon, firms are barricading their IT systems, forgetting that employees may themselves be exposing their companies to risk. "People are always the weakest link".[199]

The growing number of threats and threat actors and the widespread and successful use of social engineering attacks would indicate that either our behavior in a cyber environment is still not sufficiently cautious or the threats are emerging in ways that are limiting our ability to counter them. OSINT is proliferating, can be agglomerated and analyzed through software, and this is a boon to the social engineer or espionage professional.

Michael Workman describes how, a "Perceived severity of threat will lead people to behave in a more cautious manner if their perception of the damage or danger increases. The reverse of this, however, is also true: when people perceive that a risk has diminished, they will behave in a less cautious manner".[200] Workman adds, "Threat assessment research…indicates that when a threat is perceived, people adjust their behavior according to the amount of risk from the threat that they are willing to accept (sometimes known as risk homeostasis). This adjustment is based on the degree or severity and costs of damage they perceive to be associated with the threat".[201] The extent to which the Covid-19 pandemic and the acceleration of working remotely will alter this perception or increased level of risk is difficult to forecast but espionage, practiced by intelligence professionals, and social engineering share some basic characteristics.

Both employ cognitive and behavioral psychology which for social engineers can include very simple techniques to pull off the 'confidence trick'. Utilizing espionage or simpler reverse 'social engineering' (using their techniques back on them) could in some cases locate potential intruders prior to the stage of developing full attack abilities and specific attack corridors. The surveillance, recruiting, managing, and manipulation of human information sources (the tradecraft of espionage) can become a significant advantage in the field of information security and cyber threat intelligence. It can aid computer forensics and the work of the intelligence and security services and police and law enforcement. Additionally, technical countermeasures in cybersecurity can help mitigate outsider threats but might not always be successful against compromised insiders.

It is worth noting that the 2019 State of the Phish survey saw information security teams "shifting to a more people-centric model by proactively identifying phishing susceptibility, measuring end user risk, and delivering regular security awareness training".[202] Security awareness that promotes behavioral change which aid healthy cyber hygiene practices by moderating innate human behaviors which are exploitable to 'social engineers' and protecting against threats and risks related to information and physical assets is to be welcomed. This is largely a social psychological effort involving dynamic skepticism related to cybersecurity and the cyber-physical environment. Nevertheless, the field of 'social engineering', contains some highly sophisticated experts who combine technological knowledge with a sharp understanding of psychological biases. This enables them to compromise businesses and governments. Various degrees of risk will remain regardless of cybersecurity and physical defenses, and are a feature of an open society.

A successful social engineering attack can bypass millions of dollars of investment in technical security to expose an organization's critical information and a nations critical infrastructure. Detection is not prevention and there are many other ways to get into organizations. This can even include through the children of employees. These can be used to gain passwords or used as leverage.[203] This is a particularly insidious example of the exploitation of SE, but it demonstrates the palpable need for holistic thinking and the integration of a variety of tiered mitigations that include integrating human-focused thinking and behavioral patterns and methodologies into the cybersecurity skill set.

Reverse social engineering and counterintelligence and counterespionage (including HUMINT) can help identify attack vectors, vertical or lateral movements within target organizations, alongside remedies and modeling using cognitive and behavioral psychology. These can be used to try and induce an attacker to desist. It could also help generate investigative routes and aid technical threat hunting if the response is joined between the social engineering side of the attack, the technical methods used, and understanding their goals.

Offensively (or defensively depending on your point of view), intelligence organizations use HUMINT virtually in cyberspace for targeting and running human sources and for infiltrating Foreign Intelligence Services, terrorist cells, and criminal gangs. Persons of interest in the real world and the organizations they work for or affiliated to are first spotted, researched, surveilled, and then targeted.[204] HUMINT or reverse social engineering (as a form of counterintelligence) can be used by large companies targeted by individuals and organized criminal groups and can aid their responses and response times. Understanding competitive intelligence/industrial espionage between companies is a useful starting point and although counterintelligence has been traditionally associated with national intelligence agencies, it clearly has applications for private companies.[205] Indeed private companies, including multinational telecommunications backbone providers like Cisco Systems, actively recruit former intelligence professionals including former Chiefs of Station.[206] Insider threats might also be lessened through the use of psychologists and by strict(er) vetting procedures with deep background checks (as in intelligence agencies like the CIA). However, given the scale of critical infrastructure and the global shortfall of trained staff across the ICT industry, this could cause more problems than it solves.

Additionally, most attackers seek to cover their tracks making attribution difficult. Attribution might not be completed in time to aid the target. Also, the global shortfall in highly trained ICT personnel means expertise might well not be available or remedial action prove not to be financially viable. This is a case where the help of law enforcement and the intelligence agencies are needed. For many firms facing a financial and reputational loss this is a decision based on a cost–benefit analysis. In the SCADA-ICS-CI environment this decision could impact life and the well-being of local and national populations making it a national security issue. This again is where government support should (ideally) be able to be leveraged. This too however depends on the capability and capacity to respond in time. With major state players in cyberspace as APTs this continues to pose significant problems. This includes Russia, who frequently uses cybercriminals as proxies. This is the subject of the next chapter and considered alongside other non and sub-state actors, including terrorists and hackers.

Notes

1. David V. Gioe, Michael S. Goodman and Alicia Wanless, 'Rebalancing Cybersecurity Imperatives: Patching the Social Layer', *Journal of Cyber Policy*, Vol. 4, Issue 1 (March 2019), pp. 117–137.
2. Snigdha Poonam, 'The Scammers Gaming India's Overcrowded Job Market' (January 2 2018), https://www.theguardian.com/news/2018/jan/02/the-scammers-gaming-indias-overcrowded-job-market, accessed 20 February 2020.
3. 'Protect Yourself from Tech Support Scams' (undated), https://support.microsoft.com/en-us/help/4013405/windows-protect-from-tech-support-scams, accessed 20 February 2020.
4. FireEye and Mandiant, 'Cybersecurity's Maginot Line: A Real-World Assessment of the Defense-in-Depth Model', p. 13. Available from http://www2.fireeye.com/rs/fireye/images/fireeye-real-world-assessment.pdf, accessed 27 September 2014.

5. 'Proofpoint State of the Phish 2019 Report', p. 22, https://www.cyqueo.com/images/download/Proofpoint-US-TR_State-Of-The-Phish-2019-CYQUEO.pdf, accessed 28 July 2019.
6. 'Verizon 2021 Data Breach Investigations Report', http://verizon.com/dbir/, p. 49.
7. 'Business Email Compromise (BEC)', https://www.trendmicro.com/vinfo/us/security/definition/business-email-compromise-(bec), accessed 3 August 2019.
8. 'Verizon 2021 Data Breach Investigations Report', http://verizon.com/dbir/, p. 49.
9. 'Verizon Data Breach Investigations Report 2022', pp. 8, 11, 13, https://www.verizon.com/business/resources/reports/2022/dbir/2022-data-breach-investigations-report-dbir.pdf, accessed 11 June 2022.
10. 'Verizon Data Breach Investigations Report 2022', p. 34.
11. Joel T. Langhill, 'Defending Against the Dragonfly Cyber Security Attacks' (December 10 2014), p. 21. Available from https://www.controlglobal.com/assets/15WPpdf/150311-Belden-DragonflyCybersecurity.pdf, accessed 5 October 2018.
12. Finding and/or uncovering an enemy agent/s or infiltration is very much second best to turning them to become 'double agents'.
13. Chan D. Lieu, 'Social Engineering—Attacking the Weakest Link'. Available from https://www.giac.org/paper/gsec/2082/social-engineering-attacking-weakest-link/103563, accessed 9 September 2016.
14. Amit Steinhart, 'The Future Is Behind Us? The Human Factor in Cyber Intelligence: Interplay Between Cyber-HUMINT, Hackers and Social Engineering'. Available from http://www.amitsteinhart.com/, accessed 12 September 2016.
15. 'Verizon's 2016 Data Breach Investigations Report Finds Cybercriminals Are Exploiting Human Nature', http://www.prnewswire.com/news-releases/verizons-2016-data-breach-investigations-report-finds-cybercriminals-are-exploiting-human-nature-300258134.html, accessed 14 September 2016.
16. See for example Christopher Hadnagy, *Social Engineering: The Art of Human Hacking* (London: Wiley, 2010), Christopher Hadnagy, *Unmasking the Social Engineer: The Human Element of Security* (London: Wiley, 2014), Kevin D. Mitnick, and William L. Simon, *The Art of Deception: Controlling the Human Element of Security* (London: Wiley, 2003), Kevin D. Mitnick, and William L. Simon, *The Art of Intrusion: The Real Stories Behind the Exploits of Hackers, Intruders*

and *Deceivers* (London: Wiley, 2005), and Michael Bazzell, *Open Source Intelligence Techniques: Resources for Searching and Analyzing Online Information* (Charleston: CreateSpace Independent Publishing Platform; 5th edition, 2016).
17. See for example Christopher Hadnagy and Michele Fincher, *Phishing Dark Waters: The Offensive and Defensive Sides of Malicious Emails* (London: Wiley, 2015).
18. Proofpoint State of the Phish 2019 Report, p. 21.
19. Stanley Milgram, 'Behavioral Study of Obedience', *Journal of Abnormal and Social Psychology*, Vol. 67, Issue 4 (October 1963), p. 371.
20. Milgram, 'Behavioral Study of Obedience', p. 371. His full findings are on pp. 371–378. Milgram later wrote a book on these findings. Stanley Milgram, *Obedience to Authority: An Experimental View* (New York: Harper and Row, 1974). See also Roger Brown, *Social Psychology: Second Edition* (New York: The Free Press, 1986). The experiment can be found on YouTube. See also Albert Bandura, 'Social Cognitive Theory: An Agentic Perspective', *Annual Review of Psychology*, Vol. 52 (2001), pp. 1–26.
21. See for example Michael A. Wallach, Nathan Kogan, and Daryl J. Bem, 'Diffusion of Responsibility and Level of Risk Taking in Groups', *Journal of Abnormal and Social Psychology*, Vol. 68, No. 3 (March 1964), pp. 263–274.
22. 'Derren Brown', http://derrenbrown.co.uk/about/, accessed 24 October 2016.
23. Derren Brown, *Absolute Magic A Model for Powerful Close-Up Performance* (London: H & R Magic Books, 2001), Derren Brown, *Direct Mindreading and Magical Artistry* (London: H & R Magic Books, 2002), Derren Brown, *Tricks of the Mind* (London: Channel 4 Books, 2006), Derren Brown, *Confessions of a Conjuror* (London: Channel 4 Books, 2010). See also Nick Kolenda, *Methods of Persuasion: How to Use Psychology to Influence Human Behavior* (No location: Kolena Entertainment, 2013), For scholarly work on these subjects see for example Robert B. Cialdini, *Influence: The Psychology of Persuasion* Revised edition (New York: Harper Business 2008), Dan Ariely, *Predictably Irrational: The Hidden Forces That Shape Our Decisions* (London: HarperCollins 2008), Noah J. Goldstein, Steve J. Martin, Robert B. Cialdini, *Yes! 50 Secrets from the Science of Persuasion* (New York: Simon & Shuster, 2008).

24. T. Melcher, D. Winter, B. Hommel, R. Pfister, P. Dechent, and O. Gruber, 'The Neural Substrate of the Ideomotor Principle Revisited: Evidence for Asymmetries in Action-effect Learning', *Neuroscience*, Vol. 231 (February 2013), pp. 13–27.
25. Ian Mann, *Hacking the Human: Social Engineering Techniques and Security Countermeasures* (Aldershot: Gower, 2008), pp. 242–244.
26. See for example Richard Bandler and John Grinder, *Frogs into Princes: Neuro Linguistic Programming* (Utah: Real People Press, 1979).
27. Gareth Roderique-Davies, 'Neuro-Linguistic Programming: Cargo Cult Psychology?', *Journal of Applied Research in Higher Education*, Vol. 1, Issue 2 (2009), pp. 58–63.
28. Doug O'Brien, 'Ericksonian Language Patterns' (July 19 2008), https://ericksonian.com/ericksonian-language-patterns, accessed 21 February 2020. For an explanation of the methods and influence of Milton H. Erickson (especially in hypnotherapy) see for example Hugh Gunnison, 'Hypnocounseling: Ericksonian Hypnosis for Counselors', *Journal of Counseling & Development*, Vol. 68, Issue 4 (March/April 1990), pp. 450–453.
29. Paul Tosey and Jane Mathison, *Neuro-Linguistic Programming: A Critical Appreciation for Managers and Developers* (Basingstoke: Palgrave Macmillan, 2008), p. 144.
30. www.levesoninquiry.org.uk/wp.../Witness-Statement-of-Ian-Hurst.pdf, accessed 20 May 2011. Quoted in R. Gerald Hughes and Kristan Stoddart, 'Hope and Fear: Intelligence and the Future of Global Security a Decade after 9/11', *Intelligence and National Security*, Vol. 27, No. 5 (October 2012), pp. 642–643.
31. www.levesoninquiry.org.uk/wp.../Witness-Statement-of-Ian-Hurst.pdf, accessed 20 May 2011.
32. https://www.social-engineer.com/about/, accessed 19 September 2016.
33. https://digitalguardian.com/blog/social-engineering-attacks-common-techniques-how-prevent-attack, accessed 19 September 2016.
34. https://theantisocialengineer.com/about-us/, accessed 19 September 2016.
35. See for example 'Hacking Humans & Social Engineering', https://www.youtube.com/watch?v=r5kd0KZ_MVs, 'DefCamp 2014—Social Engineering, or "Hacking People"', https://www.youtube.com/watch?v=JAOTRgWdPTU, 'Defcon 21—Social Engineering: The Gentleman Thief', https://www.youtube.com/watch?v=1kkOKvPrdZ4, all accessed 19 September 2016.

36. Ian Mann, *Hacking the Human 2* (Whitley Bay: Consilience Media, 2013) and Ian Mann's earlier book *Hacking the Human*, Christopher Hadnagy, *Unmasking the Social Engineer*.
37. http://www.exabeam.com/ and https://www.crowdstrike.com/, both accessed 25 October 2016.
38. See for example Monika Aring, 'Background Paper Prepared for the Education for All Global Monitoring Report 2012 Youth and Skills: Putting Education to Work Report on Skills Gaps' (2012), http://unesdoc.unesco.org/images/0021/002178/217874e.pdf, accessed 25 October 2016.
39. Josh Fruhlinger, 'What Is Phishing? How This Cyber Attack Works and How to Prevent It' (May 9 2019), https://www.csoonline.com/article/2117843/what-is-phishing-how-this-cyber-attack-works-and-how-to-prevent-it.html, accessed 11 July 2019.
40. 'Proofpoint State of the Phish 2019 Report', p. 8.
41. 'Proofpoint State of the Phish 2019 Report', p. 9. Other factors can include class, gender, sexual orientation, ethnicity, and education.
42. 'Proofpoint State of the Phish 2019 Report', p. 12.
43. Fruhlinger, 'What Is Phishing?'.
44. Fruhlinger, 'What Is Phishing?'.
45. Fruhlinger, 'What Is Phishing?'.
46. Noa Bar-Yosef, 'The Structure of a Cybercrime Organization—Hackers Have Supply Chains Too!' (September 23 2010), https://www.securityweek.com/structure-crybercrime-organization-hackers-have-supply-chains-too, accessed 11 July 2019.
47. Bar-Yosef, 'The Structure of a Cybercrime Organization'.
48. 'Verizon 2019 Data Breach Investigation Report', p. 51, https://enterprise.verizon.com/resources/reports/2019-data-breach-investigations-report.pdf, accessed 13 July 2019.
49. 'Verizon 2021 Data Breach Investigations Report', http://verizon.com/dbir/, p. 52.
50. 'Verizon Data Breach Investigations Report 2022', pp. 33–35.
51. 'Proofpoint State of the Phish 2019 Report', p. 17.
52. 'Proofpoint State of the Phish 2019 Report', p. 19.
53. 'Common Features of Phishing Emails' (undated), https://www.phishing.org/what-is-phishing, accessed 6 August 2019.
54. 'Proofpoint State of the Phish 2019 Report', pp. 13–14.
55. Simple countermeasures to business practise issues are already well established. See for example Peter O. Okenyi and Thomas J. Owens,

'On The Anatomy of Human Hacking', *Information Systems Security*, Vol. 16, Issue 6 (December 2007), pp. 302–314.
56. 'Global Energy Cyberattacks: "Night Dragon"' (February 10 2011), p. 3, https://securingtomorrow.mcafee.com/wp-content/uploads/2011/02/McAfee_NightDragon_wp_draft_to_customersv1-1.pdf, accessed 7 February 2019.
57. 'Increase in Insider Threat Cases Highlight Significant Risks to Business Networks and Proprietary Information' (September 23 2014), https://www.ic3.gov/media/2014/140923.aspx, accessed 9 September 2018.
58. Rory Cellan-Jones, 'Hacking the House: Do MPs Care About Cyber-Security?' (December 3 2017), https://www.bbc.co.uk/news/technology-42217017, accessed 17 March 2020.
59. 'Increase in Insider Threat Cases'.
60. Anonymous, 'Marcus Hutchins Spared US Jail Sentence over Malware Charges' (July 26 2019), https://www.bbc.co.uk/news/technology-49127569, accessed 2 August 2019.
61. 'Social Engineering', https://null-byte.wonderhowto.com/how-to/social-engineering/, accessed 2 August 2019.
62. 'The Social Engineering Framework', https://www.social-engineer.org/framework/se-tools/computer-based/social-engineer-toolkit-set/, accessed 2 August 2019.
63. 'Metasploit The World's Most Used Penetration Testing Framework', https://www.metasploit.com/, accessed 2 August 2019.
64. J.M. Porup, 'What Is Metasploit? And How to Use This Popular Hacking Tool Metasploit is a Widely Used Penetration Testing Tool That Makes Hacking Way Easier Than It Used to Be. It Has Become an Indispensable Tool for Both Red Team and Blue Team' (March 25 2019), https://www.csoonline.com/article/3379117/what-is-metasploit-and-how-to-use-this-popular-hacking-tool.html, accessed 2 August 2019.
65. Porup, 'What Is Metasploit?'.
66. A hacking technique that allows an attacker to authenticate to a remote server or service through passwords which are stored not as clear text but as a hash value (a unique identifier used in data storage and retrieval applications).
67. Porup, 'What Is Metasploit?'.
68. Sylvain Leroux, 'Kali Linux Review: Not Everyone's Cup of Tea' (May 9 2019), https://itsfoss.com/kali-linux-review/, accessed 2 August 2019.

69. 'Hacker Tools Top 10' (December 12 2018), https://www.concise-courses.com/hacking-tools/top-ten/, accessed 2 August 2019.
70. 'Maltego CE', https://www.paterva.com/buy/maltego-clients/maltego-ce.php, accessed 2 August 2019.
71. 'Maltego CE'.
72. Raghu Nallani Chakravartula, 'What Is Enumeration?' (February 28 2018), https://resources.infosecinstitute.com/what-is-enumeration/#gref, accessed 3 August 2019.
73. 'Maltego CE', https://www.paterva.com/buy/maltego-clients/maltego-ce.php, accessed 2 August 2019.
74. Charu C. Aggarwal, *Social Network Data Analytics* (New York: Springer, 2011), pp. 1–15 and Judith S. Kleinfeld, 'The Small World Problem', *Society*, Vol. 39, Issue 2 (January/February 2002), pp. 61–66.
75. Jeffrey Travers and Stanley Milgram, 'An Experimental Study of the Small World Problem', *Sociometry*, Vol. 32, No. 4 (December 1969), pp. 425–441. These issues are further detailed in Chapter 6.
76. Judith S. Kleinfeld, 'Could It Be a Big World After All? The "Six Degrees of Separation" Myth' (April 12 2001), https://www.cs.princeton.edu/~chazelle/courses/BIB/big-world.htm, accessed 21 February 2020. Very little systematic study has taken place on the sociology of hacking in different cultures. What has been studied, including in this book and its companion volumes, is their targets and tactics. Whilst these books, drawing on related research, has attempted to better understand the 'sociology of hacking', especially in Russia and China, and drawing on their underlying culture, history and theories, this can be extended and developed far more. The TTPs offer few clues for generating better sociological or psychological understanding.
77. https://inequality.org/facts/wealth-inequality/, accessed 21 February 2020.
78. 'Website Mirroring | Website Hacking #2' (April 15 2019), https://www.allabouthack.com/2019/04/website-mirroring-website-hacking-2.html, accessed 3 August 2019. YouTube contains many tutorials to mirror websites. Many are based on the tool HTTrack.
79. 'Burp Suite Editions', https://portswigger.net/burp and 'Web Application vs Website: What Suits Your Business Better' (June 10 2018), https://www.cleveroad.com/blog/what-is-the-difference-between-website-and-web-application-choose-what-fits-your-business, both accessed 3 August 2019.

80. Dan Ryan, 'Watch How Easy It Is for Your RFID Card to be Cloned By Hackers [Video]' (January 9 2018), https://insights.identicard.com/blog/watch-how-easy-it-is-for-your-rfid-card-to-be-cloned-by-hackers-video, accessed 3 August 2019.
81. Web/Internet advertisements containing malware.
82. Ryan Heartfield and George Loukas, 'A Taxonomy of Attacks and a Survey of Defence Mechanisms for Semantic Social Engineering Attacks', *ACM Computing Surveys (CSUR)*, Vol. 48, Issue 3 (February 2016), pp. 1–41.
83. See for example Thomas R. Peltier, 'Social Engineering: Concepts and Solutions', *Information Systems Security*, Vol. 15, Issue 5 (2006), pp. 13–21, Xin (Robert) Luo, Richard Brody, Alessandro Seazzu and Stephen Burd, 'Social Engineering: The Neglected Human Factor for Information Security Management', *Information Resources Management Journal*, Vol. 24, No. 3 (July–September 2011), pp. 1–8 and Pekka Tetri and Jukka Vuorinen, 'Dissecting Social Engineering', *Behaviour and Information Technology*, Vol. 32, Issue 10 (2013), pp. 1014–1023.
84. '2021 Data Breach Investigations Report', http://verizon.com/dbir/, pp. 50, 52.
85. 'Verizon Data Breach Investigations Report 2022', p. 34.
86. Sam Jones, 'The Spy Who Liked Me: Britain's Changing Secret Service' (September 29 2016), https://www.ft.com/content/b239dc22-855c-11e6-a29c-6e7d9515ad15, accessed 17 June 2019. Quoted in Kyle Cunliffe, 'The Art of Science and Betrayal', unpublished PhD thesis, Aberystwyth University (2019), p. 5.
87. Jones, 'The Spy Who Liked Me: Britain's Changing Secret Service'. See also Danny Steed, 'Disrupting the Second Oldest Profession the Impact of Cyber on Intelligence', in Miriam Dunn Cavelty and Andreas Wenger (eds.), *Cyber Security Politics Socio-Technical Transformations and Political Fragmentation* (New York: Routledge, 2022), pp. 205–219.
88. (U//FOUO) U.S. Army Intelligence Analysis Manual (November 19 2017), https://publicintelligence.net/us-army-intelligence-analysis/, accessed 9 August 2019.
89. Joel Brenner, *America the Vulnerable: Inside the New Threat Matrix of Digital Espionage, Crime, and Warfare* (New York: Penguin Press, 2011), p. 167.
90. Part of my book *Russia's Cyber Offensive Against the West*.

91. Pierre Thomas, Jack Cloherty and Jason Ryan, 'How the FBI Busted Anna Chapman and the Russian Spy Ring' (November 1 2011), http://abcnews.go.com/blogs/politics/2011/11/how-the-fbi-busted-anna-chapman-and-the-russian-spy-ring/, accessed 29 September 2016.
92. 'MI6 'C' Speech on Fourth Generation Espionage' (December 3 2018), https://www.gov.uk/government/speeches/mi6-c-speech-on-fourth-generation-espionage, accessed 7 August 2019.
93. Sir David Omand, Jamie Bartlett, and Carl Miller, 'Introducing Social Media Intelligence (SOCMINT)', *Intelligence and National Security*, Vol. 27, Issue 6 (December 2012), pp. 801–823.
94. Jeff Wise, 'Disrupting Cyberwar with Open Source Intelligence' (October 19 2018), https://www.hpe.com/us/en/insights/articles/disrupting-cyberwar-with-open-source-intelligence-1810.html and Moritz Rakuszitzky, 'Third Suspect in Skripal Poisoning Identified as Denis Sergeev, High-Ranking GRU Officer' (February 14 2019), https://www.bellingcat.com/news/uk-and-europe/2019/02/14/third-suspect-in-skripal-poisoning-identified-as-denis-sergeev-high-ranking-gru-officer/, both accessed 9 August 2019.
95. Larry M. Wortzel, *The Chinese People's Liberation Army and Information Warfare* (Carlisle, PA: United States Army War College Press, 2014), p. 2.
96. Aaron F. Brantly, 'Aesop's Wolves: The Deceptive Appearance of Espionage and Attacks in Cyberspace', *Intelligence and National Security*, Vol. 31, No. 5 (July 2016), p. 675. Background to "Assessing the Russian Activities and Intentions in Recent US Elections": The Analytic Process and Cyber Incident Attribution, 6 January 2017, p. 5, https://www.dni.gov/files/documents/ICA_2017_01.pdf, accessed 2 September 2019.
97. Brantly, 'Aesop's Wolves', p. 675.
98. Avi-Tal and David Siman-Tov, 'HUMINT in the Cybernetic Era: Gaming in Two Worlds', *Military and Strategic Affairs*, Vol. 7, No. 3 (2015), p. 98.
99. On the process of intelligence analysis and the work of intelligence analysts see for example Julian Richards, *The Art and Science of Intelligence Analysis* (Oxford: Oxford University Press, 2011).
100. For detailed information on the intelligence cycle see Mark Phythian, *Understanding the Intelligence Cycle* (Abingdon: Routledge, 2013).

101. Vaughan Sherman, 'How Accurate are Bourne and Bond? As an Ex-CIA Officer' (December 6 2017), https://m.huffpost.com/us/entry/2451700/amp, and Anonymous, 'What Are Spies Really Like?', (April 2 2012), https://www.bbc.co.uk/news/magazine-17560253, both accessed 7 August 2019.
102. 'MI6 'C' Speech on Fourth Generation Espionage'.
103. Jones, 'The Spy Who Liked Me: Britain's Changing Secret Service'. See also Kristan Stoddart, 'Edward Snowden and PRISM: Negotiating the Post 9/11 "Surveillance State"', in Tobias T. Gibson and Kurt W. Jefferson (eds.), *Contextualizing Security: A Reader* (Athens, GA: University of Georgia Press, 2022).
104. Brenner, *America the Vulnerable*, pp. 82–90.
105. Robert Wallace, 'A Time for Counterespionage', in Jennifer E. Sims and Burton Gerber (eds.), *Vaults, Mirrors, & Masks: Rediscovering U.S. Counterintelligence* (Washington, DC: Georgetown University Press, 2008), p. 113.
106. Jones, 'The Spy Who Liked Me: Britain's Changing Secret Service'.
107. Anonymous, "Serious' Hack Attacks from China Targeting UK Firms' (April 4 2017), http://www.bbc.co.uk/news/technology-39478975, accessed 4 April 2017.
108. Joel Brenner, 'How Obama Fell Short on Cybersecurity: Under the President's Proposals, We'll Remain America the Vulnerable' (January 21 2015), http://www.politico.com/magazine/story/2015/01/state-of-the-union-cybersecurity-obama-114411.html#ixzz3PjrwwEnf, accessed 24 January 2015. Brenner's wider views on cybersecurity can be found in Joel Brenner, *Glass Houses: Privacy, Secrecy and Cyber Insecurity in a Transparent World* (New York: Penguin 2013).
109. 'MI6 'C' Speech on Fourth Generation Espionage'.
110. Also commonly referred to variously as Computer Network Attack (CNA), Computer Network Defense (CND), Computer Network Exploitation (CNE), Computer Network Operations (OCO), and Computer Network Reconnaissance (CNR).
111. Brantly, 'Aesop's Wolves', p. 675. Clark and Konrad, on which Brantly bases this argument, found their ideas of an 'offensive advantage' on economic models and game theory. They argue "A defending player needs to successfully defend all fronts, and an attacker needs to win at only one. Multiple fronts result in a considerable disadvantage for the defending player, and even if there is a defense advantage at each of them, the payoff of the defending player is zero if the number of fronts

is large. With some positive probability, in the equilibrium defending players surrender without expending effort". Derek J. Clark and Kai A. Konrad, 'Asymmetric Conflict: Weakest Link Against Best Shot', *Journal of Conflict Resolution*, Vol. 51, No. 3 (June 2007), p. 457, pp. 457–469.
112. For a description and examples of covert action see Loch K. Johnson and James J. Wirtz (eds.), *Intelligence: The Secret World of Spies An Anthology* Third Edition (Oxford: Oxford University Press, 2011), pp. 225–285.
113. Dan Goodin, 'Massive US-Planned Cyberattack Against Iran Went Well Beyond Stuxnet' (February 17 2016), http://arstechnica.co.uk/tech-policy/2016/02/massive-us-planned-cyberattack-against-iran-went-well-beyond-stuxnet/, accessed 1 November 2016.
114. Jeff Goldman, 'FBI, DHS Warn of Surge in Insider Threats from Disgruntled Employees' (September 25 2014), https://www.esecurityplanet.com/network-security/fbi-dhs-warn-of-surge-in-insider-threats-from-disgruntled-employees.html, accessed 8 September 2018.
115. 'Insider Threat—Cyber', https://www.dhs.gov/insider-threat-cyber, accessed 9 September 2018.
116. 'Insider Threat—Cyber'.
117. The male equivalent is known as a 'raven'. Greg Henderson, 'Marine Sgt. Clayton Lonetree Was Convicted of Espionage and...' (August 21 1987), https://www.upi.com/Archives/1987/08/21/Marine-Sgt-Clayton-Lonetree-was-convicted-of-espionage-and/8612556500687/, 'A Chronicle of KGB Seduction, Sexual Entrapment, Incompetence and CIA Arrogance', https://www.youtube.com/watch?v=AkVVWu686VY, both accessed 21 February 2020.
118. Jane Wakefield, 'Office Intruder 'Steals' Data' (May 6 2009), http://news.bbc.co.uk/1/hi/technology/7843206.stm, accessed 22 June 2015.
119. Wakefield, 'Office Intruder 'Steals' Data'.
120. Wakefield, 'Office Intruder 'Steals' Data'.
121. Wakefield, 'Office Intruder 'Steals' Data'.
122. Colin Greenlees, talk given at Digital Wales, Celtic Manor, UK, 4–6 June 2015 attended by author.
123. 'Bump Keys' can also be used. 'Pick a Lock in Seconds with a Bump Key' 14 April 2010), https://www.youtube.com/watch?v=WpH_t0u5Ybg, accessed 16 September 2018. This video, uploaded by 'Scam School' (who have 1.94 million subscribers) and dedicated to social

engineering, has been viewed 21 million times. Based on previous visits to the site, this is a rapidly growing number.
124. 'The PlugBot: Hardware Botnet Research Project', https://www.redteamsecure.com/the-plugbot-hardware-botnet-research-project/, accessed 16 September 2018. It is open source code available on the developer platform Github. https://github.com/redteamsecurity/PlugBot-Plug, accessed 16 September 2018.
125. 'Hackers Love This Tiny Box' (September 9 2016), https://www.youtube.com/watch?v=r0jBLzmrH9w&feature=youtu.be, accessed 17 September 2018.
126. redteamsecuritytraining/PlugBot-Plug, https://github.com/redteamsecuritytraining/PlugBot-Plug, accessed 21 February 2020.
127. 'Watch Hackers Break into the US Power Grid' (May 11 2016), https://www.youtube.com/watch?v=pL9q2lOZ1Fw&feature=youtu.be, accessed 16 September 2018.
128. Lucian Constantin, 'Researchers Show Ways to Bypass Home and Office Security Systems Many Door Sensors, Motion Detectors and Security Keypads Can Be Bypassed Using Simple Techniques, Researchers from Bishop Fox Said' (July 31 2013), https://www.csoonline.com/article/2133815/physical-security/researchers-show-ways-to-bypass-home-and-office-security-systems.html, accessed 17 September 2018.
129. 'Cyber Threat and Vulnerability Analysis of the U.S. Electric Sector Prepared by: Mission Support Center Idaho National Laboratory August 2016', p. 11, https://www.energy.gov/sites/prod/files/2017/01/f34/Cyber%20Threat%20and%20Vulnerability%20Analysis%20of%20the%20U.S.%20Electric%20Sector.pdf, accessed 12 August 2019.
130. Lauren J. Sullins, '"Phishing" for a Solution: Domestic and International Approaches to Decreasing Online Identity Theft', in Indira Carr (ed.), *Computer Crime* (London: Routledge, 2007). See also Lauren Silverman, 'There Are Plenty of RFID-Blocking Products, But Do You Need Them?' (July 4 2017), https://www.npr.org/sections/alltechconsidered/2017/07/04/535518514/there-are-plenty-of-rfid-blocking-products-but-do-you-need-them?t=1537186154889, accessed 17 September 2018.
131. 'How to Clone a Security Badge in Seconds' (May 23 2016), https://www.youtube.com/watch?v=cxxnuofREcM, accessed 17 September 2018.
132. 'Watch Hackers Break into the US Power Grid' (May 11 2016).

133. This question is worthy of systematic study.
134. George Wrenn, 'Cyber Security and Critical Infrastructure MIT Industrial Liaison Program (ILP)' (June 9 2015), https://www.youtube.com/watch?v=JCbme19f7yQ, accessed 17 September 2018.
135. Alex Marquardt, 'New Report Reveals Apparent Plot Against Electrical Grid' (November 4 2021), https://edition.cnn.com/videos/politics/2021/11/04/drone-threat-power-grid-new-details-marquardt-lead-pkg-vpx.cnn, accessed 9 November 2021.
136. 'Red Teaming vs Penetration Testing vs Vulnerability Scanning vs Vulnerability Assessments' (undated), https://penconsultants.com/home/red-teaming-vs-penetration-testing-vs-vulnerability-scanning-vs-vulnerability-assessments/, accessed 22 February 2020.
137. George Wrenn, 'Cyber Security and Critical Infrastructure MIT Industrial Liaison Program (ILP)'.
138. Cyber Threat and Vulnerability Analysis of the U.S. Electric Sector, p. 1.
139. U.S. Charges Russian GRU Officers with International Hacking and Related Influence and Disinformation Operations, https://www.justice.gov/opa/pr/us-charges-russian-gru-officers-international-hacking-and-related-influence-and, accessed 7 January 2019.
140. Anonymous, 'How the Dutch Foiled Russian 'Cyber-Attack' on OPCW' (October 4 2018), https://www.bbc.co.uk/news/world-europe-45747472, accessed 7 January 2018.
141. Anonymous, 'Major Power Cut Leaves Large Parts of England Without Electricity the National Grid Said Two Generators Went Down and the Issue Is Now Resolved but Commuters Have Been Left Stranded' (August 10 2019), https://news.sky.com/story/large-parts-of-london-and-south-east-without-electricity-after-power-cut-11781338 and Anonymous, 'UK Power Cut: National Grid Promises to Learn Lessons from Blackout' (August 10 2019), https://www.bbc.co.uk/news/uk-49302996, accessed 11 August 2019.
142. Remarks made by a number of informed commentators, including at Digital Wales 2015, Celtic Manor UK, 8–9 June 2015 attended by author.
143. '2016 Internet Security Threat Report', available from https://www.symantec.com/en/uk/security-center/threat-report, accessed 2 November 2016.

144. 'Global Study Reveals Businesses and Countries Vulnerable Due to Shortage of Cybersecurity Talent', 26 July 2016, https://newsroom.intel.com/news-releases/global-study-reveals-businesses-countries-vulnerable-due-to-shortage-cybersecurity-talent/, accessed 2 November 2016.
145. William Crumpler and James A. Lewis, 'The Cybersecurity Workforce Gap' (January 2019), https://csis-website-prod.s3.amazonaws.com/s3fs-public/publication/190129_Crumpler_Cybersecurity_FINAL.pdf, accessed 3 August 2021.
146. (ISC)² A Resilient Cybersecurity Profession Charts the Path Forward, (ISC)² Cybersecurity Workforce Study, 2021, p. 3. https://www.isc2.org//-/media/ISC2/Research/2021/ISC2-Cybersecurity-Workforce-Study-2021.ashx, accessed 13 June 2022.
147. 'Global Study Reveals Businesses and Countries Vulnerable Due to Shortage of Cybersecurity Talent', 26 July 2016.
148. 'Cyber Threat and Vulnerability Analysis of the U.S. Electric Sector', p. 16.
149. (ISC)² Cybersecurity Workforce Study, 2021, pp. 27–33.
150. 'Dawn Kawamoto, 'Number of CISOs Rose 15% This Year' (June 5 2017), https://www.darkreading.com/careers-and-people/number-of-cisos-rose-15--this-year/d/d-id/1329050 and Max Metzger, 'CISO Salaries May Soon Hit £1 Million—But Few Qualified for Top Roles' (May 22 2017), https://www.scmagazine.com/news/network-security/ciso-salaries-may-soon-hit-1-million-but-few-qualified-for-top-roles, both accessed 22 July 2019.
151. Graham Burton, 'World's Largest Companies Recruiting CISOs for Board-level Roles' (May 30 2014), https://www.computing.co.uk/ctg/news/2347513/worlds-largest-companies-recruiting-cisos-for-board-level-roles, accessed 22 July 2019.
152. Metzger, 'CISO Salaries May Soon Hit £1 Million'.
153. Metzger, 'CISO Salaries May Soon Hit £1 Million'.
154. 'State of Cybersecurity 2019', p. 3, http://m.isaca.org/Knowledge-Center/Research/Documents/cyber/state-of-cybersecurity-2019-part-1_res_eng_0319a.pdf, accessed 22 July 2019.
155. 'State of Cybersecurity 2019', pp. 3–4, 11–13.
156. 'State of Cybersecurity 2019', p. 13. See also Joe Burton and Simona R. Soare, 'Understanding the Strategic Implications of the Weaponization of Artificial Intelligence', in Tomáš Minárik, Siim Alatalu, Stefano Biondi, Massimiliano Signoretti, Ihsan Tolga and Gábor Visky (eds.),

2019 11th International Conference on Cyber Conflict: Silent Battle (Tallinn: CCD COE Publications, 2019), pp. 249–266.
157. 'State of Cybersecurity 2019', p. 13.
158. 'State of Cybersecurity 2019', pp. 14–16.
159. 'State of Cybersecurity 2019', p. 10.
160. 'State of Cybersecurity 2019', p. 18.
161. ISACA, 'State of Cybersecurity 2021 Part 1: Global Update on Workforce Efforts, Resources and Budgets', p. 7, https://www.isaca.org/bookstore/bookstore-wht_papers-digital/whpsc211, accessed 5 August 2021.
162. Steve Morgan, 'Cybersecurity Labor Crunch to Hit 3.5 Million Unfilled Jobs by 2021' (June 6 2017), https://www.csoonline.com/article/3200024/cybersecurity-labor-crunch-to-hit-35-million-unfilled-jobs-by-2021.html, accessed 22 July 2019.
163. 'Cyber Threat and Vulnerability Analysis of the U.S. Electric Sector', p. 2.
164. Formerly the Global Information Security Workforce (GISW) study. Brian NeSmith, 'The Cybersecurity Talent Gap Is An Industry Crisis' (August 9 2018), https://www.forbes.com/sites/forbestechcouncil/2018/08/09/the-cybersecurity-talent-gap-is-an-industry-crisis/#2fdfc6a4a6b3 and '(ISC)² Report Finds Cybersecurity Workforce Gap Has Increased to More Than 2.9 Million Globally' (October 17 2018), https://www.isc2.org/News-and-Events/Press-Room/Posts/2018/10/17/ISC2-Report-Finds-Cybersecurity-Workforce-Gap-Has-Increased-to-More-Than-2-9-Million-Globally, both accessed 22 July 2019.
165. 'OECD Employment Outlook 2019', https://www.oecd.org/employment/outlook/, accessed 22 July 2019.
166. 'Skills for Jobs', p. 5, https://www.oecdskillsforjobsdatabase.org/data/Skills%20SfJ_PDF%20for%20WEBSITE%20final.pdf, accessed 22 July 2019.
167. 'The Future of Jobs Report 2018', http://www3.weforum.org/docs/WEF_Future_of_Jobs_2018.pdf, accessed 22 July 2019.
168. ISACA, 'State of Cybersecurity 2021 Part 1', p. 7.
169. INTERPOL, 'Global Landscape on Covid-19 Cyberthreat', https://www.interpol.int/Crimes/Cybercrime/COVID-19-cyberthreats, accessed 5 August 2021.
170. ISACA 'State of Cybersecurity 2021 Part 1', p. 7.
171. Dana A. Shea, 'Critical Infrastructure: Control Systems and the Terrorist Threat', Congressional Research Service, 21 February 2003.

Available from http://fas.org/irp/crs/RL31534.pdf, accessed 15 July 2014. See also Michael Warner, 'Cyber-Security: A Pre-history', *Intelligence and National Security*, Vol. 27, No. 5 (October 2012), pp. 781–799. On U.S. efforts in this realm, see Myriam Dunn Cavelty, *Cyber-Security and Threat Politics: US Efforts to Secure the Information Age* (London: Routledge 2008) and Jason Healey (ed.), *A Fierce Domain: Conflict in Cyberspace 1986–2002* (Vienna, VGN: CSSA/Atlantic Council, 2013), pp. 14–88.

172. 'Public Law No: 115-278 (11/16/2018) Cybersecurity and Infrastructure Security Agency Act of 2018', https://www.congress.gov/bill/115th-congress/house-bill/3359, accessed 22 February 2020.
173. 'About Us', https://www.us-cert.gov/about-us, accessed 23 July 2019.
174. 'National Cybersecurity and Communications Integration Center', https://www.dhs.gov/cisa/national-cybersecurity-communications-integration-center and 'About Us', https://www.us-cert.gov/about-us, both accessed 23 July 2019.
175. Cynthia Brumfield, 'What Is the CISA? How the New Federal Agency Protects Critical Infrastructure from Cyber Threats' (July 1 2019), https://www.csoonline.com/article/3405580/what-is-the-cisa-how-the-new-federal-agency-protects-critical-infrastructure-from-cyber-threats.html. See also https://uscode.house.gov/statutes/pl/115/278.pdf and https://www.cisa.gov/, both accessed 22 February 2020.
176. Kristan Stoddart, Kevin Jones, Hugh Soulsby, Andrew Blyth, Peter Eden, Peter Burnap and Yulia Cherdantseva, 'Live Free or Die Hard: U.S.–UK Cybersecurity Policies', *Political Science Quarterly*, Vol. 131, No. 4 (Winter 2016), pp. 812–820.
177. Baiba Kaskina, Edgars Taurins, and Andrea Dufkova (ENISA), 'CSIRT Capabilities How to Assess Maturity? Guidelines for National and Governmental CSIRTs' (December 2015), p. 9, https://www.enisa.europa.eu/publications/csirt-capabilities/at_download/fullReport, accessed 30 March 2020. More on ENISA's approach to 'maturity' can be found in Andrea Dufková (ed.), (ENISA), 'CERT Community Recognition Mechanisms and Schemes' (November 2013), https://www.enisa.europa.eu/publications/cert-community-recognition-mechanisms-and-schemes, pp. 12–40.
178. Kaskina, Taurins, and Dufková (ENISA), 'CSIRT Capabilities How to Assess Maturity?', pp. 9–10.
179. Kaskina, Taurins, and Dufkova (ENISA), 'CSIRT Capabilities How to Assess Maturity?', p. 13.

180. 'The National Cyber Security Centre', https://www.ncsc.gov.uk/, accessed 23 July 2019.
181. Darren Pauli, 'CERT Australia Rebuffs Ex-staff Criticism' (February 8 2013), https://www.itnews.com.au/news/cert-australia-rebuffs-ex-staff-criticism-331618, Geetha Nandikotkur, 'CERT-In Says Hacking Declining, But Critics Express Doubts Do the Latest Statistics Reflect Reality?' (December 18 2018), https://www.bankinfosecurity.asia/cert-in-says-hacking-declining-but-critics-express-doubts-a-11868 and Daniel Binns, 'After Heartbleed, Can Britain's New Cyber Emergency Response Team Beat the Hackers' (April 15 2014), https://metro.co.uk/2014/04/15/after-heartbleed-can-britains-new-cyber-emergency-response-team-beat-the-hackers-4698848/. See also Stuart Madnick, Xitong Li and Nazli Choucri, 'Experiences and Challenges with using CERT Data to Analyze International Cyber Security', Working Paper CISL# 2009-13 September 2009, https://pdfs.semanticscholar.org/d4a5/c681807b6f38b2dbcd4cea5894c12011cc81.pdf, all accessed 23 July 2019.
182. This might well be indicated by the April 2016 disclosures on the activities of the Panamanian law firm, Mossack Fonseca. See for example, Anonymous, 'Panama Papers Q&A: What Is the Scandal About?', http://www.bbc.co.uk/news/world-35954224, accessed 5 April 2016.
183. 'Communication from the Commission on a European Programme for Critical Infrastructure Protection' (December 12 2006), https://eur-lex.europa.eu/LexUriServ/LexUriServ.do?uri=COM:2006:0786:FIN:EN:PDF, 'Critical Infrastructure Protection' (undated), https://ec.europa.eu/jrc/en/research-topic/critical-infrastructure-protection, both accessed 22 February 2020. Detailed work on this 'patchwork quilt' was conducted by the author for the Airbus project 'SCADA systems and the Cyber Security Lifecycle'. As of writing it is unpublished. How Britain's withdrawal from the EU could affect intelligence sharing was not part of the work.
184. Kārlis Podiņš and Kenneth Geers, 'Aladdin's Lamp: The Theft and Re-weaponization of Malicious Code', in Tomáš Minárik, Raik Jakschis and Lauri Lindström (eds.), *2018 10th International Conference on Cyber Conflict CyCon X: Maximising Effects* (Tallinn: CCD COE Publications, 2018), pp. 187–203.
185. Cynthia Harvey, 'Top Cybersecurity Companies of 2018' (August 22 2018), https://www.esecurityplanet.com/products/top-cybersecurity-companies-2018.html, accessed 12 July 2018.

186. 'Cyber Threat and Vulnerability Analysis of the U.S. Electric Sector', p. 16.
187. 'Cybersecurity Framework', https://www.nist.gov/cyberframework, accessed 22 July 2019.
188. 'ISACA Produces New Audit Program Based on NIST Framework' (January 10 2017), http://www.isaca.org/About-ISACA/Press-room/News-Releases/2017/Pages/ISACA-Produces-New-Audit-Program-Based-on-NIST-Framework.aspx, accessed 22 July 2019.
189. 'About NCI', https://www.nationalisacs.org/about-nci, accessed 22 July 2019.
190. 'About ISACs', https://www.nationalisacs.org/about-isacs and 'Member ISACS', https://www.nationalisacs.org/member-isacs, both accessed 22 July 2019.
191. 'Information Sharing and Analysis Centers', https://www.enisa.europa.eu/topics/national-cyber-security-strategies/information-sharing. See also Information Sharing and Analysis Centres (ISACs) Cooperative models, https://www.enisa.europa.eu/publications/information-sharing-and-analysis-center-isacs-cooperative-models, both accessed 22 July 2019.
192. https://www.isac.io/, accessed 22 July 2019.
193. 'About ISA', https://isalliance.org/about-isa/, accessed 22 July 2019.
194. https://www.ncfta.net/, accessed 22 July 2019.
195. 'Cyber Operations Tracker', https://www.cfr.org/interactive/cyber-operations, accessed 22 July 2019.
196. https://icspa.org/about-us/, http://www.iaac.org.uk/about/, both accessed 22 July 2019.
197. https://www.ukcybersecuritycouncil.org.uk/, accessed 13 June 2022.
198. Daniel Kapellmann and Rhyner Washburn, 'Call to Action: Mobilizing Community Discussion to Improve Information-Sharing About Vulnerabilities in Industrial Control Systems and Critical Infrastructure', in Minárik, Alatalu, Biondi, Signoretti, Tolga and Visky (eds.), *Silent Battle*, pp. 37–60.
199. Wakefield, 'Office Intruder 'Steals' Data'.
200. Michael Workman, 'Gaining Access with Social Engineering: An Empirical Study of the Threat', *Information Systems Security*, Vol. 16, Issue 6 (2007), p. 317.
201. Workman, 'Gaining Access with Social Engineering', p. 317.
202. 'Proofpoint State of the Phish 2019 Report', p. 12.

203. Remarks made under Chatham House Rules at the Insider Threat Summit, Monterey, California, 29 March–1 April 2016 attended by author.
204. Melissa Russano, Fadia M. Narchet, Steven M. Kleinman, and Christian A. Meissner, 'Structured Interviews of Experienced HUMINT Interrogators', *Applied Cognitive Psychology*, Vol. 28, Issue 6 (November/December 2014), pp. 847–859. This is part of a Special Issue on information gathering in law enforcement and intelligence settings and contains much that is of value and interest in terms of hacking the human.
205. Anonymous, 'Ethical Collection of Competitive Intelligence', https://www.competitivefutures.com/ethical-collection-competitive-intelligence/, accessed 23 February 2020.
206. As with former military personnel, some struggle to readjust to civilian life and "crash and burn" in the private sector. Art Keller, 'What Do Former CIA Spies Do When They Quit the Spy Game?' (October 12 2012), https://www.forbes.com/sites/realspin/2012/10/12/what-do-former-cia-spies-do-when-they-quit-the-spy-game/, https://www.afio.com/14_careers.htm, accessed 23 February 2020.

6

Non and Sub-State Actors: Cybercrime, Terrorism, and Hackers

Introduction

The actors behind cyber intrusions and breaches are both foreign and domestic. Foreign groups can contain domestic actors and domestic groups foreign actors.[1] These groups range from 'script kiddies', who are (predominantly) young people engaging in illegal activities ranging from probing organizations to distributed denial of service (DDoS) attacks. These can be 'lone wolves' who act alone or in concert or competition. This can include 'patriotic hackers' or collectives such as the 'hacktivist' group 'Anonymous' and 'leaktivists' who use organizations like Wikileaks, Bellingcat, and the Investigative Consortium of Investigative Journalists (ICIJ), to leak confidential information. The most virulent actors are sophisticated state-run and intelligence-led Advanced Persistent Threats (APTs) hostile to other nation-states. Most of these cyberattackers will target the weakest links in the fence; the so-called low-hanging fruit, as these represent easier entry points. These are frequently where campaigns begin.

Cybercriminals, and organized criminal gangs which encompass non-state hackers, can also be leveraged by nation-states as proxy forces and 'privateers'. The 2018 U.S. Intelligence World Threat Assessment warned that "The FBI assesses that US losses from cybercrime in 2016 exceeded $1.3 billion, and some industry experts predict such losses could cost the global economy $6 trillion by 2021".[2] The 2019 assessment warned that "financially motivated cyber criminals…could increasingly disrupt US critical infrastructure in the health care, financial, government, and emergency service sectors, based on the patterns of activities against these sectors in the last few years".[3] The 2021 assessment that:

> Cyber threats from nation states and their surrogates will remain acute. Foreign states use cyber operations to steal information, influence populations, and damage industry, including physical and digital critical infrastructure. Although an increasing number of countries and nonstate actors have these capabilities, we remain most concerned about Russia, China, Iran, and North Korea. Many skilled foreign cybercriminals targeting the United States maintain mutually beneficial relationships with these and other countries that offer them safe haven or benefit from their activity.[4]

These actors are not just criminal. This includes what are currently latent threats to critical infrastructure. In a 2016 report into 'Cyber Threat and Vulnerability Analysis of the U.S. Electric Sector', Idaho National Laboratory recorded that across multiple fronts, threat actors from nation-states including Russia, China, and Iran, as well as non-state actors "pose varying threats to the power grid". These include terrorists and hacktivist groups and "A determined, well-funded, capable threat actor with the appropriate attack vector can succeed to varying levels depending on what defenses are in place".[5] The 2020 Department of Homeland Security (DHS) threat assessment singled out China's ability "to threaten and potentially disrupt U.S. critical infrastructure" especially "critical manufacturing, defense-industrial base, energy, healthcare, and transportation sectors". This included through Chinese ICT companies and the rollout of 5G. In addition, "Cybercriminals increasingly will target U.S. critical infrastructure to generate profit" mainly through ransomware.[6] Many are determined, well-funded, and capable and there are a wide range of hackers out there with varying technical skills and motivations.

Outsider Threats, Insider Threats, and Target Spotting

For those looking from outside the gates or at the gates themselves, targeting an individual, group, or organization can provide valuable or compromising information. This can begin online or physically near a target. Online observation can begin with basic Internet searches on Google (or Russia's Yandex and China's Baidu), through business platforms such as LinkedIn, social media such as Facebook or Twitter, or on company websites. Many companies thrive on publicizing their business and employees in turn can be keen to maximize their profile. This can be used to identify the most promising organizations and individuals and help determine what approach methods have the greatest chances of success.

These activities can lead to information that is of value to the social engineer or to an intelligence agency.[7] This is a simple means of harvesting valuable Open Source Intelligence (OSINT) prior to or during target surveillance (and later as a possible gauge to aftereffects). Social engineers and their counterparts in the intelligence world have become adept at 'spotting' these targets. They target individuals or organizations who have, or appear to have, access to information or are attractive targets for the access they can provide. These targets can be 'cultivated' over time through a variety of means prior to a direct approach (especially of individual people). A riskier 'cold' approach can be made but the most valuable intelligence assets are not targeted or recruited but 'walk-ins'. These offer their services or volunteer information. OSINT can again be used to provide background information to corroborate their stories, identify them and their associations—although false trails can be laid.

'Walk-ins' are often 'rogue insiders'. The most common reasons for rogue insiders are: financial considerations/greed; revenge, grudges or dissatisfaction; blackmail or entrapment; national pride or corporate competitiveness; emotional involvement with someone hostile to the target organization or country; naiveté; ideology; or whistleblowing. They could also be false flag approaches or part of an exercise or operation, including a counterintelligence operation to lure hostile actors.

Prior to the movement and travel restrictions brought about by the 2020–2022 Covid-19 pandemic, Verizon's 2019 Data Breach Investigations Report found that most threats in the *public sector* were espionage related with 29 percent for financial gain.[8] Verizon's 2019 Insider Threat Report saw that most cybercrime, *both public and private sector*, was for financial gain with a significant percentage from insiders or partners. Insiders also cited grudges and fear as motivating factors.[9] These and other motives can be readily exploited by criminals, organized crime gangs, or by corporate and state espionage. Verizon's 2022 Data Breach Investigations Report (DBIR) recorded that around 4 in 5 breaches were due to cybercrime, with espionage No. 2. Only 2 percent was due to ideology or grudges.[10]

Accurate information-intelligence on a target, target organization, the culture of that organization, the business practices of the target, and their host country or national affiliation increases the opportunities for infiltration. The same can be said of an individual target's social, political, religious, or financial status and other related factors. This biographical accumulation and analysis can also be used for the purposes of entrapment. Entrapment can take several forms but the most common are financial or sexual. However, this form of 'coercive recruitment' is far from a guarantee of success as resentment to blackmail or exploitation can also lead people to turn themselves in.

Insiders can include "disgruntled or former employees" attempting "to extort their employer for financial gain" in a variety of ways.[11] The financial penalties for companies exposed to 'rogue insiders' range from thousands to millions of dollars through exposure to ransomware or wiperware or to employ cybersecurity professionals, install network countermeasures, or cover legal fees. This is combined with "loss of revenue and/or customers, and the purchase of credit monitoring services for employees and customers affected by a data breach".[12] Much more is at stake for companies in the defense-industrial base and critical infrastructure.

The Department of Homeland Security (DHS) has stated, "The exploitation of business networks and servers by disgruntled and/or former employees has resulted in several significant FBI investigations in which individuals used their access to destroy data, steal proprietary software, obtain customer information, purchase unauthorized goods and services using customer accounts, and gain a competitive edge at a new company".[13] Methods of exfiltration can include cloud services like Dropbox and personal e-mail accounts as well as walking out with USB sticks and SD cards. As major leaks have shown, such as that perpetrated by Edward Snowden, even a single trusted insider has the capacity to do great harm to an organization. More than one 'rogue insider' or mole multiplies the risk and potential harms. Snowden showed that even the most security-focused organization is vulnerable.

Even relatively low-level personnel can know a great deal about the people and processes within an organization or have access to sensitive areas, including security guards and cleaners. They also often have daily access to people higher up in an organization or are able to gain access to secure locations.[14] Whether knowingly or unwittingly they can provide valuable access and information-intelligence at the gateway and within the walls. Mitigation is easier said than done. Degrees of trust are a fundamental social commodity both in interpersonal terms in social situations and in the workplace. Those who live in liberal democracies are also used to freedom to travel and open movement and professionals (including academics) can be invited onto restricted or secure sites. Recognizing this is useful to a social engineer and although insider threat detection and prevention programs are in place, they are not foolproof.[15] Hackers and crackers have become adept at social engineering (SE).

Hackers, Hacking Groups, and Social Engineering

Hackers and crackers might be motivated by ideas of peer recognition, fame, or infamy, often coupled with a desire to explore, learn, and test their limits of where and what they can access and breach. Monetary gain has also become a prime motivation. With the growth of the Internet from the mid-late 1990s hackers began to see increasing opportunities present themselves through criminal activities. These include financial fraud, identity theft, extortion, intellectual property theft, and denial of service/DDoS attacks. The growth of the Internet and its low entry points have enabled (often self-taught) individuals to become successful hackers/crackers and form hacking and cracking collectives. These need not have met personally (or stay anonymous) and can communicate solely online/remotely.

As Chapter 5 detailed, SE can allow hackers to bypass external and internal defensive security such as firewalls and Demilitarized Zones and security mechanisms designed to prevent access. SE can obtain information such as passwords and grant privileges to access data and compromise systems. There are often rich digital footprints to track targets and the environment is increasingly target rich as ever-increasing numbers of people, companies, products, and services expand their online presence. These opportunities are set to grow as the 'Internet of Things' and SMART technologies expand into wider use through wearable tech and further embedded into 'smart homes', 'smart grids', 'smart cities', and countrywide adoption. Security-by-design even if matured and widely adopted cannot always account for human error or accident or well-crafted SE exploits.[16] SE shares many of the same characteristics as Human Intelligence (HUMINT) and espionage, practiced by intelligence professionals. HUMINT is often more sophisticated, but both employ both cognitive and behavioral psychology which can include very simple techniques or 'confidence tricks'.

This includes pretexting. In the United States in 2015/2016 a disturbing example of pretexting took place at the National Security Agency (NSA) and Central Intelligence Agency (CIA) using exactly these techniques. In this case a script kiddie hacking group calling themselves 'Crackas With Attitude' accessed the e-mail accounts of the Director of the CIA, John O. Brennan, the Director of National Intelligence, James R. Clapper, and a number of other senior officials. Evidence submitted by the Federal Bureau of Investigation (FBI) indicates that the hackers from this group conned technical support staff into providing them with the means to access their personal e-mail accounts.

The BBC reported they "posed as technicians from ISPs [Internet Service Providers] and other service companies to get passwords re-set so they could take over accounts and get at federal computer systems".[17] Part of their pretexting scam involved telephone calls made by one or more British teenagers who were members of 'Crackas With Attitude'. They used this access to leak details of 29,000 FBI and Homeland Security employees onto the Internet.[18] An American member of this group contacted the *New York Post* in October 2015 to "brag about his exploits". He said that the documents were stored as attachments to around 40 e-mails that he read after breaking into Brennan's account on 12 October 2015. The group used pretexting to dupe Verizon technical support (Brennan's private ISP) into disclosing his personal information. Using this, they then deceived AOL into resetting his password. This gave them access to Brennan's e-mail. The same member of the group also claimed to have listened to the voicemails of Jeh Johnson, then Director of Homeland Security, and to have screenshotted billing information through pretexting his Comcast ISP account.[19]

At one level it is clearly poor cyber hygiene to use personal e-mail accounts for anything sensitive or private, especially when holding responsible positions. The same can be said of any use of unencrypted e-mail. These are issues that dogged Hillary Clinton's bid for the White House dating from her use of personal e-mail for government work when she served as Secretary of State.[20] At the same time, as one inside source told the *New York Post*, "[The] problem with these older-generation guys is that they don't know anything about cybersecurity".[21] At another level it is concerning that ISPs provided personal information to unauthorized users, in this case 'script kiddies', through simple pretexting and equally simple social engineering methods. A member of 'Crackas With Attitude' who spoke with *Wired* magazine showed how easy pretexting was.[22] It was not unlike a carefully crafted phishing attack—another attack vector that can utilize social engineering.[23] These can be used to extort money or to harvest or intercept exploitable information-intelligence. The UK leader of the group was given a two-year jail sentence when he turned 18, another U.S. member a five-year prison sentence.[24]

Social Network Analysis

Social engineering by criminal gangs and non- and sub-state hackers, embrace various forms of data gathering and analysis using OSINT as gateways to a target. The more professional CYBERINT carried out by intelligence agencies (as well as in police investigations) also leverages OSINT and a variety of other intelligence gathering and exploitation methods discussed in other chapters. The value of this target profiling can be extended through Social Networking Analysis (SNA) which is useful in examining social structures. SNA now embraces information contained on social media such as Facebook and Twitter as well as sites for professional networking such as LinkedIn.[25]

SNA speaks to Stanley Milgram experiments of six degrees of separation described in Chapter 5.[26] Aggarwal argues "While such hypotheses have largely remained conjectures over the last few decades, the development of online social networks has made it possible to test such hypotheses at least in an online setting. This is also referred to as the *small world* phenomenon".[27] In the online world geographical remoteness is not an obstacle and 'six degrees' can diminish class, racial, cultural, financial, and other social barriers. They are also useful in finding drawbridges and tunnels to enter protected castles.

The idea of 'six degrees of separation' and 'small world' phenomena are described in an interesting article by the psychologist Judith S. Kleinfeld. Kleinfeld delved into Milgram's original papers at Yale University and the literature on 'six degrees' and 'small world' studies. Kleinfeld concluded, "We are used to thinking of "six" as a small number, but in terms of practical connections, "six" may be a huge number indeed. Nothing is so useful as a good problem. How we are connected to each other remains an eternally fascinating mystery…and a researchable one."[28] Kleinfeld's study showed that the notion of 'six degrees' and 'small world' might be a myth began by Milgram himself which has become perpetuated by popular culture and academics, especially in the field of mathematics.[29]

Despite these valid questions regarding social connections and social connectivity, online interactions enabled by the growth of the Internet and now the IoT are increasing. As they increase, they are arguably also flattening the hierarchy of social, financial, temporal, religious, racial, and class-based obstacles in the social chains that link us or decouple us (as Kleinfeld found in her 2002 study).[30] These are as apparent in the real world and in the effects the online world has upon the real world. Whether these bonds are strong or weak, the data that can be mined from social networks, social media, and from a wider variety of sources and standpoints (such as SNA, telecommunications metadata, and Big Data analytics) is considerable. This can produce highly detailed and highly accurate 'patterns of life' both for individuals and the social networks, groups, and nations they are affiliated to.[31] The boundaries of these rivers of data might well settle upon the capacity to accurately interpret our complex patterns of life derived from the information-intelligence we generate.

SNA for example can be built up through an initial analysis of dyadic ties between individuals. These ties can be familial, romantic, through work and leisure, or via individuals with mutual interests. These can be layered around other social interactions between individual actors and groups. SNA can be a highly effective tool for looking at social structures and social hierarchies horizontally and vertically. SNA utilizes network and graph theories as well as models and these networks can be readily visualized through sociograms which show nodes as points with ties represented as lines as seen in Fig. 6.1.

Characterizing social relationships and social interactions as nodes which form a networked structure comprised of individuals, actors, or agents[32] provides for representational tools for analytic deductions or insights to be made and relationships (actual or professed) to be easily mapped, studied, and interpreted. This is constructed through a series of ties, edges, or links illustrating the relationships or interactions that form these connections. SNA services are widely available commercially.[33]

Visual representations of these social networks that are produced through SNA might best be visualized as a complex family tree which can move through time and space. It can capture family and romantic relationships, friendship and acquaintances, employment and work relationships as well as terrorist connections or criminal activities. The use and utilization of social media networks such as Facebook or search engines such as Google, using analytics, metadata, coupled with Big Data can all be used to provide extremely useful and deep-reaching social mapping for information-intelligence. Facebook, Google, and many other firms across the ICT industry also use them, especially for marketing products and services.[34] As well as commercial activities they

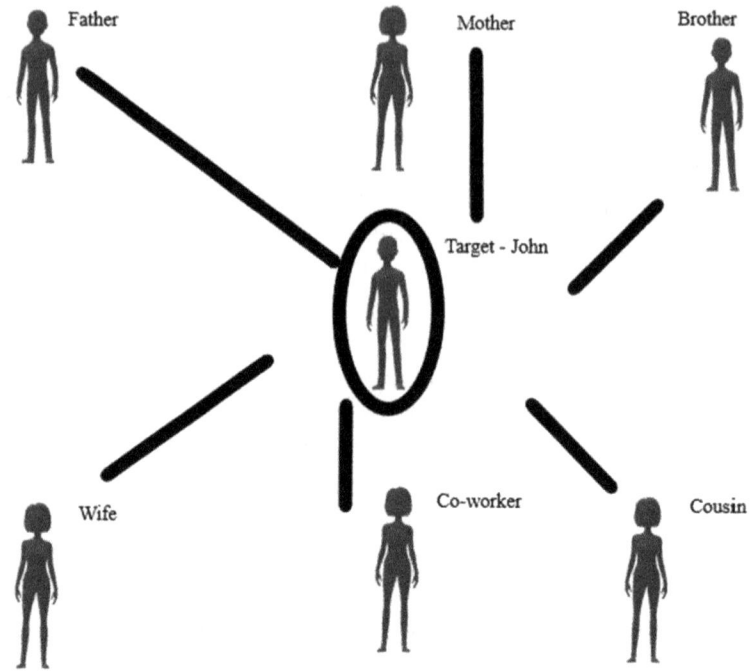

Fig. 6.1 Simple SNA

have also been used in intelligence programs such as the NSA-led PRISM programs and in law enforcement and policing.[35]

Social network analysis is most widely used in sociology but has significant coverage and use in Computer Science and fields involving social group dynamics. As Wasserman and Faust describe:

> The pioneers of social network analysis came from sociology and social psychology (for example, Moreno, Cartwright, Newcomb, Bavelas) and anthropology (Barnes, Mitchell). In fact, many people attribute the first use of the term "social network" to Barnes (1954). The notion of a network of relations linking social entities, or of webs of ties among social units emanating through society, has found wide expression throughout the social sciences…Its utility is great, and the problems that can be answered with it are numerous, spanning a broad range of disciplines.[36]

Its uses include anthropology, geography, medicine and biology, economics, social and behavioral psychology, and as an end user consumer tool. It can also be a useful tool for mapping disease transmission, population changes and migratory patterns, which can be layered with data on ecology, climatology, or physical or human geography.[37] As Christina Prell explains:

> The growing popularity of this approach coincides with the growing awareness in our society of the interdependencies and complexities of social systems in particular, but also ecological ones as well; social policy makers, for example, are increasingly turning their attention to the role of networks for understanding a range of problems on the local, regional, societal and even global level…The literature on social networks is vast: there are three journals that focus entirely on social networks – *Connections*, *Journal of Social Structure* and *Social Networks* – and there are a number of books that focus on the techniques, concepts, theories and mathematics that form the field.[38]

The ties that bind individuals to social groups and the social networks formed from these ties and interdependencies can be very useful tools for security, law enforcement, and intelligence professionals. For the amateur social engineer, encompassing both the curious and malicious/criminal, these networks and nodes are also apparent as they target individuals and organizations; including through tools like Maltego. The purpose, capacity, and methodology able to be brought to bear will vary considerably but many of their goals will be similar in an attempt to hack the human. SNA methodology, and the sophistication able to be employed, can provide shortcuts or pathways to target individuals, groups, and organizations. This includes increasing ways of 'spear phishing' targets.[39] In so doing common patterns that might otherwise be obscured or not readily apparent can be observed and analytical and methodological traits and tactics can be identified and refined.[40]

SNA as a Law Enforcement and Intelligence Tool

It is always a possibility that the hunter can also become the hunted. These same tools can also be used against those observing, interpreting, and targeting.[41] For example, in cases of industrial espionage, terrorism, and organized crime, social engineering can be flagged to police/law enforcement or intelligence agencies. Traffic analysis from devices like mobile/cell phone transmissions based on geodetic locations and data publicly broadcast can also be a useful instrument in both professional and 'amateur' toolkits. Politicians looking to regulate or gain backdoors into encrypted smartphone apps like WhatsApp or Telegram also use these same applications (as do criminals and terrorists).

Whilst the tech industry, including Facebook, Google, and Twitter works with law enforcement, balancing their business model of user participation and engagement (to drive advertising) with privacy and free speech laws (especially in liberal democracies like the U.S.) against the needs and desires of police and law enforcement to be able to interrogate the data they hold has been a source of tension.[42] For police and law enforcement, the data these companies hold or can provide on individuals and the people they are in contact with can be extremely useful for the prosecution and defense of crimes and terrorism. It can aid investigations, down to fine grain detail on individuals, matched to much wider analytical tools including SNA tools and methodologies to be able to tackle crime and even predict crime (predictive policing). This carries great benefits for policing but comes with the potential for misuse and misidentification.[43]

It may be that "Political conversations, both online and offline, occur most often between people with close personal ties—spouses, close friends, and relatives". But "How common are informal political discussions on social media? How often do such discussions occur across partisan boundaries? Do these cross-cutting discussions occur primarily via existing relationships or via "weak ties"—for example, friends of friends?". This is a debate which extends beyond social media and into other areas of the online world and into offline discussion and discourse analysis.[44]

Civic (and civil) engagement are features of representative democracy and for the most part mainstream politics is conducted without threats to personal safety. However, online death threats to politicians and public figures are not uncommon. Those in the public eye run the risk of physical violence or even assassination. This risk has "risen dramatically since the early 1970s" and "reflective of the emergence of a new wave of terrorist groups, radical and universal ideologies operating on a global scale, and a growing willingness by oppressive regimes to use assassinations as a tool in their treatment of political opposition". Moreover, "while most assassinations of government officials were perpetrated by sub-state violent groups, most assassinations of opposition leaders were initiated by ruling political elites or their proxies".[45]

Whilst authoritarian and kleptocratic states like Russia have used this as part of their toolbox, a wider question affecting democracies is "whether online platforms are contributing to political polarization or serving to dampen its most corrosive effects" and a breakdown in tolerance and civility which is crossing into the real world.[46] The events of January 6 2021 is the most visceral reminder of this corrosion on America's body politic when the capitol building was stormed in Washington DC as Congress sought to certify Joe Biden's election victory.[47] Social media, cell phone and bodycam footage, and CCTV have each been part of the evidence presented in court against those prosecuted.[48] The events of January 6 also lends itself to studying modern group dynamics and increasing beliefs in conspiracies and shared or discordant realities.[49]

In addition, whistleblowing sites like Wikileaks are, for some, adding to these effects. The debate between whistleblowing and spying has been brought into focus by Julian Assange and WikiLeaks as well as Edward Snowden. In between lie questions relating to surveillance and accountability and cyber offense versus cyber defense typified by PRISM and the activities of the NSA. Protection of citizens from harms by national security agencies and law enforcement from criminals, terrorists, sub-state, and state actors needs to be balanced against privacy concerns, freedom of speech, and a free press.[50] SNA is an extremely useful tool for counterterrorism.

Terrorism

What has so far been seen in terms of cyber terrorism is online radicalization, support, and coordination. There are many examples. In October 2015, Kosovo national Ardit Ferizi was arrested in Malaysia for supplying Junaid Hussain with the personal data of 1,351 American soldiers. Hussain subsequently posted this on Twitter. According to the Department of Justice, Ferizi obtained this data by hacking into the computer system of an American company—not a protected military network.[51] Hussain, a British national of Pakistani decent and one of the infamous ISIS 'Beatles' had already been jailed as a teenager in 2012 for hacking then British Prime Minister Tony Blair's special advisor and 'phone bombing' the UK national anti-terrorism hotline. He was killed in a drone strike in Syria in 2015.[52] A year later Ferizi, received a 20-year sentence after being extradited from Malaysia to the United States.[53]

Hussain had also used Twitter and Surespot, an encrypted chat/messenger app, to radicalize others and as a key member of the self-styled CyberCaliphate taught others to hack bank accounts. Those he radicalized included Elton Simpson and Nadir Soofi, who attacked a 'Draw Muhammad' cartoon contest in Garland, Texas, in May 2015 and Usaamah Abdullah Rahim who was planning to attack the contest's main organizer but "instead attempted to attack members of the Boston police force". He also used Surespot to message Junead Khan who was arrested before he could attack U.S. military personnel in Britain; addresses for whom were allegedly provided by Hussain who went by the moniker Abu Hussain al Britani.[54]

For 'lone wolves' like Khan, Mohamed Bouhlel, who drove a truck into pedestrians in Nice, France or Adrian Ajao (aka Khalid Masood) who killed four people by driving a rental car into pedestrians on London's Westminster Bridge and then stabbed a police officer, the Internet provides a radicalization and support system, and smart phones are their mobile gateways. Although terrorists might be physically isolated in a geographic location, psychologically they can connect with their like-minded ideological comrades in an online chatroom, or through encrypted apps on their smartphone. In this way technology has influenced societal behavior and can be a lethal cocktail. Terrorists and terrorist sympathizers can also draw on these platforms and the 'Dark Net' for propaganda and for sharing, supporting, and nourishing beliefs and knowledge.[55] This said, after 2018 "IS' online capabilities reflect the overall collapse of the physical caliphate, previously the central pillar of its project".[56]

Extreme right-wing terrorists like Anders Brevik and Brenton Tarrant have also used the Internet and smartphones in similar ways to distribute their self-styled manifestos and views and, in Tarrant's case, to use Facebook to livestream his attacks on two mosques in Christchurch, New Zealand in March 2019 where he murdered 51 people. "Telegram remains the platform of choice" for communications among most of these individuals and groups and "for terrorist sympathisers, who continue to exploit its advantageous encryption and file-sharing capabilities".[57]

Apps and social media platforms like Facebook and Twitter can also be used to broadcast their views or to disseminate these tragic events—both as they unfold and afterward for propaganda or to entice 'drive-by' users to view the content.[58] These images and video can (and has) also been used by foreign organizations like the Russian Internet Research Agency (the most active of all sub-state actors through to the present) to divide societies and fuel intolerance. These actors do this because they understand the importance of the social layer and how that environment can be used to shape and influence target audiences. The purposes can be as varied as citizens contemplating how or whether to vote or lone wolves deliberating jihad. Through social media, apps, and smartphones exist ecosystems and producers of disinformation, propaganda, and destabilizing narratives.[59] The capacity of Internet-enabled platforms to radicalize or support existing confirmation bias in these respects picks at the knots that bind societies.[60] However, so far this is the only real uses cyber has been put to by terrorists and this is why then Director of the FBI, James B. Comey, said to the cybersecurity industry in 2016:

> At the bottom of the stack, which may surprise you, we would put terrorists. The reason they're at the bottom of our stack is terrorist organizations around the world, especially the group that calls itself the Islamic State, are proficient at using the Internet to spread their message of hate, to

recruit, to communicate for operational purposes. They are literally able to buzz in the pockets of fellow travelers or would-be terrorists 24 hours a day. And that has an enormous impact on the FBI's counterterrorism work. But what we don't see them doing yet, and I underline yet, is moving towards and developing the capability for computer intrusions. But logic tells us that has to be the future of terrorism. As we make it harder and harder for them to get physically into this country to kill people and to do damage, surely they are going to turn to try to come in as a photon and do damage through the Internet.[61]

Similar points have been made by America's allies.[62] At the far end of the cybersecurity spectrum lies the potential for a terrorist attack on critical infrastructure to be a mass casualty event. This could well be difficult to prevent should well-organized transnational terrorist groups go down this road. A 2017 U.S. Department of Defense report noted that terrorist attacks might not be deterrable even by "the promise of severe punishment". Instead it advocated prevention/pre-emption and an emphasis on defense.[63] They also noted that "Virtually any actor with substantial resources can now develop or buy the capability to attack elements of U.S. critical infrastructure with cyber weapons. North Korea, Iran, and terrorist groups have strong motivation to purchase such capabilities where possible, and to develop their own substantially improved attack capabilities".[64]

The 2019 U.S. intelligence World Threat Assessment said that "*The growing availability and use of publicly and commercially available cyber tools is increasing the overall volume of unattributed cyber activity around the world.* The use of these tools increases the risk of misattributions and misdirected responses by both governments and the private sector".[65] OSINT and CYBERINT could provide an overview and detail on emerging threats, but terrorist cells can be difficult to track. Should they be identified, ELINT, SATINT, and HUMINT can be leveraged. As Tal and Siman-Tov assess, "There is no substitute for HUMINT if the intelligence picture is to be comprehended as a whole, especially for organizations like ISIS and Hamas, which have a low cyber signature; total reliance on technological intelligence in obtaining accurate information about them is inadequate and questionable".[66]

Mitigating mass casualty cyber enabled terrorist threats might be "at the bottom of the stack" but its demonstrable capability for remote destruction, the media impact this would generate, and the potential it would highlight are significant factors for agencies engaged in counterterrorism.[67] A mass casualty event by cyberattacking critical infrastructure would be complex and time consuming to achieve but it would not be significantly inhibited by cost and the barriers to entry are eminently surmountable. This is also recognized by companies like Boeing which has a full-time security and counterterrorism (CT) team (including looking the radicalization process).[68]

Encryption and the Risk of 'Going Dark'

The use of The Onion Router (TOR), a widely used encrypted Internet browser, and proxy servers compounds CYBERINT problems for intelligence agencies and police in the United States, and around the world. This acts as one of the barriers for identification/attribution and prosecution, as does the growing use of encryption by major technology companies (developed in part as a reaction to the Snowden disclosures).[69] Indeed, the use of end-to-end encryption have meant that the surveillance activities of the intelligence community for eavesdropping on electronic communications have been made more difficult.[70]

The encryption debate was brought into sharp public focus during the first half of 2016 by the dispute between the FBI and Apple over unlocking or hacking the iPhone of Syed Rizwan Farook, the Islamic State-inspired terrorist who, along with his wife Tashfeen Malik, killed 14 people and wounded 22 others during an attack in San Bernardino, California, in December 2015. Apple has cooperated with government agencies in the past, including on the NSA's PRISM program, but as technology journalist Kim Zetter noted, "the government wants a way to access data on gadgets, even when those devices use secure encryption to keep it private".[71]

This has implications beyond Apple and the FBI. If the FBI had legally compelled Apple to comply with this request, "it would also set a precedent for other countries to follow and ask Apple to provide their authorities with the same software tool."[72] Authoritarian states can apply the same arguments as the FBI but those seen as political dissidents as well as terrorists. If Apple and other technology providers that offer end-to-end encryption resist, they could be denied access to that market. This could make life difficult for them and their customers. In this particular case, Apple resisted the FBI's request, forcing the FBI to pay around $1 million to a third party to unlock Farook's iPhone.[73]

At the same time, the FBI's rationale is relatively straightforward. Comey repeatedly raised concerns that law enforcement (as well as the intelligence community) in the United States could be "going dark"—a problem also recognized by Europol.[74] This means that they are unable to access encrypted devices and encrypted communications despite having the legal and constitutional authority to do so.[75] This valid argument could also be made by more authoritarian states like Russia and China and other regimes that do not hold the same views of free speech as the United States.[76] Encrypted communications are also used by criminals, including organized crime gangs and cybercriminals. They are also used by journalists, dissidents, and whistleblowers and even sophisticated encryption has been broken.[77] It is also not preventing scammers from contacting us.

State-Backed/State-Sanctioned Cybercrime

Many of us has had phishing e-mails offering too-good-to-be true investment opportunities and the like. If these phishing activities are successful these can be used as a gateway to access our passwords and empty our bank accounts, gather our personal and business contacts, or to spy on us through our webcams and the listening devices contained in our computers, tablets, and phones. The amount of personal data we hold on fixed and mobile devices and on the cloud is growing.[78] For victims of cybercrime, the first port of call might well be law enforcement, our banks, and ultimately intelligence agencies such as the NSA. They will need to access and analyze Internet data to investigate and to build a case. Criminality and other hostile threats are fueled by both low entry points to the Internet, and growing levels of technical knowledge in a spectrum encompassing multiplying threat actors.

Cybercrime is a growth industry which to criminals offers rich rewards for relatively low risk. In this murky world there is evidence that nations including Russia have leveraged them for state-run intelligence activities. Former FBI official Leo Taddeo felt in "no doubt [that] Russia uses these criminal organizations to mask their state-sponsored intelligence and military operations".[79] In November 2014 Admiral Michael S. Rogers, then head of the NSA and U.S. Cyber Command, stated:

> what we had traditionally seen in the criminal sector was criminal actors, gangs, groups, penetrating systems and trying to steal information that they then could sell or use to generate revenue. So credit card information, selling personal information on—there's actually a market out there to sell personal information on individuals. They had been stealing - we had been watching them and observing them stealing data associated with generating revenue. The next trend that I think we're going to see…[is] criminal actors now engaging not just in the theft of information designed to generate revenue but also potentially as a surrogate for other groups, other nations. Because I'm watching nation states attempt to obscure…their fingerprints. And one of the ways to do that is to use surrogate groups to attempt to execute these things for you. It's one reason…while we're watching criminal actors start to use some of the tools that we historically have seen nation states using now, you're starting to see criminal gangs in some instances using those tools, which suggests to us that increasingly in some scenarios we're going to see more linkages between the nation state and some of these groups. That's a troubling development for us.[80]

The drivers for this can be financial incentives or pressures from governments to penetrate systems as demonstrated by Russia. This can be either penetration from the outside, to target an insider, or a combination of both. Their motive has traditionally been to be to monetarize their activities through blackmail or extortion but Russia, and likely other nations, have leveraged them (often under threats of prosecution) to do their bidding. Chinese cybercrime, in direct contrast to Russia, is mostly indigenous.[81] What Chinese hackers excel at, and frequently labeled cybercrime, is cyberespionage.[82]

Cybercriminals and States

Cybercrime threat actors can be found in organized criminal gangs, hacker groups, extremist, and terrorist groups and can be operating independently or collectively in hostile or unstable states.[83] Cybercrime is bringing new sellers to the market (unseen by software vendors) and they can have a big payday. According to Northrop–Grumman, these create "plausible scenarios…in which freelance hackers are hired by state entities to perform network exploitation on a contract basis.[84] These could be 'script kiddies' all the way through to fully organized groups affiliated to nation-states. States do not have a monopoly of coding capabilities, including for the next series of Stuxnet's, BlackEnergy and Industroyer, although they have much greater scale and resources. This poses a proliferation problem.[85]

In addition, with cyberwar(fare) programs being run by states through their intelligence agencies, some states are now building and buying code using the black market/'Dark Net'. It is alleged 'guns for hire' are being told not only what to write but also how to write it against specific targets. This includes data at rest (e.g. stored locally on hard drives) and data in motion (transited over the Internet or on Intranets). Some of these attacks can be both criminal and espionage.[86] Nations can also leak advanced tools onto the mass market. This includes Russian tools provided to the 'Shadow Brokers', a group suspected of affiliation to a nation-state operating out of either Russia or China.[87]

The most likely suspect is Russia and the Latvian-based news organization Meduza claims, "Russian programmers on the job market are invited to work at…research institutes with connections to "the agency" [the FSB]".[88] Whilst employed by the FSB, some will also continue to engage in cybercrime (which may be state sanctioned).[89] Klimsburg, writing in 2011, argued that much of this activity is conducted quite openly with the Russian hacker scene fiercely competitive and marked by venality with a focus on material (monetary) benefits. They also have social support networks (online and offline) and their main aim is stealing from the West and where prosecutions are dropped as well as hackers co-opted into Russian government service.[90] This situation is not exclusive to Russia and both cybercriminals and APTs have gone after banks and money transfer systems including SWIFT, as well as repositories of cryptocurrencies.[91]

Northrop–Grumman's 2012 study recorded that, "Non-state criminal operators and state-sponsored professional intelligence or military actors typically operate in the same environment and sometime against similar classes of targets. This is an overlap that poses attribution challenges".[92] This illustrates that the intermix between state, sub-state, or non-state actors can be blurry. The Tallinn Manual 2.0 also indicates some, (but not all) of the difficulties of accounting for the role played by cybercriminals and terrorists under the Law of Armed Conflict (LOAC).[93] Individuals and entities operating at the sub-state level remain open to obscuration, interpretation, and abuse by states.[94] The 'Dark Net' is a breeding ground for these actors and illegal activities of all descriptions, and all manner of illegal goods and services.[95]

'Dark Net' Markets

Vendors on the 'Dark Net' sell anything from firearms to hitmen for hire, illegal drugs delivered by post, stolen data, credit card numbers, cloned credit cards and ATM skimmers, to stolen identities, and fake documentation, as well as malware and exploits. They also host illegal videos and images. This includes serious child exploitation, murder, and torture. 'Dark net' sites do get closed through international and interagency cooperation between national and supranational international law enforcement agencies including national police forces and bodies like Europol and the FBI. But mounting prosecutions across transnational borders is both technically and politically difficult. Even when sites are closed down others often spring up in their place.[96] Closing the main sites and prosecuting their administrators is merely cutting the head of the hydra.[97] If and when they are closed, prosecutions depend on whether they are state-protected 'privateers' and if extradition agreements are in place and acted on.

'Black Hat' tools and services for sale on the 'Dark Net' include those for tailored malware attacks, phishing campaigns, industrial espionage, and insider information. These sell for around the following: remote logins for corporate networks $2–$30; targeted attack on companies $4,500; targeted attack on individuals $2,000; phishing kits $40; with espionage and insider trading $1,000–$15,000.[98] McAfee's December 2018 threat report stated that the takedowns of underground forums on the 'Dark Net' have had little effect on the criminal hacker scene and "many other underground markets have eagerly filled the gaps".[99] Moreover, "With the services on offer, the effectiveness of cybercriminals is increasing".[100] Despite the closure of 'marketplaces' like the infamous Silk Road in October 2013 by the FBI, and Hansa and AlphaBay between 2016–2017 in a multinational law enforcement effort led by Dutch police, "Competing marketplaces…eagerly filled the gap".[101] These included Silk Road 2.0, Dream Market, Wall Street Market, Olympus Market, Valhalla, and xDedic—all of which too were eventually shut down too.[102]

These "marketplaces, generally accessible via TOR, focus on selling narcotics and other illicit goods…offer hacking tools, hackers for hire, and data records. The accessibility of these marketplaces to a large public makes them a force to be reckoned with. Stolen digital data, which drives much of the profits, will continue to be a key motivator".[103] Their existence is also enabled through 'bulletproof hosting', where companies rent server space with 'no questions asked' in countries with light restrictions on content hosting. Users cover their tracks through TOR and VPNs. It is estimated by Europol that $1 billion was traded on the 'Dark Net' during 2019 (mainly through cryptocurrencies).[104]

Users of the 'Dark Net' increasingly communicate through encrypted apps like Telegram and Wickr, and trade continues largely through cryptocurrencies like Monero and Bitcoin (which remains the main cryptocurrency on the 'Dark Net'). The operational security of these marketplaces has matured as the result of takedowns by law enforcement as they have become more aware of the techniques being used against them. This includes vendors employing free penetration testing services and automating Pretty Good Privacy (PGP) encryption. According to Europol's 2021 IOCTA report, what has evolved is sales of wiperware/ransomware as-a-service alongside a greater exposure to data on individuals and companies. Between 2019–2020 known ransomware payments grew by 300 percent. The contact point between buyers and sellers was increasingly through encrypted apps, with Telegram and Wickr usernames listed as a point of contact.[105]

Organized Crime, Ransomware, and the 'Dark Net'

Cryptocurrencies are the payment of choice for ransomware. Ransomware has evolved from mass and untargeted phishing e-mails to well-crafted and plausible spear phishing e-mails against socially engineered targets.[106] Despite some notable successes by law enforcement bodies like Europol and vigilance by individuals and education of cyber hygiene, for many cybercriminals ransomware is not only the most widespread but also the most lucrative form of cyberattack.[107] It is a growth industry which again offers rich rewards for relatively low risk.[108]

This is evidenced by the lack of visible prosecutions, especially if criminals or criminal gangs are operating in areas where prosecutions or extraditions are likely to be difficult. According to the U.S. Secret Service's appendix to Verizon's 2022 DBIR, "This past year clearly demonstrated the increasing impact ransomware is having on businesses, critical infrastructure and national security. The most prolific ransomware networks are Russian-speaking, though this crime is not limited to one country or region. According to one industry estimate, 74% of ransomware payments were Russian affiliated". Because destructive wiperware is now also being masked as ransomware, "coupled with the limited cooperation of some states in countering ransomware", this represents "a growing risk which blurs distinctions [between] politically and financially motivated cybercrimes".[109] The gateway to ransomware and wiperware is regularly through social engineering.

If they are successful against their targets the most common methods of extracting ransom are to encrypt company data with the means to decrypt them provided only if the ransom is paid. As Bryan Lee of Palo Alto Networks noted, "Ransomware, specifically cryptographic ransomware, has quickly become one of the greatest cyberthreats facing organizations around the world. This criminal business model has proven to be highly effective in generating revenue for cyber adversaries in addition to causing significant operational impact to affected organizations. It is largely victim agnostic, spanning the globe and affecting all major industry verticals".[110]

There is nothing off limits to criminals. This includes hospitals who have been forced to shut down systems and turn away patients whilst U.S. cities and municipalities have also been increasing subjects of ransomware attacks.[111] IBM reported in 2021 that ransomware was responsible for 23 percent of incidents. Those targeted included manufacturing, the energy sector (oil and gas especially), and law firms. Those behind the attacks would also threaten to leak sensitive data if payment was not made. These are not idle threats with auctions for data dumps taking place on the 'Dark Net'. Most sellers are criminal gangs, but this can include foreign governments and even hacktivists.[112] This said, between 2015–2019, publicly disclosed hacktivist attacks saw a "sharp decrease" whilst "cryptojacking attacks and attacks on critical infrastructure organizations" rose.[113]

Another growth area is sales of Common Vulnerabilities and Exposures (CVE) combined with browser exploit kits for ransomware. These pieces of malware come in prepacked form under names like RIG, Grandsoft, Fallout, and GandCrab with "cybercriminals…eager to weaponize both new and old vulnerabilities".[114] Exploitation of these vulnerabilities, which currently profit-driven, are topics of active discussion on 'Dark Net' forums. Furthermore, there is related interest in the development of ransomware-as-a-service. McAfee has found evidence that "developers are forming strategic partnerships with other essential services, such as crypter services and exploit kits, to better service their customers and increase infection rates".[115] Crypter services develop obfuscation malware designed to elude anti-malware security products.

File encrypting ransomware includes GandCrab and Scarab. Signposting its origins, GandCrab is designed not to infect Russian systems. These and other ransomware variants and their associated tools are actively being developed. End users will wait for new releases, and these are being tailored to stay ahead of cybersecurity industry responses. They are designed to be persistent, evade existing defenses, and escalate privileges. This allows attackers to move laterally and vertically into breached systems and networks.[116] Corporate payers are often regarded as ATMs with just over a quarter finding their files decrypted after paying the ransom.[117] This is the thin end of the wedge. A large but unknown number of cases go unreported for fear of reputational damage and stock price falls (among other concerns).

On the 'Dark Net' there are Remote Desktop Protocol (RDP) shops selling access to hacked computer systems "to potentially cripple cities and bring down major companies" ranging in size from as little as 15 to over 40,000 RDP connections on the Russian-based Ultimate Anonymity Service.[118] RDP is a legitimate tool designed by Microsoft to help system administrators access remote computers (similar to TeamViewer—popular with telephone scammers) which is now being exploited by organized cybercriminals. Compromises through RDP can misdirect investigations and "By leveraging RDP, an attacker need not create a sophisticated phishing campaign, invest in malware obfuscation, use an exploit kit, or worry about antimalware defenses. Once attackers gain access, they are in the system". Moreover, "By accessing a system via RDP, attackers can obtain almost all data stored on a system. This…can be used for identity theft, account takeovers, credit card fraud, and extortion, etc."[119]

Windows 2008 and 2012 server hacks are most widely offered (indicating ongoing breaches into organizations) but Operating Systems from Windows XP to Windows 10 were also on offer as is Windows Embedded Standard (now Windows IOT). This is widely used by 'thin clients' used by "hotel kiosk systems, announcement boards, point-of-sale (POS) systems, and even parking meters among others" which "are often overlooked and not commonly updated, making them an ideal backdoor target for an attacker". Access to "multiple government systems" is also "being sold worldwide, including those linked to the

United States, and dozens of connections linked to health care institutions, from hospitals and nursing homes to suppliers of medical equipment".[120]

Even more worryingly, McAfee has found that "access linked to security and building automation systems of a major international airport could be bought for only US$10…This is definitely not something you want to discover on a Russian underground RDP shop, but the story gets worse". They also discovered the "airport's automated transit system, the passenger transport system that connects terminals…might be openly accessible from the Internet". Exploiting RDPs requires "no zero-day exploit, elaborate phishing campaign, or watering hole attack".[121] This kind of attack compounds problems for police and law enforcement and their frequent partners in the intelligence agencies.

Internationally, the FBI, Interpol and Europol require not only the resources to fight cybercrime but also sufficient funding. The former head of Europol, Rob Wainwright, has also said that intelligence sharing between like-minded governments is also a high priority.[122] Wainwright's successor, Catherine De Bolle, said in the foreword to Europol's 2020 annual Internet Organized Crime Threat Assessment (IOCTA) "Cybercrime remains among the most dynamic forms of crime encountered by law enforcement in the EU".[123] In addition to established *modi operandi*, new methods are being found. Ransomware masquerading as wipers whose intent is data destruction through encryption is one.

WannaCry and Petya/NotPetya

This change was demonstrated by the WannaCry ransomware in May 2017. This had the footprint of a nation-state. WannaCry spread rapidly and infected around 200,000 computers in 150 countries worldwide. The worst effects were seen in Russia, Ukraine, India, and Taiwan and badly hit Britain's National Health Service (NHS) which led them to temporarily revert to paper record keeping.[124] WannaCry used a vulnerability in legacy Windows Operating Systems to spread (especially Windows XP). Although a patch was rapidly issued by Microsoft, legacy systems remain a problem area which is difficult to remedy—even for the U.S. government.[125]

WannaCry used an exploit in Windows known EternalBlue which was developed by the NSA and then repurposed for WannaCry by North Korea after it was hacked by the 'Shadow Brokers'.[126] This highlights two things. First state actors—those acting under political direction or sanctioned to operate by the state and able to be 'plausibly' denied—are engaging in activities normally associated with cybercriminals. Second, that state and non-state actors can rewire code which can be repurposed for their own ends and able to be used against those who first developed it, pointing to the proliferation problem of malware for cyber offense—and cybercrime.[127]

While initially masked as ransomware, not unlike WannaCry a month earlier,[128] the Petya/NotPetya attack, which appeared in June 2017, was designed to make systems, databases, and machines unusable. Although Petya/NotPetya propagated globally, Ukraine was the "epicenter" with "as many as 60 percent of the systems" there effected.[129] Unlike WannaCry, Petya/NotPetya "was built to destroy, not extort" using both open-source code and a variant of the EternalBlue exploit stolen from the NSA (for which there was the existing patch from Microsoft).[130] It was designed to look like a WannaCry variant and claimed to have encrypted data whilst demanding ransomware payments in Bitcoin. However, it was wiperware—the files could not have been decrypted once Petya was installed.[131]

Almost immediately Western intelligence began to point the finger at Russia.[132] In January 2018 *The Washington Post* reported the CIA had attributed NotPetya to Russia's GRU.[133] In October 2020 the Justice Department indicted six members of the GRU's Military Unit 74,455 (part of the APT 'Fancy Bear'/APT28).[134] While 60% of the victims were in Ukraine, further suggesting a politically motivated attack, a number of innocent bystanders were hit globally. The economic effects were remarkable. FedEx, the logistics company, has said that some of its information was lost forever and it took over a month to restore normal functioning leading to a $300 million loss.[135] Similarly, shipping conglomerate Maersk reported it affected global shipping, taking them a month to restore systems. They have also estimated the damage to be up to $300 million.[136] Saint Gobain, Reckitt Beckinser, and several other companies were also impacted as Petya/NotPetya spread to at least 65 countries.[137]

Both used the same exploits stolen from the NSA in the 'Shadow Brokers' attack, EternalBlue, and DoublePulsar. These exploits were also used by China in "several disparate campaigns, likely at the hands of multiple separate groups".[138] CrowdStrike analyzed that "The rapid incorporation of all of these exploits into China-based operations suggests these adversaries may have access to a centralized dissemination channel for tools and exploits. It is also possible that China was already aware of some or all of these vulnerabilities".[139] This is supported by the cybersecurity industry which "has suggested that the Chinese National Vulnerability Database (CNNVD) is a loose cover for the Ministry of State Security (MSS) and provides early access of vulnerabilities to China's intelligence services before publicly reporting them".[140] This is another illustration of the blurring of lines between state, sub-state, non-state, and proxy actors. It might also conceivably offer evidence of coordination.

The Cloak of Attribution: The Use of Proxy Actors by States

Tim Maurer in *Cyber Mercenaries* argues that "States are only one subset of a larger group of actors with significant offensive cyber capabilities".[141] Maurer quoted a U.S. Secret Service officer who believed that "Many of the [non-state] actors that we look at on a daily and weekly basis have capabilities that actually exceed the capabilities of most nation-states". Maurer describes that "states use these proxies for a wide variety of purposes not limited to projected power abroad".[142] This is recognized by U.S. military with the Joint Chiefs of Staff declaring, "Nation-states may conduct operations directly or may outsource them to third parties, including front companies, patriotic hackers, or other surrogates, to achieve their objectives".[143] This includes cybercriminal gangs.

Maurer recognizes that whilst states might pursue different models for controlling actors detached from the state "they also face a common challenge and have an interest in balancing the benefits of proxy relationships with the cost and increased risk of escalation".[144] Escalation control is made more difficult by proxies and privateer freelancing, including what Maurer terms 'cyber mercenaries' in Russia and China.[145] The use of these proxies and privateers is to deflect and diffuse state responsibilities. This diminishes the international Rule of Law. This also has direct implications for the Law of Armed Conflict and considerations over what constitutes direct participation in hostilities by civilians.[146]

So long as they are used for plausibly deniable acts and campaigns by states like Russia and China the use of these proxies and privateers poses challenges for the West and their friends and allies in the non-Western world.[147] According to Mark Galeotti, in Russia during the 2010s: "Some of the cybercriminals and cybersecurity experts that the gangs employ also work for the government. Most, however, do not. Hackers themselves rarely fit the model of organized crime, as their structures are generally collectives. Instead of becoming members of gangs, they tend to be outside consultants, hired for specific jobs".[148] This relationship has elements of co-dependency, but this is an unequal and asymmetric relationship. Their use could also lead to unintended escalation and harm state-on-state relations, complicating signaling and diplomacy, and escalation control. At worst, should a proxy cyberattack cross the threshold with an act considered an 'armed attack' under the Law on Armed Conflict this could lead to return fire or militarized conflict.

Proxies also undercut cyber deterrence strategies, combined with a failure of diplomatic initiatives aimed at preventing attacks on economic, political, and critical infrastructure targets. Proving state responsibility has been problematic. It remains far too easy for intelligence agencies, from where most state level cyber intrusions and breaches are run or orchestrated, to hide behind the cloak provided by cybercriminals and malicious hackers through the 'attribution problem'. The attribution problem is well recognized and although a number of individuals have faced prosecution, many escape judicial proceedings or disappear undetected. At the same time, advancements in technical computer forensics are beginning to show the holes in this cloak. In trying to uphold the Rule of Law, legal remedies such as prosecution and extradition remain problematic. Publicly naming states accused of cyberattacks or harboring individuals perpetrating them has hitherto not shamed them into stopping.

This said, cybercrime has increasingly been prosecuted across national jurisdictions in some high-profile cases. This includes Ovidiu-Ionut Nicola-Roman, a Romanian who was the first person to be extradited to the United States for phishing, and Marcel Lazar (Guccifer), another Romanian who hacked the e-mail accounts of several senior U.S. officials which began the investigation into Hillary Clinton's use of a private e-mail server when Secretary of State. Many more are deemed not worth prosecuting with extradition proceedings lengthy, costly, and difficult or unrealistic. For citizens of Russia, China, or North Korea no extradition treaty exists with the United States and any state can resist or refuse extradition. Even for allied or friendly nations prosecutions and extraditions are subject to lengthy appeals procedures and fraught with difficulties.[149] The case of Wikileaks founder Julian Assange who faces extradition from the UK to the U.S. is a visible reminder of the difficulties involved.[150] There has instead been a drip effect stemming from these legal impediments which has seen cases of major cybercrimes become unprosecutable.[151] Beyond cybercrime, a number also purport to hack for patriotic reasons at the sub-state level.[152] China is known to use these 'patriotic hackers' at their universities and many study or work abroad.[153]

Conclusion

The different threats and threat actors in this chapter reflect the gestalt environment of cybersecurity. Each are part of the whole and offer different manifestations of threats against the West. Despite high levels of investment in technical cybersecurity products and services, these threats and threat actors continue to grow in eye watering numbers and continue to metastasize.[154] The global shortfall in personnel involved in cybersecurity is a factor but a greater factor is that the Internet has become a breeding ground for online communities.[155]

Hacking can be self-taught at the cost of an Internet connection and laptop making the barriers to entry very low (technical proficiency notwithstanding) and social engineering is easily accomplished in a target rich environment filled with actionable OSINT. This means that cybercrime, including those seeking to take this as a route to monetarize and develop an existing skill set, lacks many of the barriers to entry and advancement as opposed to the competitive world of the global cybersecurity market. Against this background, the promise of high returns for limited risk is co-opting people into criminality. Turning them away from criminality relies on the belief in and adherence to the Rule of Law in a wide number of states.

Lawlessness and the willingness of states like Russia to use cybercriminals as proxy forces or 'privateers' is a corrupt and corrupting practice. This leads organized criminals and criminal gangs to potentially easy gains without undue fear of prosecution or extradition. The 'Dark Net' is a haven for these activities where it is used to hone and share skills and exploits and sell what they have stolen/hacked.[156] It is also where they can self-organize into hacking groups with crimes like ransomware offering big rewards for low risk. For these communities, malware provides the keys to the cupboard.

Against the West, evidence points to Russia and former Soviet states allied or affiliated to Russia as a 'safe space' for criminal activities. This is directed outwards. Cybercrime in China is, in contrast, largely domestic. China has not embraced 'cyber privateers'. Instead, the tighter grip the Chinese Communist Party and Ministry of State Security hold on Internet access and the levels of domestic monitoring lead instead more towards cooperation between the state and the 'Black Hats'.[157] In the West too, whilst cybercriminals are prosecuted, there have been many cases where they are given light sentences and treated as a 'white collar' criminals. A number, like Kevin Mitnick and Marcus Hutchins now works for the cybersecurity industry. The admixture of criminals, including those turned from 'poacher turned gamekeeper' like Mitnick and Hutchins, is evident in Russia and China except they are more likely to continue being 'Black Hats' working for or with the state.

It is unclear how many are involved in cybercrime and the numbers working on malware development, but investigative tools like Social Network Analysis can aid police, law enforcement, and the intelligence agencies (as it can criminals and terrorists who practice social engineering). Government and industry cybersecurity threat reports show however that the global scope of cybercrime and awareness of it is increasing. Ben Hammersley's thought provoking BBC TV series 'Cyber Crimes' has also highlighted in some depth the types of crime, complexities, and sophistication involved in some cases as well as some of the prosecutional difficulties involved.[158] When and whether crimes such as drug trafficking committed using 'Dark Net' marketplaces through to potentially devastating acts such as an attack on critical infrastructure committed by a cyber 'gun for hire' through the 'Dark Web' can be deterred or prosecuted requires international cooperation at the global level. Current cooperation takes place between friends and allies in an environment where states are both targets and attackers.

Although technical attribution has advanced, many 'Black Hats' cover their tracks through TOR and VPNs and communicate on encrypted channels, devices, and apps. These same tools are used by terrorists to communicate and coordinate locally or at distance. They are also used when searching for propaganda or instructions for carrying out attacks. This is how 'lone wolves' are 'radicalized' allied to their 'echo chambers' on the Internet drawing vulnerable or impressionable people in; especially the young. Different 'echo chambers' draw people into cybercrime. This points to the need to leverage social and behavioral psychology to examine "motivations as varied as financial gain and political/ideological protest" which "are often instigated by groups, as opposed to individuals acting alone".[159]

Views and collectives form quickly but how long they last, whether they have longevity, and what binds them and breaks them need systematic study. Added into this melting pot is the intermix between sub-state criminality and state intelligence services, particularly evident in Russia, as in the case of WannaCry, Petya, and NotPetya. At the sub-state level in China meanwhile, overt research at Chinese universities is being married to covert intelligence gathering by the state with cyberespionage by the commercial sector. This adds layers of complexity to an already complex environment defined at least as much by anarchy as the Rule of Law. This is one of several major issues debated in the conclusion which begins by reflecting on the demonstrative potential for cyberwar through attacking critical infrastructure.

Notes

1. Lucas Kello, *The Virtual Weapon and International Order* (New Haven, CT: Yale University Press, 2017), pp. 160–194.
2. Worldwide Threat Assessment of the US Intelligence Community Daniel R. Coats 13 February 2018, p. 14, https://www.dni.gov/files/documents/Newsroom/Testimonies/2018-ATA---Unclassified-SSCI.pdf, accessed 13 December 2018.
3. Worldwide Threat Assessment of the US Intelligence Community Daniel R. Coats Director of National Intelligence Senate Select Committee on Intelligence (29 January 2019), p. 6, https://www.dni.gov/files/ODNI/documents/2019-ATA-SFR---SSCI.pdf, accessed 7 July 2019. See also Verizon 2019 Data Breach Investigation Report, pp. 44–45, https://enterprise.verizon.com/resources/reports/2019-data-breach-investigations-report.pdf, accessed 13 July 2019.
4. Annual Threat Assessment of the US Intelligence Community, Office of the Director of National Intelligence (April 9, 2021), p. 20, https://www.dni.gov/files/ODNI/documents/assessments/ATA-2021-Unclassified-Report.pdf, accessed 6 August 2021.
5. Cyber Threat and Vulnerability Analysis of the U.S. Electric Sector Prepared by: Mission Support Center Idaho National Laboratory August 2016, p. ii, https://www.energy.gov/sites/prod/files/2017/01/f34/Cyber%20Threat%20and%20Vulnerability%20Analysis%20of%20the%20U.S.%20Electric%20Sector.pdf, accessed 17 April 2019.
6. Homeland Threat Assessment October 2020, pp. 8–9, https://www.dhs.gov/sites/default/files/publications/2020_10_06_homeland-threat-assessment.pdf, accessed 6 August 2021.
7. Jane Wakefield, 'Office Intruder 'Steals' Data', http://news.bbc.co.uk/1/hi/technology/7843206.stm, accessed 22 June 2015.

8. Verizon, 2019 Data Breach Investigations Report, pp. 55, https://www.fbcinc.com/e/docitcyber/presentations/Verizon2019DataBreakInvestigationsReport.pdf, accessed 6 April 2020.
9. Verizon, 2019 Insider Threat Report Executive Summary, p. 6, https://enterprise.verizon.com/resources/executivebriefs/insider-threat-report-executive-summary.pdf, accessed 11 April 2020.
10. Verizon Data Breach Investigations Report 2022, pp. 8, 11, 13, 71–72, https://www.verizon.com/business/resources/reports/2022/dbir/2022-data-breach-investigations-report-dbir.pdf, accessed 11 June 2022.
11. Homeland Security Public Service Announcement, Increase in Insider Threat Cases Highlight Significant Risks to Business Networks and Proprietary Information (September 23 2014), https://www.ic3.gov/media/2014/140923.aspx, accessed 9 September 2018.
12. Homeland Security Public Service Announcement, Increase in Insider Threat Cases (September 23 2014).
13. Homeland Security Public Service Announcement, Increase in Insider Threat Cases (September 23 2014).
14. These issues were placed under active study by the U.S. Department of Defense in 2017. Department of Defense Defense Science Board Task Force on Cyber Deterrence February 2017, p. 20, https://www.armed-services.senate.gov/imo/media/doc/DSB%20CD%20Report%202017-02-27-17_v18_Final-Cleared%20Security%20Review.pdf, accessed 1 December 2018.
15. This includes U.S. government and critical infrastructure sites. Executive Order 13587—Structural Reforms to Improve the Security of Classified Networks and the Responsible Sharing and Safeguarding of Classified Information, 7 October 2016, https://www.whitehouse.gov/the-press-office/2011/10/07/executive-order-13587-structural-reforms-improve-security-classified-net. See also, https://www.sei.cmu.edu/research-capabilities/all-work/display.cfm?customel_datapageid_4050=21232, both accessed 14 November 2016.
16. Michael Workman, 'Gaining Access with Social Engineering: An Empirical Study of the Threat', *Information Systems Security*, Vol. 16, Issue 6 (2007), pp. 315–331.
17. Anonymous, 'Arrests Over Hacks of CIA and FBI Staff' (September 9 2016), http://www.bbc.co.uk/news/technology-37316615, accessed 10 September 2016.

18. Anonymous, 'Two Years for Teen 'Cyber Terrorist' who Targeted US Officials' (20 April 2018), https://www.bbc.co.uk/news/uk-england-leicestershire-43840075, accessed 29 September 2019.
19. Philip Messing, Jamie Schram, and Bruce Golding, 'Teen Says He Hacked CIA Director's AOL Account' (October 18 2015), http://nypost.com/2015/10/18/stoner-high-school-student-says-he-hacked-the-cia/, accessed 10 September 2016.
20. See for example Anthony Zurcher, 'Hillary Clinton's 'Emailgate' Diced and Sliced' (19 September 2016), http://www.bbc.co.uk/news/world-us-canada-31806907, accessed 24 October 2016.
21. Messing, Schram, and Golding, 'Teen Says He Hacked CIA Director's AOL Account' (October 18 2015).
22. Kim Zetter, 'Teen Who Hacked CIA Director's Email Tells How He Did It' (October 19 2015), https://www.wired.com/2015/10/hacker-who-broke-into-cia-director-john-brennan-email-tells-how-he-did-it/, accessed 10 September 2016.
23. Rachna Dhamija, J.D. Tygar, and Marti Hearst, 'Why Phishing Works', *Proceedings of the SIGCHI Conference on Human Factors In Computing Systems* (April 2006), pp. 581–590 and Tom N. Jagatic, Nathaniel A. Johnson, Markus Jakobsson and Filippo Menczer, 'Social Phishing', *Communications of the ACM*, Vol. 50, No.10 (October 2007), pp. 94–100.
24. 'Anonymous, 'Arrests Over Hacks of CIA and FBI Staff' (September 9 2016), Lorenzo Franceschi-Bicchierai, 'Teen Who Hacked Ex-CIA Director John Brennan Gets Sentenced to 2 Years of Prison' (April 20 2018), https://www.vice.com/en_us/article/pax87v/kane-gamble-crackas-with-attitude-cwa-sentence-prison, Thomas Brewster, 'Sources: Martin Shkreli Thrown In Solitary Confinement After Claims He Ran Company From Prison' (April 1 2019), https://www.forbes.com/sites/thomasbrewster/2019/04/01/sources-martin-shkreli-thrown-in-solitary-confinement-after-claims-he-ran-company-from-prison/, both accessed 1 January 2019.
25. Charu C. Aggarwal, *Social Network Data Analytics* (New York: Springer, 2011), esp. pp. 1–15.
26. Aggarwal, *Social Network Data Analytics*, p. 2.
27. Aggarwal, *Social Network Data Analytics*, p. 2.
28. Judith S. Kleinfeld, 'The Small World Problem', *Society*, Vol. 39, Issue 2 (January/February 2002), p. 66.
29. Kleinfeld, 'The Small World Problem', p. 62.

30. Kleinfeld, 'The Small World Problem', pp. 61–66.
31. This is unless misinformation or disinformation is purposely distributed for the purposes of confusion, investigation, or entrapment.
32. In the sense of human/social agency.
33. See for example https://www.sometics.com/en/sociogram, accessed 17 September 2016.
34. David Lazer, Ryan Kennedy, Gary King, Alessandro Vespignani, 'The Parable of Google Flu: Traps in Big Data Analysis', *Science*, Vol. 343, Issue 6176 (14 March 2014), pp. 1203–1205.
35. Barton Gellman and Laura Poitras, 'U.S., British Intelligence Mining Data from Nine U.S. Internet Companies in Broad Secret Program' (June 6 2013), http://www.washingtonpost.com/investigations/us-intelligence-mining-data-from-nine-us-internet-companies-in-broad-secret-program/2013/06/06/3a0c0da8-cebf-11e2-8845-d970ccb04497_story.html, accessed February 15, 2016.
36. Stanley Wasserman and Katherine Faust, *Social Network Analysis: Methods and Applications* (Cambridge: Cambridge University Press, 1994), p. 10.
37. See for example Wullianallur Raghupathi and Viju Raghupathi, 'Big Data Analytics in Healthcare: Promise and Potential', *Health Information Science and Systems*, Vol. 2, No. 3 (February 2014), pp. 1–10, and Ashwin Belle, Raghuram Thiagarajan, S. M. Soroushmehr, Fatemeh Navidi, Daniel A. Beard, and Kayvan Najarian, 'Big Data Analytics in Healthcare', *BioMed Research International* (July 2015), pp. 1–16. James B. Heffernan, Patricia A. Soranno, Michael J. Angilletta, Lauren B. Buckley, Daniel S. Gruner, Tim H. Keitt, James R. Kellner Daniel S Gruner, John S. Kominoski, Adrian V. Rocha, Jingfeng Xiao, Tamara K. Harms, Simon J. Goring, Lauren E. Koenig, William H. McDowell, Heather Powell, Andrew D. Richardson, Craig A. Stow, Rodrigo Vargas, and Kathleen C. Weathers, 'Macrosystems Ecology: Understanding Ecological Patterns and Processes at Continental Scales', *Frontiers in Ecology and the Environment*, Vol. 12, No. 1 (February 2014), pp. 5–14. Pierre Deville, Catherine Linard, Samuel Martin, Marius Gilbert, Forrest R. Stevens, Andrea E. Gaughan, Vincent D. Blondel, and Andrew J. Tatem, 'Dynamic Population Mapping Using Mobile Phone Data', *Proceedings of the National Academy of Sciences*, Vol. 111, No. 45 (November 2014), pp. 15888–15893.
38. Christina Prell, *Social Network Analysis: History, Theory and Methodology* (London: Sage, 2011), p. 2.

39. Anonymous, "Serious' Hack Attacks from China Targeting UK Firms' (4 April 2017), http://www.bbc.co.uk/news/technology-39478975, accessed 4 April 2017.
40. For more on SNA see for example David Easley and John Kleinberg, *Networks, Crowds, and Markets: Reasoning About a Highly Connected World* (Cambridge: Cambridge University Press, 2010), Charles Kadushin, *Understanding Social Networks: Theories, Concepts, And Findings* (Oxford: Oxford University Press, 2011), Stephen P. Borgatti, Martin G. Everett and Jeffrey C. Johnson, *Analyzing Social Networks* (London: Sage, 2013), Matthew A. Russell, *Mining the Social Web: Data Mining Facebook, Twitter, LinkedIn, Google +, GitHub, and More* (Cambridge: O'Reilly, Second Edition 2013),
41. Alicia Bargar, Janis Butkevics, Stephanie Pitts and Ian McCulloh, 'Challenges and Opportunities to Counter Information Operations Through Social Network Analysis and Theory', in Tomáš Minárik, Siim Alatalu, Stefano Biondi, Massimiliano Signoretti, Ihsan Tolga and Gábor Visky (eds.), *2019 11th International Conference on Cyber Conflict: Silent Battle* (Tallinn: CCD COE Publications, 2019), pp. 231–248.
42. Anonymous, 'Cooperation or Resistance?: The Role of Tech Companies in Government Surveillance' (April 10 2018), https://harvardlawreview.org/2018/04/cooperation-or-resistance-the-role-of-tech-companies-in-government-surveillance/, accessed 27 January 2020.
43. John S. Hollywood, Michael J. D. Vermeer, Dulani Woods, Sean E. Goodison and Brian A. Jackson, 'Using Social Media and Social Network Analysis in Law Enforcement Creating a Research Agenda, Including Business Cases, Protections, and Technology Needs' (2018), https://www.rand.org/pubs/research_reports/RR2301.html?adbid=1024409625238491136&adbpl=tw&adbpr=22545453&adbsc=social_20180731_2462281, and Jon Fasman, 'I Know What You'll Do Next Summer More Data and Surveillance are Transforming Justice Systems' (June 2 2018), https://www.economist.com/technology-quarterly/2018-05-02/justice, both accessed 25 September 2019.
44. Joshua A. Tucker, Andrew Guess, Pablo Barberá, Cristian Vaccari, Alexandra Siegel, Sergey Sanovich, Denis Stukal, and Brendan Nyhan, 'Social Media, Political Polarization, and Political Disinformation: A

Review of the Scientific Literature' (March 2018), p. 9, https://www.hewlett.org/wp-content/uploads/2018/03/Social-Media-Political-Polarization-and-Political-Disinformation-Literature-Review.pdf, accessed 17 November 2019.

45. Arie Perliger, 'The Causes and Impact of Political Assassinations', *CTC Sentinel* [Combatting Terrorism Center, West Point], Vol. 8, Issue 1 (January 2015), https://ctc.usma.edu/the-causes-and-impact-of-political-assassinations/, accessed 17 November 2019.

46. Tucker, et al., 'Social Media, Political Polarization, and Political Disinformation', p. 9.

47. Christopher Wray, Director, Federal Bureau of Investigation, Examining the January 6 Attack on the U.S. Capitol *Statement for the Record* (June 15 2021), https://www.fbi.gov/news/testimony/examining-the-january-6-attack-on-the-us-capitol-wray-061521, accessed 14 June 2022.

48. Marshall Cohen, 'Evidence Shows Capitol Rioters Brutally Attacked Police with Flagpoles, Fire Extinguishers and Fists' (January 21 2021), https://edition.cnn.com/2021/01/21/politics/capitol-rioters-attacking-police/index.html, https://www.politico.com/news/2022/01/04/jan-6-insurrection-sentencing-tracker-526091, both accessed 14 June 2022.

49. Helen C. Harton, Mathew Gunderson, Martin J. Bourgeois, "'I'll Be There With You": Social Influence and Cultural Emergence at the Capitol on January 6', *Group Dynamics: Theory, Research, and Practice*, Vol. 26, (forthcoming 2022). Carl A. Grant & Paul D. Grant, 'A Failure to Educate: January 6, 2021 and the Banality of Evil', *The Educational Forum*, Vol. 86, No. 2 (March 2022), pp. 109–124.

50. An interesting case of a Canadian journalist working for the TV station Viceland and Canadian authorities and their views on his contacts with an Islamic State insurgent, highlights some of the difficulties between press freedom and counter terrorism. Rachel Browne, 'Canadian Judge Orders VICE News Journalist to Hand Over Digital Messages', https://news.vice.com/article/canadian-judge-orders-vice-news-journalist-to-hand-over-digital-messages, accessed 3 April 2017. Similar reasoning provided by this journalist, Ben Makuch, and his organization were also used by Glenn Greenwald and other journalists in their dealings with Edward Snowden.

51. Maura Conway, 'Reality Check: Assessing the (Un)likelihood of Cyberterrorism', in Thomas M. Chen, Lee Jarvis, Stuart Macdonald

(eds.), *Cyberterrorism. Understanding, Assessment, and Response* (New York: Springer, 2014), p. 103–122.
52. Gianluca Mezzofiore, 'Team Poison's Junaid Hussain Jailed for Tony Blair Hack and Phone Bombing Anti-Terror Hotline' (July 2 2014), https://www.ibtimes.co.uk/team-poison-phone-bomb-hacker-anti-terror-367660, accessed 3 January 2020.
53. ISIL-Linked Kosovo Hacker Sentenced to 20 Years in Prison, September 23 2016, https://www.justice.gov/opa/pr/isil-linked-kosovo-hacker-sentenced-20-years-prison, accessed 10 August 2021.
54. https://www.counterextremism.com/extremists/junaid-hussain, https://www.counterextremism.com/extremists/usaamah-abdullah-rahim, https://www.counterextremism.com/extremists/elton-simpson.
55. David V. Gioe, Michael S. Goodman and Alicia Wanless, 'Rebalancing Cybersecurity Imperatives: Patching the Social Layer', *Journal of Cyber Policy*, Vol. 4, No. 1 (March 2019), p. 127.
56. IOCTA Internet Organized Crime Threat Assessment (2019), p. 48, https://www.europol.europa.eu/iocta-report, accessed 7 January 2020.
57. IOCTA Internet Organized Crime Threat Assessment (2019), pp. 48–49.
58. Laura Bliss, 'What Facebook isn't Telling us About its Fight Against Online Abuse' (May 21 2018), http://theconversation.com/what-facebook-isnt-telling-us-about-its-fight-against-online-abuse-96818, accessed 30 September 2019.
59. Gioe, Goodman, Wanless, 'Rebalancing Cybersecurity Imperatives: Patching the Social Layer', pp. 128–129.
60. Twitter is frequently the medium of choice to send anonymized death threats to political figures. Elizabeth Arnold, Danya Bazaraa, Josh Thomas, 'Labour MP Jess Phillips Shares Death Threat Letter Sent to Her as She Slams PM' (25 September 2019), https://www.mirror.co.uk/news/politics/labour-mp-jess-phillips-shares-20205763 and Will Oremus, 'The Pipe Bomb Suspect Appears to Have Tweeted Death Threats. Twitter Saw No Problem' (October 26 2018), https://slate.com/technology/2018/10/twitter-account-linked-to-cesar-sayoc-made-death-threats-and-twitter-declined-to-suspend-him.html, accessed 26 September 2019.
61. 'Inside the FBI: Director Comey Addresses Cyber Security Experts' (September 2 2016), https://www.fbi.gov/audio-repository/inside-podcast-comey-cyber-speech-090216.mp3/view, accessed 22 November 2018.

62. See for example, Chancellor's Speech to GCHQ on Cyber Security (17 November 2015), https://www.gov.uk/government/speeches/chancellors-speech-to-gchq-on-cyber-security, accessed 20 March 2016.
63. Department of Defense Defense Science Board Task Force on Cyber Deterrence February 2017, pp. 4, 11.
64. Department of Defense Defense Science Board Task Force on Cyber Deterrence February 2017, p. 27. See also Martin Rudner, 'Cyber-Threats to Critical National Infrastructure: An Intelligence Challenge', *International Journal of Intelligence and Counterintelligence*, Vol. 26, Issue 3, (May 2013), pp. 453–481.
65. Emphasis in original. Worldwide Threat Assessment (29 January 2019), p. 7.
66. Avi Tal and David Siman-Tov, 'HUMINT in the Cybernetic Era: Gaming in Two Worlds', *Military and Strategic Affairs*, Vol. 7, No. 3 (December 2015), p. 96.
67. Also noteworthy is the work that is being undertaken by the Global Internet Forum to Counter Terrorism (GIFCT) and Tech Against Terrorism. https://gifct.org/, https://www.techagainstterrorism.org/, both accessed 30 June 2022.
68. Remarks made under Chatham House Rules at the 'Insider Threat Summit', Monterey, California, 29 March-1 April 2016 attended by author.
69. See, for example, Joe Miller, 'Google and Apple to Introduce Default Encryption' (19 September 2014), http://www.bbc.co.uk/news/technology-29276955 and Anonymous, 'Tor Project Makes Efforts to Debug Dark Web' (23 July 2014), http://www.bbc.co.uk/news/technology-28447023, both accessed 31 March 2015.
70. See for example Conor Friedersdorf, 'How Dangerous Is End-to-End Encryption?' (14 July 2015), http://www.theatlantic.com/politics/archive/2015/07/nsa-encryption-ungoverned-spaces/398423/, accessed 16 February 2016.
71. Kim Zetter, 'Apple's FBI Battle Is Complicated. Here's What's Really Going On' (18 February 2016), accessed at https://www.wired.com/2016/02/apples-fbi-battle-is-complicated-heres-whats-really-going-on/, accessed 29 May 2016.
72. Zetter, 'Apple's FBI Battle Is Complicated' (February 18 2016).

73. Mark Hosenball, 'FBI Paid under $1 Million to Unlock San Bernardino iPhone: Sources' (4 May 2016), http://www.reuters.com/article/us-apple-encryption-idUSKCN0XQ032, accessed 29 May 2016.
74. Anonymous, 'Europol Chief Warns on Computer Encryption' (29 March 2015), http://www.bbc.co.uk/news/technology-32087919, accessed 29 May 2016.
75. Amy Hess, Executive Assistant Director, Science and Technology Branch, Federal Bureau of Investigation, statement before the House Committee on Energy and Commerce, Subcommittee on Oversight and Investigation, 19 April 2016, https://www.fbi.gov/news/testimony/deciphering-the-debate-over-encryption, accessed 29 May 2016.
76. Lulu Yilun Chen, 'Telegram Traces Massive Cyber Attack to China During Hong Kong Protests' (June 13 2019), https://www.bloomberg.com/news/articles/2019-06-13/telegram-traces-cyber-attack-to-china-amid-hong-kong-protests, accessed 4 January 2020.
77. Bill Marczak, John Scott-Railton, Sarah McKune, Bahr Abdul Razzak, and Ron Deibert, 'Hide and Seek Tracking NSO Group's Pegasus Spyware to Operations in 45 Countries' (September 18 2018), The Citizen Lab, Munk School of Global Affairs, University of Toronto, https://tspace.library.utoronto.ca/bitstream/1807/95391/1/Report%23113--hide%20and%20seek.pdf, accessed 14 June 2022.
78. There are a plethora of examples of what is possible to do with the right toolkit and expertise and demonstrations at Defcon, the premium hacker conference in the United States, are food for thought. See for example 'Real Future: What Happens When You Dare Expert Hackers To Hack You (Episode 8)' (24 February 2016) https://www.youtube.com/watch?v=bjYhmX_OUQQ, accessed 15 March 2016.
79. Patrick Howell O'Neill, 'Russia's Rise to Cyberwar Superpower' (February 29 2020), https://www.dailydot.com/debug/russia-cyberwar-cyberattack-dnc-breach-history/, accessed 9 April 2020.
80. Hearing of the House (Select) Intelligence Committee Subject: "Cybersecurity Threats: The Way Forward". November 20, 2014, https://www.nsa.gov/news-features/speeches-testimonies/Article/1620360/hearing-of-the-house-select-intelligence-committee-subject-cybersecurity-threat/, accessed 21 January 2019.
81. Jon R. Lindsay, 'The Impact of China on Cybersecurity: Fiction and Friction', *International Security*, Vol. 39. No. 3. (Winter 2014/15), p. 18. William T. Hagestad II, *Chinese Cyber Crime 2016: Hacking*

Underground in the People's Republic of China 2nd Edition (Scotts Valley, CA: CreateSpace Independent Publishing, 2016). Marin Ivezic, 'Cybercrime in China – a Growing Threat for the Chinese Economy' (February 2 2017), https://cyberkinetic.com/cybersecurity-cyber-risk/chinese-cybercrime/, accessed 11 August 2021.
82. This is the subject of a separate book.
83. Views expressed under Chatham House Rules at the BT SASIG Conference on the Insider Threat London, 8 May 2015 attended by author.
84. Bryan Krekel, Patton Adams, and George Bakos, *Occupying the Information High Ground: Chinese Capabilities for Computer Network Operations and Cyber Espionage* Prepared for the U.S.-China Economic and Security Review Commission by Northrop Grumman Corp (McLean, VA: Northrop Grumman Corporation, March 7, 2012), p. 97, https://info.publicintelligence.net/USCC-ChinaCyberEspionage.pdf, accessed 16 April 2019.
85. Views expressed at CyCon 2015, Tallinn, Estonia, 27–28 May 2015, under Chatham House Rules attended by author.
86. Views expressed at CyCon 2015, Tallinn, Estonia, 27–28 May 2015, under Chatham House Rules attended by author.
87. Bruce Schneier, 'Who Are the Shadow Brokers? What is—and isn't—Known About the Mysterious Hackers Leaking National Security Agency secrets' (May 23 2017), https://www.theatlantic.com/technology/archive/2017/05/shadow-brokers/527778/, accessed 8 January 2020.
88. Sasha Baranovskaya, 'Moscow's Cyber-Defense How the Russian Government Plans to Protect the Country from the Coming Cyberwar' (19 July 2017), https://meduza.io/en/feature/2017/07/19/moscow-s-cyber-defense, accessed 28 December 2017.
89. Alexander Martin, 'Cyber Criminal Charged with Stealing £76 m While Working for Russian Intelligence' (December 5 2019), https://news.sky.com/story/cyber-criminal-charged-with-stealing-76m-while-working-for-russian-intelligence-11878896, accessed 4 January 2020.
90. Alexander Klimburg, 'Mobilising Cyber Power', *Survival*, Vol. 53, No. 1 (February–March 2011), pp. 50–51.
91. IOCTA Internet Organized Crime Threat Assessment (2019), pp. 25, 26.
92. Krekel, Adams and Bakos, *Occupying the Information High Ground*, p. 94.

93. Michael N. Schmitt (ed.), *Tallinn Manual 2.0 on the International Law Applicable to Cyber Operations* (New York: Cambridge University Press, 2017), pp. 379–396.
94. Schmitt (ed.), *Tallinn Manual 2.0*, pp. 87–92.
95. 'Double Blow to Dark Web Market Places' (May 3 2019), https://www.europol.europa.eu/newsroom/news/double-blow-to-dark-web-marketplaces, accessed 4 January 2020.
96. The 'surface web' also hosts illegal content. This includes on restricted or 'gated' sites on public forums and on the main social media sites. This also permits criminals and cybercriminals to sell goods and services.
97. Anonymous, 'Cyber-Thieves Turn to 'Invisible Net' to Set Up Attacks' (June 7 2019), https://www.bbc.co.uk/news/technology-47319971, accessed 4 January 2020.
98. Anonymous, 'Cyber-Thieves Turn to 'Invisible Net' to Set Up Attacks' (June 7 2019).
99. McAfee Labs Threats Report (December 2018), p. 2, https://www.mcafee.com/enterprise/en-us/assets/reports/rp-quarterly-threats-dec-2018.pdf, accessed 25 July 2019.
100. McAfee Labs Threats Report (December 2018), p. 2.
101. McAfee Labs Threats Report (December 2018), p. 4 and Andy Greenberg, 'Operation Bayonet: Inside the Sting That Hijacked an Entire Dark Web Drug Market Dutch Police Detail for the First Time How They Secretly Hijacked Hansa, Europe's Most Popular Dark Web Market' (August 3 2018), https://www.wired.com/story/hansa-dutch-police-sting-operation/?verso=true, accessed 25 July 2019.
102. Charlie Osborne, 'Failed Student Jailed for Silk Road, Dark Web Drug Profiteering The Operator of Silk Road 2.0 was Unemployed and Yet Lived the High Life' (April 15 2019), https://www.zdnet.com/article/failed-student-jailed-for-silk-road-dark-web-drug-profiteering/, Liv Rowley, 'Sweet Dream(s): An Examination of Instability in the Darknet Markets' (May 10 2019), https://www.blueliv.com/blog/research/sweet-dreams-instability-in-the-darknet-markets/, both accessed 25 June 2019. IOCTA Internet Organized Crime Threat Assessment (2019), p. 44.
103. McAfee Labs Threats Report (December 2018), p. 4.

104. IOCTA Internet Organized Crime Threat Assessment (2019), pp. 44–46. Norton, 'What is Bulletproof Hosting?' (undated), https://us.norton.com/internetsecurity-emerging-threats-what-is-bulletproof-hosting.html, accessed 8 January 2020.
105. IOCTA Internet Organized Crime Threat Assessment 2021, pp. 21–22, 34–38, https://www.europol.europa.eu/cms/sites/default/files/documents/internet_organised_crime_threat_assessment_iocta_2021.pdf, accessed 15 June 2022.
106. See for example Anonymous, 'Cyber-Thieves Cash in from Malware' (June 9 2015), http://www.bbc.co.uk/news/technology-33048949, and 'How Ransomware Infects Computers', https://www.mcafee.com/enterprise/en-us/security-awareness/ransomware/ransomware-infections.html, both accessed 10 November 2016.
107. IOCTA Internet Organized Crime Threat Assessment (2019), p, 4.
108. 2021 Data Breach Investigations Report, http://verizon.com/dbir/, pp. 55–57, 79–83. Verizon Data Breach Investigations Report 2022, pp. 100–103.
109. Verizon Data Breach Investigations Report 2022, (Appendix D), p. 100.
110. Bryan Lee (Unit 42), 'Ransomware: Unlocking the Lucrative Criminal Business Model', p. 2. Available from https://www.paloaltonetworks.com/apps/pan/public/downloadResource?pagePath=/content/pan/en_US/resources/research/ransomware-report, accessed 10 November 2016.
111. Anna Beahm, 'DCH Health System 'Closed to All But Most Critical' New Patients Due to Ransomware Attack' (October 2 2019), https://www.al.com/news/2019/10/dch-health-system-closed-to-all-but-most-critical-new-patients-due-to-ransomware-attack.html, accessed 2 October 2019. IOCTA Internet Organized Crime Threat Assessment (2019), pp. 15–17, Kenneth Kraszewski, 'SamSam and the Silent Battle of Atlanta', in Minárik, Alatalu, Biondi, Signoretti, Tolga and Visky (eds.), *2019 11th International Conference on Cyber Conflict: Silent Battle*, pp. 291–306.
112. IBM Security, X-Force Threat Intelligence Index 2021, pp. 8, 10, 30–33, 39–44, 47, https://www.ibm.com/downloads/cas/M1X3B7QG, accessed 8 January 2021.

113. Camille Singleton, 'The Decline of Hacktivism: Attacks Drop 95 Percent Since 2015' (May 16 2019), https://securityintelligence.com/posts/the-decline-of-hacktivism-attacks-drop-95-percent-since-2015/. See also Michael Mabee, 'Recent Critical Infrastructure Attacks Expose Our Vulnerability—And the Need For Change' (December 28 2020), https://centerforsecuritypolicy.org/recent-critical-infrastructure-attacks-expose-our-vulnerability-and-the-need-for-change/, and the 'Secure the Grid' (STG) Coalition, https://securethegrid.com/, accessed 8 January 2021.
114. McAfee Labs Threats Report (December 2018), p. 5.
115. McAfee Labs Threats Report (December 2018), pp. 5–6.
116. McAfee Labs Threats Report (December 2018), pp. 8–9.
117. D.H. Kass, Riviera Beach, Florida Ransomware Attack; City Pays $600,000' (June 20 2019), https://www.msspalert.com/cybersecurity-breaches-and-attacks/ransomware/riviera-beach-florida-malware-attack/, accessed 4 January 2020.
118. John Fokker, 'Organizations Leave Backdoors Open to Cheap Remote Desktop Protocol Attacks' (July 11 2018), https://securingtomorrow.mcafee.com/other-blogs/mcafee-labs/organizations-leave-backdoors-open-to-cheap-remote-desktop-protocol-attacks/, accessed 25 July 2019.
119. Fokker, 'Organizations Leave Backdoors Open' (July 11 2018).
120. Fokker, 'Organizations Leave Backdoors Open' (July 11 2018).
121. Fokker, 'Organizations Leave Backdoors Open' (July 11 2018).
122. Martin Caudron, 'How Europol Became a Center Point for the F.B.I.' (April 27 2018), https://medium.com/euintheus/how-europol-became-a-center-point-for-the-f-b-i-2ccc96f105bb, Dick Carozza, 'Sir Rob Wainwright Encourages 'Powering Up' Information, Intelligence Sharing' (June 18 2018), https://www.fraudconferencenews.com/home/2018/6/18/sir-wainwright-encourages-powering-up-information-intelligence-sharing, both accessed 8 January 2020.
123. IOCTA Internet Organized Crime Threat Assessment 2020, https://www.europol.europa.eu/activities-services/main-reports/internet-organised-crime-threat-assessment-iocta-2020, p, 4.
124. Abi Tyas Tunggal, 'What is the WannaCry Ransomware Attack?' (December 13 2019), https://www.upguard.com/blog/wannacry, accessed 9 January 2019. The author experienced directly the effects of WannaCry in a British NHS hospital at the time and saw the effects

firsthand on front line services. This included ambulances queuing and treatment backlogs and delays.
125. Federal Agencies' Reliance on Outdated and Unsupported Information Technology: A Ticking Time Bomb (May 25 2016), https://oversight.house.gov/hearing/federal-agencies-reliance-on-outdated-and-unsupported-information-technology-a-ticking-time-bomb/, accessed 3 September 2018.
126. Tunggal, 'What is the WannaCry Ransomware Attack?' (December 13 2019).
127. Dan Goodin, 'Stolen NSA Hacking Tools were Used in the Wild 14 months Before Shadow Brokers Leak' (May 7 2019), https://arstechnica.com/information-technology/2019/05/stolen-nsa-hacking-tools-were-used-in-the-wild-14-months-before-shadow-brokers-leak/, Shaun Nichols, 'Remember Those Stolen 'NSA Exploits' Leaked Online by the Shadow Brokers? The Chinese Had Them a Year Before' (May 7 2019), https://www.theregister.co.uk/2019/05/07/equation_group_tools/, both accessed 9 January 2019.
128. Anonymous/Gordon Corera, 'Cyber-attack: US and UK blame North Korea for WannaCry' (December 19 2017), http://www.bbc.co.uk/news/world-us-canada-42407488, accessed January 20 2018.
129. Russell Brandom, 'The Petya Ransomware is Starting to Look Like a Cyberattack in Disguise. The Ransomware that Wasn't' (June 28 2018), https://www.theverge.com/2017/6/28/15888632/petya-goldeneye-ransomware-cyberattack-ukraine-russia, accessed June 20 2018.
130. Iain Thomson, 'Everything You Need to Know About the Petya, er, NotPetya Nasty Trashing PCs Worldwide. This isn't Ransomware—It's Merry Chaos' (June 28 2017), https://www.theregister.co.uk/2017/06/28/petya_notpetya_ransomware/, accessed Jun 20 2018.
131. 'New Petya / NotPetya / ExPetr Ransomware Outbreak' (June 28 2017), https://www.kaspersky.com/blog/new-ransomware-epidemics/17314/, accessed January 20 2018.
132. Sam Jones, 'Finger Points at Russian State Over Petya Hack Attack Tactics of Those Behind Malware Match Kremlin Playbook, Say Analysts' (June 30 2017), https://www.ft.com/content/f300ad84-5d9d-11e7-b553-e2df1b0c3220, accessed June 20 2018.
133. Ellen Nakashima, 'Russian Military was Behind 'NotPetya' Cyberattack in Ukraine, CIA Concludes', *The Washington Post*, January 12 2018.
134. https://www.justice.gov/opa/press-release/file/1328521/download, pp. 16–23.

135. 'FedEx Corp. Reports First Quarter Earnings' (September 19 2017), http://about.van.fedex.com/newsroom/fedex-corp-reports-first-quarter-earnings-2/, accessed January 20 2018.
136. Jonathan Saul, 'Global Shipping Feels Fallout from Maersk Cyber Attack' (June 29 2017), https://www.reuters.com/article/us-cyber-attack-maersk/global-shipping-feels-fallout-from-maersk-cyber-attack-idUSKBN19K2LE, accessed January 20 2018.
137. Brett Molina, Jon Swartz and Rachel Sandler, 'Petya Cyberattack Spreads to 65 Countries' (28 June 2017), https://www.usatoday.com/story/tech/talkingtech/2017/06/28/petya-cyberattack-spreads-65-countries/435016001/, accessed January 20 2018.
138. Crowd Strike, 2018 Global Threat Report Blurring The Lines Between Statecraft and Tradecraft, p. 4. Available from https://www.crowdstrike.com/resources/reports/2018-crowdstrike-global-threat-report-blurring-the-lines-between-statecraft-and-tradecraft/, accessed 31 March 2020. Another exploit called Eternal Romance might also have been used in Not Petya though this might be another name for Double Pulsar. https://www.justice.gov/opa/press-release/file/1328521/download, pp. 20–23.
139. Crowd Strike, 2018 Global Threat Report Blurring The Lines Between Statecraft and Tradecraft, p. 4.
140. Crowd Strike, 2018 Global Threat Report Blurring The Lines Between Statecraft and Tradecraft, p. 4.
141. Tim Maurer, *Cyber Mercenaries: The State, Hackers, and Power* (Cambridge: Cambridge University Press, 2018), p. x.
142. Maurer, *Cyber Mercenaries*, pp. xi–xii.
143. Joint Publication 3-12 Cyberspace Operations Joint Chiefs of Staff (8 June 2018), p. I–11, https://www.jcs.mil/Portals/36/Documents/Doctrine/pubs/jp3_12.pdf?ver=2018-07-16-134954-150, accessed 6 April 2020.
144. Maurer, *Cyber Mercenaries*, p. xii.
145. Maurer, *Cyber Mercenaries*, pp. 94–120.
146. Tassilo V. P. Singer, 'Participation in Hostilities Due to Autonomous Cyber Weapons', in Henry Rõigas, Raik Jakschis, Lauri Lindström, Tomáš Minárik (eds.), *2017 9th International Conference on Cyber Conflict Defending the Core* (Tallinn CCD COE Publications, 2017), pp. 121–134.
147. Maurer, *Cyber Mercenaries*, pp. 123–157.

148. Mark Galeotti, 'Gangster's Paradise: How Organised Crime Took Over Russia' (March 23 2018), https://www.theguardian.com/news/2018/mar/23/how-organised-crime-took-over-russia-vory-super-mafia, accessed 25 February 2020.
149. See for example the cases of Gary McKinnon, Lauri Love, Roger Clark and Kim Dotcom.
150. Anonymous, 'Julian Assange: Campaigner or Attention-Seeker?' (July 30 2018), https://www.bbc.co.uk/news/world-11047811, accessed 18 October 2018.
151. E.F.G. Ajayi, 'Challenges to Enforcement of Cyber-Crimes Laws and Policy', *Journal of Internet and Information Systems*, Vol. 6, No. 1 (August 2016), pp. 1–12.
152. Tobias Feakin, 'Enter the Cyber Dragon Understanding Chinese intelligence agencies' cyber capabilities', Australian Strategic Policy Institute Special Report, Issue 50 (June 2013), p. 4, https://www.aspi.org.au/report/special-report-enter-cyber-dragon-understanding-chinese-intelligence-agencies-cyber. Bart Parys, 'The KeyBoys are Back in Town' (November 2 2017),
    https://www.pwc.co.uk/issues/cyber-security-data-privacy/research/the-keyboys-are-back-in-town.html, Wun Nan, 'From Hackers to Entrepreneurs: The Sino-US Cyberwar Veterans Going Straight' (August 21 2013), https://www.scmp.com/news/china/article/1298200/hackers-entrepreneurs-sino-us-cyberwar-veterans-going-straight, Ken Dunham and Jim Melnick, '"Wicked Rose" and the NCPH Hacking Group', https://krebsonsecurity.com/wp-content/uploads/2012/11/WickedRose_andNCPH.pdf, all accessed 9 January 2020.
153. Alex Joske, 'Picking Flowers, Making Honey The Chinese Military's Collaboration with Foreign Universities, ASPI International Cyber Policy Centre, Policy Brief Report No. 10/2018, https://s3-ap-southeast-2.amazonaws.com/ad-aspi/2018-10/Picking%20flowers%2C%20making%20honey_0.pdf?H5sGNaWXqMgTG_2F2yZTQwDw6OyNfH.u, accessed 22 April 2019.
154. 2016 Internet Security Threat Report, available from https://www.symantec.com/en/uk/security-center/threat-report, accessed 2 November 2016. Krekel, Adams and Bakos, *Occupying the Information High Ground*, p. 96.

155. Global Study Reveals Businesses and Countries Vulnerable Due to Shortage of Cybersecurity Talent, 26 July 2016, https://newsroom.intel.com/news-releases/global-study-reveals-businesses-countries-vulnerable-due-shortage-cybersecurity-talent/, accessed 2 November 2016.
156. Technical tools are being developed to combat and map activities on the 'Dark Net', Matthias Schäfer, Martin Strohmeier, Marc Liechti, Markus Fuchs, Markus Engel and Vincent Lenders, 'BlackWidow: Monitoring the Dark Web for Cyber Security Information', in Minárik, Alatalu, Biondi, Signoretti, Tolga and Visky (eds.), *2019 11th International Conference on Cyber Conflict: Silent Battle*, pp. 499–520.
157. Paul Mozur and Chris Buckley, 'Spies for Hire: China's New Breed of Hackers Blends Espionage and Entrepreneurship' (August 26 2021), https://www.nytimes.com/2021/08/26/technology/china-hackers.html. Recorded Future/Insikt Group, 'Illegal Activities Endure on China's Dark Web Despite Strict Internet Control' (October 5 2021), https://go.recordedfuture.com/hubfs/reports/cta-2021-1005.pdf, accessed 10 October 2021.
158. Kate Whitehead, 'New BBC Series Offers a Peek into the World of Cybercrime', https://www.scmp.com/lifestyle/technology/article/1629989/new-bbc-series-offers-peek-world-cybercrime, accessed 30 March 2015. In addition, there is considerable literature on cybercrime. See for example Misha Glenny, *Dark Market Cyberthieves Cybercops and You* (London: Bodley Head, 2011) and Misha Glenny, *Dark Market How Hackers Became the New Mafia* (London: Random House, 2011).
159. John McAlaney, Helen Thackray and Jacqui Taylor, 'The Social Psychology of Cybersecurity', *The Psychologist*, Vol. 29, No. 9 (September 2016), p. 686.

7

Conclusion

At a 2014 hearing of the U.S. House Intelligence Committee its chair, Michael J. Rogers, noted that discussions of cyberwar and the "hypothetical dangers of a cyber Pearl Harbor" have "become a bit of a cliché in cybersecurity circles". At the same time Rogers warned that threats to critical infrastructure, including electrical power and financial networks, "is actually becoming less hypothetical every day". Although there have been no damaging attacks on U.S. critical infrastructure yet he cautioned, "I wouldn't take much comfort in that".[1] These dangers might remain hypothetical but this book has set out to systematically demonstrate the potential for cyberwarfare should cyber be unchained and its uses unconstrained against critical infrastructure targets. This possibility exists because of our high reliance on technology. The potential uses of cyberwarfare are unlike other modes of warfare which are founded on the use of (military) force and threats of use, including in hybrid modes of conflict. John Arquilla is disquieted by this potential lamenting in 2011 that "As one of the progenitors of cyberwar, I am mortified by the eagerness of many to use this mode of conflict against other societies".[2]

Cyberattacks by nations thus far have not escalated to war but the escalatory potential is clearly evident. Notwithstanding the potential, "real-world capabilities have already been demonstrated". These include the BlackEnergy and Industroyer campaigns that Russia conducted against Ukraine in December 2015 and again in 2016 which "confirmed the potential for cyber attacks to inhibit and disrupt electric system operations by compromising a SCADA system at host and device levels" followed by Industroyer 2.0 which targeted high-voltage electrical substations in Ukraine following their invasion in February 2022.[3] This means attacks on field devices that control everything from electrical generation and distribution to water treatment and transport can be hit as well as the systems designed to control them. This, like the cyberattack on the Natanz nuclear processing plant in Iran using Stuxnet, and the earlier Aurora experiment are examples of cyber-kinetic attacks that effect machinery in the real world.[4]

However, the U.S. cybersecurity industry is also a world leader and defensive security products and services can raise the bar against even the most concerted Advanced Persistent Threats (APTs). It remains the lead innovator across multiple ICT domains and home to Silicon Valley with Microsoft, Apple, and Facebook (now rebranded as Meta) based there. Moreover, the Snowden revelations also revealed some of the extent of the capabilities of the National Security Agency (NSA) as well as Britain's Government Communications Headquarters (GCHQ) allied to their Five Eyes partners in Canada, Australia, and New Zealand. To this can be added the cybersecurity defenses across the European Union (EU) and in Western militaries. Hughes and Colarik, in their interesting study of cyberwar definitions note:

> The strategic value of cyberspace rests both in the infrastructure itself and in the information that is being globally stored, transmitted, and shared. This massive infrastructure moves across State borders—sovereign areas of controlled space. It also traverses those expanses that are open to all nations; international waters and orbital pathways. The data and information flowing through this infrastructure comprises many of the forms of communication that individuals, nation States and sub and supra State organizations use on a daily basis to conduct the transactions

underpinning twenty first century society. Any deliberate disruption of this infrastructure or the information it contains is likely to be harmful to States, citizens, and international stability. Accordingly, governments across the world are expanding their security doctrines to include the defense—and in some cases the exploitation—of cyberspace.[5]

Bearing in mind the harm that can be inflicted on civilian society through cyberattack, a growing number of militaries now view cyberspace as the fifth domain of warfare alongside land, sea, air, and space. Thus far, only a relatively small number of cyberattacks with clear state-on-state dimensions have been seen. This debate began with denial of service attacks on Estonia in 2007.

On Cyberwarfare

The much discussed cyberattack on Estonia, like many other cyber campaigns and events, fell (substantially) beneath the threshold of an armed attack. This is the first principle a state can use against another state to declare war under the United Nations (UN) Charter and international law by invoking the right to self-defense. The potential and demonstrated capabilities of cyberwarfare against critical infrastructure challenges this definition which presupposes armed attack to be a military use of force. This was the subject of Chapter 2, 'On Cyberwarfare'.

What constitutes an armed attack and the right to self-defense is central to debates in NATO's Tallinn process (which NATO sponsors and oversees) culminating in the Tallinn manuals and discussed at NATO's annual cyber conference (CyCon) in Tallinn, Estonia. This debate, and the expert legal advice it offers, are tied into Just War Theory/Tradition. This encompasses *jus ad bellum* (the justice of war), *jus in bello* (justice in war), and also *jus post bellum* (justice after war). *Jus in bello* has over time, especially after the two World Wars helped embed rules of war through international treaties/international law. This includes conceptions of proportionality and discrimination; definitions of combatants and non-combatants/civilians (a complicating factor as cyberspace contains an array of sub and non-state 'combatants'); the non-targeting of civilians; avoidance of 'collateral damage' to non-military targets; and rights of individual and collective self-defense. International law, built around the Tallinn process, has also attempted to define just cause, the pre-emptive use of force and its uses as a last resort, and helps inform military Rules of Engagement and the use of cyber as offensive 'weapons' and what are permissible defensive uses. The Tallinn Manuals provide a good indicator of how complex and subjective these issues are as they relate to cyberspace.[6]

Complicating this is that state-on-state cyberattacks are also closely aligned to, and run by, state intelligence agencies. Espionage, including cyberespionage, is legal under international law. These issues were discussed at length in Chapter 2, which used a Clausewitzian framework to question the nature and character of cyberwar. It also looked at the present and the future potential for cyberwar to better understand it—if war and conflict are the continuation of politics by other means.[7] It did so with a view to NATO, and other states allied or aligned to the West, and a standpoint that sees the United States as the lead nation under challenge from China, and to a differing and lesser extent Russia. This 'security architecture' encompasses both Western and non-Western nations such as Australia, New Zealand, South Korea, and Japan which ring Asia–Pacific enabled through America's network of formal alliances and defensive and extended deterrence commitments.

There are underlying motives that can be used to explain the conduct and character of cyberwar and cyberwarfare. Clarke and Knake in their 2010 book *Cyber War* opined that "The perception that cyberspace is a "domain" where fighting takes place, a domain that the U.S. must "dominate,", pervades America military thinking on the subject of cyber war".[8] They add that the Pentagon sees not only that "cyber war is real" but "the almost reverential way in which it is discussed as the keystone holding up the edifice of modern war-fighting capability".[9]

This places the emphasis on the use of offensive cyber to obtain superiority and deny adversaries seeking to gain the initiative. For Russia this includes civilian targets which most other nations consider to be 'off-limits' (thus far) in a 'peacetime' scenario. Tim Mauer in his 2018 book, *Cyber Mercenaries*, questioned what if "the long discussion about cyber war obscured other potential ways in which international stability can be and is being undermined through offensive cyber operations?...[can] cyber capabilities be used to help empower politicians that would make...escalation less likely?".[10] Another way is to demonstrate the potential of cyber as a state disabler and potential war winner on its own, especially allied to military force.

Attacking Critical Infrastructure

Arquilla wrote in 2011 that "cyberwar as a form of strategic attack on national infrastructures quickly overshadowed our original vision".[11] More than a decade later, the vision of what cyberwar is and might be remain blurry. The nature and extent of a major and concerted attack on critical infrastructure would currently have to cross the threshold of 'armed attack' to be defined as an act of war. Chapter 3, 'Cyberwar: Attacking Critical Infrastructure', set out to demonstrate how this is feasible. It did so by setting out how critical infrastructure systems are automated and how and why they are computer controlled. It began by detailing two real-world proof-of-concept cases, the 2007 Aurora test conducted by Idaho National Laboratory (INL) and the 2010 use of the Stuxnet worm against Iran's Natanz nuclear processing plant. It then when onto show the implications of both.

It was followed by a sector-by-sector breakdown of threats covering electrical generation and distribution; water treatment and sanitation; dams; the oil and gas industry; chemical plants; ports and logistics; merchant shipping; and the transport sectors of road and rail and civil aviation. The most important of these—because it powers all sectors—is the electrical sector, but each showed a demonstrable threat to life through cyberattack alongside the potential for local and widespread disruption and/or destruction. This chapter showed that used to its full potential cyberwar has the capacity to be highly destructive and qualify as the use of armed force.

Chapter 3 also looked at the underpinnings of these sectors—Supervisory Control and Data Acquisition (SCADA) systems and their control of field devices that form Industrial Control Systems (ICS). It concluded with some good news. Cybersecurity awareness is increasing, and cyber threat intelligence sharing is improving between government and private industry. However, this is accompanied by the bad news that there are increasing threats and threat actors; including evolving state-based APTs. Terrorists are not, as yet (and hopefully never will), using cyber as a method of attacking critical infrastructure but that risk exists.

Whether this is directed upon a nation like the Russian attacks upon Estonia in 2007, Georgia in 2008, and Ukraine in 2015/6 and 2022 or part of a broader campaign would shape the parameters of cyberwar(fare). This is part of the 'fog of cyberwar'. The capabilities, resource base, and will of the actor(s) involved would determine responses, the options open, and dictate the speed of any recovery. So far, what we have seen qualifies as cyberespionage and cybercrime (mainly through ransomware and wiperware) against the sectors that make up critical infrastructure alongside attacks on the democratic process (most notably election interference) and the financial services sector. Election interference and economic espionage are covered in two follow-on books, with the former a method of Social Psychological Warfare. Both are a large-scale and persistent phenomena.

Cyberwar(fare) in contrast, through its modes and possible uses, has the capacity to cross the threshold into 'armed attack' and qualify as the 'use of force'. Whilst these are the conventional definitions and legal benchmarks to invoke the right of self-defense and a declaration of war, they do not fit well with all the other effects cyber, as the fifth domain of warfare, can accomplish. This may be one of the outcomes of cyber as a Revolution in Military Affairs (RMA). As Galeotti describes in relation to Russian activities, "the whole debate about hybrid war is really two debates intertwined: about the strategic challenge from an embittered and embattled Russia, and the changing nature of war in the modern age".[12] Theoretically informed discussions of these topics and the wider discourse on cybersecurity writ large remain somewhat embryonic and polarized. This book has attempted to move these debates forward and increase understanding.

Pinprick Attacks and First Strike

Much of this conceptual confusion is a result of information warfare, influence operations, and cyberespionage by states becoming subsumed into the discourse of cyberwar(fare) without being deconstructed. Conceptualizing this as 'hybrid warfare' may be a better fit unless 'hybrid warfare' is being used as part of long-term tactics or stratagems to engage in full-scale cyberwarfare alone to bring about disruption, destruction, or collapse or as tools of coercion. This embraces the utility of cyber as a precursor to politico-military use of force, or as a prelude to armed invasion. As Czosseck, Ottis and Talihärm argue, "Politically motivated actors can cover the entire spectrum of cyber attack, from high-profile strikes against critical infrastructure, to millions of pinprick attacks that can weaken the state over a long period of time".[13] A stark government commissioned study by INL in 2016 pointed out:

> It is likely that many more cyber incidents occur than are reported. Known attacks against the energy sector often follow a phased pattern that focuses on discovery, capture, and exfiltration of data, which generally does not produce tangible or immediately detectable consequences. However, if an attacker's goal is to "degrade, disrupt, deny, [or] destroy" utility operations, prior reconnaissance and established access provide launch points for destructive payloads (malware). No lasting damage–physical, cyber-physical, or otherwise–to U.S. utilities as a result of a cyber attack has yet been reported publicly, but known cyber attacks and campaigns targeting U.S. electric utilities have been highly publicized.[14]

Depending on the actors involved, almost invariably state versus state but also plausible in regional conflicts or disputes, and the scale of cyber capability and cyber dependency of each actor, cyberattacks might be a war winning factor alone in a situation of a 'quick grab' as Russia undertook when it annexed the Crimea from Ukraine in 2014. In the longer term, hostilities a which appear to be a 'million pinprick attacks' can grow in strength or be used to degrade or weaken a state over time. There are few test cases to go on where cyber has been used for kinetic or military effects but the capacity for disruption, destruction, or collapse exists. However, this is context dependent and also related to the capacity and will of the actors involved—with Ukraine's response to Russia's invasion a case in point.

Any initiative to strike first is bound by the speed of attacks and because inaction could lead to vulnerable targets being hardened against attack or made inaccessible. This logic of a pre-emptive first strike used to cripple a nation's war-fighting ability reflects the decades old debate over nuclear strategy which has been ported over to emerging cyber strategy. This is particularly in the U.S. where deterrence was built on the potential use of demonstrated capabilities, survivable second-strike forces, and declaratory policies. At the beginning of the 2010s, within the Department of Defense's strategy lay the recognition that America was vulnerable to the kind of cyber first-strike it was advocating "because to do otherwise would put other kinds of American dominance at risk" allied to the reality that the U.S. is more cyber dependent than their adversaries.[15] At that time, Stuxnet had only just been discovered and Edward Snowden's leaks of the NSA's monitoring and offensive cyber capabilities was yet to happen.

As a defender, America is deploying a continental rather than 'whole-of-nation' response and, as David E. Sanger articulates: "The challenge is to think about how to defend a civilian infrastructure that the United States government does not control, and private networks where companies and American citizens often don't want their government lurking—even for the purpose of defending them".[16] Sanger bemoaned that in contrast to the first nuclear age, there is an absence of grand strategic debates around cyberwarfare. He attributes this to more players on the field, political division in the U.S., as well as over classification of cyber weapons by the national intelligence apparatus. Sanger also makes a case for declaring capabilities and living "within some limits" as a means of deterrence and stabilization.[17] This has merit.

Cybersecurity Defenses: Risk Management and Legacy Systems

Across the three interrelated layers of cyberspace: the logical layer comprising data, the physical layer of equipment, and the social layer (people), vulnerabilities and risks are evident. Systems are primarily designed to ensure Confidentiality, Integrity, and Availability of systems and data (the CIA triad). For critical infrastructure, the CIA triad is reversed. Always-on availability is prioritized because the systems are running essential processes which keep sites operating. This includes power and water on demand. ICS and SCADA systems depend on the data they depend on to be uncorrupted and trustworthy—the Integrity leg of the triad. This data is essential for the safe functioning of electrical generation and distribution and temperature and pressure among many other processes. For other industries like financial services, confidentiality and integrity are paramount.

Chapter 4 began by breaking down the terminology used in cybersecurity and by Information Security (Infosec) professionals. It then details common technical attack methods known as tactics, techniques, and procedures (TTPs) which aid forensic analysis of the actor or actors involved. It went onto detail defensive systems and technologies. Chapter 4 also details how legacy systems can be used to gain access. Across critical infrastructure this includes hardware and software which date from the 1970s and 80s. The U.S. Department of Defense even operates a critical nuclear command and control system from the 1960s. They were designed to ensure performance and reliability. At this time ARPANET, the forerunner of today's Internet, was only in its infancy and cybersecurity vulnerabilities were only exploitable through physical access. Now the modern world depends on the Internet and connectivity, including wireless signals, and the development of 4G and 5G communications—developments which will be ongoing. Alongside routes in through the Internet, this now permits remote access to critical infrastructure sites. Sites can also be accessed 'over the fence' or on-site using equipment which can be bought commercially off-the-shelf.

The history of modern computing tells us that despite cybersecurity barriers rising, driven by the growing awareness of threats and threat actors, there will always be bugs, glitches, mis-designs, and backdoors. Security-by-design will raise the barriers against malicious actors but there will always be ghosts in the machine. As hardware and software is gradually replaced, legacy systems will remain. New builds will become old builds and could do so more rapidly than in the past given Moore's

Law (the maxim that computer capacity doubles every two years) and the onwards and accelerating drive of ICT innovation and maturation.

This means that residual risk will remain a factor. In addition, Ted G. Lewis offers a different approach to the problems of ICT dependency and residual risk. Lewis, who spent his career in industry and academia, suggests that "deeper analysis suggests that collapse is built into infrastructure, itself, because of its structural complexity. Risk is in the way systems are "wired together" and operated. Fragility-the opposite of complexity-is a symptom of complexity…modern society is responsible for the fragility of the very infrastructure it depends on for survival".[18]

Lewis adds, "most infrastructure has emerged as a complex system subject to unpredictable behavior when stressed…And construction of most sectors is largely accidental or emergent" with structural dependency on networked connectivity.[19] Lewis explained that small incidents can expand and enlarge and produce unseen catastrophe because, much like the economic crashes of 1929 and 2008, the faults are in the system itself. They are the kindling that can spread the fire. For the electric sector this came in the report by INL in 2016.

They pointed out the importance of maintaining the delivery of electricity "above security" as the electric grid modernizes and adopts new technology. Increasing automation is needed to "keep up with exponential increases in electricity demand and consumption". Automation has also allowed the industry to reduce staff levels whilst improving reliability and the expectancy of power on demand. This means increasing requirements for real-time data and "even broader automation and control capabilities".[20] These issues are not confined to America and are global in scope. Communications tools and hyper-connectivity enable 'globalization' and development and the 'Internet of things' (IoT) continues to expand with the advent of a wide variety of SMART devices and increasing remote tools and automation. This is also providing a wide attack surface for an ever-increasing number of threat actors who also use human nature to exploit these underlying insecurities.

Hacking the Human

This is explored in Chapter 5, 'Hacking the Human'. This analyzes social engineering (SE) methods and practices which are used to exploit the link thought weakest in the cybersecurity fence—people. The various methods of SE are examined as are the cognitive and behavioral triggers they are designed to exploit. These are most widely, and most successfully, used for spear phishing. This uses Open Source Intelligence (OSINT), including corporate websites and social media, to design e-mails (the main gateway) targeting individuals and companies. Chapter 5 showed how SE can be mitigated but also how spear phishing tools and websites make it a very widespread and successful attack vector.

Technical solutions to cyber security procedures, such as firewalls, secure hardware products, Intrusion Detection Systems and the like are always going to be only steps towards security and not an end point so long as human–machine interaction takes place. Solutions including access privileges, organizational security, and segmentation can all be bypassed either by design and ingenuity, resources, and/or chance. Physical security is also not a solution either in itself or in isolation. It is also the case that "Many users [might] have little to no understanding of their responsibility to protect information assets. It is critical that businesses understand the value of a security awareness program and make a commitment to closing the awareness gap".[21]

Security awareness that promotes behavioral change and aids healthy cyber hygiene practices by reducing innate human behaviors which are exploitable to 'social engineers' and protects against threats and risks related to information and physical assets is to be welcomed. This is largely a social psychological effort involving dynamic skepticism regarding the cyber physical environment. The field of social engineering contains some highly sophisticated actors who combine technological knowledge and a sharp understanding of psychological biases. This enables them to seriously compete against cybersecurity providers, and the law enforcement and state intelligence organizations who track them. In this environment, companies and suppliers are forced to accept various degrees of risk to their businesses. A successful social engineering attack can bypass millions of dollars of investment in technical security to expose an organization's critical information.

Chapter 5 also showed how and why, the electric sector is probably the most tempting target for cyberattack against critical infrastructure. INL pointed out in their 2016 study that as demand increases or decreases, mostly in known patterns related to seasonal weather and times of day, production is increased or decreased to meet electrical generation needs. This requires "system load stability" which is supplied and balanced by a wide number and type of power generation including nuclear, coal fired, and renewable energy sources. At the local level "control loops do not depend on large scale geographically dispersed control systems and therefore the cyber attack surface is considered to be more focused on gaining access to the local control system".[22] This is where RedTeam Security, described in Chapter 5, successfully targeted as a means to improve security.

This is also possible by remote access and the introduction of malware through a wide variety of injection types, or by "pivoting through a trusted communication path in the organization's corporate networks".[23] E-mail injection is the major threat vector but "Control system cyber security is further impaired by poor institutional cyber hygiene such as weak or no password usage, outdated or unpatched software, and even poor physical security". This means that "Attack vectors have evolved to exploit poor personnel awareness of threats to cyber security more often than attacking hardware directly".[24]

Detection is not a solution by itself, and prevention of external threats is limited by human fallibility and technology (both of which are guards). There are many other ways to get into organizations including through the children of employees and through them to gain home network access, passwords, and for coercion.[25] This is a particularly insidious example of the exploitation of SE but it demonstrates the palpable need for the integration of different, human-factor thinking patterns and methodologies into the cyber security skill set. Concerningly, PricewaterhouseCoopers' 2018 Global State of Information Security Survey found that less than half of their respondents had implemented recommended good or best practice. This included vulnerability and threat assessments, penetration testing or active monitoring and analysis such as Cyber Threat Intelligence (CTI). They also found most of the corporate world pushes cybersecurity beneath board level rather than directing it through a board level Chief Information Security Officer (CISO). Limited awareness at board level increases risk and reduces resilience.[26]

Reducing Risk

In many cases of SE there are measures that can help provide safeguards but against sophisticated and targeted techniques, especially those that prey upon individuals and human fallibility, this can be difficult to stop entirely. Mitigation practices include not providing personal or account information unless you initiated the exchange. If an unsolicited exchange has taken place and raised suspicions, it is a good practice to report it up the chain of command or to the IT department within an organization. If deemed serious, this can also be flagged to police and law enforcement bodies, or by directly flagging soliciting attempts or exchanges to appropriate state intelligence agencies. These are practices that should be encouraged but should be appropriate to the risk otherwise already overworked professionals will be overwhelmed, including in cybersecurity where there is a shortfall of qualified staff globally.

Guarding against SE and spear phishing is a first line of defense and one where good computer hygiene, education, and best practices can help. These can be encouraged and nurtured as a measure of 'common sense'. The greatest threat to a social engineering attack can be naivety or lack of awareness or deficient evaluation. Although best practice is a process that needs to be evolved over time, encryption and strong user authentication with robust regulation remain important considerations for critical infrastructure owner-operators with "relationships to governments...a key factor in how they handle security".[27] A joint McAfee/CSIS study from 2011 recorded that China was then further ahead in locating and adopting 'best practices' "well ahead of the United States, the UK and Australia, the next highest rated countries". Furthermore. "The sectors with the highest security adoption rates were banking and energy. Water/sewage had the lowest rate of any sector".[28] All critical infrastructure is regulated (even if lightly) and governments can intervene for reasons of national security "regardless of the efficiency provided by a market. Indeed, even free market economists accept that governments might need to protect industries for reasons of national security".[29]

However, even with help and support from governments it is difficult to guard against 'spear phishing' which involves targeted attacks against individuals and businesses. Effective social engineering, as opposed to' cold' attacks or mass e-mails (spam) designed to trawl for victims, is enabled by the kind of information-intelligence gathering empowered by the Internet and the use of botnets. Targeted attacks can be achieved through simple searches, searching for social media profiles online, or in more advanced cases by purchasing individual or company data and studying their targets. This can include bogus friend requests which look plausible or enticing for business opportunities or for personal advancement, or out of curiosity or innocence. Some SE attacks can also emanate from trusted sources or appear to come from trusted sources (a man-in-the-middle attack). Websites, both open source and 'Dark Net' sites, offer advice on SE attacks and software such as Metasploit and Kali Linux is available for anyone to download.

Risk Management and Resilience

Cyberattacks are frequently designed to lay hidden and the harm they can do can remain latent until activated. Discovery has been problematic but as intrusions have increased this has got better. Still, threat assessments and risk management appear to remain problematic areas which only receive priority post-intrusion or breach and not as a proactive business strategy. This applies to real-time alerts, threat analysis, and threat prioritization. These can give attackers advantages against new, existing, or unmitigated vulnerabilities in both hardware and software. Within critical infrastructure this "lack of preparation" is built on the "unpredictability of progressively complicated, layered control systems. The increasing complexity of devices used by utilities and other critical infrastructure elements in everyday operations…paralleled by an increase in attack surfaces and vulnerabilities".[30]

Therefore, the resilience, the capacity to resist attacks, recover from or overcome their effects and prevent being breached or compromised become central to reducing risk. This is particularly important for systems that large sections of the economy rely on including banking and finance.[31] This requires raising awareness and capacity building, proactive and tailored risk management and establishing or elevating meaningful security standards. Baseline security such as anti-virus and firewalls should be regularly updated and properly configured, and good and skeptical cyber hygiene practices should be to treat approaches in the same way you would a knock on the door.

Beyond this, especially for critical infrastructure operators, good and best practices include the National Institute of Standards and Technology (NIST) Cybersecurity Framework. This has five pillars: protect, detect, identify, respond, and recover.[32] Detection includes monitoring and (desirably) comprehensive risk management including "crisis and response planning, including understanding the need to notify and cooperate with authorities" and information sharing.[33] Baseline standards and sensible security practices are inexpensive. Corporate investment into cyber/IT security and risk management needs to be seen by boardrooms (and increasing small to medium sized businesses and enterprises, especially those in the supply chain to larger firms), as a proactive insurance and not as costs to be avoided (especially staff costs). Technical countermeasures in computer security can help mitigate against outsider threats but might not always be successful against compromised insiders.

States as Advanced Persistent Threats

Despite these measures, protecting against highly organized and well-resourced actors such as Russia and China and their highly orchestrated intelligence apparatuses remains problematic. States represent Advanced Persistent Threats (APTs) in cyberspace and whilst they too use social engineering and similar TTPs to cybercriminals they also use other more sophisticated means and can buy-in cybercriminal expertise, zero-days, or tools.

For state intelligence operatives, part of the process of intelligence gathering and espionage practices include honey pots/traps,[34] bribery, and blackmail—including through *kompromat* (compromising information), threats and intimidation, or the use of force.[35] Corporate intrusions and breaches are useful for long-term economic, political, and military gain and private e-mail hacks (which are a feature of cybercrime and hacktivism) can, especially for states, also yield information for *kompromat* on individuals or organizations which can be exploited or blackmailed now or in the future. They can be useful for providing access and exploitation or even background intelligence to form 'sleeper cells' or implant moles. These practices run a gamut of techniques ranging from the use of OSINT to the dark arts of espionage. False or misleading information, including deep fakes, can also be used to target and exploit individuals and organizations.

State-sponsored attacks combine off-the-shelf tools and custom tools, and their cyberespionage campaigns are known to have been patiently carried out for years. No criminal organization will do this, only states. This forms part of a much broader series of activities against companies, organizations, and individuals and can include direct state-on-state cyberespionage. This occurred in the Office of Personnel Management (OPM) breach by China between 2013–15.[36] For states like China and Russia, this opens the possibility of 'kill switches' created through back door access years in the making. Cyberattacks/weapons can be installed years before any hostilities ready to be activated, used as a contingency, or for coercion. This includes through targeting or acquiring companies in supply chains. The global nature of business practices and the complexity of supply chains makes this very difficult to stop. This is also apparent through the manufacture of chips and firmware (software hardwired into chips at the manufacturing stage).[37]

These vendors are vulnerable to compromise despite stringent security procedures, which include staff vetting. There are many possible ways to exploit this pathway including corporate mergers and acquisitions of civilian companies with military contracts, by purchasing human or technological potential or intellectual property (IP), or by blackmailing or bribing individuals. The tradecraft of espionage is steeped in examples of each of these in cases of corporate and industrial espionage or to seek competitive advantage, and in state-run or state-sponsored activities (which also attempt to penetrate governments).

Chapter 5 also looked at physical security and the potential to compromise trusted insiders. In addition, it questions the benefits and hazards of 'full disclosure' of security vulnerabilities and the activities of 'white hat' penetration testing. It also called for increased investment in Science, Technology, Engineering and Mathematics (STEM) subjects to deal with the shortage of people in cybersecurity and underinvestment in cybersecurity across industry and by Western governments. This includes bolstering Computer Emergency Response Teams (CERTs) and government agencies tasked with cybersecurity. These rely on OSINT and forensic analysis by private sector providers and the CTI they provide to their clients and wider public, allied to industry and private self-help organizations.

Admiral Michael S. Rogers, when commander of U.S. Cyber Command and Director of the NSA, testified to the Senate Armed Services Committee in May 2017 that for America "cyber is one area we have to acknowledge that we have peer competitors who have every bit as much capacity and capability as we do".[38] Rogers added, "Most of our critical infrastructure providers are doing their best to better secure their networks. But if they get attacked by an adversary with the resources and capabilities of a nation state like China or Russia or Iran, it certainly isn't a fair fight".[39]

Russia has demonstrated its capabilities against critical infrastructure, China has not but has at least as much capability as Russia and much greater capacity. Iran and North Korea have much lower levels of capability and capacity, and in respect to North Korea, a focus on ransomware. However, whilst the capacity might exist, be demonstrated and demonstrable, and intentions appear clear and imminent, this is also an area where intentions and political will can change over time or be pressured into change through diplomacy, negotiation, and (at worst) through crisis management. Capabilities and intentions can also be discerned through espionage/cyberespionage but attended by problems and risks.[40] Under the presidency of Vladimir Putin, Russia has not responded to calls from the West to change direction. In respect to cyberespionage, neither has Xi Jinping the President of the People's Republic of China.

The U.S. Intelligence Community and a 'Whole of Nation' Effort

Joined up thinking on combatting threats in cyberspace, whether 'known knowns' or 'unknown unknowns' is generally framed as a whole-of-government approach with cooperation from private industry.[41] In America this is directed from the president down under the U.S. Code of Laws whilst policy and guidance is usually provided to the president by a team of National Security Advisors, but the president can choose his (or her) own course.

Presidential decision-making is also aided by the National Security Council (NSC), made up of senior advisers from the military and intelligence branches of government. Among others, this includes the Under Secretary of Defense for Policy (USD(P)), who is the Secretary of Defense's Principal Cyber Advisor, and the Deputy Assistant Secretary of Defense for Cyber Policy. The Director of the National Security Agency supports decision-making with cybersecurity guidance part of the remit of the wider DoD.[42] For government, NSA support includes: "target development assistance; situational awareness and attack; sensing and warning; threat analysis; internal threat hunting; red-teaming and security assist visits; communications monitoring; forensics; linguist support; and other specialized support, as authorized".[43] The NSA's analysts alongside partner agencies, of which there are many with a variety of roles and expertise, dissect and interpret the data as part of all-source intelligence collection and analysis. Briefings are reports that are then produced which are discussed within the NSA and shared with other agencies. These are then channeled and discussed through the chain of command before they reach the very highest levels of government.

The Director of the Defense Intelligence Agency (DIA), a three-star military appointment which rotates between Army, Navy, Air Force, and Marines approximately every three years, is responsible for the timely delivery of military intelligence from across the U.S. military and into the policymaking machinery. The DIAs Director is "a principal adviser to the Secretary of Defense and to the Chairman of the Joint Chiefs of Staff on matters of military intelligence".[44] Like the NSA, they are tasked with providing all-source intelligence for Cyberspace Operations (CO). The DIA is the prime contractor of counterintelligence for the DoD whose aim is "cyberspace superiority" by providing "worldwide cyberspace...situational awareness and coordination".[45] The DIA also helps protect the Department of Defense Information Network (DODIN) and has overall responsibility for the Joint Worldwide Intelligence Communications System which is "a strategic, secure, high-capacity telecommunications network serving the IC with voice, data, and video services".[46]

The DIA also has responsibility for setting intelligence priorities and targets across the military, including for cyber operations, and enabling "interoperability between tactical, theater, and national intelligence-related systems and between intelligence-related systems and the tactical, theater, and national elements of the DODIN".[47] DC3, a department which operates out of the U.S. Air Force (USAF), supports the DoD as a center of excellence in the areas of "forensics; cyberspace investigative training; research, development, test and evaluation; and cyberspace vulnerability analysis for DODIN protection". It acts for law enforcement, the Intelligence Community (IC), and for specialized counterintelligence and counterterrorism organizations.[48] The Department of Homeland Security (DHS) and the Department of Justice (DOJ) also have domestic responsibilities for national protection and investigation.

This 'whole-of-government' approach has been hampered by political division in the U.S. exacerbated by the covert action of Russian interference in U.S. politics, especially in the 2016 presidential election.[49] Currently, this form of covert action has not been reciprocated by the United States in Russia. The Joint Chiefs of Staff (JCS) look instead to 'cyberspace exploitation' which "includes activities to gain intelligence and support operational preparation of the environment for current and future operations through actions such as gaining and maintaining access to networks, systems, and nodes of military value; maneuvering to positions of advantage; and positioning cyberspace capabilities to facilitate follow-on actions". These activities aid situational awareness, helps discover vulnerabilities, target development, and "the planning, execution, and assessment of military operations". For the JCS offensive cyber operations should possess the abilities to deny, degrade, disrupt, destroy, and manipulate.[50]

The JCS and the wider DoD are the responsibility of the Secretary of Defense, currently General Lloyd J. Austin III. Austin will be aware, like his predecessor under the Trump administration, Mark T. Esper, of the need to find, recruit, develop, and retain cyber talent in the U.S. military and IC because there is "a select pool of people that can do that and they have a lot of other incentives to go work outside the military".[51] These incentives have seen the DoD facing "tremendous challenges in recruiting and retaining trained and experienced cybersecurity professionals" due to more lucrative employment in Silicon Valley and the wider commercial sector.[52]

What does not help are cross-cutting responsibilities for cybersecurity between the State Department, Defense Department, and DHS added to which is that critical infrastructure is largely owned and operated by the private sector outside of direct government control. Multi-agency responsibilities across cyber undercut national strategy and this includes the role of the Office of the Director of National Intelligence (DNI) whose current incumbent is Avril D. Haines. They too have a significant remit in U.S. cyber policy.[53] Ongoing partisan political division is a further serious impediment to a whole-of-nation response. To understand the scale of their task and the extent the West is on the defensive, the U.S. Department of Defense two-year long study of cyber deterrence concluded:

> Russia and China are increasing their already substantial capabilities to hold U.S. critical infrastructure at risk by cyber targeting of inherently vulnerable ICT and industrial control system (ICS) architectures. In the face of these ongoing efforts, the U.S. Government and the private sector should continue to intensify their efforts to defend and boost the cyber resilience of U.S. critical civilian infrastructure. However, even with sustained improvements, such progress will not be adequate to deny Russia and China the ability to unleash catastrophic cyber attacks on the United States, given their massive resources, and capabilities-at-scale (e.g., intelligence apparatus, ability to influence supply chains, and ability

to introduce and sustain vulnerabilities) to dedicate to their objectives. Barring major unforeseen breakthroughs in the cyber defense of U.S. civilian critical infrastructure, the United States will not be able to prevent large-scale and potentially catastrophic cyber attacks by Russia or China; for the foreseeable future, we will have to rely heavily on deterrence by cost imposition. In bolstering our cyber deterrence posture relative to major powers, the United States must account for another reality: over the coming years, Russia and China will also be working to increase their ability through cyberattack (and other means) to delay, disorganize, disrupt, and where possible negate U.S. military capabilities. Such cyberattacks may target military systems specifically, or the civilian critical infrastructure on which civil and military activities depend. An attack on military systems might result in U.S. guns, missiles, and bombs failing to fire or detonate or being directed against our own troops; or food, water, ammo, and fuel not arriving when or where needed; or the loss of position/navigation ability or other critical warfighter enablers. Moreover, the successful combination of these attacks could severely undermine the credibility of the U.S. military's ability to both protect the homeland and fulfill our extended deterrence commitments.[54]

Zugzwang

Invoking Clausewitz's ideas of war as an outgrowth of politics, Kalevi J. Holsti argues that "Policymakers, whether princes, kings, Politbureaus, cabinets, or presidents, generally seek to defend, extend, or achieve certain known objectives. When these purposes and the means to implement them are incompatible with the values and interests of other actors, the probability of the use of force increases".[55] Related to this reasoning, structural issues within the international system involving power, perception, and prevention leads to "numerous examples of the "Peloponnesian syndrome". This is "where states go to war preventively because they face an impending hegemony or preponderance of power by their main rival". This notwithstanding, "there are many more instances when no such stakes are involved... since clearly not all power contests end in war".[56] These remain politico-military choices which can be acted on, ameliorated diplomatically, or defended through joining alliances like NATO.

Through to the late twentieth-century wars have been solely fought on land, at sea, and in the air. Now space provides commanding heights and cyber has become the fifth domain of warfare. It has become a component of the other four domains for militaries but where cyberespionage campaigns are run largely by intelligence agencies. These campaigns are at their most potent if used against critical infrastructure.

Concerted movement against critical infrastructure by major powers against one another would be akin to the chess term *zugzwang*. This is where a player has to make a move but this results in a serious, often decisive, disadvantage. It is a move compelled by the rules of the game but one which, knowing the result, a player does not want to make. This gives cyberwar entropic characteristics. This has been briefly demonstrated by Russia against Ukraine in the years preceding their 2022 invasion. This included temporarily taking down part of their electrical grid (which they again attempted in 2022). Even before the invasion, this demonstrated Russia's leadership do not consider civilian targets 'off-limits'. Russia's APT 28 and 29 ('Fancy Bear' and Cozy Bear') are two of the most virulent threat actors in cyberspace.

Far from having its claws clipped by diplomatic expulsions and economic sanctions, the Russian bear under Putin remains emboldened and, potentially along with China, a revisionist power. If this is, as Paul Kennedy suggests, part of another cycle of *The Rise and Fall of the Great Powers*, we would be wise to pay attention, especially if China's rise leads to Western, especially American, overstretch; whether alone or in coordination with Russia.[57] For China and America the danger of Thucydides Trap, where a rising power challenges a hegemon, remains.[58]

According to Chinese General Dai Qingmin: "Whoever controls information and controls networks will have the whole world at his feet…".[59] This is no longer just a theory. Chinese theorizing of what we conceive of in the West as cyberwarfare has greatly advanced from the time of the first Gulf War in 1991 and allied to their capacity and capabilities for cyberwarfare against critical infrastructure is a major concern.[60] With the development of the Internet and growing global connectivity, cybersecurity concerns have become an accepted social and international norm. Chris Demchak testified in 2017 that cyberspace has "created and spread a new form of system-vs-system 'cybered conflict'".

She added, "China's scale and its ever growing skill in this form of conflict challenge democratic societies' influence over the interstate system...However, the democratic civil societies have themselves equally to blame for much of the cyber threats emergent today due to their own blinders and failure to act early and collectively.[61]

This is part of the reason we are in the middle of a cyber arms race (an arms race that includes Machine Learning, Artificial Intelligence, and quantum computing) and if improved organizing principles cannot be established, then a "Wild Wild West" style anarchy within the World Wide Web might prevail for the foreseeable future. Although we hope we never see a 'Cybergeddon' scenario emerge, there might instead be lower-level attacks to disrupt elements of critical infrastructure such as oil refineries, chemical plants, or transport. A lower order surprise attack is also possible whilst James Clapper noted "we must be prepared for a large-scale strike—a "Cyber Pearl Harbor" scenario or "Cyber 9/11".[62] However, given the globalized and interconnected nature of national economies, an attack of this nature might not only affect only the target state, sector, or company. A spillover or cascade effect cannot be ruled out. This is particularly true in terms of the financial and energy sectors. Russia and China proposed a peacetime non-aggression pact in the UN as a basis for discussion in 2015, supported by the UN's Group of Governmental Experts. This might be worth revisiting because without attempts at consensus and norms building, we will see more of the same or worse even if this would likely dissolve if war or a warlike scenario developed.[63]

Cyberwarfare raises complex and interrelated issues tied into societal needs, social order, state and societal resiliency, international law, international order, the security dilemma, development, and a wide range of other factors.[64] Understanding cross-sectorial and cross-border dependencies work alongside mitigation and recovery plans, especially in terms of critical infrastructure. At the far end of the scale, it is worth pondering how resilient and prepared we are in the West should some of the scenarios in this book emerge and what would lead to social breakdown?

Our responses and ability to respond to a cyberwar targeting critical infrastructure between peer adversaries like Russia and China would dictate the ability and capacity to recover. Whilst these scenarios were modeled in nuclear war planning during the Cold War, this has not been put to the test since World War II.[65] Two centuries ago the great moral philosopher Immanuel Kant wrote, "there mere cessation of hostilities is no guarantee of continued peaceful relations, and unless this guarantee is given by every individual to his neighbor—which can only be done in a state of society regulated by law—one man is at liberty to challenge another and treat him as an enemy".[66]

Notes

1. Hearing of the House (Select) Intelligence Committee Subject: "Cybersecurity Threats: The Way Forward", November 20 2014, https://www.nsa.gov/news-features/speeches-testimonies/Article/1620360/hearing-of-the-house-select-intelligence-committee-subject-cybersecurity-threat/, accessed 16 January 2019.
2. John Arquilla, 'The Computer Mouse that Roared: Cyberwar in the Twenty-First Century', *Brown Journal of World Affairs*, Vol. xviii, Issue 1 (Fall/Winter 2011), p. 32.
3. Cyber Threat and Vulnerability Analysis of the U.S. Electric Sector Prepared by: Mission Support Center. Idaho National Laboratory August 2016, p. 6, https://www.energy.gov/sites/prod/files/2017/01/f34/Cyber%20Threat%20and%20Vulnerability%20Analysis%20of%20the%20U.S.%20Electric%20Sector.pdf, accessed 12 August 2019. 'Industroyer2: Industroyer reloaded' (April 12 2022), https://www.welivesecurity.com/2022/04/12/industroyer2-industroyer-reloaded/, accessed 20 May 2022.
4. Even now the United States has neither confirmed nor denied it was responsible. This is a well-rehearsed defense for intelligence-led covert action.
5. Daniel Hughes and Andrew Colarik, 'The Hierarchy of Cyber War Definitions', in G.A. Wang et al. (eds.), Pacific Asia Workshop on Intelligence and Security (2017), p. 15. Available from https://pdfs.semanticscholar.org/34c5/8f3a28f836bd78352381e9f6054dd78f374d.pdf, accessed 13 September 2018.
6. Michael N. Schmitt (ed.), *Tallinn Manual 2.0* on the International Law Applicable to Cyber Operations (New York: Cambridge University Press, 2017).

7. Carl von Clausewitz, *On War*, edited and translated by Michael Howard and Peter Paret, (Oxford: Oxford University Press, 1997), pp. 20–21, 27–29.
8. Richard A. Clarke and Robert K. Knake, *Cyber War The Next Threat to National Security and What To Do About It* (New York: HarperCollins, 2010), p. 44.
9. Clarke and Knake, *Cyber War*, p. 44.
10. Tim Maurer, *Cyber Mercenaries: The State, Hackers, and Power* (Cambridge: Cambridge University Press, 2018), p. xii.
11. Arquilla, 'The Computer Mouse that Roared', p. 32.
12. Mark Galeotti, '(Mis)Understanding Russia's two 'hybrid wars'', *Critique and Humanism*, Vol. 59, Issue 1 (2018), https://www.eurozine.com/misunderstanding-russias-two-hybrid-wars/, accessed 23 December 2018.
13. Christian Czosseck, Rain Ottis and Anna-Maria Talihärm, 'Estonia After the 2007 Cyber Attacks: Legal, Strategic and Organisational Changes in Cyber Security', *International Journal of Cyber Warfare and Terrorism*, Vol. 1, Issue 1 (January-March 2011), pp. 61–62.
14. Cyber Threat and Vulnerability Analysis of the U.S. Electric Sector, p. 4.
15. Clarke and Knake, *Cyber War*, p. 46.
16. David E. Sanger, 'The age of cyberwar is here. We can't keep citizens out of the debate', *The Guardian*, July 28 2018.
17. Sanger, 'The age of cyberwar is here'.
18. Ted G. Lewis, *Critical Infrastructure Protection in Homeland Security: Defending a Nation* Second Edition (Hoboken, NJ: Wiley, 2015), p. xv.
19. Lewis, *Critical Infrastructure Protection in Homeland Security*, p. xv.
20. Cyber Threat and Vulnerability Analysis of the U.S. Electric Sector, p. 1.
21. Peter O. Okenyi and Thomas J. Owens, 'On The Anatomy of Human Hacking', *Information Systems Security*, Vol. 16, Issue 6 (December 2007), p. 314.
22. Cyber Threat and Vulnerability Analysis of the U.S. Electric Sector, p. 9.
23. Cyber Threat and Vulnerability Analysis of the U.S. Electric Sector, p. 9.
24. Cyber Threat and Vulnerability Analysis of the U.S. Electric Sector, p. 16.
25. Remarks made under Chatham House Rules at the 'Insider Threat Summit', Monterey, California, 29 March-1 April 2016 attended by author.

26. 'Strengthening digital society against cyber shocks', https://www.pwc.com/us/en/services/consulting/cybersecurity/library/information-security-survey/strengthening-digital-society-against-cyber-shocks.html, accessed 13 August 2019.
27. McAfee/CSIS, 'In the Crossfire Critical Infrastructure in the Age of Cyber War', p. 39, https://www.govexec.com/pdfs/012810j1.pdf, accessed 26 July 2019.
28. McAfee/CSIS, 'In the Crossfire Critical Infrastructure in the Age of Cyber War', p. 19.
29. Vinod K. Aggarwal and Andrew W. Reddie, 'Comparative industrial policy and cybersecurity: A framework for analysis', *Journal of Cyber Policy*, Vol. 3, Issue 3, (December 2018), p. 296.
30. Cyber Threat and Vulnerability Analysis of the U.S. Electric Sector, pp. 4, 15. See also Yulia Cherdantseva, Pete Burnap, Simin Nadjm-Tehrani and Kevin Jones, 'A configurable dependency model of a SCADA system for goal-oriented risk assessment'. *Applied Sciences*, Vol. 12 Issue 10 (May 2022), 4880, pp. 1–29.
31. 'Cyber Resilience', https://www.iaac.org.uk/cyber-resilience/, accessed 27 September 2019.
32. 'NIST Releases Version 1.1 of its Popular Cybersecurity Framework' (April 16 2018), https://www.nist.gov/news-events/news/2018/04/nist-releases-version-11-its-popular-cybersecurity-framework, accessed 17 August 2019.
33. McAfee/CSIS, 'In the Crossfire Critical Infrastructure in the Age of Cyber War', pp. 37–39.
34. In a cyber security context this might be considered to be a form of disinformation—a practice widely used in espionage activities.
35. Samuel D. Porteous, 'Economic espionage: Issues arising from increased government involvement with the private sector', *Intelligence and National Security*, Vol. 9, Issue 4 (October 1994), pp. 735–752.
36. Committee on Oversight and Government Reform U.S. House of Representatives 114th Congress The OPM Data Breach: How the Government Jeopardized Our National Security for More than a Generation September 7 2016, pp. 42–50, https://www.cylance.com/content/dam/cylance/pdfs/reports/The-OPM-Data-Breach-How-the-Government-Jeopardized-Our-National-Security-for-More-than-a-Generation.pdf, accessed 27 April 2019.

37. See for example Wyatt Andrews, 'Some U.S. military parts imported from China' (June 9 2017), https://www.cbsnews.com/news/some-us-military-parts-imported-from-china/, accessed 5 September 2017, Dave Lee, "Foreshadow' attack affects Intel chips' (August 15 2018), https://www.bbc.co.uk/news/technology-45191697, accessed 1 October 2019.
38. 'Cybersecurity Threats and Defense Strategy' (May 9 2017), https://www.c-span.org/video/?428023-1/nsa-director-rogers-russia-poses-thr eat-congressional-elections&start=1166, accessed 27 September 2018.
39. Hearing of the House (Select) Intelligence Committee Subject: "Cybersecurity Threats: The Way Forward".
40. Martin Libicki, 'Drawing Inferences from Cyber Espionage', in Tomáš Minárik, Raik Jakschis and Lauri Lindström (eds.), *2018 10th International Conference on Cyber Conflict CyCon X: Maximising Effects* (Tallinn: CCD COE Publications, 2018), pp. 109–122.
41. National Cyber Strategy of the United States of America September 2018, https://www.whitehouse.gov/wp-content/uploads/2018/09/National-Cyber-Strategy.pdf, accessed 22 November 2018. John P. Carlin, 'Detect, Disrupt, Deter: A Whole-of-Government Approach to National Security Cyber Threats', *Harvard National Security Journal*, Vol. 7, Issue 2 (June 2016), pp. 391–436.
42. Also serving as chief of the Central Security Service and, through to 2018, director of USCYBERCOM.
43. Joint Publication 3–12 Cyberspace Operations Joint Chiefs of Staff (8 June 2018), p. III-9.
44. 'Leadership', http://www.dia.mil/About/Leadership/, accessed 27 November 2018.
45. Joint Publication 3–12 Cyberspace Operations Joint Chiefs of Staff (8 June 2018), p. III-9.
46. Joint Publication 3–12 Cyberspace Operations Joint Chiefs of Staff (8 June 2018), p. III-9.
47. Joint Publication 3–12 Cyberspace Operations Joint Chiefs of Staff (8 June 2018), p. III-9.
48. Joint Publication 3–12 Cyberspace Operations Joint Chiefs of Staff (8 June 2018), p. III-10.
49. A subject covered in-depth in my book *Russia's Cyber Offensive Against the West* and in Special Counsel Robert S. Mueller, III, Report On The Investigation Into Russian Interference In The 2016 Presidential Election Volume I of II (March 2019), https://www.justice.gov/storage/report.pdf, accessed 2 September 2019.

50. Joint Publication 3–12 Cyberspace Operations Joint Chiefs of Staff (8 June 2018), p. II-7.
51. Remarks by Secretary Esper at the Air Force Association's 2019 Air, Space & Cyber Conference, National Harbor, Maryland (September 18 2019), https://www.defense.gov/Newsroom/Transcripts/Transcript/Article/1964448/remarks-by-secretary-esper-at-the-air-force-associations-2019-air-space-cyber-c/, accessed 27 January 2020.
52. Arthur MacDougall and Michael Myers, 'Pentagon faces array of challenges in retaining cybersecurity personnel' (June 9 2018), https://thehill.com/opinion/cybersecurity/391426-pentagon-faces-array-of-challenges-in-retaining-cybersecurity-personnel, accessed 23 November 2018.
53. https://www.dni.gov/, accessed 22 November 2018. On the roles and responsibilities of the DNI and the wider U.S. intelligence community see also R. Gerald Hughes and Kristan Stoddart, 'Hope and Fear: Intelligence and the Future of Global Security a Decade after 9/11', *Intelligence and National Security*, Vol. 27, Issue 5 (October 2012), pp. 625–652 and Loch K. Johnson, 'A Conversation with James R. Clapper, Jr., The Director Of National Intelligence in the United States, Intelligence and National Security', *Intelligence and National Security*, Vol. 30, Issue 1 (January 2015), pp. 1–25.
54. Department of Defense Defense Science Board Task Force on Cyber Deterrence February 2017, p. 17.
55. Kalevi J. Holsti, *Major Texts on War, the State, Peace, and International Order* (London: Springer Nature, 2016), p. 14.
56. Holsti, *Major Texts on War, the State, Peace, and International Order*, pp. 15–16.
57. Paul Kennedy, *The Rise and Fall of the Great Powers: Economic Change and Military Conflict from 1500–2000* (London: Fontana, 1989).
58. Graham Allison, *Destined for War: Can America and China Escape Thucydides' Trap?* (New York: Houghton Mifflin Harcourt, 2017).
59. Timothy L. Thomas, *Decoding the Virtual Dragon Critical Evolutions in the Science and Philosophy of China's Information Operations and Military Strategy* (Fort Leavenworth, KS: Foreign Military Studies Office, 2007), p. 106.
60. Detailed in my book *China and its Embrace of Offensive Cyberespionage*.
61. Key Trends across a Maturing Cyberspace affecting U.S. and China Future Influences in a Rising deeply Cybered, Conflictual, and Post-Western World Dr. Chris C. Demchak Testimony before

Hearing on China's Information Controls, Global Media Influence, and Cyber Warfare Strategy Panel 3: Beijing's Views on Norms in Cyberspace and China's Cyber Warfare Strategy U.S.-China Economic and Security Review Commission Washington, DC 4 May 2017, p. 5, https://www.uscc.gov/sites/default/files/Chris%20Demchak%20May%204th%202017%20USCC%20testimony.pdf, accessed 3 April 2020.

62. Remarks as delivered by The Honorable James R. Clapper Director of National Intelligence "National Intelligence, North Korea, and the National Cyber Discussion" International Conference on Cyber Security Fordham University, January 7 2015 https://www.dni.gov/index.php/newsroom/speeches-interviews/speeches-interviews-2015/item/1156-remarks-as-delivered-by-dni-james-r-clapper-on-national-intelligence-north-korea-and-the-national-cyber-discussion-at-the-international-conference-on-cyber-security, accessed 28 September 2019.

63. Martin Libicki, 'The Coming of Cyber Espionage Norms', in Henry Rõigas, Raik Jakschis, Lauri Lindström, Tomáš Minárik (eds.), *2017 9th International Conference on Cyber Conflict Defending the Core* (Tallinn CCD COE Publications, 2017), pp. 14–17.

64. Jacqueline Eggenschwiler, 'Big tech's push for norms to tackle uncertainty in cyberspace', in Miriam Dunn Cavelty and Andreas Wenger (eds.), *Cyber Security Politics Socio-Technical Transformations and Political Fragmentation* (New York: Routledge, 2022), pp. 186–204.

65. Desmond Ball and Jeffrey Richelson (eds.), *Strategic Nuclear Targeting* (Ithaca: Cornell University Press, 1986).

66. Immanuel Kant, *Perpetual Pace: A Philosophical* Essay (originally published, 1795), (New York: Cosimo, 2010), p. 9.

Bibliography

Books

Adamsky, D., *The Culture of Military Innovation: The Impact of Cultural Factors on the Revolution in Military Affairs in Russia, the US, and Israel* (Stanford: Stanford University Press, 2010).
Aggarwal, C.C., *Social Network Data Analytics* (New York: Springer, 2011).
Allison, G., *Destined for War: Can America and China Escape Thucydides' Trap?* (New York: Houghton Mifflin Harcourt, 2017).
Ariely, D., *Predictably Irrational: The Hidden Forces That Shape Our Decisions* (London: HarperCollins, 2008).
Arquilla, J., and Ronfeldt, D., *In Athena's Camp: Preparing for Conflict in the Information Age* (Santa Monica: RAND Corporation, 1997).
Bacon, P., and Hobson, C. (eds.), *Responding to the 2011 Earthquake, Tsunami and Fukushima Nuclear Crisis* (Abingdon: Routledge, 2014).
Ball, D., and Richelson, J. (eds.), *Strategic Nuclear Targeting* (Ithaca: Cornell University Press, 1986).
Bandler, R., and Grinder, J., *Frogs into Princes: Neuro Linguistic Programming* (Utah: Real People Press, 1979).

Bazzell, M., *Open Source Intelligence Techniques: Resources for Searching and Analyzing Online Information* Fifth Edition (Charleston: CreateSpace Independent Publishing Platform, 2016).
Betz, D.J., and Stevens, T., *Cyberspace and the State Towards a Strategy for Cyber-Power* (New York: Routledge, 2011).
Borgatti S.P., Everett, M.G., and Johnson, J.C., *Analyzing Social Networks* (London: Sage, 2013).
Brangetto, P., Maybaum, M., and Stinissen, J. (eds.), *2014 6th International Conference on Cyber Conflict* (Tallinn: CCD COE Publications, 2014).
Brenner, J., *America the Vulnerable: Inside the New Threat Matrix of Digital Espionage, Crime, and Warfare* (New York: Penguin Press, 2011).
Brenner, J., *Glass Houses: Privacy, Secrecy and Cyber Insecurity in a Transparent World* (New York: Penguin, 2013).
Brown, D., *Absolute Magic: A Model for Powerful Close-Up Performance* (London: H & R Magic Books, 2001).
Brown, D., *Direct Mindreading and Magical Artistry* (London: H & R Magic Books, 2002).
Brown, D., *Tricks of the Mind* (London: Channel 4 Books, 2006).
Brown, D., *Confessions of a Conjuror* (London: Channel 4 Books, 2010).
Brown, R., *Social Psychology: Second Edition* (New York: The Free Press, 1986).
Browne, A. (ed.), *Neural Network Perspectives on Cognition and Adaptive Robotics* (Bristol: IOC Publishing, 1997).
Brunnée, J., and Toope, S.J., *Legitimacy and Legality in International Law* (Cambridge: Cambridge University Press, 2010).
Buzan, B., Wæver, O., and de Wilde, J., *Security: A New Framework for Analysis* (London: Lynne Rienner, 1997).
Carr, I. (ed.), *Computer Crime* (London: Routledge, 2007).
Carr, J., *Inside Cyber Warfare: Mapping the Cyber Underworld* (Sebastopol, CA: O'Reilly Media, 2010).
Cavelty, M.D., *Cyber-Security and Threat Politics: US Efforts to Secure the Information Age* (London: Routledge, 2008).
Cavelty, M.D., and Wenger A. (eds.), *Cyber Security Politics Socio-Technical Transformations and Political Fragmentation* (New York: Routledge, 2022).
Chen, T.M., Jarvis, L., and Macdonald, S. (eds.), *Cyberterrorism: Understanding, Assessment, and Response* (New York: Springer, 2014).
Cialdini, R.B., *Influence: The Psychology of Persuasion* Revised Edition (New York: HarperBusiness, 2008).
Clarke, R.A., and Knake, R.K., *Cyber War: The Next Threat to National Security and What to Do About It* (New York: HarperCollins, 2010).

Comfort, L., *Designing Resilience: Preparing for Extreme Events* (Pittsburgh: University of Pittsburgh Press, 2010).
Czosseck, C., and Geers, K. (eds.), *The Virtual Battlefield: Perspectives on Cyber Warfare* (Amsterdam: IOS Press, 2009).
Dai Qingmin, *Wangdian Yiti Zhan Yinlun* (*Introduction to Integrated Network and Electronic Warfare*) (Beijing: PLA Press, 2002).
d'Amato, C. et al. (eds.), *The Semantic Web—ISWC 2017 16th International Semantic Web Conference, Vienna, Austria, October 21–25, 2017, Proceedings, Part II* (Cham, Switzerland: Springer, 2017).
Demchak, C.C., *Wars of Disruption and Resilience: Cybered Conflict, Power and National Security* (Athens: University of Georgia Press, 2011).
Easley, D., and Kleinberg, J., *Networks, Crowds, and Markets: Reasoning About a Highly Connected World* (Cambridge: Cambridge University Press, 2010).
Fidler, D.P., *The Snowden Reader* (Bloomington: Indiana University Press, 2015).
Fridman, O., *Russian 'Hybrid Warfare': Resurgence and Politicisation* (London: Hurst, 2018).
Futter, A., *Hacking the Bomb: Cyber Threats and Nuclear Weapons* (Washington, DC: Georgetown University Press, 2018).
Geers, K., *Strategic Cyber Security* (Tallinn: CCD COE Publications, 2011).
Ghattas, K., *Black Wave: Saudi Arabia, Iran, and the Forty-Year Rivalry That Unraveled Culture, Religion, and Collective Memory in the Middle East* (New York: Henry Holt, 2020).
Glenny, M., *Dark Market Cyberthieves Cybercops and You* (London: Bodley Head, 2011)
Glenny, M., *Dark Market How Hackers Became the New Mafia* (London: Random House, 2011).
Goldstein, N.K., Martin, S.J., and Cialdini, R.B., *Yes! 50 Secrets from the Science of Persuasion* (New York: Simon & Shuster, 2008).
Gray, C.S., *Strategy for Chaos: Revolutions in Military Affairs and the Evidence of History* (London: Frank Cass, 2002).
Gray, C.S., *Irregular Enemies and the Essence of Strategy: Can the American Way of War Adapt?* (Carlisle Barracks, PA: Strategic Studies Institute, U.S. Army War College, 2006).
Greenwald, G., *No Place to Hide: Edward Snowden, the NSA and the Surveillance State* (New York: Hamish Hamilton, 2014).
Haddow, G., Bullock, J., and Coppola, D.P., *Introduction to Emergency Management* Fifth Edition (Waltham, MA: Butterworth-Heinemann, 2013).

Hadnagy, C., *Social Engineering: The Art of Human Hacking* (London: Wiley, 2010).
Hadnagy, C., *Unmasking the Social Engineer: The Human Element of Security* (London: Wiley, 2014).
Hadnagy, C., and Fincher, M., *Phishing Dark Waters: The Offensive and Defensive Sides of Malicious Emails* (London: Wiley, 2015).
Hagestad II, W.T., *Chinese Cyber Crime 2016: Hacking Underground in the People's Republic of China* Second Edition (Scotts Valley, CA: CreateSpace Independent Publishing, 2016).
Haskew, M.E., *Aircraft Carriers: The Illustrated History of the World's Most Important Warships* (Minneapolis, MA: Zenith Press, 2016).
Healey, J. (ed.), *A Fierce Domain: Conflict in Cyberspace 1986–2002* (Vienna, VGN: CSSA/Atlantic Council, 2013).
Henderson, J., and Ferguson, A., *International Partnership in Russia Conclusions from the Oil and Gas Industry* (London: Palgrave Macmillan, 2014).
Hoffman, F.G., *Conflict in the 21st Century: The Rise of Hybrid Wars* (Arlington, VA: Potomac Institute for Policy Studies, 2007).
Holsti, K.J., *Major Texts on War, the State, Peace, and International Order* (London: Springer Nature, 2016).
Howard, P.N., *Pax Technica: How the Internet of Things May Set Us Free or Lock Us Up* (New Haven, CT: Yale University Press, 2015).
Hurd, I., *After Anarchy: Legitimacy and Power in the United Nations Security Council* (Princeton, NJ: Priceton University Press, 2007).
Iancu, N., Fortuna, A., Barna, C., and Teodor, M. (eds.), *Countering Hybrid Threats: Lessons Learned from Ukraine* (Amsterdam: IOS Press, 2016).
Inkster, N., *China's Cyber Power* (Abingdon: Routledge, 2016).
Jasper, S., *Strategic Cyber Deterrence* (Lanham, MD: Rowman & Littlefield, 2017).
Johnson, L.K., and Wirtz, J.J. (eds.), *Intelligence: The Secret World of Spies an Anthology* Third Edition (Oxford: Oxford University Press, 2011).
Kadushin, C., *Understanding Social Networks: Theories, Concepts, and Findings* (Oxford: Oxford University Press, 2011).
Kant, I., *Perpetual Pace: A Philosophical* Essay (originally published, 1795) (New York: Cosimo, 2010).
Kello, L., *The Virtual Weapon and International Order* (New Haven, CT: Yale University Press, 2017).
Kennedy, P., *The Rise and Fall of the Great Powers: Economic Change and Military Conflict from 1500–2000* (London: Fontana, 1989).

Keohane, R.O., and Nye, J.S., *Power and Interdependence: World Politics in Transition* (Boston: Little, Brown & Co., 1977).

Klimburg, A., *The Darkening Web: The War for Cyberspace* (New York: Penguin, 2017).

Knapp, E.D., and Langill, J.T., *Industrial Network Security: Securing Critical Infrastructure Networks for Smart Grid, SCADA, and Other Industrial Control Systems* Second Edition (New York: Elsevier, 2015).

Knox, M., and Murray, W., *The Dynamics of Military Revolution 1300–2050* (Cambridge: Cambridge University Press, 2001).

Kolenda, N., *Methods of Persuasion: How to Use Psychology to Influence Human Behavior* (No Location: Kolena Entertainment, 2013).

Kont, M., Pihelgas, M., Wojtkowiak, J., Trinberg, L., and Osula, A.M., *Insider Threat Detection Study* (Tallinn: CCD COE Publications, 2015).

Krutz, R.L., *Securing SCADA Systems* (Indianapolis, IN: Wiley, 2006).

Kurzweil, R., *The Singularity Is Near: When Humans Transcend Biology* (New York: Penguin, 2005).

Lewis, T.G., *Critical Infrastructure Protection in Homeland Security: Defending a Nation* Second Edition (Hoboken, NJ: Wiley, 2015).

Libicki, M.C., *Conquest in Cyberspace: National Security and Information Warfare* (New York: Cambridge University Press, 2007).

Lin, H., *Cyber Threats and Nuclear Weapons* (Stanford: Stanford University Press, 2021).

Mann, I., *Hacking the Human: Social Engineering Techniques and Security Countermeasures* (Aldershot: Gower, 2008).

Mann, I., *Hacking the Human 2* (Whitley Bay: Consilience Media, 2013).

Maurer, T., *Cyber Mercenaries: The State, Hackers, and Power* (Cambridge: Cambridge University Press, 2018).

Maybaum, M., Osula, A.-M., and Lindström, L. (eds.), *2015 7th International Conference on Cyber Conflict: Architectures in Cyberspace* (Tallinn: NATO Cooperative Cyber Defence Centre of Excellence, 2015).

Milgram, S., *Obedience to Authority: An Experimental View* (New York: Harper and Row, 1974).

Minárik, T., Alatalu, S., Biondi, S., Signoretti, M., Tolga, I., and Visky, G. (eds.), *2019 11th International Conference on Cyber Conflict: Silent Battle* (Tallinn: CCD COE Publications, 2019).

Minárik, T., Jakschis, R., and Lindström, L. (eds.), *2018 10th International Conference on Cyber Conflict CyCon X: Maximising Effects* (Tallinn: CCD COE Publications, 2018).

Mitnick, K.D., and Simon, W.L., *The Art of Deception: Controlling the Human Element of Security* (London: Wiley, 2003).

Mitnick, K.D., and Simon, W.L., *The Art of Intrusion: The Real Stories Behind the Exploits of Hackers, Intruders and Deceivers* (Indianapolis, IN: Wiley, 2005).

Murray, W., and Mansoor, P.R. (eds.), *Hybrid Warfare: Fighting Complex Opponents from the Ancient World to the Present* (New York: Cambridge University Press, 2012).

Nascimento, A.C.A., and Barreto, P. (eds.), *9th International Conference, ICITS 2016 Tacoma, WA, USA, August 9–12, 2016 Revised Selected Papers* (Cham: Springer, 2016).

Omand, D., *Securing the State* (London: Hurst, 2010).

Osula, A.M., and Rõigas H. (eds.), *International Cyber Norms Legal, Policy & Industry Perspectives* (Tallinn: CCD COE Publications, 2016).

Pacepa, I.M., and Rychlak, R.J., *Disinformation: Former Spy Chief Reveals Secret Strategies for Undermining Freedom, Attacking Religion, and Promoting Terrorism* (Washington, DC: WND Books, 2013).

Phillips, B.D., *Disaster Recovery* Second Edition (Boca Raton, FL: Taylor and Francis, 2016).

Phythian, M., *Understanding the Intelligence Cycle* (Abingdon: Routledge, 2013).

Pissanidis, N., Rõigas, H., and Veenendaal, M. (eds.), *2016 8th International Conference on Cyber Conflict Cyber Power* (Tallinn: CCD COE Publications, 2016).

Podins, K., Stinissen, J., and Maybaum, M. (eds.), *5th International Conference on Cyber Conflict Proceedings* (Tallinn: CCD COE Publications, 2013).

Prell, C., *Social Network Analysis: History, Theory and Methodology* (London: Sage, 2011).

Reason, J., *Human Error* (Cambridge: Cambridge University Press, 1990).

Richards, J., *Cyber-War: The Anatomy of the Global Security Threat* (London: Palgrave Macmillan, 2014).

Richards, J., *The Art and Science of Intelligence Analysis* (Oxford: Oxford University Press, 2011).

Rid, T., *Cyber War Will Not Take Place* (London: Hurst, 2013).

Rodrigue, J.-P., *The Geography of Transport Systems* (New York: Routledge, 2013).

Rõigas, H., Jakschis, R., Lindström, L., and Minárik, T. (eds.), *2017 9th International Conference on Cyber Conflict Defending the Core* (Tallinn: CCD COE Publications, 2017).

Ross, A., *The Industries of the Future* (New York: Simon & Shuster, 2016).
Russell, M.A., *Mining the Social Web: Data Mining Facebook, Twitter, LinkedIn, Google+, GitHub, and More* Second Edition (Cambridge: O'Reilly, 2013).
Sanger, D.E., *The Perfect Weapon War, Sabotage, and Fear in the Cyber Age* (London: Scribe, 2018).
Scanlon, M., and Le-Khac, N.H. (eds.), *Proceedings of 16th European Conference on Cyber Warfare and Security* (Reading: Academic Conferences Publishing International, 2017).
Schmitt, M.N. (ed.), *Tallinn Manual on the International Law Applicable to Cyber Warfare* (Cambridge: Cambridge University Press, 2013).
Schmitt, M.N. (ed.), *Tallinn Manual 2.0 on the International Law Applicable to Cyber Operations* Second Edition (Cambridge: Cambridge University Press, 2017).
Sherman, V., 'How Accurate Are Bourne and Bond? As an Ex-CIA Officer' (December 6 2017), https://m.huffpost.com/us/entry/2451700/amp.
Sims, J.E., and Gerber, B. (eds.), *Vaults, Mirrors, & Masks: Rediscovering U.S. Counterintelligence* (Washington, DC: Georgetown University Press, 2008).
Singer, P.W., and Friedman, A., *Cybersecurity and Cyberwar: What Everyone Needs to Know* (Oxford: Oxford University Press, 2014).
Slim, H., *Killing Civilians: Method, Madness, and Morality in War* (Oxford: Oxford University Press, 2010).
Stoddart, K., *Losing an Empire and Finding a Role: Britain, the USA, NATO and Nuclear Weapons 1964–1970* (Palgrave Macmillan, 2012).
Stoddart, K., *The Sword and the Shield: Britain, America, NATO and Nuclear Weapons 1970–1976* (Palgrave Macmillan, 2014).
Stoddart, K., *Facing Down the Soviet Union: Britain, the USA, NATO and Nuclear Weapons 1976–1983* (Palgrave Macmillan, 2014).
Thomas, T.L., *Decoding the Virtual Dragon Critical Evolutions in the Science and Philosophy of China's Information Operations and Military Strategy* (Fort Leavenworth, KS: Foreign Military Studies Office, 2007).
Thompson, S.G., and Ghanea-Hercock, R. (eds.), *Defence Applications of Multi-Agent Systems* (DAMAS 2005) (Berlin: Springer, 2006).
Tosey, P., and Mathison, J., *Neuro-Linguistic Programming: A Critical Appreciation for Managers and Developers* (Basingstoke: Palgrave Macmillan, 2008).
Tzu, S., *The Art of War*, translated and annotated by Samuel B. Griffith (Oxford: Oxford University Press, 1963).
van Creveld, M., *The Transformation of War* (New York: Free Press, 1991).

Van Puyvelde, D., and Brantly, A.F., *Cybersecurity Politics, Governance and Conflict in Cyberspace* (Cambridge: Polity Press, 2019).
Van Wie Davis, E., *Shadow Warfare: Cyberwar Policy in the United States, Russia and China* (Lanham, MD: Rowman & Littlefield, 2021).
von Clausewitz, C., *On War*, edited and translated by Michael Howard and Peter Paret (Ware: Wordsworth Editions Limited, 1997).
von Clausewitz, C., *Principles of War*, translated by Hans Wilhelm Gatkze (Harrisburg, PA: Military Service Publishing, 1942).
Walzer, M., *Just and Unjust Wars: A Moral Argument with Historical Illustrations* Fifth Edition (New York: Basic Books, 2015).
Wasserman, S., and Faust, K., *Social Network Analysis: Methods and Applications* (Cambridge: Cambridge University Press, 1994).
Watts, B.D., *Clausewitzian Friction and Future War* Revised Edition, McNair Paper 68 (Washington, DC: National Defense University, 2004).
Weiss, J. *Protecting Industrial Control Systems from Electronic Threats* (New York: Momentum Press, 2010).
Wendt, A., *Social Theory of World Politics* (Cambridge: Cambridge University Press, 1999).
Wortzel, L.M., *The Chinese People's Liberation Army and Information Warfare* (Carlisle, PA: United States Army War College Press, 2014).
Wright, Q., *A Study of War* (Chicago: Chicago University Press, 1942).
Wyciszkiewicz, E. (ed.), *Geopolitics of Pipelines Energy Interdependence and Inter-state Relations in the Post-Soviet Area* (Warsaw: Polski Instytut Spraw Miedzynarodowych, 2009).
Young, S., and Aitel, D., *The Hacker's Handbook: The Strategy Behind Breaking into and Defending Networks* (Boca Raton: Auerbach, 2003).
Ziolkowski, K. (ed.), *Peacetime Regime for State Activities in Cyberspace: International Law, International Relations and Diplomacy* (Tallinn: NATO CCD COE Publications, 2013).

Book Chapters

Abbasi, A. et al., 'Descriptive Analytics: Examining Expert Hackers in Web Forums', in *2014 IEEE Joint Intelligence and Security Informatics Conference*, https://ieeexplore.ieee.org/stamp/stamp.jsp?arnumber=6975554.
Alagic, G., Broadbent, A., Fefferman, B., Gagliardoni, T., Schaffner, C., and St. Jules, M., 'Computational Security of Quantum Encryption', in Nascimento and Barreto (eds.), *9th International Conference, ICITS 2016*.

Anuta, C., 'Old and New in Hybrid Warfare', in Iancu et al. (eds.), *Countering Hybrid Threats*.

Applegate, S.D., and Stavrou, A., 'Towards a Cyber Conflict Taxonomy', in Podins et al. (eds.), *5th International Conference on Cyber Conflict Proceedings*.

Arquilla, J., and Ronfeldt, D., 'Cyberwar Is Coming!', in Arquilla and Ronfeldt, *In Athena's Camp*.

Bargar, A., Butkevics, J., Pitts, S., and McCulloh, I., 'Challenges and Opportunities to Counter Information Operations Through Social Network Analysis and Theory', in Minárik et al. (eds.), *Silent Battle*.

Bonfanti, M.E., Artificial Intelligence and the Offense-Defense Balance in Cyber Security', in Cavelty and Wenger (eds.), *Cyber Security Politics*.

Boothby, B., 'Law, Ethics and Cyber Warfare', in Glorioso, L., and Osula, A.-M. (eds.), *1st Workshop on Ethics of Cyber Conflict Proceedings*, Tallinn, CCDCOE (2014), https://ccdcoe.org/uploads/2018/10/2013ethics-workshop-proceedings.pdf.

Brantly, A.F., 'Battling the Bear: Ukraine's Approach to National Cyber and Information Security', in Cavelty and Wenger (eds.), *Cyber Security Politics*.

Brantly, A.F., 'The Cyber Deterrence Problem', in Minárik et al. (eds.), *CyCon X: Maximising Effects*.

Burton, J., and Soare, S.R., 'Understanding the Strategic Implications of the Weaponization of Artificial Intelligence', in Minárik et al. (eds.), *Silent Battle*.

Conway, M., 'Reality Check: Assessing the (Un)likelihood of Cyberterrorism', in Chen et al. (eds.), *Cyberterrorism: Understanding, Assessment, and Response*.

Creery, A.A., and Byres, E.J., 'Industrial Cybersecurity for Power System and SCADA Networks', in *Proceedings of the IEEE Petroleum and Chemical Industries Conference, Institute of Electrical and Electronics Engineers, Denver* (September 2007), https://www.tofinosecurity.com/professional/industrial-cybersecurity-power-system-and-scada-networks.

Dhamija, R., Tygar, J.D., and Hearst, M., 'Why Phishing Works', in *Proceedings of the SIGCHI Conference on Human Factors in Computing Systems* (April 2006).

Diamond, J., 'Early Patriotic Hacking', in Healey (ed.), *A Fierce Domain: Conflict in Cyberspace 1986–2002*.

Ducheine, P., and van Haaster, J., 'Fighting Power, Targeting and Cyber Operations' in Brangetto et al., *2014 6th International Conference on Cyber Conflict*.

Durante, M., 'Violence, Just Cyber War and Information', in Glorioso and Osula (eds.), *1st Workshop on Ethics of Cyber Conflict Proceedings, Tallinn*, CCDCOE (2014).

D'Urso, M., 'The Cyber-Combatant: A New Status for a New Warrior', in Glorioso and Osula (eds.), *1st Workshop on Ethics of Cyber Conflict Proceedings, Tallinn*, CCDCOE (2014).

Eggenschwiler, J., 'Big Tech's Push for Norms to Tackle Uncertainty in Cyberspace', in Cavelty and Wenger (eds.), *Cyber Security Politics*.

Giles, K., and Hagestad II, W., 'Divided by a Common Language: Cyber Definitions in Chinese, Russian and English' in Podins et al. (eds.), *5th International Conference on Cyber Conflict Proceedings*.

Hallaq, B., Somer, T., Osula, A.-M., Ngo, K., and Mitchener-Nissen, T., 'Artificial Intelligence Within the Military Domain and Cyber Warfare', in Scanlon and Le-Khac (eds.), *Proceedings of 16th European Conference on Cyber Warfare and Security*.

Healey, J., and Jenkins, N., 'Rough-and-Ready: A Policy Framework to Determine if Cyber Deterrence Is Working or Failing', in Minárik et al. (eds.), *Silent Battle*.

Hodgson, Q.E., 'Understanding and Countering Cyber Coercion', in Minárik et al. (eds), *CyCon X: Maximising Effects*.

Hughes, R., 'Towards a Global Regime for Cyber Warfare', in Czosseck and Geers (eds.), *The Virtual Battlefield*.

Hughes, D., and Colarik, A., 'The Hierarchy of Cyber War Definitions', in Wang, Chau, and Chen (eds.), *Pacific Asia Workshop on Intelligence and Security* (2017). Available from https://pdfs.semanticscholar.org/34c5/8f3a28f83 6bd78352381e9f6054dd78f374d.pdf.

Janicke, H., Siewe, F., Jones, K., Cau, A., and Zedan, H., 'Analysis and Run-Time Verification of Dynamic Security Policies', in Thompson and Ghanea-Hercock (eds.), *Defence Applications of Multi-Agent Systems*.

Janicke, H. et al., 'A Compositional Event & Time-Based Policy Model', in *Seventh IEEE International Workshop on Policies for Distributed Systems and Networks (POLICY'06)*, 2006.

Kapellmann, D., and Washburn, R., 'Call to Action: Mobilizing Community Discussion to Improve Information-Sharing About Vulnerabilities in Industrial Control Systems and Critical Infrastructure', in Minárik et al. (eds.), *Silent Battle*.

Kingston, J., 'Mismanaging Risk and the Fukushima Nuclear Crisis', in Bacon and Hobson (eds.), *Responding to the 2011 Earthquake*.

Koch, R., 'Hidden in the Shadow: The Dark Web—A Growing Risk for Military Operations?', in Minárik et al. (eds.), *Silent Battle*.
Kong, J.Y., Kim, K.G., and Lim, J.I., 'The All-Purpose Sword: North Korea's Cyber Operations and Strategies', in Minárik et al. (eds.), *Silent Battle*.
Kosseff, J., 'The Contours of 'Defend Forward' Under International Law', in Minárik et al. (eds.), *Silent Battle*.
Kraszewski, K., 'SamSam and the Silent Battle of Atlanta', in Minárik, et al. (eds.), *Silent Battle*.
Kuerbis, B., Badiei, F., Grindal, K., and Mueller, M., 'Understanding Transnational Cyber Attribution Moving from "Whodunit" to Who Did It', in Cavelty and Wenger (eds.), *Cyber Security Politics*.
Lehmann, J. et al., 'Distributed Semantic Analytics Using the SANSA Stack', in d'Amato et al. (eds.), *The Semantic Web*.
Lewis, J.A., 'The Role of Offensive Cyber Operations in NATO's Collective Defence', Tallinn Paper 8 (2015), https://ccdcoe.org/uploads/2018/10/TP_08_2015_0.pdf.
Libicki, M., 'Drawing Inferences from Cyber Espionage', in Minárik et al. (eds), *CyCon X: Maximising Effects*.
Libicki, M., 'The Coming of Cyber Espionage Norms' in Rõigas et al. (eds.), *Defending the Core*.
Lindsay, J.R., 'Quantum Computing and Classical Politics: The Ambiguity of advantage in signals intelligence', in Cavelty and Wenger (eds.), *Cyber Security Politics*.
Mačák, K., 'From the Vanishing Point Back to the Core: The Impact of the Development of the Cyber Law of War on General International Law' in Rõigas et al. (eds.), *Defending the Core*.
Madnick, S., Li, X., and Choucri, N., 'Experiences and Challenges with Using CERT Data to Analyze International Cyber Security', Working Paper CISL# 2009–13 September 2009, https://pdfs.semanticscholar.org/d4a5/c681807b6f38b2dbcd4cea5894c12011cc81.pdf.
Mansoor, P.R., 'Introduction: Hybrid Warfare in History', in Murray and Mansoor (eds.), *Hybrid Warfare*.
Maynard, P., McLaughlin, K., and Haberler, B., 'Towards Understanding Man-in-the-Middle Attacks on IEC 60870-5-104 SCADA Networks'. Paper presented at International Symposium for ICS & SCADA Cyber Security Research (ICS-CSR), St Polten, Austria (2014), http://www.qub.ac.uk/sites/CSIT/ACEpublications/2016Papers/Fileroupload,734096,en.pdf.
Moore, D., 'Targeting Technology: Mapping Military', in Minárik et al. (eds), *CyCon X: Maximising Effects*.

Naik, N. et al., 'Fuzzy Hashing Aided Enhanced YARA Rules for Malware Triaging', in *2020 IEEE Symposium Series on Computational Intelligence (SSCI)*.

Özkan, B.E., and Bulkan, S., 'Hidden Risks to Cyberspace Security from Obsolete COTS Software', in Minárik, et al. (eds.), *Silent Battle*.

Podiņš, K., and Geers, K., 'Aladdin's Lamp: The Theft and Re-weaponization of Malicious Code', in Minárik et al. (eds.), *CyCon X: Maximising Effects*.

Rivera, J., 'Achieving Cyberdeterrence and the Ability of Small States to Hold Large States at Risk', in Maybaum et al. (eds.), *Architectures in Cyberspace*.

Satasiya, D., and Rupal, R.D., 'Analysis of Software Defined Network Firewall (SDF)', in *2016 International Conference on Wireless Communications, Signal Processing and Networking (WiSPNET)*, https://ieeexplore.ieee.org/abstract/document/7566125/authors#authors.

Schäfer et al., 'BlackWidow: Monitoring the Dark Web for Cyber Security Information', in Minárik et al. (eds.), *Silent Battle*.

Sharkey, N.E., and Heemskerk, J.N.H., 'The Neural Mind and the Robot', in Browne (ed.), *Neural Network Perspectives*.

Sharma, A., 'Cyber Wars: A Paradigm Shift from Means to Ends', in Czosseck and Geers (eds.), *The Virtual Battlefield*.

Singer, T.V.P., 'Participation in Hostilities Due to Autonomous Cyber Weapons', in Rõigas et al. (eds.), *Defending the Core*.

Smeets, M., and Lin, H.S., 'Offensive Cyber Capabilities: To What Ends?', in Minárik et al. (eds.), *CyCon X: Maximising Effects*.

Soltan, S., Mittal, P., and Poor, H.V., 'BlackIoT: IoT Botnet of High Wattage Devices Can Disrupt the Power Grid', in *Proceedings of the 27th USENIX Security Symposium* (August 15–17, 2018, Baltimore, MD, USA), pp. 15–32, https://www.usenix.org/conference/usenixsecurity18/presentation/soltan.

Stockburger, P.Z., 'Control and Capabilities Test: Toward a New Lex Specialis Governing State Responsibility for Third Party Cyber Incidents' in Rõigas et al. (eds.), *Defending the Core*.

Stoddart, K., 'Edward Snowden and PRISM: Negotiating the Post 9/11 "Surveillance State"', in Gibson and Jefferson (eds.), *Contextualizing Security: A Reader*.

Strohmeier, M. et al., 'Assessing the Impact of Aviation Security on Cyber Power', in Pissanidis et al. (eds.), *2016 8th International Conference on Cyber Conflict Cyber Power*.

Sullins, L.J., '"Phishing" for a Solution: Domestic and International Approaches to Decreasing Online Identity Theft', in Carr (ed.), *Computer Crime*.

Trimble, D., Monken, J., and Sand, A.F.L., 'A Framework for Cybersecurity Assessments of Critical Port Infrastructure', in *2017 International Conference on Cyber Conflict (CyCon U.S.)*, 2017, https://ieeexplore.ieee.org/document/8167506/authors#authors.

Upadhyaya, R., and Jain, A., 'Cyber Ethics and Cyber Crime: A Deep Delved Study into Legality, Ransomware, Underground Web and Bitcoin Wallet', in *2016 International Conference on Computing, Communication and Automation (ICCCA)*, 2016, https://ieeexplore.ieee.org/document/7813706/similar#similar.

van der Meer, S., 'Cyber Warfare and Nuclear Weapons: Game-Changing Consequences?', in Meier and Suh (eds.), Reviving Nuclear Disarmament. Paths Towards a Joint Enterprise, SWP Working Paper, FG03 No. 6 (December 2016), https://www.swp-berlin.org/en/publication/reviving-nuclear-disarmament/.

Wallace, R., 'A Time for Counterespionage', in Sims and Gerber (eds.), *Vaults, Mirrors, & Masks*.

Yin, Z. et al., 'How Do Fixes Become Bugs?', in *Proceedings of the 19th ACM SIGSOFT Symposium and the 13th European Conference on Foundations of Software Engineering* (September 5–9 2011), http://opera.ucsd.edu/paper/fse11.pdf.

Zhu, B., Joseph, A., and Sastry, S., 'A Taxonomy of Cyber Attacks on SCADA Systems', in *2011 IEEE International Conferences on Internet of Things, and Cyber, Physical and Social Computing*, https://ieeexplore.ieee.org/stamp/stamp.jsp?tp=&arnumber=6142258.

Journal Articles

Aggarwal, V.K., and Reddie, A.W., 'Comparative Industrial Policy and Cybersecurity: A Framework for Analysis', *Journal of Cyber Policy*, Vol. 3, Issue 3 (December 2018).

Ajayi, E.F.G., 'Challenges to Enforcement of Cyber-Crimes Laws and Policy', *Journal of Internet and Information Systems*, Vol. 6, Issue 1 (August 2016).

Ani, U.P.D., He, H., and Tiwari, A., 'Review of Cybersecurity Issues in Industrial Critical Infrastructure: Manufacturing in Perspective', *Journal of Cyber Security Technology*, Vol. 1, Issue 1 (December 2016).

Armbruster, G., Endicott-Popovsky, B., and Whittington, J., 'Threats to Municipal Information Systems Posed by Aging Infrastructure', *International Journal of Critical Infrastructure Protection*, Vol. 6, Issues 3–4 (December 2013).

Arquilla, J., and Ronfeldt, D., 'Cyberwar Is Coming!', *Comparative Strategy*, Vol. 12, Issue 2 (Spring 1993).

Arquilla, J., 'The Computer Mouse That Roared: Cyberwar in the Twenty-First Century', *Brown Journal of World Affairs*, Vol. 18, Issue 1 (Fall/Winter 2011).

Arquilla, J., 'Twenty Years of Cyberwar', *Journal of Military Ethics*, Vol. 12, Issue 1 (April 2013).

Bandura, A., 'Social Cognitive Theory: An Agentic Perspective', *Annual Review of Psychology*, Vol. 52 (2001).

Bechtol Jr., B.E., 'North Korea and Support to Terrorism: An Evolving History', *Journal of Strategic Security*, Vol. 3, Issue 2 (Summer 2010).

Belle, A. et al., 'Big Data Analytics in Healthcare', *BioMed Research International* (July 2015).

Betz, D., 'Cyberpower in Strategic Affairs: Neither Unthinkable nor Blessed', *Journal of Strategic Studies*, Vol. 35, Issue 5 (October 2012).

Blank, S., and Kim, Y., 'Economic Warfare a la Russe: The Energy Weapon and Russian National Security Strategy', *The Journal of East Asian Affairs*, Vol. 30, Issue 1 (Spring/Summer 2016).

Blum, A.L., and Langley, P., 'Selection of Relevant Features and Examples in Machine Learning', *Artificial Intelligence*, Vol. 97 (1997).

Brantly, A.F., 'Aesop's Wolves: The Deceptive Appearance of Espionage and Attacks in Cyberspace', *Intelligence and National Security*, Vol. 31, Issue 5 (July 2016).

Brenner, J., 'Correspondence: Debating the Chinese Cyber Threat', *International Security*, Vol. 40, Issue 1 (July 2015).

Bronk, C., and Tikk-Ringas, E., 'The Cyber Attack on Saudi Aramco', *Survival*, Vol. 55, Issue 2 (April–May 2013).

Buchanan, B., and Cunningham, F.S., 'Preparing the Cyber Battlefield: Assessing a Novel Escalation Risk in a Sino-American Crisis', *Texas National Security Review*, Vol. 3, Issue 4 (Fall 2020).

Carlin, J.P., 'Detect, Disrupt, Deter: A Whole-of-Government Approach to National Security Cyber Threats', *Harvard National Security Journal*, Vol. 7, Issue 2 (June 2016).

Chekov, A.D., Makarycheva, A.V., Solomentseva, A.M., Suchkov, M.A., and Sushentsov, A.A., 'War of the Future: A View from Russia', *Survival*, Vol. 61, Issue 6 (December 2019–January 2020).

Cherdantseva, Y. et al., 'A Review of Cyber Security Risk Assessment Methods for SCADA Systems', *Computers & Security*, Vol. 56 (February 2016).

Cherdantseva, Y., Burnap, P., Nadjm-Tehrani, S., and Jones, K., 'A Configurable Dependency Model of a SCADA System for Goal-Oriented Risk Assessment', *Applied Sciences*, Vol. 12, Issue 10 (May 2022), 4880.

Chestnut, S., 'Illicit Activity and Proliferation North Korean Smuggling Networks', *International Security*, Vol. 32, Issue 1 (Summer 2007).

Clark, D.J., and Konrad, K.A., 'Asymmetric Conflict: Weakest Link Against Best Shot', *Journal of Conflict Resolution*, Vol. 51, Issue 3 (June 2007).

Comert, G. et al., 'Modeling Cyber Attacks at Intelligent Traffic Signals', *Transportation Research Record: Journal of the Transportation Research Board*, Vol. 2672, Issue 1 (December 2018).

Czosseck, C., Ottis, R., and Talihärm, A.M., 'Estonia After the 2007 Cyber Attacks: Legal, Strategic and Organisational Changes in Cyber Security', *International Journal of Cyber Warfare and Terrorism*, Vol. 1, Issue 1 (January–March 2011).

Danks, D., and Danks, J.H., 'The Moral Permissibility of Automated Responses During Cyberwarfare', *Journal of Military Ethics*, Vol. 12, Issue 1 (April 2013).

Deville, P. et al., 'Dynamic Population Mapping Using Mobile Phone Data', *Proceedings of the National Academy of Sciences*, Vol. 111, Issue 45 (November 2014).

Dimitrov, D.V., 'Medical Internet of Things and Big Data in Healthcare', *Healthcare Informatics Research*, Vol. 22, No. 3 (July 2016).

Duddu, V., 'A Survey of Adversarial Machine Learning in Cyber Warfare', *Defence Science Journal*, Vol. 68, Issue 4 (July 2018).

Dunn Cavelty, Myriam, 'Cyber-Terror—Looming Threat or Phantom Menace? The Framing of the US Cyber-Threat Debate', *Journal of Information Technology and Politics*, Vol. 4, Issue 1 (April 2008).

Ernst, J.M., and Michaels, A.J., 'Framework for Evaluating the Severity of Cybervulnerability of a Traffic Cabinet', *Transportation Research Record: Journal of the Transportation Research Board*, Vol. 2619, Issue 1 (January 2017).

Eshelman, R., and Derrick, D., 'Relying on the Kindness of Machines? The Security Threat of Artificial Agents', *Joint Forces Quarterly*, Vol. 77 (April 2015).

Evertsz, R., Ritter, F.E., Russell, S., and Shepherdson, D., 'Modeling Rules of Engagement in Computer Generated Forces', in *Proceedings of the 16th Conference on Behavior Representation in Modeling and Simulation* (Orlando, FL: University of Central Florida, 2007).

Feng, Y. et al., 'Vulnerability of Traffic Control System Under Cyberattacks with Falsified Data', *Transportation Research: Record Journal of the Transportation Research Board*, Vol. 2672, Issue 1 (December 2018).

Fischerkeller, M.P., and Harknett, R.J., 'Deterrence Is Not a Credible Strategy for Cyberspace', *Orbis*, Vol. 61, Issue 3 (Summer 2017).

Flournoy, M., and Sulmeyer, M., 'Battlefield Internet: A Plan for Securing Cyberspace', *Foreign Affairs*, Vol. 97, Issue 5 (September/October 2018).

Galeotti, M., '(Mis)Understanding Russia's Two 'Hybrid Wars'', *Critique & Humanism*, Vol. 59, Issue 1 (2018). Reprinted in https://www.eurozine.com/misunderstanding-russias-two-hybrid-wars/?pdf.

Gartzke, E., 'The Myth of Cyberwar Bringing War in Cyberspace Back Down to Earth', *International Security*, Vol. 38, Issue 2 (Fall 2013).

Gioe, D.V., Goodman, M.S., and Wanless, A., 'Rebalancing Cybersecurity Imperatives: Patching the Social Layer', *Journal of Cyber Policy*, Vol. 4, Issue 1 (March 2019).

Goldsmith, J., 'How Cyber Changes the Laws of War', *European Journal of International Law*, Vol. 24, Issue 1 (February 2013).

Grant, C.A, and Grant, P.D., 'A Failure to Educate: January 6, 2021 and the Banality of Evil', *The Educational Forum*, Vol. 86, Issue 2 (March 2022).

Gray, C.S., 'Strategic Culture as Context: The First Generation of Theory Strikes Back', *Review of International Studies*, Vol. 25, Issue 1 (January 1999).

Green, L.C., 'Cicero and Clausewitz or Quincy Wright: The Interplay of Law and War', *United States Airforce Academy Journal of Legal Studies*, Vol. 9 (1998–1999).

Gunnison, H., 'Hypnocounseling: Ericksonian Hypnosis for Counselors', *Journal of Counseling & Development*, Vol. 68, Issue 4 (March/April 1990).

Harrer, G.A., 'Cicero on Peace and War', *The Classical Journal*, Vol. 14, Issue 1 (October 1918).

Harton, H.C., Gunderson, M., and Bourgeois, M.J, '"I'll be There with You": Social Influence and Cultural Emergence at the Capitol on January 6', *Group Dynamics: Theory, Research, and Practice*, Vol. 26 (forthcoming 2022).

Heartfield, R., and Loukas, G., 'A Taxonomy of Attacks and a Survey of Defence Mechanisms for Semantic Social Engineering Attacks', *ACM Computing Surveys (CSUR)*, Vol. 48, Issue 3 (February 2016).

Heffernan, J.B. et al., 'Macrosystems Ecology: Understanding Ecological Patterns and Processes at Continental Scales', *Frontiers in Ecology and the Environment*, Vol. 12, Issue 1 (February 2014).

Hoffman, F.G., 'Hybrid Warfare and Challenges', *Joint Forces Quarterly*, Issue 52 (Spring 2009).

Holt, T.J., and Lampke, E., 'Exploring Stolen Data Markets Online: Products and Market Forces', *Criminal Justice Studies*, Vol. 23, Issue 1 (January 2010).

Holt, T.J., Strumsky, D., Smirnova, O., and Kilger, M., 'Examining the Social Networks of Malware Writers and Hackers', *International Journal of Cyber Criminology*, Vol. 6, Issue 1 (June 2012).

Howard-Hassmann, R.E., 'State-Induced Famine and Penal Starvation in North Korea', *Genocide Studies and Prevention*, Vol. 7, Issue 2 (August/December 2012).

Hughes, R.G., and Stoddart, K., 'Hope and Fear: Intelligence and the Future of Global Security a Decade After 9/11', *Intelligence and National Security*, Vol. 27, Issue 5 (October 2012).

Iansiti, M., and Lakhani, K.R., 'The Truth About Blockchain', *Harvard Business Review*, January–February 2017, p. 4, https://hbr.org/2017/01/the-truth-about-blockchain.

Jagatic, T.N., Johnson, N.A., Jakobsson, M., and Menczer, F., 'Social Phishing', *Communications of the ACM*, Vol. 50, Issue 10 (October 2007).

Johnston, A.I., 'Thinking About Strategic Culture', *International Security*, Vol. 19, Issue 4 (Spring 1995).

Johnson, L.K., 'A Conversation with James R. Clapper, Jr., The Director of National Intelligence in the United States, Intelligence and National Security', *Intelligence and National Security*, Vol. 30, Issue 1 (January 2015).

Junio, T.J., 'How Probable Is Cyber War? Bringing IR Theory Back in to the Cyber Conflict Debate', *Journal of Strategic Studies*, Vol. 36, Issue 1 (February 2013).

Kaiser, R., 'The Birth of Cyberwar', *Political Geography*, Vol. 46 (2015).

Kello, L., 'The Meaning of the Cyber Revolution Perils to Theory and Statecraft', *International Security*, Vol. 38, Issue 2 (Fall 2013).

Keohane, R.O., and Nye, J.S., 'Power and Interdependence Revisited', *International Organization*, Vol. 41, Issue 4 (Autumn 1987).

Khraisat, A., and Alazab, A., 'A Critical Review of Intrusion Detection Systems in the Internet of Things: Techniques, Deployment Strategy, Validation Strategy, Attacks, Public Datasets and Challenges', *Cybersecurity*, Vol. 4, Issue 18 (March 2021).

Khraisat, A., Gondal, I., Vamplew, P., and Kamruzzaman, J., Survey of Intrusion Detection Systems: Techniques, Datasets and Challenges, *Cybersecurity*, Vol. 2, Issue 20 (July 2019).

Kleinfeld, J.S., 'The Small World Problem', *Society*, Vol. 39, Issue 2 (January/February 2002).

Kliem, T., 'You Can't Cyber in Here, This Is the War Room! A Rejection of the Effects Doctrine on Cyberwar and the Use of Force in International Law', *Journal on the Use of Force and International Law*, Vol. 4, Issue 2 (June 2017).

Klimburg, A., 'Mobilising Cyber Power', *Survival*, Vol. 53, Issue 1 (February–March 2011).

Kranenbarg, M.W., Holt, T.J., and van der Ham, J., 'Don't Shoot the Messenger! A Criminological and Computer Science Perspective on Coordinated Vulnerability Disclosure', *Crime Science*, Vol. 7, Issue 16 (December 2018).

Langer, R., 'Stuxnet's Secret Twin', *Foreign Policy* (19 November 2013).

Lanskoy, M., and Myles-Primakoff, D., 'Power and Plunder in Putin's Russia', *Journal of Democracy*, Vol. 29, Issue 1 (January 2018).

Lazer, D. et al., 'The Parable of Google Flu: Traps in Big Data Analysis', *Science*, Vol. 343, Issue 6176 (14 March 2014).

Libicki, M., 'The Nature of Strategic Instability in Cyberspace', *Brown Journal of World Affairs*, Vol. 18, Issue 1 (Fall/Winter 2011).

Libicki, M.C., 'Cyberspace Is Not a Warfighting Domain', *Journal of Law and Policy for the Information Society*, Vol. 8, Issue 2 (Fall 2012).

Liff, A.P., 'The Proliferation of Cyberwarfare Capabilities and Interstate War, Redux: Liff Responds to Junio', *Journal of Strategic Studies*, Vol. 36, Issue 1 (February 2013).

Lindsay, J.R., 'The Impact of China on Cybersecurity: Fiction and Friction', *International Security*, Vol. 39. Issue 3 (Winter 2014/2015).

Lindsay, J.R., and Kello, L., 'Correspondence: A Cyber Disagreement', *International Security*, Vol. 39, Issue 2 (Fall 2014).

Luiijf, E., 'Why Are We So Unconsciously Insecure?', *International Journal of Critical Infrastructure Protection*, Vol. 6, Issues 3–4 (December 2013).

Luo, X. et al., 'Social Engineering: The Neglected Human Factor for Information Security Management', *Information Resources Management Journal*, Vol. 24, Issue 3 (July–September 2011).

Lykou, G., Anagnostopoulou, A., and Gritzalis, D., 'Smart Airport Cybersecurity: Threat Mitigation and Cyber Resilience Controls', *Sensors*, Vol. 19, Issue 19 (January 2019).

McAlaney, J., Thackray, H., and Taylor, J., 'The Social Psychology of Cybersecurity', *The Psychologist*, Vol. 29, Issue 9 (September 2016).

Meakins, J., 'A Zero-Sum Game: The Zero-Day Market in 2018', *Journal of Cyber Policy*, Vol. 4, Issue 1 (January 2019).

Melcher, T. et al., 'The Neural Substrate of the Ideomotor Principle Revisited: Evidence for Asymmetries in Action-Effect Learning', *Neuroscience*, Vol. 231 (February 2013).

Mikail, E.H., and Aytekin, C.A., 'The Communications and Internet Revolution in International Relations', *Open Journal of Political Science*, Issue 6 (September 2016).

Milgram, S., 'Behavioral Study of Obedience', *Journal of Abnormal and Social Psychology*, Vol. 67, Issue 4 (October 1963).

Monarch, B., 'Black Start: The Risk of Grid Failure from a Cyber Attack and the Policies Needed to Prepare for It', *Journal of Energy and Natural Resources Law*, Vol. 38, Issue 2 (April 2020).

Murdoch, W.J. et al., 'Definitions, Methods, and Applications in Interpretable Machine Learning', *Proceedings of the National Academy of Sciences (PNAS)*, Vol. 116, Issue 44 (October 2019).

Nicholson, A. et al., 'SCADA Security in the Light of Cyber-Warfare', *Computers & Security*, Vol. 31, Issue 4 (June 2012).

Okenyi, P.O., and Owens, T.J., 'On the Anatomy of Human Hacking', *Information Systems Security*, Vol. 16, Issue 6 (December 2007).

Omand, D., Bartlett, J., and Miller, C., 'Introducing Social Media Intelligence (SOCMINT), *Intelligence and National Security*, Vol. 27, Issue 6 (December 2012).

Peltier, T.R., 'Social Engineering: Concepts and Solutions', *Information Systems Security*, Vol. 15, Issue 5 (2006).

Perliger, A., 'The Causes and Impact of Political Assassinations', *CTC Sentinel* [Combatting Terrorism Center, West Point], Vol. 8, Issue 1 (January 2015), https://ctc.usma.edu/the-causes-and-impact-of-political-assassinations/.

Porteous, S.D., 'Economic Espionage: Issues Arising from Increased Government Involvement with the Private Sector', *Intelligence and National Security*, Vol. 9, Issue 4 (October 1994).

Pun, D., 'Rethinking Espionage in the Modern Era', *Chicago Journal of Internal Law*, Vol. 18, Issue 1 (Summer 2017).

Raghupathi, W., and Raghupathi, V., 'Big Data Analytics in Healthcare: Promise and Potential', *Health Information Science and Systems*, Vol. 2, Issue 3 (February 2014).

Ravndal, J.A., 'Anders Behring Breivik's Use of the Internet and Social Media', *Journal EXIT-Deutschland*, Vol. 2 (2013).

Reason, J., 'Human Error: Models and Management', *British Medical Journal*, Vol. 320 (March 2000).

Redya, V., K. Chatrapati, S., and Kamalesh, V.N., 'Paper on Types of Firewall and Design Principles', *International Journal of Science and Research*, Vol. 6, Issue 14 (2015).

Richardson, J., 'Stuxnet as Cyberwarfare: Applying the Law of War to the Virtual Battlefield', *Journal of Information Technology and Privacy Law*, Vol. 29, Issue 1 (Fall 2011).

Rid, T., 'An Imperfect Weapon', *Survival*, Vol. 60, Issue 5 (2018).

Robinson, M., Jones, K., and Janicke, H., 'Cyber Warfare: Issues and Challenges', *Computers & Security*, Vol. 49 (March 2015).

Roderique-Davies, G., 'Neuro-Linguistic Programming: Cargo Cult Psychology?', *Journal of Applied Research in Higher Education*, Vol. 1, Issue 2 (2009).

Rogoff, M.A., and Collins Jr., E., 'The Caroline Incident and the Development of International Law', *Brooklyn Journal of International Law*, Vol. 16, Issue 3 (1990).

Rudner, M., 'Cyber-Threats to Critical National Infrastructure: An Intelligence Challenge', *International Journal of Intelligence and Counterintelligence*, Vol. 26, Issue 3 (May 2013).

Russano, M. et al., 'Structured Interviews of Experienced HUMINT Interrogators', *Applied Cognitive Psychology*, Vol. 28, Issue 6 (November/December 2014).

Ruzicka, J., 'Failed Securitization: Why It Matters', *Polity*, Vol. 51, Issue 2 (April 2019).

Salisbury, D., 'North Korea's Missile Programme and Supply Side Controls: Lessons for Countering Illicit Procurement', *The RUSI Journal*, Vol. 163, Issue 4 (August/September 2018).

Saygin, A.P., Cicekli, I., and Akman, I., 'Turing Test: 50 Years Later', *Minds and Machines*, Vol. 10, Issue 4 (2000).

Schmitt, M.N., '21st Century Conflict: Can the Law Survive?', *Melbourne Journal of International Law*, Vol. 8, Issue 2 (October 2007).

Schmitt, M.N., 'The Law of Cyber Warfare: Quo Vadis?', *Stanford Law & Policy Review*, Vol. 25 (2014).

Shaalan, A.M., 'Adopting Measures to Reduce Power Outages', *Electrical & Computer Engineering: An International Journal (ECIJ)*, Vol. 7, Issue 1/2 (June 2018).

Shahi, A., and Abdoh-Tabrizi, E., 'Iran's 2019–2020 Demonstrations: The Changing Dynamics of Political Protests in Iran', *Asian Affairs*, Vol. 51, Issue 1 (January 2020).

Sharkey, N.E., 'The Evitability of Autonomous Robot Warfare', *International Review of the Red Cross*, Vol. 94, Issue 886 (Summer 2012).

Shea, T.C., 'Post-Soviet Maskirovka, Cold War Nostalgia, and Peacetime Engagement', *Military Review*, Vol. 82, Issue 3 (May–June 2002).

Siboni, G., Abramski, L., and Sapir, G., 'Iran's Activity in Cyberspace: Identifying Patterns and Understanding the Strategy', *Cyber, Intelligence, and Security*, Vol. 4, Issue 1 (March 2020).

Simpson, E., 'Clausewitz's Theory of War and Victory in Contemporary Conflict', *Parameters*, Vol. 47, Issue 4 (Winter 2017–2018).

Soldatov, A., and Borogan, I., 'Russia's Surveillance State', *World Policy Journal*, Vol. 30, Issue 23 (Fall 2013).

Solovyeva, A., Hynek, N., 'Going Beyond the "Killer Robots" Debate: Six Dilemmas Autonomous Weapon Systems Raise', *Central European Journal of International and Security Studies*, Vol. 12, Issue 3 (September 2018).

Srivastava, P., and Khan, R., 'A Review Paper on Cloud Computing', *International Journals of Advanced Research in Computer Science and Software Engineering*, Vol. 8, Issue 6 (June 2018).

Steinhart, A., 'The Future Is Behind Us? The Human Factor in Cyber Intelligence: Interplay Between Cyber-HUMINT, Hackers and Social Engineering'. Available from http://www.amitsteinhart.com/.

Stevens, T., 'Knowledge in the Grey Zone: AI and Cybersecurity', *Digital War*, Vol. 1, Issue 1 (December 2020).

Stockburger, P.Z., 'Known Unknowns: State Cyber Operations, Cyber Warfare, and the Jus Ad Bellum', *American University International Law Review*, Vol. 31, Issue 4 (2016).

Stoddart, K., Jones, K., Soulsby, H., Blyth, A., Eden, P., Burnap, P., and Cherdantseva, Y., 'Live Free or Die Hard: U.S.–UK Cybersecurity Policies', *Political Science Quarterly*, Vol. 131, Issue 4 (Winter 2016).

Sulmeyer, M., 'How the U.S. Can Play Cyber-Offense', *Foreign Affairs* (March 22 2018).

Tal, A., and Siman-Tov, D., 'HUMINT in the Cybernetic Era: Gaming in Two Worlds', *Military and Strategic Affairs*, Vol. 7, Issue 3 (2015).

Taneski, G.V.N., and Dojchinovski, M., 'The Danger of "Hybrid Warfare" from a Sophisticated Adversary: The Russian "Hybridity" in the Ukrainian Conflict', *Defense and Security Analysis*, Vol. 33, Issue 4 (December 2017).

Tang, S., 'The Security Dilemma: A Conceptual Analysis', *Security Studies*, Vol. 18, Issue 3 (October 2009).

Tetri, P., and Vuorinen, J., 'Dissecting Social Engineering', *Behaviour and Information Technology*, Vol. 32, Issue 10 (2013).

Thoben, K.-D., Wiesner, S., and Wuest, T., '"Instrustrie 4.0" and Smart Manufacturing—A Review of Research Issues and Application Examples', *International Journal of Automation Technology*, Vol. 11, Issue 1 (2017).

Timmers, P., 'The European Union's Cybersecurity Industrial Policy', *Journal of Cyber Policy*, Vol. 3, Issue 3 (December 2018).

Travers, J., and Milgram, S., 'An Experimental Study of the Small World Problem', *Sociometry*, Vol. 32, Issue 4 (December 1969).

Valeriano, B., Jensen B., and Maness, R.C., *Cyber Strategy: The Evolving Character of Power and Coercion* (New York: Oxford University Press, 2018).

van Creveld, M., 'The Transformation of War Revisited', *Small Wars & Insurgencies*, Vol. 13, Issue 2 (Summer 2002).

Vinge, V., 'The Coming Technological Singularity, *Whole Earth Review*, Issue 81 (Winter 1993).

Wallace, D.A., McCarthy, A.H., Visger, M., 'Peeling Back the Onion of Cyber Espionage After Tallinn 2.0', *Maryland Law Review*, Vol. 78, Issue 2 (2019).

Wallach, M.A., Kogan, N., and Bem, D.J., 'Diffusion of Responsibility and Level of Risk Taking in Groups', *Journal of Abnormal and Social Psychology*, Vol. 68, Issue 3 (March 1964).

Walsh, T., 'The Singularity May Never Be Near', *AI Magazine*, Vol. 38, Issue 3 (Fall 2017).

Warner, M., 'Cyber-Security: A Pre-history', *Intelligence and National Security*, Vol. 27, Issue 5 (October 2012).

Wendt, A., 'Anarchy Is What States Make of It: The Social Construction of Power Politics', *International Organization*, Vol. 46, Issue 2 (Spring 1992).

Wigell, M., and Vihma, A., 'Geopolitics Versus Geoeconomics: The Case of Russia's Geostrategy and Its Effects on the EU', *International Affairs*, Vol. 92, Issue 3 (May 2016).

William S. Lind, Keith Nightengale, John F. Schmitt, Joseph W. Sutton, Gary I. Wilson, 'The Changing Face of War: Into the Fourth Generation', *Marine Corps Gazette*, 22–26 (October 1989).

Wilkie, R., 'Hybrid Warfare: Something Old, Not Something New', *Air & Space Power Journal*, Vol. 23, Issue 4 (Winter 2009).

Wilson, P., 'The Myth of International Humanitarian Law', *International Affairs*, Vol. 93, Issue 3 (May 2017).

Workman, M., 'Gaining Access with Social Engineering: An Empirical Study of the Threat', *Information Systems Security*, Vol. 16, Issue 6 (2007).

Yalcintas, A., and Alizadeh, N., 'Digital Protectionism and National Planning in the Age of the Internet: The Case of Iran', *Journal of Institutional Economics*, Vol. 16, Issue 4 (August 2020).

Zaidi, W.H., 'Stages of War, Stages of Man: Quincy Wright and the Liberal Internationalist Study of War', *The International History Review*, Vol. 40, Issue 2 (March 2018).

Zimba, A., Wang, Z., and Chen, H., 'Multi-Stage Crypto Ransomware Attacks: A New Emerging Cyber Threat to Critical Infrastructure and Industrial Control Systems', *ICT Express*, Vol. 4, Issue 1 (March 2018).

Mainstream Media Sources (Newspaper Reports, Magazines, and Periodicals)

Aaro, D., 'Massive Electrical Failure Cuts Power to Argentina and Uruguay' (June 16 2019), https://www.foxnews.com/world/massive-electrical-failure-cuts-power-to-argentina-and-uruguay.

Andrews, W., 'Some U.S. Military Parts Imported from China' (June 9 2017), https://www.cbsnews.com/news/some-us-military-parts-imported-from-china/.

Anonymous, '70mn Cyberattacks, Mostly Foreign, Targeted Russia's Critical Infrastructure in 2016 – FSB' (January 25 2017), https://www.rt.com/news/374973-cyber-attacks-russian-infrastructure/.

Anonymous, 'Arrests Over Hacks of CIA and FBI Staff' (September 9 2016), http://www.bbc.co.uk/news/technology-37316615.

Anonymous, 'Blank Screens at Bristol Airport After Cyber Attack' (September 16 2018), https://www.itv.com/news/westcountry/2018-09-16/blank-screens-at-bristol-airport-after-cyber-attack/.

Anonymous, 'British Airways: Suspect Code That Hacked Fliers 'Found'' (September 11 2018), https://www.bbc.co.uk/news/technology-45481976.

Anonymous, 'Christmas Ransomware Attack Hit New York Airport Servers' (January 10 2020), https://apnews.com/article/fbefe0ccdfac9279df8c8170 68482b1b.

Anonymous, 'Cooperation or Resistance?: The Role of Tech Companies in Government Surveillance' (April 10 2018), https://harvardlawreview.org/2018/04/cooperation-or-resistance-the-role-of-tech-companies-in-government-surveillance/.

Bibliography

Anonymous, 'Copycat Coders Create 'Vulnerable' Apps' (October 7 2019), https://www.bbc.co.uk/news/technology-49960387.

Anonymous, 'Cyber-Thieves Cash in from Malware' (June 9 2015), http://www.bbc.co.uk/news/technology-33048949.

Anonymous, 'Cyber-Thieves Turn to 'Invisible Net' to Set Up Attacks' (June 7 2019), https://www.bbc.co.uk/news/technology-47319971.

Anonymous, 'Daniel Kelley: The Teen Behind the Cybercrime Screen' (June 10 2019), https://www.bbc.co.uk/news/uk-wales-48120428.

Anonymous, 'Europol Chief Warns on Computer Encryption' (March 29 2015), http://www.bbc.co.uk/news/technology-32087919.

Anonymous, 'How the Dutch Foiled Russian 'Cyber-Attack' on OPCW' (October 4 2018), https://www.bbc.co.uk/news/world-europe-45747472.

Anonymous, 'Iran Says Key Natanz Nuclear Facility Hit by 'Sabotage'', https://www.bbc.co.uk/news/world-middle-east-56708778.

Anonymous, 'Julian Assange: Campaigner or Attention-Seeker?' (July 30 2018), https://www.bbc.co.uk/news/world-11047811.

Anonymous, 'Major Cyberattack on UK Infrastructure Is 'When, Not If'' (January 23 2018), https://news.sky.com/story/major-cyberattack-on-uk-infrastructure-is-when-not-if-11219026.

Anonymous, 'Major Power Cut Leaves Large Parts of England Without Electricity: The National Grid Said Two Generators Went Down and the Issue Is Now Resolved But Commuters Have Been Left Stranded' (August 10 2019), https://news.sky.com/story/large-parts-of-london-and-south-east-without-electricity-after-power-cut-11781338.

Anonymous, 'Marcus Hutchins Spared US Jail Sentence Over Malware Charges' (July 26 2019), https://www.bbc.co.uk/news/technology-49127569.

Anonymous, 'Panama Papers Q&A: What Is the Scandal About?', http://www.bbc.co.uk/news/world-35954224.

Anonymous, 'Rosaviatsiya Extends Temporary Closure of 11 Airports in Southern Russia' (June 4 2022), https://azeritimes.com/2022/06/04/rosaviatsiya-extends-temporary-closure-of-11-airports-in-southern-russia-2/.

Anonymous, 'Russian Bots Rigged Voice Kids TV Talent Show Result' (May 16 2019), https://www.bbc.co.uk/news/world-europe-48293196.

Anonymous, 'SBU Thwarts Cyber Attack from Russia Against Chlorine Station in Dnipropetrovsk Region' (July 11 2018), https://en.interfax.com.ua/news/general/517337.html.

Anonymous, ''Serious' Hack Attacks from China Targeting UK Firms' (April 4 2017), http://www.bbc.co.uk/news/technology-39478975.

Anonymous, 'Tor Project Makes Efforts to Debug Dark Web' (July 23 2014), http://www.bbc.co.uk/news/technology-28447023.

Anonymous, 'Two Years for Teen 'Cyber Terrorist' Who Targeted US Officials' (April 20 2018), https://www.bbc.co.uk/news/uk-england-leicestershire-438 40075.

Anonymous, 'UK Power Cut: National Grid Promises to Learn Lessons from Blackout' (August 10 2019), https://www.bbc.co.uk/news/uk-49302996.

Anonymous, 'What Are Spies Really Like?' (April 2 2012), https://www.bbc.co.uk/news/magazine-17560253.

Anonymous/Gordon Corera, 'Cyber-Attack: US and UK Blame North Korea for WannaCry' (December 19 2017), http://www.bbc.co.uk/news/world-us-canada-42407488.

Arnold, E., Bazaraa, D., and Thomas, J., 'Labour MP Jess Phillips Shares Death Threat Letter Sent to Her as She Slams PM' (September 25 2019), https://www.mirror.co.uk/news/politics/labour-mp-jess-phillips-shares-20205763.

Atwood, K., and Cohen, Z., 'US in Contact with Zelensky Through Secure Satellite Phone' (March 1 2022), https://edition.cnn.com/europe/live-news/ukraine-russia-putin-news-03-02-22/h_6b5c8062541ddb6c36dd43ca7039 1608.

Barkin, N., 'Exclusive: Five Eyes Intelligence Alliance Builds Coalition to Counter China' (October 12 2018), https://www.reuters.com/article/us-china-fiveeyes-idUSKCN1MM0GH.

Barnes, J.E., and Goldman, A., 'Captured, Killed or Compromised: C.I.A. Admits to Losing Dozens of Informants' (October 5 2021), https://www.nytimes.com/2021/10/05/us/politics/cia-informants-killed-captured.html.

Batchelor, T., 'Tracking Apps That Reveal Location of British Warships Spark Security Fears' (February 5 2018), https://www.independent.co.uk/news/uk/home-news/royal-navy-tracking-app-warship-nato-russia-china-military-security-a8191896.html.

Berwick, A., and Wilson, T., 'How Crypto Giant Binance Became a Hub for Hackers, Fraudsters and Drug Traffickers' (June 6 2022), https://www.reuters.com/investigates/special-report/fintech-crypto-binance-dirtymoney/.

Beuth, P. et al., 'Merkel and the Fancy Bear' (May 12 2017), https://www.zeit.de/digital/2017-05/cyberattack-bundestag-angela-merkel-fancy-bear-hacker-russia.

Binns, D., 'After Heartbleed, Can Britain's New Cyber Emergency Response Team Beat the Hackers' (April 15 2014), https://metro.co.uk/2014/04/15/after-heartbleed-can-britains-new-cyber-emergency-response-team-beat-the-hackers-4698848/.

Brennan, R.J., 'Cyber Attack on Small Illinois Water Treatment Plant Has Serious Implications: Security Expert' (November 21 2011), https://www.thestar.com/news/world/2011/11/21/cyber_attack_on_small_illinois_water_treatment_plant_has_serious_implications_security_expert.html.

Brewster, T., 'Sources: Martin Shkreli Thrown in Solitary Confinement After Claims He Ran Company From Prison' (April 1 2019), https://www.forbes.com/sites/thomasbrewster/2019/04/01/sources-martin-shkreli-thrown-in-solitary-confinement-after-claims-he-ran-company-from-prison/.

Browne, R., 'Canadian Judge Orders VICE News Journalist to Hand Over Digital Messages', https://news.vice.com/article/canadian-judge-orders-vice-news-journalist-to-hand-over-digital-messages.

Burgess, M., 'What Is the Internet of Things? WIRED Explains From Hairbrushes to Scales, Consumer and Industrial Devices Are Having Chips Inserted into Them to Collect and Communicate Data' (February 16 2018), https://www.wired.co.uk/article/internet-of-things-what-is-explained-iot.

Cadwalladr, C., 'Arron Banks, Brexit and the Russia Connection' (June 16 2018), https://www.theguardian.com/uk-news/2018/jun/16/arron-banks-nigel-farage-leave-brexit-russia-connection.

Carlin, J., 'A Farewell to Arms' (May 1 1997), http://archive.wired.com/wired/archive/5.05/netizen.html.

Carman, A., 'Hammertoss Malware Represents Culmination of 'Best Practices' for Cyber Attackers' (July 29 2015), https://www.scmagazine.com/home/security-news/hammertoss-malware-represents-culmination-of-best-practices-for-cyber-attackers/.

Cellan-Jones, R., 'Hacking the House: Do MPs Care About Cyber-Security?' (December 3 2017), https://www.bbc.co.uk/news/technology-42217017.

Chen, L.Y., 'Telegram Traces Massive Cyber Attack to China During Hong Kong Protests' (June 13 2019), https://www.bloomberg.com/news/articles/2019-06-13/telegram-traces-cyber-attack-to-china-amid-hong-kong-pro tests.

Chuang, T., 'Cyber Attack on CDOT Computers Estimated to Cost Up to $1.5 Million So Far', https://www.denverpost.com/2018/04/05/samsam-ransomware-cdot-cost/.

Cohen, M., 'Evidence Shows Capitol Rioters Brutally Attacked Police with Flagpoles, Fire Extinguishers and Fists' (January 21 2021), https://edition.cnn.com/2021/01/21/politics/capitol-rioters-attacking-police/index.html.

Cohen, Z., 'Intelligence Officials Ask Congress Not to Hold Threats Hearings After Angering Trump Last Year' (January 16 2020), https://edition.cnn.com/2020/01/16/politics/us-intelligence-officials-world-wide-threats-hearing-testimony/index.html.

Collinson, S., 'Ransomware Attacks Saddle Biden with Grave National Security Crisis' (June 7 2021), https://edition.cnn.com/2021/06/07/politics/president-joe-biden-cyber-attacks-russia-putin-trump-economy/index.html.

Conger, K., and Satariano, A., 'Volunteer Hackers Converge on Ukraine Conflict with No One in Charge (March 4 2022), https://www.nytimes.com/2022/03/04/technology/ukraine-russia-hackers.html.

Culbertson, A., 'Half of European Flights Face Delay After Computer Failure' (April 4 2018), https://news.sky.com/story/half-of-european-flights-face-delay-after-computer-failure-11315397.

'Cybersecurity Threats and Defense Strategy' (May 9 2017), https://www.c-span.org/video/?428023-1/nsa-director-rogers-russia-poses-threat-congressional-elections&start=1166.

Davies, P., 'Cyber Espionage Is Key to Russia's Invasion of Ukraine. The International Community Is Fighting Back' (March 9 2022), https://www.euronews.com/next/2022/03/09/cyberespionage-is-key-to-russia-s-invasion-of-ukraine-the-international-community-is-fight.

Dreyfuss, E., 'US Weapons Systems Are Easy Cyberattack Targets, New Report Finds a New Report Says the Department of Defense "Likely Has an Entire Generation of Systems That Were Designed and Built Without Adequately Considering Cybersecurity."' (October 10 2018), https://www.wired.com/story/us-weapons-systems-easy-cyberattack-targets/.

Fasman, J., 'I Know What You'll Do Next Summer More Data and Surveillance Are Transforming Justice Systems' (June 2 2018), https://www.economist.com/technology-quarterly/2018-05-02/justice.

Franceschi-Bicchierai, L., 'Teen Who Hacked Ex-CIA Director John Brennan Gets Sentenced to 2 Years of Prison' (April 20 2018), https://www.vice.com/en_us/article/pax87v/kane-gamble-crackas-with-attitude-cwa-sentence-prison.

Galeotti, M.,, 'Gangster's Paradise: How Organised Crime Took Over Russia' (March 23 2018), https://www.theguardian.com/news/2018/mar/23/how-organised-crime-took-over-russia-vory-super-mafia.

Gellman, B., 'Cyber-Attacks by Al Qaeda Feared' (June 27 2002), https://www.washingtonpost.com/archive/politics/2002/06/27/cyber-attacks-by-al-qaeda-feared/5d9d6b05-fe79-432f-8245-7c8e9bb45813/.

Gellman, B., and Poitras, L., 'U.S., British Intelligence Mining Data from Nine U.S. Internet Companies in Broad Secret Program' (June 6 2013), http://www.washingtonpost.com/investigations/us-intelligence-mining-data-from-nine-us-internet-companies-in-broad-secret-program/2013/06/06/3a0c0da8-cebf-11e2-8845-d970ccb04497_story.html.

Glassberg, J., 'What You Need to Know About 'Drive-By' Cyber Attacks' (February 4 2015), https://www.foxbusiness.com/features/what-you-need-to-know-about-drive-by-cyber-attacks.

Gold, H., 'We Know Who Is Attacking Us and We Know How to Get Even, Says Israel's Cyber Defense Chief' (December 4 2021), https://edition.cnn.com/2021/12/04/middleeast/israel-cyberattack-intl-cmd/index.html.

Greenberg, A., 'Iranian Hackers Launch a New US-Targeted Campaign as Tensions Mount Three Cybersecurity Firms Have Identified Phishing Attacks Stemming from Iran—That May Lay the Groundwork for Something More Destructive' (June 20 2021), https://www.wired.com/story/iran-hackers-us-phishing-tensions/.

Greenberg, A., 'Mind the Gap: This Researcher Steals Data With Noise, Light, and Magnets' (February 7 2018), https://www.wired.com/story/air-gap-researcher-mordechai-guri/.

Greenberg, A., 'Operation Bayonet: Inside the Sting That Hijacked an Entire Dark Web Drug Market Dutch Police Detail for the First Time How They Secretly Hijacked Hansa, Europe's Most Popular Dark Web Market' (August 3 2018), https://www.wired.com/story/hansa-dutch-police-sting-operation/?verso=true.

Greenberg, A., 'Russia's Sandworm Hackers Attempted a Third Blackout in Ukraine' (April 12 2022), https://www.wired.com/story/sandworm-russia-ukraine-blackout-gru/.

Groenfeldt, T., 'Insiders Pose a Serious Threat to Corporate Information' (May 8 2014), http://www.forbes.com/sites/tomgroenfeldt/2014/05/08/insiders-pose-a-serious-threat-to-corporate-information/.

Hall, K., 'Cyber Security Is a Board-Level Issue, Warn Government Spooks' (September 5 2012), https://www.computerweekly.com/news/2240162676/Companies-must-tackle-cyber-threats-warn-government-spooks.

Hardy, E., 'Failure to Launch: Stock Market Open Delayed by Nearly Two Hours as Technical Issue at LSE Stalls FTSE Share Trading' (August 16 2019), https://www.thisismoney.co.uk/money/markets/article-7363411/Failure-launch-UK-share-trading-delayed-nearly-2-hours-technical-issue-LSE.html.

Haynes, D., 'Iran's Secret Cyber Files' (Undated), https://news.sky.com/story/irans-secret-cyber-files-on-how-cargo-ships-and-petrol-stations-could-be-attacked-12364871.

Haynes, D., 'Ukraine: 'Massive Cyber Attack' Shuts Down Government Websites' (January 14 2022), https://news.sky.com/story/ukraine-says-massive-cyber-attack-has-shut-down-government-websites-12515487.

Hosenball, M., 'FBI Paid Under $1 Million to Unlock San Bernardino iPhone: Sources' (May 4 2016), http://www.reuters.com/article/us-apple-encryption-idUSKCN0XQ032.

House Homeland Security Committee Hearing on the Colonial Pipeline Cyber Attack (June 9 2021), https://www.c-span.org/video/?512332-1/colonial-pipeline-ceo-joseph-blount-testifies-house-homeland-security-committee.

Hubbard, B., Karasz, P., and Reed, S., 'Two Major Saudi Oil Installations Hit by Drone Strike, and U.S. Blames Iran (September 14 2019), https://www.nytimes.com/2019/09/14/world/middleeast/saudi-arabia-refineries-drone-attack.html.

Jones, S., 'Finger Points at Russian State Over Petya Hack Attack Tactics of Those Behind Malware Match Kremlin Playbook, Say Analysts' (June 30 2017), https://www.ft.com/content/f300ad84-5d9d-11e7-b553-e2df1b0c3220.

Jones, S., 'Licensed to Hack: The Rise of the Cyber Privateer Russia Is Increasingly Using Criminal Proxies, Say Western Intelligence Officials' (March 16 2017), https://www.ft.com/content/21be48ec-0a48-11e7-97d1-5e720a26771b.

Jones, S., 'The Spy Who Liked Me: Britain's Changing Secret Service' (September 29 2016), https://www.ft.com/content/b239dc22-855c-11e6-a29c-6e7d9515ad15.

Keller, A., 'What Do Former CIA Spies Do When They Quit the Spy Game?' (October 12 2012), https://www.forbes.com/sites/realspin/2012/10/12/what-do-former-cia-spies-do-when-they-quit-the-spy-game/, https://www.afio.com/14_careers.htm.

Kiger, P.J., "American Blackout': Four Major Real-Life Threats to the Electric Grid' (October 25 2013), https://www.nationalgeographic.com/environment/great-energy-challenge/2013/american-blackout-four-major-real-life-threats-to-the-electric-grid/.

Kovachich, L., 'Russia Flirts with Internet Sovereignty China Specialist Leonid Kovachich on How Russia Might Overtake China in Internet Censorship' (February 1 2019), https://www.themoscowtimes.com/2019/02/01/russia-flirts-with-internet-sovereignty-op-ed-a64369.

Krebbs, C., 'The cyber warfare predicted in Ukraine may be yet to come', https://www.ft.com/content/2938a3cd-1825-4013-8219-4ee6342e20ca.

Kuznetzov, N., 'Russia's Energy Sector Set to Thrive in 2017' (January 11 2017), https://www.forbes.com/sites/nikolaikuznetsov/2017/01/11/russias-energy-sector-could-thrive-in-2017/#7b62ef7b9595.

Larson, S., 'Is Russian Social Media Meddling 'Cyberwarfare'?' (November 3 2017), https://money.cnn.com/2017/11/03/technology/business/russian-social-media-info-ops-cyberwar/index.html.

Lee, D., "Foreshadow' Attack Affects Intel Chips' (August 15 2018), https://www.bbc.co.uk/news/technology-45191697.

Lyngaas, S., 'US Warns That Iranian Government-Sponsored Hackers Are Targeting Key US Infrastructure' (November 17 2021), https://edition.cnn.com/2021/11/17/politics/us-iran-hackers-warning/index.html.

Macalister, T., 'Piper Alpha Disaster: How 167 Oil Rig Workers Died' (July 4 2013), https://www.theguardian.com/business/2013/jul/04/piper-alpha-disaster-167-oil-rig.

Macalister, T., 'Who Would Get the Oil Revenues If Scotland Became Independent?', *The Guardian*, March 2 2012.

Mackie, T., 'Revealed: New Era of State Sponsored Hacking Can Turn Oil Rigs Into 'Bomb' That Can Kill' (February 18 2018), https://www.express.co.uk/news/world/920437/computer-hacker-cyber-hack-saudi-arabia-cyber-criminals-oil-rigs.

Marquardt, A., 'New Report Reveals Apparent Plot Against Electrical Grid' (November 4 2021), https://edition.cnn.com/videos/politics/2021/11/04/drone-threat-power-grid-new-details-marquardt-lead-pkg-vpx.cnn.

Marr, B., 'A Very Brief History of Blockchain Technology Everyone Should Read (February 16 2018), https://www.forbes.com/sites/bernardmarr/2018/02/16/a-very-brief-history-of-blockchain-technology-everyone-should-read/#17ef6b77bc47.

Martin, A., 'Cyber Criminal Charged with Stealing £76m While Working for Russian Intelligence' (December 5 2019), https://news.sky.com/story/cyber-criminal-charged-with-stealing-76m-while-working-for-russian-intelligence-11878896.

Meserve, J., 'Sources: Staged Cyber Attack Reveals Vulnerability in Power Grid' (September 26 2007), http://edition.cnn.com/2007/US/09/26/power.at.risk/.

Messing, P., Schram, J., and Golding, B., 'Teen Says He Hacked CIA Director's AOL Account' (October 18 2015), http://nypost.com/2015/10/18/stoner-high-school-student-says-he-hacked-the-cia/.

Mezzofiore, G., 'Team Poison's Junaid Hussain Jailed for Tony Blair Hack and Phone Bombing Anti-Terror Hotline' (July 2 2014), https://www.ibtimes.co.uk/team-poison-phone-bomb-hacker-anti-terror-367660.

Miller, J., 'Google and Apple to Introduce Default Encryption' (September 19 2014), http://www.bbc.co.uk/news/technology-29276955.

Miller, J., 'Intel Vows to Stop Using 'Conflict Minerals' in New Chips', http://www.bbc.co.uk/news/technology-25636001.

Miller, G., Nakashima, E., and Entous, A., 'Obama's Secret Struggle to Punish Russia for Putin's Election Assault' (June 23 2017), https://www.washingtonpost.com/graphics/2017/world/national-security/obama-putin-election-hacking/.

Milner, A., 'Mosul Dam: Why the Battle for Water Matters in Iraq' (August 18 2014), http://www.bbc.co.uk/news/world-middle-east-28772478.

Molina, B., Swartz, J., and Sandler, R., 'Petya Cyberattack Spreads to 65 Countries' (June 28 2017), https://www.usatoday.com/story/tech/talkingtech/2017/06/28/petya-cyberattack-spreads-65-countries/435016001/.

Mozur, P., and Buckley, C., 'Spies for Hire: China's New Breed of Hackers Blends Espionage and Entrepreneurship' (August 26 2021), https://www.nytimes.com/2021/08/26/technology/china-hackers.html.

Myre, G., 'How Does Ukraine Keep Intercepting Russian Military Communications?' (April 26 2022), https://www.npr.org/2022/04/26/1094656395/how-does-ukraine-keep-intercepting-russian-military-communications.

Nakashima, E., 'Russian Military Was Behind 'NotPetya' Cyberattack in Ukraine, CIA Concludes', *The Washington Post*, January 12 2018.

Nan, W., 'From Hackers to Entrepreneurs: The Sino-US Cyberwar Veterans Going Straight' (August 21 2013), https://www.scmp.com/news/china/article/1298200/hackers-entrepreneurs-sino-us-cyberwar-veterans-going-straight.

NeSmith, B., 'The Cybersecurity Talent Gap Is an Industry Crisis' (August 9 2018), https://www.forbes.com/sites/forbestechcouncil/2018/08/09/the-cybersecurity-talent-gap-is-an-industry-crisis/#2fdfc6a4a6b3.

Newman, L.H., 'A Cisco Router Bug Has Massive Global Implications' (May 13 2019), https://www.wired.com/story/cisco-router-bug-secure-boot-trust-anchor/.

Newman, L.H., 'Menacing Malware Shows the Dangers of Industrial Systems Sabotage' (January 18 2018), https://www.wired.com/story/triton-malware-dangers-industrial-system-sabotage/.

Ng, A., 'These 6 Charts Show How Sanctions Are Crushing Iran's Economy' (March 22 2021), https://www.cnbc.com/2021/03/23/these-6-charts-show-how-sanctions-are-crushing-irans-economy.html.

'Offshore Oil Fields' (Source: https://www.bbc.co.uk/news/10298342, June 14 2010).

Ong, D., 'Russian General Brutally Dies in Ukraine: 'Collected His Guts…Back in His Belly'' (April 27 2022), https://www.ibtimes.com/russian-general-brutally-dies-ukraine-collected-his-gutsback-his-belly-3488076.

Opie, R., 'Adelaide Teen Hacked into Apple Twice Hoping the Tech Giant Would Offer Him a Job' (May 27 2019), https://www.abc.net.au/news/2019-05-27/adelaide-teenager-hacked-into-apple-twice-in-two-years/11152492.

PA/MEHR News Agency, 'Iran Loses $150 Billion a Year Due to Brain Drain' (January 8, 2014), https://en.mehrnews.com/news/101558/Iran-loses-150-billion-a-year-due-to-brain-drain.

Patel, B., 'Legacy systems are problems for boardrooms not computer geeks' (February 1 2017), https://www.ft.com/content/5bf9de84-d665-11e6-944b-e7eb37a6aa8e.

Patrikarakos, D., 'Social Media Networks Are the Handmaiden to Dangerous Propaganda' (November 2 2017), https://time.com/5008076/nyc-terror-attack-isis-facebook-russia/.

PBS-NOVA, 'CyberWar Threat' http://www.pbs.org/wgbh/nova/military/cyberwar-threat.html.

PBS-NOVA, 'CyberWar Threat', https://www.youtube.com/watch?v=DAl7Cre3BeI.

Peachey, P., 'Director of Europol: 'Top Computer Graduates Are Being Lured into Cybercrime' (December 29 2014), https://www.independent.co.uk/news/uk/crime/director-of-europol-top-computer-graduates-are-being-lured-into-cybercrime-9948990.html.

Perlroth, N., and Krauss, C., 'A Cyberattack in Saudi Arabia Had a Deadly Goal. Experts Fear Another Try' (March 15 2018), https://www.nytimes.com/2018/03/15/technology/saudi-arabia-hacks-cyberattacks.html.

Perlroth, N., and Sanger, D.E., 'Cyberattacks Put Russian Fingers on the Switch at Power Plants, U.S. Says' (March 15 2018), *New York Times*, March 15 2018. https://www.politico.com/news/2022/01/04/jan-6-insurrection-sentencing-tracker-526091

Poonam, S., 'The Scammers Gaming India's Overcrowded Job Market' (January 2 2018), https://www.theguardian.com/news/2018/jan/02/the-scammers-gaming-indias-overcrowded-job-market.

'Pro-Russian Hacking Group Hits Back at Anonymous – RT' (February 2022), https://www.rt.com/russia/551080-killnet-hackers-anonymous-retaliation-russia/.

Radio Farda, 'Iran's Intelligence Minister Boasts of Wide-Ranging Successes' (April 20 2019), https://en.radiofarda.com/a/iran-s-intelligence-minister-boasts-of-wide-ranging-successes/29892972.html.

Rahmani, M., 'Iran Passive Defense Org. Role Vital in Foiling New Threats', https://en.mehrnews.com/news/164314/Iran-Passive-Defense-Org-role-vital-in-foiling-new-threats.

Sanger, D.E., 'The Age of Cyberwar Is Here. We Can't Keep Citizens Out of the Debate', *The Guardian*, July 28 2018.

Sanger, D.E., and Barnes, J.E., 'U.S. and Britain Help Ukraine Prepare for Potential Russian Cyberassault' (December 20 2021), https://www.nytimes.com/2021/12/20/us/politics/russia-ukraine-cyberattacks.html.

Saul, J., 'Global Shipping Feels Fallout from Maersk Cyber Attack' (June 29 2017), https://www.reuters.com/article/us-cyber-attack-maersk/global-shipping-feels-fallout-from-maersk-cyber-attack-idUSKBN19K2LE.

Schechner, S., 'Ukraine's 'IT Army' Has Hundreds of Thousands of Hackers, Kyiv Says' (March 4 2022), https://www.wsj.com/livecoverage/russia-ukraine-latest-news-2022-03-04/card/ukraine-s-it-army-has-hundreds-of-thousands-of-hackers-kyiv-says-RfpGa5zmLtavrot27OWX.

Shanker, T., 'Panetta Warns of Dire Threat of Cyberattack on U.S.' (October 11 2012), http://www.nytimes.com/2012/10/12/world/panetta-warns-of-dire-threat-of-cyberattack.html?pagewanted=all&_r=0.

Siddiqui, F., 'Cyberattack on San Francisco Transit Agency Prompts Senate Questions for Metro' (January 9 2017), https://www.washingtonpost.com/news/dr-gridlock/wp/2017/01/09/cyberattack-on-san-francisco-transit-agency-prompts-senate-questions-for-metro/.

Siddiqui, F., 'System Glitch Leaves Metro's Nerve Center Unable to Control Its Tracks, Wrecking Morning Commute', *The Washington Post*, January 5 2017.

Silverman, L., 'There Are Plenty of RFID-Blocking Products, But Do You Need Them?' (July 4 2017), https://www.npr.org/sections/alltechconsidered/2017/07/04/535518514/there-are-plenty-of-rfid-blocking-products-but-do-you-need-them?t=1537186154889.

Solon, O., 'What Happens If You Play Along with a Microsoft 'Tech Support' Scam?', https://www.wired.co.uk/article/malwarebytes.

Staff, T., 'Screens at Iran Airport Said Hacked with Anti-Regime Messages' (May 25 2018), https://www.timesofisrael.com/screens-at-iran-airport-said-hacked-with-anti-regime-messages/.

Stone, M., 'Iran Claims It Has 'Missile Cities' as 6 Incidents Prompt Theories It Is Under Attack' (July 6 2020), https://news.sky.com/story/coincidence-or-attack-what-is-behind-the-six-curious-incidents-in-iran-12021907.

Szary, W., Auchard, E., 'Polish Airline, Hit by Cyber Attack, Says All Carriers Are at Risk' (June 22 2015), https://www.reuters.com/article/us-poland-lot-cybercrime-idUKKBN0P21DC20150622.

Thomas, P., Cloherty, J., and Ryan, R., 'How the FBI Busted Anna Chapman and the Russian Spy Ring' (November 1 2011), http://abcnews.go.com/blogs/politics/2011/11/how-the-fbi-busted-anna-chapman-and-the-russian-spy-ring/.

Thomson, M., 'Iranian Cyber Attack on New York Dam Shows Future of War' (March 24 2016), https://time.com/4270728/iran-cyber-attack-dam-fbi/.

Tidy, J., 'European Oil Facilities Hit by Cyber-Attacks' (February 3 2022), https://www.bbc.co.uk/news/technology-60250956.

Tidy, J., 'Ukraine: EU Deploys Cyber Rapid-Response Team (February 22 2022), https://www.bbc.co.uk/news/technology-60484979.

Tidy, J., 'Predatory Sparrow: Who Are the Hackers Who Say They Started a Fire in Iran?' (July 11 2022), https://www.bbc.co.uk/news/technology-62072480.

'Verizon's 2016 Data Breach Investigations Report Finds Cybercriminals Are Exploiting Human Nature', http://www.prnewswire.com/news-releases/verizons-2016-data-breach-investigations-report-finds-cybercriminals-are-exploiting-human-nature-300258134.html.

Volpicelli, G.M., 'Russia Is Facing a Tech Worker Exodus', https://www.wired.com/story/russian-techies-exodus-ukraine/.

Volz, D., 'Researchers Link Cyberattack on Saudi Petrochemical Plant to Russia' (October 23 2018), https://www.wsj.com/articles/u-s-researchers-link-cyberattack-on-saudi-petrochemical-plant-to-russia-1540322439.

Wakefield, J., 'Office Intruder 'Steals' Data', http://news.bbc.co.uk/1/hi/technology/7843206.stm, (May 6 2009).

Ward, M., 'How to Hack a Nation's Infrastructure' (May 20 2013), https://www.bbc.co.uk/news/technology-22524274.

Ward, M., 'XP - The Operating System That Will Not Die' (March 5 2014), http://www.bbc.co.uk/news/technology-26432473.

Ward, M., 'Why Some Computer Viruses Refuse to Die' (August 14 2018), https://www.bbc.co.uk/news/technology-44564709.

'What Are 'Rare Earths' Used for?' (March 13 2012), http://www.bbc.co.uk/news/world-17357863.
Whitehead, K., 'New BBC Series Offers a Peek into the World of Cybercrime', https://www.scmp.com/lifestyle/technology/article/1629989/new-bbc-series-offers-peek-world-cybercrime.
Winer, S., "Dutch Mole' Planted Stuxnet Virus in Iran Nuclear Site on Behalf of CIA, Mossad' (September 3 2019), https://www.timesofisrael.com/dutch-mole-planted-infamous-stuxnet-virus-in-iran-nuclear-site-report/.
Zetter, K., 'Exclusive: Comedy of Errors Led to False 'Water-Pump' Hack Report' (November 30 2011), https://www.wired.com/2011/11/water-pump-hack-mystery-solved/.
Zetter, K., 'Did a U.S. Government Lab Help Israel Develop Stuxnet?' (January 17 2011), https://www.wired.com/2011/01/inl-and-stuxnet/.
Zetter, K., 'Apple's FBI Battle Is Complicated. Here's What's Really Going On' (February 18 2016), https://www.wired.com/2016/02/apples-fbi-battle-is-complicated-heres-whats-really-going-on/.
Zetter, K., 'Teen Who Hacked CIA Director's Email Tells How He Did It' (October 19 2015), https://www.wired.com/2015/10/hacker-who-broke-into-cia-director-john-brennan-email-tells-how-he-did-it/.
Zinets, N., 'Ukraine Charges Russia with New Cyber Attacks on Infrastructure' (February 15 2017), http://www.reuters.com/article/us-ukraine-crisis-cyber/ukraine-charges-russia-with-new-cyber-attacks-on-infrastructure-idUSKBN15U2CN.
Zurcher, A., 'Hillary Clinton's 'Emailgate' Diced and Sliced' (September 19 2016), http://www.bbc.co.uk/news/world-us-canada-31806907.

Government, Intergovernmental, and Non-governmental Sources

'2019-005-Eastern Mediterranean and Red Seas-GPS Interference', https://www.maritime.dot.gov/content/2019-005-eastern-mediterranean-and-red-seas-gps-interference.
'3314 (XXIX). Definition of Aggression', https://documents-dds-ny.un.org/doc/RESOLUTION/GEN/NR0/739/16/IMG/NR073916.pdf?OpenElement.
'About ISACs', https://www.nationalisacs.org/about-isacs.
'About NCI', https://www.nationalisacs.org/about-nci.
'About Us', https://www.us-cert.gov/about-us.

Bibliography

Abrams, M., and Weiss, J., 'Malicious Control System Cyber Security Attack Case Study–Maroochy Water Services, Australia', http://csrc.nist.gov/groups/SMA/fisma/ics/documents/Maroochy-Water-Services-Case-Study_report.pdf.

'Achieve and Maintain Cyberspace Superiority Command Vision for US Cyber Command', https://www.cybercom.mil/Portals/56/Documents/USCYBERCOM%20Vision%20April%202018.pdf?ver=2018-06-14-152556-010.

'Advance Policy Questions for Lieutenant General Paul Nakasone, USA Nominee for Commander, U.S. Cyber Command and Director, National Security Agency/Chief, Central Security Service', https://www.armed-services.senate.gov/imo/media/doc/Nakasone_APQs_03-01-18.pdf.

'Alert (AA20-006A) Potential for Iranian Cyber Response to U.S. Military Strike in Baghdad' (January 6 2020), https://us-cert.cisa.gov/ncas/alerts/aa20-006a.

'Alert (AA20-106A) Guidance on the North Korean Cyber Threat' (June 23 2020), https://www.cisa.gov/uscert/ncas/alerts/aa20-106a.

'Alert (AA22-057A) Update: Destructive Malware Targeting Organizations in Ukraine' (April 28 2022), https://www.cisa.gov/uscert/ncas/alerts/aa22-057a.

'AA22-110A U.S. Support for Connectivity and Cybersecurity in Ukraine' (May 10 2022), https://www.state.gov/u-s-support-for-connectivity-and-cybersecurity-in-ukraine/.

'Alert (TA17-293A) Advanced Persistent Threat Activity Targeting Energy and Other Critical Infrastructure Sectors' (October 20 2017, updated March 15 2018), https://www.us-cert.gov/ncas/alerts/TA17-293A.

'All Groups from Iran', https://apt.thaicert.or.th/cgi-bin/listgroups.cgi. This Site Has Now Moved to https://apt.etda.or.th/cgi-bin/listgroups.cgi.

'All Groups from North Korea' (Regularly Updated), https://apt.thaicert.or.th/cgi-bin/listgroups.cgi?c=North%2520Korea&v=&s=&m=&x=.

'Annual Threat Assessment of the US Intelligence Community, Office of the Director of National Intelligence' (April 9 2021), https://www.dni.gov/files/ODNI/documents/assessments/ATA-2021-Unclassified-Report.pdf.

'Annual Threat Assessment of the US Intelligence Community, Office of the Director of National Intelligence' (February 2022), https://www.dni.gov/files/ODNI/documents/assessments/ATA-2022-Unclassified-Report.pdf.

'APT Group: Chafer, APT 39', https://apt.thaicert.or.th/cgi-bin/showcard.cgi?g=Chafer%2C%20APT%2039&n=1.

'APT Group: APT 33, Elfin, Magnallium', https://apt.thaicert.or.th/cgi-bin/showcard.cgi?g=APT%2033%2C%20Elfin%2C%20Magnallium&n=1.

'APT Group: Clever Kitten', https://apt.thaicert.or.th/cgi-bin/showcard.cgi?g=Clever%20Kitten&n=1.

'APT Group: Domestic Kitten', https://apt.thaicert.or.th/cgi-bin/showcard.cgi?g=Domestic%20Kitten&n=1.

'APT Group: Flying Kitten, Ajax Security Team', https://apt.thaicert.or.th/cgi-bin/showcard.cgi?g=Flying%20Kitten%2C%20Ajax%20Security%20Team&n=1.

'APT Group: Kimsuky, Velvet Chollima', https://apt.thaicert.or.th/cgi-bin/showcard.cgi?g=Kimsuky%2C%20Velvet%20Chollima&n=1.

'APT Group: Lazarus Group, Hidden Cobra, Labyrinth Chollima', https://apt.thaicert.or.th/cgi-bin/showcard.cgi?g=Lazarus%20Group%2C%20Hidden%20Cobra%2C%20Labyrinth%20Chollima&n=1.

'APT Group: Leafminer, Raspite', https://apt.thaicert.or.th/cgi-bin/showcard.cgi?g=Leafminer%2C%20Raspite&n=1.

'APT Group: Magic Hound, APT 35, Cobalt Gypsy, Charming Kitten', https://apt.thaicert.or.th/cgi-bin/showcard.cgi?g=Magic%20Hound%2C%20APT%2035%2C%20Cobalt%20Gypsy%2C%20Charming%20Kitten&n=1.

'APT Group: Madi', https://apt.thaicert.or.th/cgi-bin/showcard.cgi?g=Madi&n=1.

'APT Group: MalKamak', https://apt.thaicert.or.th/cgi-bin/showcard.cgi?g=MalKamak&n=1.

'APT Group: Parisite, Fox Kitten, Pioneer Kitten', https://apt.thaicert.or.th/cgi-bin/showcard.cgi?g=Parisite%2C%20Fox%20Kitten%2C%20Pioneer%20Kitten&n=1.

'APT Group: Reaper, APT 37, Ricochet Chollima, ScarCruft', https://apt.thaicert.or.th/cgi-bin/showcard.cgi?g=Reaper%2C%20APT%2037%2C%20Ricochet%20Chollima%2C%20ScarCruft&n=1.

'APT Group: Rocket Kitten, Newscaster, NewsBeef', https://apt.thaicert.or.th/cgi-bin/showcard.cgi?g=Rocket%20Kitten%2C%20Newscaster%2C%20NewsBeef&n=1.

Aring, M., 'Background Paper Prepared for the Education for All Global Monitoring Report 2012 Youth and Skills: Putting Education to Work Report on Skills Gaps' (2012), http://unesdoc.unesco.org/images/0021/002178/217874e.pdf, https://www.atlanticcouncil.org/news/transcripts/transcript-nato-head-jens-stoltenberg-on-russian-aggression-ukraines-capabilities-and-expanding-the-alliance/ (January 28 2022).

'Article 2(1)–(5)', https://legal.un.org/repertory/art2.shtml.

'Article 274.1. of the Russian Criminal Code', http://www.unodc.org/documents/organized-crime/cybercrime/cybercrime-april-2018/SUSHCHIK_Item_3.pdf.

'Audit of the Federal Bureau of Investigation's Implementation of Its Next Generation Cyber Initiative' (July 2015), https://oig.justice.gov/reports/2015/a1529.pdf.

'Automated Tackling of Disinformation Study Panel for the Future of Science and Technology European Science-Media Hub' (March 2019), http://www.europarl.europa.eu/RegData/etudes/STUD/2019/624278/EPRS_STU(2019)624278_EN.pdf.

'Autonomous Ships: Regulatory Scoping Exercise Completed' (May 25 2021), https://www.imo.org/en/MediaCentre/PressBriefings/pages/MASSRSE2021.aspx.

'Background to "Assessing the Russian Activities and Intentions in Recent US Elections": The Analytic Process and Cyber Incident Attribution' (January 6 2017), https://www.dni.gov/files/documents/ICA_2017_01.pdf.

Baram, G., and Wechsler, O., 'Cyber Threats to Space Systems Current Risks and the Role of NATO' (June 2020), https://www.japcc.org/read-aheads/joint-air-space-power-conference-2020-read-ahead/.

Broom, D., 'What Is the EU Doing to End Its Reliance on Russian Energy?' (April 26 2022), https://www.weforum.org/agenda/2022/04/europe-russia-energy-alternatives/.

'Buffer Overflow', https://www.enisa.europa.eu/topics/csirts-in-europe/glossary/buffer-overflow, accessed 21 July 2019.

Campbell, R.J., 'Cybersecurity Issues for the Bulk Power System', Congressional Research Service (June 10 2015), available from https://fas.org/sgp/crs/misc/R43989.pdf.

'Chancellor's Speech to GCHQ on Cyber Security' (November 17 2015), https://www.gov.uk/government/speeches/chancellors-speech-to-gchq-on-cyber-security, https://www.cisa.gov/.

'Christopher Wray, Director, Federal Bureau of Investigation, Examining the January 6 Attack on the U.S. Capitol', *Statement for the Record* (June 15 2021), https://www.fbi.gov/news/testimony/examining-the-january-6-attack-on-the-us-capitol-wray-061521.

'CISA, Russian State-Sponsored and Criminal Cyber Threats to Critical Infrastructure', https://www.cisa.gov/uscert/sites/default/files/publications/AA22-110A_Joint_CSA_Russian_State-Sponsored_and_Criminal_Cyber_Threats_to_Critical_Infrastructure_4_20_22_Final.pdf.

'Committee on Armed Services United States Senate Hearing to Receive Testimony on Cyber Policy, Strategy and Organization Thursday' (May 11 2017), https://www.armed-services.senate.gov/imo/media/doc/17-45_05-11-17.pdf.

'Committee on Armed Services, United States Senate, Nominations, Thursday' (March 1 2018), https://www.armed-services.senate.gov/imo/media/doc/18-19_03-01-18.pdf.

'Committee on Oversight and Government Reform U.S. House of Representatives 114th Congress The OPM Data Breach: How the Government Jeopardized Our National Security for More than a Generation' (September 7 2016), https://www.cylance.com/content/dam/cylance/pdfs/reports/The-OPM-Data-Breach-How-the-Government-Jeopardized-Our-National-Security-for-More-than-a-Generation.pdf.

'Communication from the Commission on a European Programme for Critical Infrastructure Protection' (December 12 2006), https://eur-lex.europa.eu/LexUriServ/LexUriServ.do?uri=COM:2006:0786:FIN:EN:PDF.

'Computer Security Resource Center (CSRC), Glossary', https://csrc.nist.gov/Glossary/?term=2856.

Council of Europe, 'Chart of Signatures and Ratifications of Treaty 185', https://www.coe.int/en/web/conventions/full-list/-/conventions/treaty/185/signatures?p_auth=Q4ufyD6q.

Council of Europe, 'Details of Treaty No. 185 Convention on Cybercrime', http://conventions.coe.int/Treaty/en/Treaties/Html/185.htm.

'Country Reports on Terrorism 2020: Democratic People's Republic of Korea', https://www.state.gov/reports/country-reports-on-terrorism-2020/democratic-peoples-republic-of-korea/.

'Critical Infrastructure Protection' (Undated), https://ec.europa.eu/jrc/en/research-topic/critical-infrastructure-protection.

'Critical Infrastructure Sectors', https://www.dhs.gov/critical-infrastructure-sectors, https://crrts.eu/index.html.

'"Cyber-Attack Against Ukrainian Critical Infrastructure," Industrial Control Systems Cyber Emergency Response Team' (February 25 2016), https://ics-cert.us-cert.gov/alerts/IR-ALERT-H-16-056-01, https://www.icao.int/cybersecurity/Pages/Working-Groups.aspx.

'Cyber Defence in the EU Preparing for Cyber Warfare? (October 2014), http://www.europarl.europa.eu/EPRS/EPRS-Briefing-542143-Cyber-defence-in-the-EU-FINAL.pdf.

'Cyber Primer, Ministry of Defence' (December 2013), http://www.securethecyber.uk/wp-content/uploads/2015/10/20140716_DCDC_Cyber_Primer_Internet_Secured-VERSION-TO-BE-USED.pdf (dead link).

'Cyber Primer' Second Edition (July 2016). https://www.gov.uk/government/uploads/system/uploads/attachment_data/file/549291/20160720-Cyber_Primer_ed_2_secured.pdf.

'Cybersecurity Framework', https://www.nist.gov/cyberframework.

'Cybersecurity Reference and Resource Guide 2018, Department of Defense' (August 22 2018), https://ics-cert.us-cert.gov/.

'Cyber's Most Wanted', https://www.fbi.gov/wanted/cyber.

'Cyber Threat and Vulnerability Analysis of the U.S. Electric Sector Prepared by: Mission Support Center Idaho National Laboratory' (August 2016), https://www.energy.gov/sites/prod/files/2017/01/f34/Cyber%20Threat%20and%20Vulnerability%20Analysis%20of%20the%20U.S.%20Electric%20Sector.pdf.

Daniel R. Coats, 'Statement for the Record Worldwide Threat Assessment of the US Intelligence Community' (January 29 2019), https://www.dni.gov/files/ODNI/documents/2019-ATA-SFR---SSCI.pdf.

'Deepwater Horizon – BP Gulf of Mexico Oil Spill', https://www.epa.gov/enforcement/deepwater-horizon-bp-gulf-mexico-oil-spill.

'Defence Minister Jaak Aaviksoo: Cyber Defense—The Unnoticed Third World War' (May 8 2008), http://www.kaitseministeerium.ee/en/news/defence-minister-jaak-aaviksoo-cyber-defense-unnoticed-third-world-war.

Department of Defense Defense Science Board Task Force on Cyber Deterrence (February 2017), https://www.armed-services.senate.gov/imo/media/doc/DSB%20CD%20Report%202017-02-27-17_v18_Final-Cleared%20Security%20Review.pdf, https://www.dni.gov/.

Director GCHQ's Speech on Global Security Amid War in Ukraine (March 31 2022), https://www.gchq.gov.uk/speech/director-gchq-global-security-amid-russia-invasion-of-ukraine.

DoD Cyber Strategy (April 2015), https://www.hsdl.org/?abstract&did=764848.

'Double Blow to Dark Web Market Places' (May 3 2019), https://www.europol.europa.eu/newsroom/news/double-blow-to-dark-web-marketplaces.

Dufková, A. (ed.), (ENISA), 'CERT Community Recognition Mechanisms and Schemes' (November 2013), https://www.enisa.europa.eu/publications/cert-community-recognition-mechanisms-and-schemes, https://www.easa.europa.eu/eccsa.

'EASA Engaged in a "Responsible Disclosure" of Cybersecurity Issues in Coordination with Operators and Authorities', https://www.easa.europa.eu/newsroom-and-events/news/easa-engaged-%E2%80%9Cresponsible-disclosure%E2%80%9D-cybersecurity-issues-coordination.

'Electricity Generation Statistics—First Results', https://ec.europa.eu/eurostat/statistics-explained/index.php?title=Electricity_generation_statistics_%E2%80%93_first_results.

'Electricity Production, Consumption and Market Overview', https://ec.europa.eu/eurostat/statistics-explained/index.php/Electricity_production,_consumption_and_market_overview#Electricity_generation, https://ec.europa.eu/eurostat/cache/infographs/energy/bloc-2c.html.

'Electronic Transactions Development Agency (ETDA)', https://apt.etda.or.th/cgi-bin/listgroups.cgi.

'ELINT: A Scientific Intelligence System', https://www.cia.gov/library/center-for-the-study-of-intelligence/kent-csi/vol2no1/html/v02i1a06p_0001.htm.

'Espionage', https://www.mi5.gov.uk/espionage.

'EU Energy in Figures Statistical Pocketbook 2018'. Available from https://publications.europa.eu/en/publication-detail/-/publication/99fc30eb-c06d-11e8-9893-01aa75ed71a1/language-en, https://eugdpr.org/, https://eur-lex.europa.eu/legal-content/EN/TXT/?uri=celex:32018R1139.

'Executive Order 13587—Structural Reforms to Improve the Security of Classified Networks and the Responsible Sharing and Safeguarding of Classified Information' (October 7 2016), https://www.whitehouse.gov/the-press-office/2011/10/07/executive-order-13587-structural-reforms-improve-security-classified-net.

'Executive Order on Improving the Nation's Cybersecurity' (May 12 2021), https://www.whitehouse.gov/briefing-room/presidential-actions/2021/05/12/executive-order-on-improving-the-nations-cybersecurity/.

'Federal Agencies' Reliance on Outdated and Unsupported Information Technology: A Ticking Time Bomb' (May 25 2016), https://oversight.house.gov/hearing/federal-agencies-reliance-on-outdated-and-unsupported-information-technology-a-ticking-time-bomb/.

'Federal Cybersecurity Risk Determination Report and Action Plan' (May 2018), https://www.whitehouse.gov/wp-content/uploads/2018/05/Cybersecurity-Risk-Determination-Report-FINAL_May-2018-Release.pdf.

'Four Russian Government Employees Charged in Two Historical Hacking Campaigns Targeting Critical Infrastructure Worldwide' (March 24 2022), https://www.justice.gov/opa/pr/four-russian-government-employees-charged-two-historical-hacking-campaigns-targeting-critical.

'GAO-19–471 Report to Congressional Requesters Information Technology Agencies Need to Develop Modernization Plans for Critical Legacy Systems' (June 2019), https://www.gao.gov/assets/700/699616.pdf.

'Global Internet Forum to Counter Terrorism (GIFCT)', https://gifct.org/.

'Global Shifts in the Energy System' (November 14 2017), https://www.iea.org/weo2017/.

Gracia, J.R., O'Connor, P.W., Markel, L.C., Shan, R., Rizy, D.T., and Tarditi, A., 'Hydropower Plants as Black Start Resources' (May 2019), https://www.energy.gov/sites/prod/files/2019/05/f62/Hydro-Black-Start_May2019.pdf.

'Gulf of Mexico Fact Sheet', https://www.eia.gov/special/gulf_of_mexico/.

'Hearing of the House (Select) Intelligence Committee Subject: "Cybersecurity Threats: The Way Forward"' (November 20 2014), https://www.nsa.gov/news-features/speeches-testimonies/Article/1620360/hearing-of-the-house-select-intelligence-committee-subject-cybersecurity-threat/.

Hess, A., 'Executive Assistant Director, Science and Technology Branch, Federal Bureau of Investigation, Statement Before the House Committee on Energy and Commerce, Subcommittee on Oversight and Investigation' (April 19 2016), https://www.fbi.gov/news/testimony/deciphering-the-debate-over-encryption.

'Hexane', https://apt.thaicert.or.th/cgi-bin/showcard.cgi?g=Hexane&n=1.

'Homeland Security Public Service Announcement, Increase in Insider Threat Cases Highlight Significant Risks to Business Networks and Proprietary Information' (September 23 2014), https://www.ic3.gov/media/2014/140923.aspx.

'Homeland Threat Assessment October 2020', https://www.dhs.gov/sites/default/files/publications/2020_10_06_homeland-threat-assessment.pdf.

'How Many Nuclear Power Plants Are in the United States, and Where Are They Located?', https://www.eia.gov/tools/faqs/faq.php?id=207&t=3.

'H.R.3202—Cyber Vulnerability Disclosure Reporting Act', https://www.congress.gov/bill/115th-congress/house-bill/3202/text.

'H.R.5220—Cyber Act of War Act of 2016114th Congress (2015–2016)', https://www.congress.gov/bill/114th-congress/house-bill/5220, https://www.icc-cpi.int/ukraine.

'ICSB-11-327-01—Illinois Water Pump Failure Report', https://www.us-cert.gov/ics/tips/ICSB-11-327-01.

'Implementing Application Whitelisting', https://www.cert.govt.nz/it-specialists/critical-controls/application-whitelisting/implementing-application-whitelisting/.

'Industrial Control Systems', https://www.us-cert.gov/ics.

'Information Sharing and Analysis Centers', https://www.enisa.europa.eu/topics/national-cyber-security-strategies/information-sharing.

'Information Sharing and Analysis Centres (ISACs) Cooperative Models', https://www.enisa.europa.eu/publications/information-sharing-and-analysis-center-isacs-cooperative-models.

'Inside the FBI: Director Comey Addresses Cyber Security Experts' (September 2 2016), https://www.fbi.gov/audio-repository/inside-podcast-comey-cyber-speech-090216.mp3/view.

'Insider Threat—Cyber', https://www.dhs.gov/insider-threat-cyber.

'International Cyber Crime Iranians Charged with Hacking U.S. Financial Sector' (March 24 2016), https://www.fbi.gov/news/stories/iranians-charged-with-hacking-us-financial-sector.

'International Law and Justice', https://www.un.org/en/sections/issues-depth/international-law-and-justice/.

INTERPOL, 'Global Landscape on Covid-19 Cyberthreat', https://www.interpol.int/Crimes/Cybercrime/COVID-19-cyberthreats.

'IOCTA Internet Organized Crime Threat Assessment 2019', https://www.europol.europa.eu/iocta-report.

'IOCTA Internet Organized Crime Threat Assessment 2020', https://www.europol.europa.eu/activities-services/main-reports/internet-organised-crime-threat-assessment-iocta-2020, https://www.isac.io/.

'IOCTA Internet Organized Crime Threat Assessment 2021', https://www.europol.europa.eu/cms/sites/default/files/documents/internet_organised_crime_threat_assessment_iocta_2021.pdf.

'Iran Cyber Threat Overview and Advisories', https://us-cert.cisa.gov/iran.

'ISACA Produces New Audit Program Based on NIST Framework' (January 10 2017), http://www.isaca.org/About-ISACA/Press-room/News-Releases/2017/Pages/ISACA-Produces-New-Audit-Program-Based-on-NIST-Framework.aspx.

'ISIL-Linked Kosovo Hacker Sentenced to 20 Years in Prison' (September 23 2016), https://www.justice.gov/opa/pr/isil-linked-kosovo-hacker-sentenced-20-years-prison, https://www.justice.gov/opa/press-release/file/1328521/download.

'Joint Cybersecurity Advisory Russian State-Sponsored and Criminal Cyber Threats to Critical Infrastructure' (April 20 2022), https://www.cisa.gov/uscert/sites/default/files/publications/AA22-110A_Joint_CSA_Russian_State-Sponsored_and_Criminal_Cyber_Threats_to_Critical_Infrastructure_4_20_22_Final.pdf.

'Joint Publication 3–12 Cyberspace Operations Joint Chiefs of Staff' (June 8 2018), https://www.jcs.mil/Portals/36/Documents/Doctrine/pubs/jp3_12.pdf?ver=2018-07-16-134954-150.

Kaskina, B., Taurins, E., and Dufkova, A. (ENISA), 'CSIRT Capabilities How to Assess Maturity? Guidelines for National and Governmental CSIRTs' (December 2015), https://www.enisa.europa.eu/publications/csirt-capabilities/at_download/fullReport.

'Key Trends Across a Maturing Cyberspace Affecting U.S. and China Future Influences in a Rising Deeply Cybered, Conflictual, and Post-Western World Dr. Chris C. Demchak Testimony Before Hearing on China's Information Controls, Global Media Influence, and Cyber Warfare Strategy Panel 3: Beijing's Views on Norms in Cyberspace and China's Cyber Warfare Strategy U.S.-China Economic and Security Review Commission Washington, DC' (May 4 2017), https://www.uscc.gov/sites/default/files/Chris%20Demchak%20May%204th%202017%20USCC%20testimony.pdf.

'Leadership', http://www.dia.mil/About/Leadership/, www.levesoninquiry.org.uk/wp.../Witness-Statement-of-Ian-Hurst.pdf

'Locked Shields', https://ccdcoe.org/exercises/locked-shields/.

'MI6 'C' Speech on Fourth Generation Espionage' (December 2018), https://www.gov.uk/government/speeches/mi6-c-speech-on-fourth-generation-espionage.

'Member ISACS', https://www.nationalisacs.org/member-isacs.

'National Counterintelligence Strategy of the United States of America 2020–2022', https://www.dni.gov/files/NCSC/documents/features/20200205-National_CI_Strategy_2020_2022.pdf.

'National Cyber Security Centre a Part of GCHQ', *Annual Review 2018 Making the UK the Safest Place to Live and Work Online*, https://www.ncsc.gov.uk/annual-review-2018/docs/ncsc_2018-annual-review.pdf.

'National Cybersecurity and Communications Integration Center', https://www.dhs.gov/cisa/national-cybersecurity-communications-integration-center.

'National Cyber Strategy of the United States of America' (September 2018), https://www.whitehouse.gov/wp-content/uploads/2018/09/National-Cyber-Strategy.pdf.

National Security Agency, 'Embracing a Zero Trust Security Model' (February 2021), https://media.defense.gov/2021/Feb/25/2002588479/-1/-1/0/CSI_EMBRACING_ZT_SECURITY_MODEL_UOO115131-21.PDF.

'National Security Strategy of the United States of America' (December 2017), https://www.whitehouse.gov/wp-content/uploads/2017/12/NSS-Final-12-18-2017-0905.pdf.

'NATO Will Defend Itself' (August 27 2022), https://www.nato.int/cps/en/natohq/news_168435.htm?selectedLocale=en.

'NCSC Mission, Vision, Goals', https://www.dni.gov/index.php/ncsc-who-we-are/ncsc-mission-vision.

'New Cyber Attack Categorisation System to Improve UK Response to Incidents' (April 12 2018), https://www.ncsc.gov.uk/news/new-cyber-attack-categorisation-system-improve-uk-response-incidents.

'NIST Releases Version 1.1 of Its Popular Cybersecurity Framework' (April 16 2018), https://www.nist.gov/news-events/news/2018/04/nist-releases-version-11-its-popular-cybersecurity-framework.

'North American Cooperation on Energy Information (NACEI) North American Infrastructure Map' (actively updated), http://nacei.org/#!/maps.

'North Korean Cyber Activity' (March 25 2021), https://www.hhs.gov/sites/default/files/dprk-cyber-espionage.pdf.

'North Korean Regime-Backed Programmer Charged', https://www.justice.gov/opa/pr/north-korean-regime-backed-programmer-charged-conspiracy-conduct-multiple-cyber-attacks-and.

'Nuclear Energy', https://ec.europa.eu/energy/en/topics/nuclear-energy.

'OECD Employment Outlook 2019', https://www.oecd.org/employment/outlook/, https://www.osha.gov/etools/electric-power/illustrated-glossary/substation.

'Presidential Executive Order on Strengthening the Cybersecurity of Federal Networks and Critical Infrastructure' (May 11 2017), https://www.whitehouse.gov/presidential-actions/presidential-executive-order-strengthening-cybersecurity-federal-networks-critical-infrastructure/.

Presidential Policy Directive/PPD-20, 'U.S. Cyber Operations Policy'. Available from https://fas.org/irp/offdocs/ppd/ppd-20.pdf.

Presidential Policy Directive 21 (PPD-21), 'Critical Infrastructure Security and Resilience', signed by President Obama in February 2013, https://obamawhitehouse.archives.gov/the-press-office/2013/02/12/presidential-policy-directive-critical-infrastructure-security-and-resil.

Prucková, M., 'Cyber Attacks and Article 5—A Note on a Blurry But Consistent Position of NATO' (Undated, 2022), https://ccdcoe.org/incyder-articles/cyber-attacks-and-article-5-a-note-on-a-blurry-but-consistent-position-of-nato/.

'Public Law No: 115-278 (11/16/2018) Cybersecurity and Infrastructure Security Agency Act of 2018', https://www.congress.gov/bill/115th-congress/house-bill/3359.

'Reference Number: JAR-16-20296A December 29, 2016 Grizzly Steppe – Russian Malicious Cyber Activity', https://www.us-cert.gov/sites/default/files/publications/JAR_16-20296A_GRIZZLY%20STEPPE-2016-1229.pdf.

'Remarks as delivered by The Honorable James R. Clapper Director of National Intelligence "National Intelligence, North Korea, and the National Cyber Discussion" International Conference on Cyber Security Fordham University' (January 7 2015), https://www.dni.gov/index.php/newsroom/speeches-interviews/speeches-interviews-2015/item/1156-remarks-as-delivered-by-dni-james-r-clapper-on-national-intelligence-north-korea-and-the-national-cyber-discussion-at-the-international-conference-on-cyber-security.

'Remarks by Secretary Esper at the Air Force Association's 2019 Air, Space & Cyber Conference, National Harbor, Maryland' (September 18 2019), https://www.defense.gov/Newsroom/Transcripts/Transcript/Article/1964448/remarks-by-secretary-esper-at-the-air-force-associations-2019-air-space-cyber-c/.

'Report of the Panel of Experts Established Pursuant to Resolution 1874 (2019)', https://www.securitycouncilreport.org/atf/cf/%7B65BFCF9B-6D27-4E9C-8CD3-CF6E4FF96FF9%7D/S_2019_691.pdf.

'Roadmap to Secure Control Systems in the Water Sector (March 2008) Developed by March 2008 Water Sector Coordinating Council Cyber Security Working Group Sponsored by American Water Works Association/Department for Homeland Security', https://www.n-dimension.com/wp-content/uploads/NDSI-WATER-CybersecurityRoadmap08-1.pdf.

Robert S. Mueller, III, 'Report On The Investigation into Russian Interference in the 2016 Presidential Election Volume I of II' (March 2019), https://www.justice.gov/storage/report.pdf.

'Rostec and the Federal Security Service of Russia Signed an Information Security Agreement', https://rostec.ru/en/news/rostec-and-the-federal-security-service-of-russia-signed-an-information-security-agreement/.

'Russia Behind Cyber-Attack with Europe-Wide Impact an Hour Before Ukraine Invasion' (May 10 2022), https://www.gov.uk/government/news/russia-behind-cyber-attack-with-europe-wide-impact-an-hour-before-ukraine-invasion, https://www.ukcybersecuritycouncil.org.uk/.

Scarfone, K., and Mell, P., 'Guide to Intrusion Detection and Prevention Systems (IDPS) (Draft) Recommendations of the National Institute of

Standards and Technology, Special Publication 800–94 Revision 1 (Draft)' (July 2012), https://csrc.nist.gov/csrc/media/publications/sp/800-94/rev-1/draft/documents/draft_sp800-94-rev1.pdf.

Shea, D.A., 'Critical Infrastructure: Control Systems and the Terrorist Threat', Congressional Research Service (February 21 2003). Available from http://fas.org/irp/crs/RL31534.pdf.

'Skills for Jobs', https://www.oecdskillsforjobsdatabase.org/data/Skills%20SfJ_PDF%20for%20WEBSITE%20final.pdf.

'Statement of Admiral Michael S. Rogers Commander United States Cyber Command Before the Senate Committee on Armed Services' (May 9 2017), https://www.armed-services.senate.gov/imo/media/doc/Rogers_05-09-17.pdf.

'Status of Iran's Nuclear Programme in Relation to the Joint Plan of Action Report by the Director General' (July 20 2015), https://www.iaea.org/sites/default/files/gov-inf-2015-15.pdf.

Stoltenberg, J., 'Stoltenberg Provides Details of NATO's Cyber Policy' (May 16 2018), https://www.atlanticcouncil.org/blogs/natosource/stoltenberg-provides-details-of-nato-s-cyber-policy/.

Stouffer, K., Pillitteri, V., Lightman, S., Abrams, M., and Hahn, A., 'NIST Special Publication 800–82 Revision 2 Guide to Industrial Control Systems (ICS) Security', https://nvlpubs.nist.gov/nistpubs/specialpublications/nist.sp.800-82r2.pdf.

'Tactics, Techniques and Procedures (TTPs)', https://csrc.nist.gov/glossary/term/Tactics-Techniques-and-Procedures.

'Tallinn Manual 2.0', https://ccdcoe.org/research/tallinn-manual/.

'Tech Against Terrorism', https://www.techagainstterrorism.org/.

'The 9/11 Commission Report', https://www.9-11commission.gov/report/911Report.pdf.

'The Directive on Security of Network and Information Systems (NIS Directive)', https://ec.europa.eu/digital-single-market/en/network-and-information-security-nis-directive.

'The EU Cybersecurity Act Brings a Strong Agency for Cybersecurity and EU-Wide Rules on Cybersecurity Certification' (June 26 2019), https://ec.europa.eu/digital-single-market/en/news/eu-cybersecurity-act-brings-strong-agency-cybersecurity-and-eu-wide-rules-cybersecurity.

'The National Cyber Security Centre', https://www.ncsc.gov.uk/.

Theohary, C.A., 'Iranian Offensive Cyberattack Capabilities' (January 13 2020), https://crsreports.congress.gov/product/pdf/IF/IF11406.

'The Price Is Rights: The Violation of the Right to an Adequate Standard of Living in the Democratic People's Republic of Korea' (May 2019), https://www.ohchr.org/Documents/Countries/KP/ThePriceIsRights_EN.pdf.

'The Tallinn Manual', https://ccdcoe.org/research/tallinn-manual/.

'The United States Department of Justice Special Counsel's Office', https://www.justice.gov/sco.

'Threat Group Cards: A Threat Actor Encyclopedia', https://apt.etda.or.th/cgi-bin/aptgroups.cgi.

'Three North Korean Military Hackers Indicted in Wide-Ranging Scheme to Commit Cyberattacks and Financial Crimes Across the Globe', https://www.justice.gov/opa/pr/three-north-korean-military-hackers-indicted-wide-ranging-scheme-commit-cyberattacks-and.

Tibbetts, J., 'Quantum Computing and Cryptography: Analysis, Risks, and Recommendations for Decisionmakers', Center for Global Security Research, Lawrence Livermore National Laboratory (August 2019), https://cgsr.llnl.gov/content/assets/docs/QuantumComputingandCryptography-20190920.pdf.

'Treasury Sanctions North Korean State-Sponsored Malicious Cyber Groups' (September 13 2019), https://home.treasury.gov/news/press-releases/sm774.

'(U//FOUO) U.S. Army Intelligence Analysis Manual' (November 19 2017), https://publicintelligence.net/us-army-intelligence-analysis/.

'U.S. Researchers Simulate Compact Fusion Power Plant Concept' (September 21 2021), https://www.energy.gov/science/fes/articles/us-researchers-simulate-compact-fusion-power-plant-concept.

'United States Senate, Cyber Posture of the Services' (Tuesday March 13 2018), https://www.armed-services.senate.gov/imo/media/doc/18-25_03-13-18.pdf.

'Uranium and Nuclear Power Facts', https://www.nrcan.gc.ca/energy/facts/uranium/20070.

US-CERT, 'Alert (TA17–293A) Advanced Persistent Threat Activity Targeting Energy and Other Critical Infrastructure Sectors' (October 23 2017), https://www.us-cert.gov/ncas/alerts/TA17-293A, https://uscode.house.gov/statutes/pl/115/278.pdf

'U.S. Charges Russian GRU Officers with International Hacking and Related Influence and Disinformation Operations', https://www.justice.gov/opa/pr/us-charges-russian-gru-officers-international-hacking-and-related-influence-and.

'U.S. Department of Energy The Water-Energy Nexus: Challenges and Opportunities' (June 2014), https://www.energy.gov/sites/prod/files/2014/07/f17/Water%20Energy%20Nexus%20Full%20Report%20July%202014.pdf.

'US Energy Information Administration' http://www.eia.gov/pub/oil_gas/natural_gas/data_publications/crude_oil_natural_gas_reserves/current/pdf/gomwaterdepth.pdf (Reproduced as Public Domain) from https://commons.wikimedia.org/w/index.php?curid=29929461.

US Government Accountability Office (GAO), 'Report to Congressional Requesters, FAA Needs to Address Weaknesses in Air Traffic Control Systems' (January 2015), https://www.gao.gov/assets/670/668169.pdf.

'Warsaw Summit Communiqué, Issued by the Heads of State and Government Participating in the Meeting of the North Atlantic Council in Warsaw' (July 8–9 2016), https://www.nato.int/cps/en/natohq/official_texts_133169.htm#cyber.

'WASD Continues to Monitor Cyber-Threat Assessments, While Ensuring Safe Delivery of Drinking Water' (February 9 2021), https://www.miamidade.gov/releases/2021-02-09-WASD-cyber-security-finalone.asp.

'What Is Electricity', https://www.nrcan.gc.ca/energy/facts/electricity/20068

'What Is U.S. Electricity Generation by Energy Source?', https://www.eia.gov/tools/faqs/faq.php?id=427&t=3, https://www.whitehouse.gov/state-of-the-union-2022/.

'Window of Exposure… A Real Problem for SCADA Systems? Recommendations for Europe on SCADA Patching', ENISA (December 2013), https://www.enisa.europa.eu/publications/window-of-exposure-a-real-problem-for-scada-systems.

'World's Largest International Live-Fire Cyber Exercise Launches in Tallinn' (Undated 2022), https://ccdcoe.org/news/2022/locked-shields-2022-exercise-to-be-launched-next-week/.

'Worldwide Threat Assessment of the US Intelligence Community Daniel R. Coats' (February 13 2018), https://www.dni.gov/files/documents/Newsroom/Testimonies/2018-ATA---Unclassified-SSCI.pdf.

'Worldwide Threat Assessment of the US Intelligence Community Daniel R. Coats Director of National Intelligence Senate Select Committee on Intelligence' (January 29 2019), https://www.dni.gov/files/ODNI/documents/2019-ATA-SFR---SSCI.pdf.

Think Tank and Non-governmental Organization Reports

Anderson, C., and Sadjadpour, K., Iran's Cyber Threat: Espionage, Sabotage, and Revenge' (2018), https://carnegieendowment.org/files/Iran_Cyber_Final_Full_v2.pdf.

Andrew, R.B., 'State-Sponsored Militias Are Coming to a Server Near You' (2013), *Foreign Policy*, http://foreignpolicy.com/2013/02/12/cyber-gang-warfare/.

Anonymous, 'How Does China's First Aircraft Carrier Stack Up?' https://chinapower.csis.org/aircraft-carrier/.

Azadi, P., Mirramezani, M., Mesgaran, M.B., 'Migration and Brain Drain from Iran', Working Paper No. 9 April 2020, Stanford Iran 2040 Project, Stanford University, https://stanford.app.box.com/s/zv18ed560o38q0sefkxx4leikz5q2cpw.

Baezner, M., 'Hotspot Analysis: Iranian Cyber-Activities in the Context of Regional Rivalries and International Tensions' (May 2019), https://www.research-collection.ethz.ch/bitstream/handle/20.500.11850/344841/1/20190507_MB_HS_IRNV1_rev.pdf.

Baylon, C., with Brunt, R., and Livingstone, D., 'Cyber Security at Civil Nuclear Facilities Understanding the Risks' (September 2015), https://www.chathamhouse.org/sites/default/files/field/field_document/20151005CyberSecurityNuclearBaylonBruntLivingstoneUpdate.pdf.

Bermudez Jr., J.S., 38 North Special Report: A New Emphasis on Operations Against South Korea', https://38north.org/wp-content/uploads/2010/06/38north_SR_Bermudez.pdf.

Buchanan, B., and Sulmeyer, M., 'Russia and Cyber Operations: Challenges and Opportunities for the Next U.S. Administration' (December 13 2016), https://carnegieendowment.org/2016/12/13/russia-and-cyber-operations-challenges-and-opportunities-for-next-u.s.-administration-pub-66433.

Buxton, J., and Bingham, T., 'The Rise and Challenge of Dark Net Drug Markets', Policy Brief 7 (January 2015), Swansea University, https://core.ac.uk/download/pdf/34722885.pdf

Clemente, D., 'Cyber Security and Global Interdependence: What Is Critical?', https://www.chathamhouse.org/publications/papers/view/189645 (February 2013).

Cohn, E., Listek, C., Schuller, H., and Travis, C., 'Reforming the U.S. Nuclear Enterprise to Account for Emerging Cyber Threats', Report prepared for the Nuclear Threat Initiative (May 2022).

Cornish, P. et al., 'Cyber Security and the UK's Critical National Infrastructure: A Chatham House Report' (September 2011), https://www.chathamhouse.org/sites/default/files/public/Research/International%20Security/r0911cyber.pdf.
'"Covert Action" Memo in Report Confrontation or Collaboration? Congress and the Intelligence Community' (July 2009), http://belfercenter.ksg.harvard.edu/publication/19149/covert_action.html.
Crumpler, W., and Lewis, J.A., 'The Cybersecurity Workforce Gap' (January 2019), https://csis-website-prod.s3.amazonaws.com/s3fs-public/publication/190129_Crumpler_Cybersecurity_FINAL.pdf.
'Customary Law', https://www.icrc.org/en/war-and-law/treaties-customary-law/customary-law.
'Cyber Operations Tracker', https://www.cfr.org/interactive/cyber-operations.
Farley, R., 'Does the US Navy Have 10 or 19 Aircraft Carriers?' (April 17 2014), https://thediplomat.com/2014/04/does-the-us-navy-have-10-or-19-aircraft-carriers/.
Farnsworth, T., 'China and Russia Submit Cyber Proposal', *Arms Control Today*, November 2 2011, https://www.armscontrol.org/act/2011_11/China_and_Russia_Submit_Cyber_Proposal.
Feakin, T., 'Enter the Cyber Dragon Understanding Chinese Intelligence Agencies' Cyber Capabilities', Australian Strategic Policy Institute Special Report, Issue 50 (June 2013), https://www.aspi.org.au/report/special-report-enter-cyber-dragon-understanding-chinese-intelligence-agencies-cyber.
Foxall, A., 'Putin's Cyberwar: Russia's Statecraft in the Fifth Domain', Russia Studies Centre Policy Paper No. 9 (2016).
'Gen. Kim Yong Chol', http://www.nkleadershipwatch.org/leadership-biographies/lt-gen-kim-yong-chol/.
Giles, K., 'The Next Phase of Russian Information Warfare', https://www.stratcomcoe.org/next-phase-russian-information-warfare-keir-giles.
Groll, E., 'Cyberattack Targets Safety System at Saudi Aramco' (December 21 2017), https://foreignpolicy.com/2017/12/21/cyber-attack-targets-safety-system-at-saudi-aramco/.
Hollywood, J.S. et al., 'Using Social Media and Social Network Analysis in Law Enforcement Creating a Research Agenda, Including Business Cases, Protections, and Technology Needs' (2018), https://www.rand.org/pubs/research_reports/RR2301.html?adbid=1024409625238491136&adbpl=tw&adbpr=22545453&adbsc=social_20180731_2462281.
Joske, A., 'Picking Flowers, Making Honey: The Chinese Military's Collaboration with Foreign Universities', ASPI International Cyber Policy Centre,

Policy Brief Report No. 10/2018, https://s3-ap-southeast-2.amazonaws.com/ad-aspi/2018-10/Picking%20flowers%2C%20making%20honey_0.pdf?H5sGNaWXqMgTG_2F2yZTQwDw6OyNfH.u.

Jun, J., LaFoy, S., and Sohn, E., 'North Korea's Cyber Operations Strategy and Responses' (December 2015), https://www.csis.org/analysis/north-korea%E2%80%99s-cyber-operations.

Kaushal, M., and Tyle, S., 'The Blockchain: What It Is and Why It Matters' (January 13 2015), https://www.brookings.edu/blog/techtank/2015/01/13/the-blockchain-what-it-is-and-why-it-matters/.

Khoshnood, A., 'The Attack on Natanz and the JCPOA', BESA Center Perspectives Paper No. 1,997, April 14, 2021, https://besacenter.org/the-attack-on-natanz-and-the-jcpoa/.

Knake, R., 'A Cyberattack on the U.S. Power Grid Contingency Planning Memorandum No. 31' (April 3 2018), https://www.cfr.org/report/cyberattack-us-power-grid.

Lin, H., 'What Would Be a Sufficiently Strong Response to Russian Hacking of the U.S. Election?' (December 31 3016), https://www.lawfareblog.com/what-would-be-sufficiently-strong-response-russian-hacking-us-election.

Marczak, B., Scott-Railton, J., McKune, S., Razzak, B.A., and Deibert, R., 'Hide and Seek Tracking NSO Group's Pegasus Spyware to Operations in 45 Countries' (September 18 2018), The Citizen Lab, Munk School of Global Affairs, University of Toronto, https://tspace.library.utoronto.ca/bitstream/1807/95391/1/Report%23113--hide%20and%20seek.pdf.

Min-Seok, K., 'The State of the North Korean Military (March 2020), https://carnegieendowment.org/2020/03/18/state-of-north-korean-military-pub-81232.

'Overview', https://www.infrastructurereportcard.org/cat-item/energy/ (2017).

Rácz, A., 'Russia's Hybrid War in Ukraine: Breaking the Enemy's Ability to Resist', Finnish Institute of International Affairs (2015), available at http://www.fiia.fi/en/publication/514/russia_s_hybrid_war_in_ukraine/.

Recorded Future/Insikt Group, 'Illegal Activities Endure on China's Dark Web Despite Strict Internet Control' (October 5 2021), https://go.recordedfuture.com/hubfs/reports/cta-2021-1005.pdf.

'Risks from Artificial Intelligence', https://www.cser.ac.uk/research/risks-from-artificial-intelligence/ (regularly updated).

Sayfayn, N., and Madnick, S., 'Cybersafety Analysis of the Maroochy Shire Sewage Spill', Working Paper CISL# 2017–09 May 2017, http://web.mit.edu/smadnick/www/wp/2017-09.pdf.

Schmitt, M., 'Expert Backgrounder: NATO Response Options to Potential Russia Cyber Attacks Understanding the Legal Framework' (February 24 2022), https://www.justsecurity.org/80347/expert-backgrounder-nato-response-options-to-potential-russia-cyber-attacks/.

Sjursen, D., 'America Is Addicted to Fighting Undeclared Wars' (May 7 2017), https://nationalinterest.org/feature/america-addicted-fighting-undeclared-wars-20535.

Stoutland, P.O., and Pitts-Kiefer, S. (Foreword by Ernest J. Moniz, Sam Nunn, and Des Browne), 'Nuclear Weapons in the New Cyber Age Report of the Cyber Nuclear Weapons Study Group' (September 2018), https://media.nti.org/documents/Cyber_report_finalsmall.pdf.

Sukhankin, S., 'The FSB: A Formidable Player in Russia's Information Security Domain' (March 27 2018), https://jamestown.org/program/fsb-formidable-player-russias-information-security-domain/.

Sulmeyer, M., 'Military Set for Cyber Attacks on Foreign Infrastructure' (April 11 2018), https://www.belfercenter.org/publication/military-set-cyber-attacks-foreign-infrastructure.

Turovsky, D., 'Moscow's Cyber-Defense' and 'Cyberwellness Profile for Russian Federation', https://www.itu.int/en/ITU-D/Cybersecurity/Documents/Country_Profiles/Russia.pdf.

Von Behr, I., Reding, A., Edwards, C., and Gribbon, L., 'Radicalisation in the Digital Era: The Use of the Internet in 15 Cases of Terrorism and Extremism', RAND Europe (2013), available from https://www.rand.org/pubs/research_reports/RR453.html.

Wechsler, O., 'The Iran-Russia Cyber Agreement and U.S. Strategy in the Middle East' (March 14 2021), https://www.cfr.org/blog/iran-russia-cyber-agreement-and-us-strategy-middle-east.

Weeks, J., 'U.S. Electrical Grid Undergoes Massive Transition to Connect to Renewables' (April 28 2010), https://www.scientificamerican.com/article/what-is-the-smart-grid/.

'What is International Humanitarian Law?', https://www.icrc.org/en/doc/assets/files/other/what_is_ihl.pdf.

Industry Reports

'2016 Internet Security Threat Report', https://www.symantec.com/en/uk/security-center/threat-report.

'Adversary Labyrinth Chollima' (Undated), https://adversary.crowdstrike.com/en-US/adversary/labyrinth-chollima/.

'APT37 (Reaper): The Overlooked North Korean Actor' (February 20 2018), https://www2.fireeye.com/rs/848-DID-242/images/rpt_APT37.pdf.

Bellamy III, W., 'New Eurocontrol Data Shows Airlines Increasingly Becoming Targets for Cyber Attacks' (July 12 2021), https://www.aviationtoday.com/2021/07/12/new-eurocontrol-data-shows-airlines-increasingly-becoming-targets-cyber-attacks/.

Biasini, N., Chen, M., Karkins, A., Khodjibaev, A., Neal, C., and Olney, M., with contributions from Korzhevin, D., 'Ukraine Campaign Delivers Defacement and Wipers, in Continued Escalation' (January 21 2022), https://blog.talosintelligence.com/2022/01/ukraine-campaign-delivers-defacement.html.

Booz Allen Hamilton, 'Bearing Witness: Uncovering the Logic Behind Russian Military Cyber Operations' (2020), https://www.boozallen.com/content/dam/boozallen_site/ccg/pdf/publications/bearing-witness-uncovering-the-logic-behind-russian-military-cyber-operations-2020.pdf.

'Cisco Annual Internet Report (2018–2023) White Paper', https://www.cisco.com/c/en/us/solutions/collateral/executive-perspectives/annual-internet-report/white-paper-c11-741490.html.

'Cisco White Paper, Cisco Advanced Malware Protection Sandboxing Capabilities', https://www.cisco.com/c/en/us/products/collateral/security/whitepaper_c78-733277.pdf.

'Crashoverride Analysis of the Threat to Electric Grid Operations', https://dragos.com/blog/crashoverride/CrashOverride-01.pdf.

CrowdStrike, '2018 Global Threat Report Blurring: The Lines Between Statecraft and Tradecraft'. https://www.crowdstrike.com/resources/reports/2018-crowdstrike-global-threat-report-blurring-the-lines-between-statecraft-and-tradecraft/.

CrowdStrike, 'Observations from the Front Lines of Threat Hunting' (October 2018), https://go.crowdstrike.com/rs/281-OBQ-266/images/Report2018OverwatchReport.pdf.

'CrowdStrike Global Threat Report', https://www.crowdstrike.com/resources/reports/2018-crowdstrike-global-threat-report-blurring-the-lines-between-statecraft-and-tradecraft/.

'CrowdStrike Report Reveals Cyber Intrusion Trends from Elite Team of Threat Hunters' (October 9 2018), https://www.crowdstrike.com/resources/news/crowdstrike-report-reveals-cyber-intrusion-trends-from-elite-team-of-threat-hunters/.

'Destructive Malware Targeting Ukrainian Organizations' (January 15 2022), https://www.microsoft.com/security/blog/2022/01/15/destructive-malware-targeting-ukrainian-organizations/.

De Zan, T., d'Amore, F., and Di Camillo, F., 'The Defence of Civilian Air Traffic Systems from Cyber Threats' (December 2015), http://www.iai.it/sites/default/files/iai1523e.pdf.

'Digital Technology and the War in Ukraine' (February 28 2022), https://blogs.microsoft.com/on-the-issues/2022/02/28/ukraine-russia-digital-war-cyberattacks/.

'Disrupting Cyberattacks Targeting Ukraine' (April 7 2022), https://blogs.microsoft.com/on-the-issues/2022/04/07/cyberattacks-ukraine-strontium-russia/.

Dunham, K., and Melnick, J., '"Wicked Rose" and the NCPH Hacking Group', https://krebsonsecurity.com/wp-content/uploads/2012/11/Wicked Rose_andNCPH.pdf.

'Energetic Bear—Crouching Yeti Kaspersky Lab Global Research and Analysis Team', https://media.kasperskycontenthub.com/wp-content/uploads/sites/43/2018/03/08080817/EB-YetiJuly2014-Public.pdf.

FireEye and Mandiant, 'Cybersecurity's Maginot Line: A Real-World Assessment of the Defense-in-Depth Model', http://www2.fireeye.com/rs/fireye/images/fireeye-real-world-assessment.pdf.

'Full Discloser of Andariel, A Subgroup of Lazarus Threat Group' (June 23 2018), https://global.ahnlab.com/global/upload/download/techreport/%5BAhnLab%5DAndariel_a_Subgroup_of_Lazarus%20(3).pdf.

Giani, A., Bitar, E., Garcia, M., McQueen, M., Khargonekar, P., and Poolla, K., 'Smart Grid Data Integrity Attacks: Characterizations and Countermeasures', Second International Conference on Smart Grid Communications (October 2011). Available from http://www5vip.inl.gov/technicalpublications/Documents/5250261.pdf.

Global Energy Cyberattacks: "Night Dragon" (February 10 2011), https://securingtomorrow.mcafee.com/wp-content/uploads/2011/02/McAfee_NightDragon_wp_draft_to_customersv1-1.pdf.

Haller, S., and Magerkurth, C., 'The Real-Time Enterprise: IoT-Enabled Business Processes', IETF IAB Workshop on Interconnecting Smart Objects with the Internet (March 2011), https://www.iab.org/wp-content/IAB-uploads/2011/03/Haller.pdf.

Hichman, R., 'Conti Ransomware Gang: An Overview', https://unit42.paloaltonetworks.com/conti-ransomware-gang/.

'Industroyer2: Industroyer Reloaded' (April 12 2022), https://www.welivesecurity.com/2022/04/12/industroyer2-industroyer-reloaded/.

Information Assurance Advisory Council, 'SMEs and the Supply Chain White Paper, June 2014'.

Insikt Group, 'How North Korea Revolutionized the Internet as a Tool for Rogue Regimes (2020)', https://go.recordedfuture.com/hubfs/reports/cta-2020-0209.pdf.

'ISACA State of Cybersecurity 2021 Part 1: Global Update on Workforce Efforts, Resources and Budgets', https://www.isaca.org/bookstore/bookstore-wht_papers-digital/whpsc211.

'(ISC)^2 A Resilient Cybersecurity Profession Charts the Path Forward (ISC)^2 Cybersecurity Workforce Study, 2021', https://www.isc2.org//-/media/ISC2/Research/2021/ISC2-Cybersecurity-Workforce-Study-2021.ashx.

'(ISC)^2 Report Finds Cybersecurity Workforce Gap Has Increased to More Than 2.9 Million Globally' (October 17 2018), https://www.isc2.org/News-and-Events/Press-Room/Posts/2018/10/17/ISC2-Report-Finds-Cybersecurity-Workforce-Gap-Has-Increased-to-More-Than-2-9-Million-Globally.

Johnson, B., Caban, D., Krotofil, M., Scali, D., Brubaker, N., and Glyer, C., 'Attackers Deploy New ICS Attack Framework "Triton" and Cause Operational Disruption to Critical Infrastructure' (December 14 2018), https://www.fireeye.com/blog/threat-research/2017/12/attackers-deploy-new-ics-attack-framework-triton.html.

Keeling, C., 'Waking Shark II: Desktop Cyber Exercise: Report to Participants' (November 12 2013), https://www.bba.org.uk/wp-content/uploads/2014/02/Banking_3192106_v_1_Waking-Shark-II-Report-v1.pdf.pdf.

'Keynote Address by Dr. Arden Bement Director, National Institute of Standards & Technology at the NSF Workshop on Critical Infrastructure Protection for SCADA & IT (As Prepared)' (October 20 2003), http://www.nist.gov/director/speeches/bement_102003.cfm.

Krekel, B., Adams, P., and Bakos, B., 'Occupying the Information High Ground: Chinese Capabilities for Computer Network Operations and Cyber Espionage', Prepared for the U.S.-China Economic and Security Review Commission by Northrop Grumman Corp (McLean, VA: Northrop Grumman Corporation, March 7 2012), https://info.publicintelligence.net/USCC-ChinaCyberEspionage.pdf.

Langhill, J.T., 'Defending Against the Dragonfly Cyber Security Attacks' (December 10 2014), https://www.controlglobal.com/assets/15WPpdf/150311-Belden-DragonflyCybersecurity.pdf.

Mandiant, APT1 Exposing One of China's Cyber Espionage Units (2013), https://www.fireeye.com/content/dam/fireeye-www/services/pdfs/mandiant-apt1-report.pdf.

'Mandiant, M-Trends (2010)', https://www.fireeye.com/current-threats/annual-threat-report/mtrends/rpt-2010-mtrends.html.

McAfee/Center for Strategic and International Studies (CSIS), 'In the Crossfire: Critical Infrastructure in the Age of Cyberwar' (2011), https://www.govexec.com/pdfs/012810j1.pdf.

'McAfee Labs Threats Report' (December 2018), https://www.mcafee.com/enterprise/en-us/assets/reports/rp-quarterly-threats-dec-2018.pdf.

'Operation Sharpshooter Campaign Targets Global Defense, Critical Infrastructure' (2018), https://www.mcafee.com/enterprise/en-us/assets/reports/rp-operation-sharpshooter.pdf.

Pernet, C., and Sela, E., 'The Spy Kittens Are Back: Rocket Kitten 2', https://documents.trendmicro.com/assets/wp/wp-the-spy-kittens-are-back.pdf.

'Proofpoint State of the Phish 2019 Report', https://www.cyqueo.com/images/download/Proofpoint-US-TR_State-Of-The-Phish-2019-CYQUEO.pdf.

'Report North Korea Cyber Activity Recorded Future Insikt Group' (June 15 2017), https://go.recordedfuture.com/hubfs/reports/north-korea-activity.pdf.

Ribeiro, A., 'Cyberattacks Continue to Extend Across Europe, BlackCat Ransomware May be Involved' (February 4 2022), https://industrialcyber.co/threats-attacks/cyberattacks-continue-to-extend-across-europe-blackcat-ransomware-may-be-involved/.

Royce, R., 'Autonomous Ships The Next Step' (2016), https://www.rolls-royce.com/~/media/Files/R/Rolls-Royce/documents/customers/marine/ship-intel/aawa-whitepaper-210616.pdf.

Smith, B., 'Defending Ukraine: Early Lessons from the Cyber War' (June 22 2022), https://blogs.microsoft.com/on-the-issues/2022/06/22/defending-ukraine-early-lessons-from-the-cyber-war/.

'State of Cybersecurity 2019', http://m.isaca.org/Knowledge-Center/Research/Documents/cyber/state-of-cybersecurity-2019-part-1_res_eng_0319a.pdf.

'The Future of Jobs Report 2018', http://www3.weforum.org/docs/WEF_Future_of_Jobs_2018.pdf.

Tucker, J.A. et al., 'Social Media, Political Polarization, and Political Disinformation: A Review of the Scientific Literature' (March 2018), https://www.hewlett.org/wp-content/uploads/2018/03/Social-Media-Political-Polarization-and-Political-Disinformation-Literature-Review.pdf.

'Ukraine: Disk-Wiping Attacks Precede Russian Invasion' (February 24 2022), https://symantec-enterprise-blogs.security.com/blogs/threat-intelligence/ukraine-wiper-malware-russia.

'Verizon, 2018 Data Breach Investigations Report Executive Summary', https://enterprise.verizon.com/resources/reports/DBIR_2018_Report_execsummary.pdf.

'Verizon, 2019 Data Breach Investigation Report', https://enterprise.verizon.com/resources/reports/2019-data-breach-investigations-report.pdf.

'Verizon, 2019 Insider Threat Report Executive Summary', https://enterprise.verizon.com/resources/executivebriefs/insider-threat-report-executive-summary.pdf.

'Verizon, 2021 Data Breach Investigations Report', http://verizon.com/dbir/.

'Verizon Data Breach Investigations Report 11th edition (2018)', https://enterprise.verizon.com/resources/reports/DBIR_2018_Report.pdf.

'Verizon Data Breach Investigations Report 2022', https://www.verizon.com/business/resources/reports/2022/dbir/2022-data-breach-investigations-report-dbir.pdf.

Wyman, O., 'Large-Scale Cyber-Attacks on the Financial Services System: A Case for Better Coordinated Response and Recovery Strategies: A White Paper to the industry' (March 28 2018), https://www.oliverwyman.com/content/dam/oliver-wyman/v2/publications/2018/march/Large-Scale-Cyber-Attacks-DTCC-2018.pdf.

In-Person Conferences and Other

2nd International Symposium for ICS & SCADA Cyber Security, 11th-12th September 2014 at University of Applied Sciences St. Pölten, Austria.

8th International Conference on Cyber Conflict, Tallinn, Estonia, 1–3 June 2016.

BT SASIG Conference on the Insider Threat London, 8 May 2015.

Conference, NATO Intelligence Fusion Centre, RAF Molesworth, 2–5 November 2015.

Cunliffe, K., 'The Art of Science and Betrayal', Unpublished PhD Thesis, Aberystwyth University (2019).

CyCon 2015, Tallinn, Estonia, 27–28 May 2015.

Cyber Security Building Resilience Reducing Risk Conference, Chatham House, London, 19–20 May 2014.

Digital Wales 2015, Celtic Manor, UK, 4–6 June 2015.

Insider Threat Summit, Monterey, California, 29 March–1 April 2016.

Websites

'A Chronicle of KGB Seduction, Sexual Entrapment, Incompetence and CIA Arrogance', https://www.youtube.com/watch?v=AkVVWu686VY.
'A Detailed Description of the Data Execution Prevention (DEP) Feature in Windows XP Service Pack 2, Windows XP Tablet PC Edition 2005, and Windows Server 2003' (July 11 2017), https://support.microsoft.com/en-gb/help/875352/a-detailed-description-of-the-data-execution-prevention-dep-feature-in.
Abdalla, N.S., Davies, P.H.J., Gustafson, K., Lomas, D., and Wagner, S., 'Intelligence and the War in Ukraine', https://warontherocks.com/2022/05/intelligence-and-the-war-in-ukraine-part-1/, https://warontherocks.com/2022/05/intelligence-and-the-war-in-ukraine-part-2/.
'About ISA', https://isalliance.org/about-isa/.
'AIS Transponders', https://www.steamshipmutual.com/publications/Articles/Articles/04_AIS_Transporters.asp.
Anonymous, 'Are Ports Prepared to Deal with Threats from Hackers?' (April 6 2018), https://piernext.portdebarcelona.cat/en/technology/are-ports-prepared-to-deal-with-threats-from-hackers/.
Anonymous, 'Cyber Threat to Civil Aviation Position Paper' (May 2 2017), https://www.eurocockpit.be/positions-publications/cyber-threat-civil-aviation.
Anonymous, 'Ethical Collection of Competitive Intelligence', https://www.competitivefutures.com/ethical-collection-competitive-intelligence/.
Anonymous, 'IMO Moves Forward to Address Autonomous Ships' (May 27 2018), https://worldmaritimenews.com/archives/253639/imo-moves-forward-to-address-autonomous-ships/.
Anonymous, 'Spoofing Presidential Alerts', https://systems.cs.colorado.edu/headlines/cmas.html.
Anonymous, 'Tests Show Ease of Hacking ECDIS, Radar and Machinery' (December 21 2017), https://www.maritime-executive.com/article/tests-show-ease-of-hacking-ecdis-radar-and-machinery.
Anonymous, 'Top 5 Cyber Attacks in the Aviation Industry' (April 16 2021), https://cnsight.io/2021/04/16/top-5-cyber-attacks-in-the-aviation-industry/.

'AppLocker' (October 16 2017), https://docs.microsoft.com/en-us/windows/security/threat-protection/windows-defender-application-control/applocker/applocker-overview.

Arghire, I., 'Researchers Dissect PowerShell Scripts Used by Russia-Linked Hackers' (May 31 2019), https://www.securityweek.com/researchers-dissect-powershell-scripts-used-russia-linked-hackers.

Ashford, W., 'Cyber Attack Warnings Highlight Need to be Prepared' (June 14 2018), https://www.computerweekly.com/news/252443085/Cyber-attack-warnings-highlight-need-to-be-prepared.

Baranovskaya, S., 'Moscow's Cyber-Defense How the Russian Government Plans to Protect the Country from the Coming Cyberwar' (July 19 2017), https://meduza.io/en/feature/2017/07/19/moscow-s-cyber-defense.

Bar-Yosef, N., 'The Structure of a Cybercrime Organization - Hackers Have Supply Chains Too!' (September 23 2010), https://www.securityweek.com/structure-crybercrime-organization-hackers-have-supply-chains-too.

Beahm, A., 'DCH Health System 'Closed to All But Most Critical' New Patients Due to Ransomware Attack' (October 2 2019), https://www.al.com/news/2019/10/dch-health-system-closed-to-all-but-most-critical-new-patients-due-to-ransomware-attack.html.

Bergal, J., 'How Hackers Could Cause Chaos on America's Roads and Railways' (April 25 2018), https://gcn.com/articles/2018/04/25/hacks-transportation-systems.aspx, https://www.blackhat.com/us-22/defcon.html.

'Blackout: The Power Outage that Left 50 Million W/o Electricity | Retro Report | *The New York Times*' (November 11 2013), https://www.youtube.com/watch?v=nd3teNgUq8E.

Bliss, L., 'What Facebook Isn't Telling Us About Its Fight Against Online Abuse' (May 21 2018), http://theconversation.com/what-facebook-isnt-telling-us-about-its-fight-against-online-abuse-96818.

'Blowout' (October 2016), https://www.texasmonthly.com/articles/deepwater-horizon-prosecution/.

Bochman, A., 'Michael Assante Holds Forth on Cybersecurity Leadership', *Smart Grid Security Blog*, August 1 2012, http://smartgridsecurity.blogspot.co.uk/2012/08/michael-assante-holds-forth-on.html.

Bott, E., 'Fact Check: Malware Did Not Bring Down a Passenger Jet (August 24 2010), https://www.zdnet.com/article/fact-check-malware-did-not-bring-down-a-passenger-jet/.

Brandon, D.R., 'Getting Serious About Security Breaches with Endpoint Protection' (October 25 2016), https://www.insight.com/en_US/learn/content/2016/10252016-getting-serious-about-endpoint-security.html.

Brandom, R., 'The Petya Ransomware Is Starting to Look Like a Cyberattack in Disguise: The Ransomware That Wasn't' (June 28 2018), https://www.theverge.com/2017/6/28/15888632/petya-goldeneye-ransomware-cyberattack-ukraine-russia.

Brenner, J., "How Obama Fell Short on Cybersecurity: Under the President's Proposals, We'll Remain America the Vulnerable' (January 21 2015), http://www.politico.com/magazine/story/2015/01/state-of-the-union-cybersecurity-obama-114411.html#ixzz3PjrwwEnf.

Brew, K., 'Watering Hole Attacks: Detecting End-User Compromise Before the Damage Is Done' (April 26 2016), https://www.alienvault.com/blogs/security-essentials/watering-hole-attacks-detecting-end-user-compromise-before-the-damage-is-done.

'Browse the Top Used Topics on GitHub', https://github.com/topics.

'Browser Extensions: Are They Worth the Risk?' (September 5 2018), https://krebsonsecurity.com/2018/09/browser-extensions-are-they-worth-the-risk/.

Buchholz, S. et al., 'What Nontechnical Government Leaders Can Do Today to be Ready for Tomorrow's Quantum World' (February 6 2020), https://www2.deloitte.com/us/en/insights/industry/public-sector/the-impact-of-quantum-technology-on-national-security.html.

'Burp Suite Editions', https://portswigger.net/burp.

Burton, G., 'World's Largest Companies Recruiting CISOs for Board-Level Roles' (May 30 2014), https://www.computing.co.uk/ctg/news/2347513/worlds-largest-companies-recruiting-cisos-for-board-level-roles.

'Business Email Compromise (BEC)', https://www.trendmicro.com/vinfo/us/security/definition/business-email-compromise-(bec).

Carozza, D., 'Sir Rob Wainwright Encourages 'Powering Up' Information, Intelligence Sharing' (June 18 2018), https://www.fraudconferencenews.com/home/2018/6/18/sir-wainwright-encourages-powering-up-information-intelligence-sharing.

Caudron, M., 'How Europol Became a Center Point for the F.B.I.' (April 27 2018), https://medium.com/euintheus/how-europol-became-a-center-point-for-the-f-b-i-2ccc96f105bb.

Cerrudo, C., 'Hacking US Traffic Control Systems' (2014), https://www.defcon.org/images/defcon-22/dc-22-presentations/Cerrudo/DEFCON-22-Cesar-Cerrudo-Hacking-Traffic-Control-Systems-UPDATED.pdf, http://chemexfranchises.co.uk/, http://www.chemsol.co.uk/.

Chakravartula, R.N., 'What Is Enumeration?' (February 28 2018), https://resources.infosecinstitute.com/what-is-enumeration/#gref.

'Charming Kitten', https://malpedia.caad.fkie.fraunhofer.de/actor/charming_kitten, https://www.counterextremism.com/extremists/junaid-hussain, https://www.counterextremism.com/extremists/usaamah-abdullah-rahim, https://www.counterextremism.com/extremists/elton-simpson

Choros, A., 'World's Most Famous Hacker Says Metadata Is an "Extremely Attractive" Target, Everything Is Hackable', http://www.cybershack.com.au/news/worlds-most-famous-hacker-says-metadata-extremely-attractive-target-everything-hackable (dead link).

Cimpanu, C., 'How US Authorities Tracked Down the North Korean Hacker Behind WannaCry US Authorities Put Together Four Years Worth of Malware Samples, Domain Names, Email and Social Media Accounts to Track Down One of the Lazarus Group Hackers' (September 6 2018), https://www.zdnet.com/article/how-us-authorities-tracked-down-the-north-korean-hacker-behind-wannacry/.

Cimpanu, C., 'Port of San Diego Suffers Cyber-Attack, Second Port in a Week After Barcelona' (September 27 2018), https://www.zdnet.com/article/port-of-san-diego-suffers-cyber-attack-second-port-in-a-week-after-barcelona/.

Cloonan, J., 'Advanced Malware Detection—Signatures vs. Behavior Analysis' (April 11 2017), https://www.infosecurity-magazine.com/opinions/malware-detection-signatures/.

Cohen, C., 'House Passes Cyber Vulnerability Disclosure Reporting Act' (January 12 2018), https://www.insideprivacy.com/united-states/congress/house-passes-cyber-vulnerability-disclosure-reporting-act/.

'Common Features of Phishing Emails' (Undated), https://www.phishing.org/what-is-phishing.

Constantin, L., 'Researchers Show Ways to Bypass Home and Office Security Systems Many Door Sensors, Motion Detectors and Security Keypads Can be Bypassed Using Simple Techniques, Researchers from Bishop Fox Said' (July 31 2013), https://www.csoonline.com/article/2133815/physical-security/researchers-show-ways-to-bypass-home-and-office-security-systems.html.

Cosgrove, L., 'How Will Blockchain Technology Change the Mining Industry' (Undated), https://www.wipro.com/natural-resources/how-will-blockchain-technology-change-the-mining-industry/.

Crawley, K., 'Insecure Key Collection on GitHub Is a Dream Come True for Cyber Attackers' (April 1 2019), https://dzone.com/articles/insecure-key-collection-on-github-is-a-dream-come, https://www.crowdstrike.com/

Cullum, B., 'China Seeks Public Comment on Critical Infrastructure Cybersecurity Review Rules' (May 29 2019), https://globaldatareview.com/article/1193471/china-seeks-public-comment-on-critical-infrastructure-cybersecurity-review-rules.

'Customary International Law', https://www.law.cornell.edu/wex/customary_international_law.

'Cyber Resilience', https://www.iaac.org.uk/cyber-resilience/.

'Cyber Security 2014', http://www.corporatelivewire.com/round-tables.html?id=cyber-security-2014.

'Cyber Threats and Nuclear Weapons: A Book Talk with Professor Herb Lin' (December 6 2021), https://www.youtube.com/watch?v=TvZvh6IbWWY.

Cynthia Brumfield, 'What Is the CISA? How the New Federal Agency Protects Critical Infrastructure from Cyber Threats' (July 1 2019), https://www.csoonline.com/article/3405580/what-is-the-cisa-how-the-new-federal-agency-protects-critical-infrastructure-from-cyber-threats.html.

'Deep Neural Network' (Undated), https://www.sciencedirect.com/topics/computer-science/deep-neural-network.

'DefCamp 2014—Social Engineering, or "Hacking People"', https://www.youtube.com/watch?v=JAOTRgWdPTU.

'Defcon 21—Social Engineering: The Gentleman Thief', https://www.youtube.com/watch?v=1kkOKvPrdZ4.

'DEF CON 25 ICS Village—Joe Weiss—Cyber Security Issues with Level 0 Through 1 Devices' (October 16 2017), https://www.youtube.com/watch?v=UgvVaniZhsk.

Demchak, C.C., and Thomas, M.L., 'Can't Sail Away from Cyber Attacks: 'Sea-Hacking' from Land' (October 15 2021), https://warontherocks.com/2021/10/cant-sail-away-from-cyber-attacks-sea-hacking-from-land/.

'Derren Brown', http://derrenbrown.co.uk/about/, https://digitalguardian.com/blog/social-engineering-attacks-common-techniques-how-prevent-attack.

'DMitry', https://tools.kali.org/information-gathering/dmitry.

'Does Cisco Sell Snort?', https://www.snort.org/faq/does-cisco-sell-snort.

Dunwoody, M., 'Dissecting One of APT29's Fileless WMI and PowerShell Backdoors (POSHSPY)' (April 3 2017), https://www.fireeye.com/blog/threat-research/2017/03/dissecting_one_ofap.html.

Ellis, C., 'NIST: Vulnerability Disclosure as a Requirement for Every Organization' (January 18 2018), https://www.bugcrowd.com/nist-vulnerability-disclosure-as-a-requirement-for-every-organization/.

Erwin, S., 'STRATCOM to Design Blueprint for Nuclear Command, Control and Communications' (March 29 2019), https://spacenews.com/stratcom-to-design-blueprint-for-nuclear-command-control-and-communications/, http://www.euticals.com/, http://www.exabeam.com/, http://www.exactearth.com/.

'Exelis Awarded US Air Force Contract to Upgrade Strategic Automated Command Control System Digital Memory Technology' (December 3 2014), https://www.harris.com/press-releases/2014/12/exelis-awarded-us-air-force-contract-to-upgrade-strategic-automated-command, http://www.exxonmobil.co.uk/UK-English/about_what_chemicals.aspx.

'FedEx Corp. Reports First Quarter Earnings' (September 19 2017), http://about.van.fedex.com/newsroom/fedex-corp-reports-first-quarter-earnings-2/.

Ferguson, T., 'SCADA and ICS: Combating the Security Risk' (January 17 2018), https://threatmanagement.info/scada_and_ics_combating_the_security_risk/.

'FireEye, APT38 Un-Usual Suspects', https://content.fireeye.com/apt/rpt-apt38.

Fisher, T., '14 Free Remote Access Software Tools' (July 12 2019), https://www.lifewire.com/free-remote-access-software-tools-2625161.

Fokker, J., 'Organizations Leave Backdoors Open to Cheap Remote Desktop Protocol Attacks' (July 11 2018), https://securingtomorrow.mcafee.com/other-blogs/mcafee-labs/organizations-leave-backdoors-open-to-cheap-remote-desktop-protocol-attacks/.

Friedersdorf, C., 'How Dangerous Is End-to-End Encryption?' (July 14 2015), http://www.theatlantic.com/politics/archive/2015/07/nsa-encryption-ungoverned-spaces/398423/.

Fruhlinger, J., 'What Is Phishing? How This Cyber Attack Works and How to Prevent It' (May 9 2019), https://www.csoonline.com/article/2117843/what-is-phishing-how-this-cyber-attack-works-and-how-to-prevent-it.html.

Gallagher, S., '"EPIC" Fail—How OPM Hackers Tapped the Mother Lode of Espionage Data' (June 22 2015), https://arstechnica.com/information-technology/2015/06/epic-fail-how-opm-hackers-tapped-the-mother-lode-of-espionage-data/.

Gallagher, S., 'South Korea Claims North Hacked Nuclear Data Hackers Stole Blueprints, Employee Data, and Threatened "Destruction" If Demands Not Met' (March 17 2015), https://arstechnica.com/information-technology/2015/03/south-korea-claims-north-hacked-nuclear-data/.

Gertz, B., 'DHS, FBI Warn Companies of Ongoing Cyber Attacks on Critical Infrastructure Russia Seen as Behind Cyber Targeting of Electric Grid, Other Public Infrastructures' (October 24 2017), https://freebeacon.com/national-security/dhs-fbi-warn-companies-ongoing-cyber-attacks-critical-infrastructure/, https://github.com/hslatman/awesome-industrial-control-system-security, https://github.com/redteamsecurity/PlugBot-Plug, https://github.com/rmusser01/Infosec_Reference/blob/master/Draft/SCADA.md, https://github.com/SANSA-Stack/SANSA-Stack.

'Global Study Reveals Businesses and Countries Vulnerable Due to Shortage of Cybersecurity Talent', 26 July 2016, https://newsroom.intel.com/news-releases/global-study-reveals-businesses-countries-vulnerable-due-shortage-cybersecurity-talent/.

Goldman, J., 'FBI, DHS Warn of Surge in Insider Threats from Disgruntled Employees' (September 25 2014), https://www.esecurityplanet.com/network-security/fbi-dhs-warn-of-surge-in-insider-threats-from-disgruntled-employees.html.

Goodin, D., 'Massive US-Planned Cyberattack Against Iran Went Well Beyond Stuxnet' (February 17 2016), http://arstechnica.co.uk/tech-policy/2016/02/massive-us-planned-cyberattack-against-iran-went-well-beyond-stuxnet/.

Goodin, D., 'Hackers Trigger Yet Another Power Outage in Ukraine for the Second Year in a Row, Hack Targets Ukraine During One of Its Coldest Months' (January 11 2017), https://arstechnica.com/information-technology/2017/01/the-new-normal-yet-another-hacker-caused-power-outage-hits-ukraine/.

Goodin, D., 'Stolen NSA Hacking Tools Were Used in the Wild 14 Months Before Shadow Brokers Leak' (May 7 2019), https://arstechnica.com/information-technology/2019/05/stolen-nsa-hacking-tools-were-used-in-the-wild-14-months-before-shadow-brokers-leak/.

Goud, N., 'Malware Attack Via Twitter' (Undated), https://www.cybersecurity-insiders.com/malware-attack-via-twitter/.

Grady, J., 'Intel Sharing Between U.S. and Ukraine 'Revolutionary' Says DIA Director' (March 18 2022), https://news.usni.org/2022/03/18/intel-sharing-between-u-s-and-ukraine-revolutionary-says-dia-director.

Guarnieri, C., and Anderson, C., 'Flying Kitten to Rocket Kitten, A Case of Ambiguity and Shared Code' (December 5 2017), https://iranthreats.github.io/resources/attribution-flying-rocket-kitten/.

Gundert, L., Chohan, S., and Lesnewich, G., 'Iran's Hacker Hierarchy Exposed How the Islamic Republic of Iran Uses Contractors and Universities to

Conduct Cyber Operations' (May 9 2018), https://www.recordedfuture.com/iran-hacker-hierarchy/.

'Hackers Love This Tiny Box' (September 9 2016), https://www.youtube.com/watch?v=r0jBLzmrH9w&feature=youtu.be.

'Hacker Tools Top 10' (December 12 2018), https://www.concise-courses.com/hacking-tools/top-ten/.

'Hacking Humans & Social Engineering', https://www.youtube.com/watch?v=r5kd0KZ_MVs.

Harvey, C., 'Top Cybersecurity Companies of 2018' (August 22 2018), https://www.esecurityplanet.com/products/top-cybersecurity-companies-2018.html.

Hazell, L., 'SCADA System Vulnerabilities So Common, Insurers Won't Insure', http://cybersecuritynews.co.uk/scada-system-vulnerabilities-so-common-insurers-wont-insure/.

Henderson, G., 'Marine Sgt. Clayton Lonetree Was Convicted of Espionage and...' (August 21 1987), https://www.upi.com/Archives/1987/08/21/Marine-Sgt-Clayton-Lonetree-was-convicted-of-espionage-and/8612556500687/.

Higgins, K.J., 'Windows XP Alive & Well in ICS/SCADA Networks' (April 10 2014), http://www.darkreading.com/informationweek-home/windows-xp-alive-and-well-in-ics-scada-networks/d/d-id/1204385.

Hill, M., 'How Security Vendors Are Aiding Ukraine' (March 2 2022), https://www.csoonline.com/article/3651685/how-security-vendors-are-aiding-ukraine.html.

Hoffman, M., and Winston, T., 'Recommendations Following the Colonial Pipeline Cyber Attack' (May 11 2021), https://www.dragos.com/blog/industry-news/recommendations-following-the-colonial-pipeline-cyber-attack/.

Holmes, M., 'Experts Say Viasat Cyber Attack Exposed Ground Terminal, Satellite Supply Chain Vulnerabilities' (April 11 2022), https://www.satellitetoday.com/cybersecurity/2022/04/11/experts-say-viasat-cyber-attacks-exposed-ground-terminal-satellite-supply-chain-vulnerabilities/.

'How to Clone a Security Badge in Seconds' (May 23 2016), https://www.youtube.com/watch?v=cxxnuofREcM.

'How to Use Software Restriction Policies in Windows Server 2003' (April 17 2018), https://support.microsoft.com/en-gb/help/324036/how-to-use-software-restriction-policies-in-windows-server-2003.

Hummert, A., 'Reviewing X Sender Headers: How to Prevent Email Spoofing from Fake Senders' (February 20 2019), https://www.alienvault.com/blogs/security-essentials/how-hackers-manipulate-email-to-defraud-you-and-your-customers, http://www.iaac.org.uk/about/.

'IBM Security, X-Force Threat Intelligence Index 2021', https://www.ibm.com/downloads/cas/M1X3B7QG, https://icspa.org/about-us/.

'Increase in Insider Threat Cases Highlight Significant Risks to Business Networks and Proprietary Information' (September 23 2014), https://www.ic3.gov/media/2014/140923.aspx, https://inequality.org/facts/wealth-inequality/

'International Convention for the Safety of Life at Sea (SOLAS), 1974', http://www.imo.org/en/About/conventions/listofconventions/pages/international-convention-for-the-safety-of-life-at-sea-(solas),-1974.aspx.

'International Cyber Security Protection Alliance', https://www.icspa.org/.

'Internet Security Alliance', http://www.isalliance.org/isa-publications/.

'In the Dark: Crucial Industries Confront Cyberattacks' (April 19 2011), https://www.mcafee.com/blogs/enterprise/in-the-dark-crucial-industries-confront-cyberattacks/.

Ivezic, M., 'Cybercrime in China—A Growing Threat for the Chinese Economy' (February 2 2017), https://cyberkinetic.com/cybersecurity-cyber-risk/chinese-cybercrime/.

Jackson, G., 'How to Actively Protect Your Website from Cyberattacks' (October 22 2018), https://www.siliconrepublic.com/enterprise/website-protection-cyberattacks.

Kan, M., 'Windows XP Will Continue Receiving Security Support in China' (March 3 2014), http://www.pcworld.com/article/2103680/chinas-windows-xp-users-to-still-get-security-support.html.

Kass, D.H., 'Riviera Beach, Florida Ransomware Attack; City Pays $600,000' (June 20 2019), https://www.msspalert.com/cybersecurity-breaches-and-attacks/ransomware/riviera-beach-florida-malware-attack/.

Kawamoto, D., 'Number of CISOs Rose 15% This Year' (June 5 2017), https://www.darkreading.com/careers-and-people/number-of-cisos-rose-15--this-year/d/d-id/1329050.

Keen, S., 'Open Source Software Exposes ICS Device Vulnerabilities to Hackers' (August 14 2018), https://www.nozominetworks.com/blog/open-source-software-exposes-ics-device-vulnerabilities-to-hackers/.

Kleinfeld, J.S., 'Could It be a Big World After All? The "Six Degrees of Separation" Myth' (April 12 2001), https://www.cs.princeton.edu/~chazelle/courses/BIB/big-world.htm.

Klijnsma, Y., 'New Insights into Energetic Bear's Watering Hole Attacks on Turkish Critical Infrastructure' (November 2 2017), https://www.riskiq.com/blog/labs/energetic-bear/.

Knapp, E., 'SCADA Mischief Episode 2: Context and Correlation' (March 6 2012), https://www.securityweek.com/scada-mischief-episode-2-context-and-correlation.

Koren, M., 'The War on Ukraine Is Testing the Myth of Elon Musk', https://www.theatlantic.com/science/archive/2022/02/elon-musk-ukraine-starlink-satellites/622954/.

Kovacs, E., 'Multiple U.S. Gas Pipeline Firms Affected by Cyberattack' (April 4 2018), https://www.icscybersecurityconference.com/multiple-u-s-gas-pipeline-firms-affected-by-cyberattack/.

Kovacs, E., 'OSVDB Shut Down Permanently' (April 7 2016), https://www.securityweek.com/osvdb-shut-down-permanently.

Kozuch, I., 'APT Group OilRig: Who They Are and What You Need to Know' (April 10 2018), https://intsights.com/blog/apt-group-oilrig-who-they-are-and-what-you-need-to-know.

Krebbs, C., and Chesney, R., 'Gray Zone, Twilight Zone, or Danger Zone? Russian Cyber and Information Operations in Ukraine' (March 18 2022), https://warontherocks.com/2022/03/gray-zone-twilight-zone-or-danger-zone-russian-cyber-and-information-operations-in-ukraine/.

Kruger, D.W., 'Maskirovka—What's in It for Us?', School of Advanced Military Studies, Fort Leavenworth Kansas (December 4 1987), https://apps.dtic.mil/dtic/tr/fulltext/u2/a190836.pdf.

'Language of Espionage', https://www.spymuseum.org/education-programs/spy-resources/language-of-espionage/.

'Leafminer: New Espionage Campaigns Targeting Middle Eastern Regions Active Attack Group Is Eager to Make Use of Available Tools, Research, and the Work of Other Threat Actors' (July 25 2018), https://symantec-enterprise-blogs.security.com/blogs/threat-intelligence/leafminer-espionage-middle-east.

Lee, B., 'OilRig' (Undated), https://attack.mitre.org/groups/G0049/.

Lee, B. (Unit 42), 'Ransomware: Unlocking the Lucrative Criminal Business Model', https://www.paloaltonetworks.com/apps/pan/public/downloadResource?pagePath=/content/pan/en_US/resources/research/ransomware-report.

Leiner, B.M. et al., 'Brief History of the Internet' (1997), https://www.internetsociety.org/internet/history-internet/brief-history-internet/.

Leroux, S., 'Kali Linux Review: Not Everyone's Cup of Tea' (May 9 2019), https://itsfoss.com/kali-linux-review/.

Lieu, C.D., 'Social Engineering—Attacking the Weakest Link'. Available from https://www.giac.org/paper/gsec/2082/social-engineering-attacking-weakest-link/103563.

'Living Off the Land Defining Fileless Attack Methods', https://www.symantec.com/content/dam/symantec/docs/security-center/white-papers/istr-living-off-the-land-and-fileless-attack-techniques-en.pdf, https://www.lockheedmartin.com/en-us/capabilities/cyber/cyber-kill-chain.html

Lord, N., 'What Is Polymorphic Malware? A Definition and Best Practices for Defending Against Polymorphic Malware' (September 11 2018), https://digitalguardian.com/blog/what-polymorphic-malware-definition-and-best-practices-defending-against-polymorphic-malware.

Lydon, B., 'Cyber Security Threats: Expert Interview with Eric Byres, Part 1', http://www.automation.com/automation-news/article/cyber-security-threats-expert-interview-with-eric-byres-part-1.

Lyngaas, S., 'Biden Administration Remains 'On Guard' for Russian Cyber-attacks Amid War in Ukraine' (March 2 2022), https://edition.cnn.com/2022/03/02/politics/blinken-russia-cyberattacks-ukraine/index.html.

Mabee, M., 'Recent Critical Infrastructure Attacks Expose Our Vulnerability—And the Need for Change' (December 28 2020), https://centerforsecuritypolicy.org/recent-critical-infrastructure-attacks-expose-our-vulnerability-and-the-need-for-change/.

MacDougall, A., and Myers, M., 'Pentagon Faces Array of Challenges in Retaining Cybersecurity Personnel' (June 9 2018), https://thehill.com/opinion/cybersecurity/391426-pentagon-faces-array-of-challenges-in-retaining-cybersecurity-personnel.

Mackenzie, H., 'Shamoon Malware and SCADA Security—What Are the Impacts?' (October 25 2012), https://www.tofinosecurity.com/blog/shamoon-malware-and-scada-security-%E2%80%93-what-are-impacts.

Malenkovich, S., 'What Is a Rootkit and How to Remove It' (March 28 2013), https://www.kaspersky.co.uk/blog/rootkit/1508/.

'Maltego CE', https://www.paterva.com/buy/maltego-clients/maltego-ce.php.

Manky, D., 'Securing OT Networks Against Rising Attacks' (March 13 2018), https://www.csoonline.com/article/3261448/security/securing-ot-networks-against-rising-attacks.html, https://www.marinetraffic.com/

Masi, A., Verma, S., Oatman, M., 'Photos: Living in the Shadow of the Bhopal Chemical Disaster' (June 2 2014), http://www.motherjones.com/environment/2014/06/photos-bhopal-india-union-carbide-sanjay-verma-pesticides-explosion.

Melnick, J., 'Top 10 Most Common Types of Cyber Attacks' (May 15 2018), https://blog.netwrix.com/2018/05/15/top-10-most-common-types-of-cyber-attacks/.
Menear, H., 'McAfee: The Shifting Threat Landscape in the Manufacturing Sector' (June 21 2019), https://www.businesschief.com/leadership/8209/McAfee:-the-shifting-threat-landscape-in-the-manufacturing-sector.
'Metasploit: The World's Most Used Penetration Testing Framework', https://www.metasploit.com/.
Metzger, M., 'CISO Salaries May Soon Hit £1 Million—But Few Qualified for Top Roles' (May 22 2017), https://www.scmagazine.com/news/network-security/ciso-salaries-may-soon-hit-1-million-but-few-qualified-for-top-roles.
Miao, Y., 'Understanding Heuristic-Based Scanning vs. Sandboxing' (July 13 2015), https://www.opswat.com/blog/understanding-heuristic-based-scanning-vs-sandboxing.
'Michael Hayden, Richard Clarke on Greatest Cyberthreats Facing America' Washington Post Live (October 6 2017), https://www.youtube.com/watch?v=FdiAQBXGsMg.
Mimoso, M., 'Patching Bash Vulnerability a Challenge for ICS, SCADA' (September 25 2014), http://threatpost.com/patching-bash-vulnerability-a-challenge-for-ics-scada.
Moore, J., 'The Application Programming Languages Undergirding Some Federal IT Systems Were New When "The Andy Griffith Show" Premiered' (May 25 2016), https://www.nextgov.com/cio-briefing/2016/05/10-oldest-it-systems-federal-government/128599/.
Morgan, S., 'Cybersecurity Labor Crunch to Hit 3.5 Million Unfilled Jobs by 2021' (June 6 2017), https://www.csoonline.com/article/3200024/cybersecurity-labor-crunch-to-hit-35-million-unfilled-jobs-by-2021.html.
Mueller, M., 'What Is Evgeny Morozov Trying to Prove? A Review of "The Net Delusion"', https://www.internetgovernance.org/2011/01/13/what-is-evgeny-morozov-trying-to-prove-a-review-of-the-net-delusion/ (January 13 2011).
Muncaster, P., 'US Gas Pipelines Hit by Cyber-Attack' (April 4 2019), https://www.infosecurity-magazine.com/news/us-gas-pipelines-hit-by-cyberattack/.
Myers, A., 'Meet CrowdStrike's Adversary of the Month for April: STARDUST CHOLLIMA' (April 6 2018), https://www.crowdstrike.com/blog/meet-crowdstrikes-adversary-of-the-month-for-april-stardust-chollima/.
Nandikotkur, G., 'CERT-In Says Hacking Declining, But Critics Express Doubts Do the Latest Statistics Reflect Reality?' (December 18 2018),

https://www.bankinfosecurity.asia/cert-in-says-hacking-declining-but-critics-express-doubts-a-11868, https://www.ncfta.net/.

Neave, R., 'Cyber Attacks: Protecting the Financial Service Sector's Data Centres' (September 9 2019), https://thefintechtimes.com/cyber-attacks-data-centres/.

Nemeth, W.J., 'Future War and Chechnya: A Case for Hybrid Warfare', MA thesis, Naval Postgraduate School, Monterey, California (June 2002), https://calhoun.nps.edu/bitstream/handle/10945/5865/02Jun_Nemeth.pdf;sequence=3.

'New Petya / NotPetya / ExPetr Ransomware Outbreak' (June 28 2017), https://www.kaspersky.com/blog/new-ransomware-epidemics/17314/.

Nichols, S., 'Remember Those Stolen 'NSA Exploits' Leaked Online by the Shadow Brokers? The Chinese Had Them a Year Before' (May 7 2019), https://www.theregister.co.uk/2019/05/07/equation_group_tools/.

Norton, 'What Is Bulletproof Hosting?' (Undated), https://us.norton.com/internetsecurity-emerging-threats-what-is-bulletproof-hosting.html.

'Number of Offshore Oil Rigs Worldwide as of January 2018 by Region', https://www.statista.com/statistics/279100/number-of-offshore-rigs-worldwide-by-region/.

Nye, J.S., 'Is Cyber the Perfect Weapon?' (July 5 2018), https://www.project-syndicate.org/commentary/deterring-cyber-attacks-and-information-warfare-by-joseph-s--nye-2018-07.

O'Brien, D., 'Ericksonian Language Patterns' (July 19 2008), https://ericksonian.com/ericksonian-language-patterns, accessed 21 February 2020.

O'Leary, J. et al., 'Insights into Iranian Cyber Espionage: APT33 Targets Aerospace and Energy Sectors and Has Ties to Destructive Malware' (September 20 2017), https://www.fireeye.com/blog/threat-research/2017/09/apt33-insights-into-iranian-cyber-espionage.html.

O'Neill, P.H., 'Russia's Rise to Cyberwar Superpower' (February 29 2020), https://www.dailydot.com/debug/russia-cyberwar-cyberattack-dnc-breach-history/.

'OilRig', https://malpedia.caad.fkie.fraunhofer.de/actor/oilrig.

'Open Source IDS Tools: Comparing Suricata, Snort, Bro (Zeek), Linux' (October 26 2018), https://www.alienvault.com/blogs/security-essentials/open-source-intrusion-detection-tools-a-quick-overview.

'Operating System Replacing Windows Given Go-ahead in Russia' (May 29 2019), https://russiabusinesstoday.com/featured/operating-system-replacing-windows-given-go-ahead-in-russia/.

Oremus, W., 'The Pipe Bomb Suspect Appears to Have Tweeted Death Threats. Twitter Saw No Problem' (October 26 2018), https://slate.com/technology/2018/10/twitter-account-linked-to-cesar-sayoc-made-death-threats-and-twitter-declined-to-suspend-him.html.

Orleans, A., 'Who Is PIONEER KITTEN?' (August 31 2020), https://www.crowdstrike.com/blog/who-is-pioneer-kitten/.

Osborne, C., 'Shamoon Data-Wiping Malware Believed to be the Work of Iranian Hackers' (December 20 2018), https://www.zdnet.com/article/shamoons-data-wiping-malware-believed-to-be-the-work-of-iranian-hackers/.

Osborne, C., 'Failed Student Jailed for Silk Road, Dark Web Drug Profiteering: The operator of Silk Road 2.0 was unemployed and yet lived the high life' (April 15 2019), https://www.zdnet.com/article/failed-student-jailed-for-silk-road-dark-web-drug-profiteering/.

Paganini, P., 'Improving SCADA System Security' (December 6 2013), http://resources.infosecinstitute.com/improving-scada-system-security/.

Paganini, P., 'Cyber Threats Against the Aviation Industry' (April 8 2014), https://resources.infosecinstitute.com/topic/cyber-threats-aviation-industry/.

Paganini, P., 'Cyber Attack on Sony Pictures Is Much More than a Data Breach—Updated' (December 8 2014), https://resources.infosecinstitute.com/topic/cyber-attack-sony-pictures-much-data-breach/.

Parys, B., 'The KeyBoys Are Back in Town' (November 2 2017), https://www.pwc.co.uk/issues/cyber-security-data-privacy/research/the-keyboys-are-back-in-town.html.

Patterson, D., 'Gallery: The Top Zero Day Dark Web Markets' (January 9 2017), https://www.techrepublic.com/pictures/gallery-the-top-zero-day-dark-web-markets/.

Pauli, D., 'CERT Australia Rebuffs Ex-staff Criticism' (February 8 2013), https://www.itnews.com.au/news/cert-australia-rebuffs-ex-staff-criticism-331618.

Pederson, P., 'Aurora Revisited—By Its Original Project Lead' (July 9 2014), https://www.langner.com/2014/07/aurora-revisited-by-its-original-project-lead/.

Pertsev, A., 'Blindsided Russia's Top Officials Were Caught Off Guard by Putin's War in Ukraine. Many of Them Want to Resign—But Can't' (March 9 2022), https://meduza.io/en/feature/2022/03/09/blindsided.

'Pick a Lock in Seconds with a Bump Key' (April 14 2010), https://www.youtube.com/watch?v=WpH_t0u5Ybg.

'Piper Alpha Platform, North Sea', http://www.offshore-technology.com/projects/piper-alpha-platform-north-sea/.

Porup, J.M., 'What Is Metasploit? And How to Use This Popular Hacking Tool Metasploit Is a Widely Used Penetration Testing Tool That Makes Hacking Way Easier Than It Used to be. It Has Become an Indispensable Tool for Both Red Team and Blue Team' (March 25 2019), https://www.csoonline.com/article/3379117/what-is-metasploit-and-how-to-use-this-popular-hacking-tool.html.

Poulsen, K., 'Kingpin: How One Hacker Took Over the Billion-Dollar Cybercrime Underground', https://www.kingpin.cc/about/.

Powell, A., 'What Might COVID Cost the U.S.? Try $16 Trillion' (November 10 2020), https://news.harvard.edu/gazette/story/2020/11/what-might-covid-cost-the-u-s-experts-eye-16-trillion/.

'Protect Yourself from Tech Support Scams' (Undated), https://support.microsoft.com/en-us/help/4013405/windows-protect-from-tech-support-scams.

'Protecting Productivity: Industrial Security for the Digital Enterprise with Claroty and McAfee' (April 12 2019), https://www.youtube.com/watch?v=ZUkkI3S8JNc.

Rakuszitzky, M., 'Third Suspect in Skripal Poisoning Identified as Denis Sergeev, High-Ranking GRU Officer' (February 14 2019), https://www.bellingcat.com/news/uk-and-europe/2019/02/14/third-suspect-in-skripal-poisoning-identified-as-denis-sergeev-high-ranking-gru-officer/.

'RASPITE' (Undated), https://www.dragos.com/threat/raspite/.

Ravindra, S., 'The Role of Blockchain in Cybersecurity' (January 8 2018), https://www.infosecurity-magazine.com/next-gen-infosec/blockchain-cybersecurity/.

'Real Future: What Happens When You Dare Expert Hackers to Hack You (Episode 8)' (February 24 2016), https://www.youtube.com/watch?v=bjYhmX_OUQQ.

'Red Teaming vs Penetration Testing vs Vulnerability Scanning vs Vulnerability Assessments' (Undated), https://penconsultants.com/home/red-teaming-vs-penetration-testing-vs-vulnerability-scanning-vs-vulnerability-assessments/.

'redteamsecuritytraining/PlugBot-Plug', https://github.com/redteamsecuritytraining/PlugBot-Plug.

'Rewire Your Industry with IBM Blockchain', https://www.ibm.com/uk-en/blockchain/industries.

Rifkin, J., 'Data Security and Breach Notification Act Would Create the First-Ever Federal Standard for Penalizing Hacks of Consumer Information'

(December 22 2018), https://govtrackinsider.com/data-security-and-breach-notification-act-would-create-the-first-ever-federal-standard-for-9842596a27ba.

'RISI Online Incident Database', https://www.risidata.com/.

'Rocket Kitten', https://malpedia.caad.fkie.fraunhofer.de/actor/rocket_kitten.

Rouse, M., 'Verizon Data Breach Investigations Report (DBIR)' (Undated), https://searchsecurity.techtarget.com/definition/Verizon-Data-Breach-Investigations-Report-DBIR.

Rowley, L., 'Sweet Dream(s): An Examination of Instability in the Darknet Markets' (May 10 2019), https://www.blueliv.com/blog/research/sweet-dreams-instability-in-the-darknet-markets/.

Ryan, D., 'Watch How Easy It Is for Your RFID Card to be Cloned By Hackers [Video]' (January 9 2018), https://insights.identicard.com/blog/watch-how-easy-it-is-for-your-rfid-card-to-be-cloned-by-hackers-video.

Sabin, S., and Cerulus, L., '3 Reasons Moscow Isn't Taking Down Ukraine's Cell Networks' (March 7 2022), https://www.politico.com/news/2022/03/07/ukraine-phones-internet-still-work-00014487, https://sansa-stack.net/.

Saxena, V., 'Description of the Difference Between HIDs & NIDs' (Undated), https://www.techwalla.com/articles/description-of-the-difference-between-hids-nids.

'SCADA Systems', http://www.engineersgarage.com/articles/scada-systems.

Schneier, B., 'Who Are the Shadow Brokers? What Is—and Isn't—Known About the Mysterious Hackers Leaking National Security Agency Secrets' (May 23 2017), https://www.theatlantic.com/technology/archive/2017/05/shadow-brokers/527778/.

'Secure the Grid' (STG) coalition, https://securethegrid.com/, https://www.sei.cmu.edu/research-capabilities/all-work/display.cfm?customel_datapageid_4050=21232.

'Shining a Light on DARKSIDE Ransomware Operations' (May 11 2021), https://www.fireeye.com/blog/threat-research/2021/05/shining-a-light-on-darkside-ransomware-operations.html, http://shipfinder.co/

'Shodan Is the World's First Search Engine for Internet-Connected Devices', https://www.shodan.io/.

Siciliano, R., 'What Is a Backdoor Threat?', https://securingtomorrow.mcafee.com/consumer/identity-protection/backdoor-threat/.

Singh, S., 'Practical Scenarios for XSS Attacks' (October 4 2018), https://pentest-tools.com/blog/xss-attacks-practical-scenarios/.

Singleton, C., 'The Decline of Hacktivism: Attacks Drop 95 Percent Since 2015' (May 16 2019), https://securityintelligence.com/posts/the-decline-of-hacktivism-attacks-drop-95-percent-since-2015/, https://www.snort.org/.
Sobczak, B., 'Hackers warn of 'Tipping Point' for Critical Infrastructure' (July 27 2017), https://www.eenews.net/stories/1060057993.
Sobers, R., '10 Must-Know Cybersecurity Statistics for 2020' (January 9 2020), https://www.varonis.com/blog/cybersecurity-statistics/, https://www.social-engineer.com/about/.
'Social Engineering', https://null-byte.wonderhowto.com/how-to/social-engineering/, https://www.sometics.com/en/sociogram
Srivastava, M., 'Russia Hammered by Pro-Ukrainian Hackers Following Invasion' (May 6 2022), https://arstechnica.com/information-technology/2022/05/russia-hammered-by-pro-ukrainian-hackers-following-invasion/2/.
Stack, T., 'Internet of Things (IoT) Data Continues to Explode Exponentially. Who Is Using That Data and How?' (February 5 2018), https://blogs.cisco.com/datacenter/internet-of-things-iot-data-continues-to-explode-exponentially-who-is-using-that-data-and-how.
'Strengthening Digital Society Against Cyber Shocks', https://www.pwc.com/us/en/services/consulting/cybersecurity/library/information-security-survey/strengthening-digital-society-against-cyber-shocks.html.
'Supercomputing and Exascale Computing Overview' (Undated), https://www.intel.com/content/www/us/en/government/exascale-supercomputing.html.
Swinhoe, D., 'The 17 Biggest Data Breaches of the 21st Century' (January 26 2018), https://www.csoonline.com/article/2130877/data-breach/the-biggest-data-breaches-of-the-21st-century.html.
Symantec, 'Stuxnet 0.5: How It Evolved' (February 26 2013), https://www.symantec.com/connect/blogs/stuxnet-05-how-it-evolved, http://www.synthite.co.uk/
'Tactics, Techniques and Procedures (TTPs) Within Cyber Threat Intelligence' (January 19 2017), https://www.optiv.com/blog/tactics-techniques-and-procedures-ttps-within-cyber-threat-intelligence, https://www.teamviewer.com/en/solutions/remote-access/#gref.
'TeamViewer and Scamming', https://community.teamviewer.com/t5/Knowledge-Base/TeamViewer-and-scamming/ta-p/4715.
'The PlugBot: Hardware Botnet Research Project', https://www.redteamsecure.com/the-plugbot-hardware-botnet-research-project/.
'The Recent Storms and Floods in the UK' (February 2014), https://www.ceh.ac.uk/sites/default/files/Recent%20Storms%20Briefing.pdf, https://theantisocialengineer.com/about-us/

'The Social Engineering Framework', https://www.social-engineer.org/framework/se-tools/computer-based/social-engineer-toolkit-set/.

'The World's Most Widely Used Host-Based Intrusion Detection System Used by Tens of Thousands of Organizations Around the World', https://www.ossec.net/.

Thompson, J., 'Refineries and Associated Plant: Three Accident Case Studies' (2013), http://www.safetyinengineering.com/FileUploads/Refineries%20and%203%20accident%20case%20studies%20v2_1370770924_2.pdf.

'TimeStomp' (Undated), https://www.offensive-security.com/metasploit-unleashed/timestomp/.

Thomson, I., 'Everything You Need to Know About the Petya, er, NotPetya Nasty Trashing PCs Worldwide This Isn't Ransomware—It's Merry Chaos' (June 28 2017), https://www.theregister.co.uk/2017/06/28/petya_notpetya_ransomware/, https://www.torproject.org/.

Tucker, P., 'Russia Will Build Its Own Internet Directory, Citing US Information Warfare' (November 28 2017), http://www.defenseone.com/technology/2017/11/russia-will-build-its-own-internet-directory-citing-us-information-warfare/142822/?oref=d-river.

Tung, L., Cisco Warning: These Routers Running IOS Have 9.9/10-Severity Security Flaw' (September 26 2019), https://www.zdnet.com/article/cisco-warning-these-routers-running-ios-have-9-910-severity-security-flaw/.

Tunggal, A.T., 'What Is the WannaCry Ransomware Attack?' (December 13 2019), https://www.upguard.com/blog/wannacry.

Turner, A., 'Mike Assante's Lasting Impact on Critical Infrastructure Security (and Me)' (June 18 2019), https://www.csoonline.com/article/3403656/mike-assantes-lasting-impact-on-critical-infrastructure-security-and-me.html.

Turovsky, D., 'Moscow's Cyber-Defense: How the Russian Government Plans to Protect the Country from the Coming Cyberwar' (July 19 2018), https://meduza.io/en/feature/2017/07/19/moscow-s-cyber-defense.

Turovsky, D., 'What Is the GRU? Who Gets Recruited to be a Spy? Why Are They Exposed So Often? Here Are the Most Important Things You Should Know About Russia's Intelligence community' (November 6 2018), https://meduza.io/en/feature/2018/11/06/what-is-the-gru-who-gets-recruited-to-be-a-spy-why-are-they-exposed-so-often.

Tuwiner, J., 'What Is Bitcoin Mining and How Does It Work?' (February 8 2019), https://www.buybitcoinworldwide.com/mining/.

'U.S. Power Grid' (Undated), http://www.earthlyissues.com/uspower.htm.

Vella, H., 'Fighting Cyber Crime in the Offshore Oil and Gas Industry' (December 13 2016), https://www.offshore-technology.com/digital-disruption/cybersecurity/featurefighting-cyber-crime-in-the-offshore-oil-and-gas-industry-5692000/.

'VERIS Overview', http://veriscommunity.net/veris-overview.html.

Waltman, C., 'Aurora: Homeland Security's Secret Project to Change How We Think About Cybersecurity' (November 14 2016), https://www.muckrock.com/news/archives/2016/nov/14/aurora-generator-test-homeland-security/.

Walker, J., 'Autonomous Ships Timeline—Comparing Rolls-Royce, Kongsberg, Yara and More' (May 29 2018), https://www.techemergence.com/autonomous-ships-timeline/, http://www.warwickchem.com/

'Watch Hackers Break into the US Power Grid' (May 11 2016), https://www.youtube.com/watch?v=pL9q2lOZ1Fw&feature=youtu.be.

'Web Application vs Website: What Suits Your Business Better' (June 10 2018), https://www.cleveroad.com/blog/what-is-the-difference-between-website-and-web-application-choose-what-fits-your-business.

'Website Mirroring | Website Hacking #2' (April 15 2019), https://www.allabouthack.com/2019/04/website-mirroring-website-hacking-2.html.

Weisburd, A., Watts, C., and Berger, J.M., 'Trolling for Trump: How Russia Is Trying to Destroy Our Democracy' (November 6 2016), https://warontherocks.com/2016/11/trolling-for-trump-how-russia-is-trying-to-destroy-our-democracy/.

'West Siberian Oil Basin', http://petroneft.com/operations/west-siberian-oil-basin/.

'What Is CANVAS?', https://canvas-project.eu/canvas/what-is-canvas/.

'What Is a Firewall?' (Updated, https://www.cisco.com/c/en_uk/products/security/firewalls/what-is-a-firewall.html.

'What Is a Honey Pot?' (September 15 2018), https://resources.infosecinstitute.com/what-is-a-honey-pot/#gref.

'What Is a Honeypot? How It Can Lure Cyberattackers' (Undated), https://us.norton.com/internetsecurity-iot-what-is-a-honeypot.html.

'What Is a Man-in-the-Middle Attack?' (Undated), https://us.norton.com/internetsecurity-wifi-what-is-a-man-in-the-middle-attack.html.

'What Is a Packet Sniffer?', http://netsecurity.about.com/od/informationresources/a/What-Is-A-Packet-Sniffer.htm.

'What Is a Zero-Day Exploit? Zero-Day Exploit: An Advanced Cyber Attack Defined' (Undated), https://www.fireeye.com/current-threats/what-is-a-zero-day-exploit.html.

'What Is the Difference Between SCADA and DCS Systems?' (March 21 2018), https://www.maderelectricinc.com/blog/what-is-the-difference-between-scada-and-dcs-systems.

'What Is the Difference: Viruses, Worms, Trojans, and Bots?' (Undated), https://www.cisco.com/c/en/us/about/security-center/virus-differences.html.

'What Is a Programmable Logic Controller', http://www.amci.com/tutorials/tutorials-what-is-programmable-logic-controller.asp.

'What Is SCADA?', http://www.dpstele.com/dpsnews/techinfo/what_is_scada.php.

'What's My IP Address?', https://www.privateinternetaccess.com/pages/whats-my-ip/, and 'Multi-hop Proxy', https://attack.mitre.org/techniques/T1188/.

'Who We Are', https://www.saudiaramco.com/en/who-we-are/overview/global-presence, http://windows.microsoft.com/en-gb/windows/end-support-help

'Windows Management Instrumentation' (May 31 2018), https://docs.microsoft.com/en-us/windows/win32/wmisdk/wmi-start-page.

'Windows PowerShell' (July 8 2013), https://msdn.microsoft.com/library/dd835506.aspx, https://www.wireshark.org/.

Wise, J., 'Disrupting Cyberwar with Open Source Intelligence' (October 19 2018), https://www.hpe.com/us/en/insights/articles/disrupting-cyberwar-with-open-source-intelligence-1810.html.

World Economic Forum on Risk and Resilience with Marsh & McLennan and Zurich Insurance Group, https://www.weforum.org/reports/the-global-risks-report-2020.

Wrenn, G., 'Cyber Security and Critical Infrastructure MIT Industrial Liaison Program (ILP)' (June 9 2015), https://www.youtube.com/watch?v=JCbme19f7yQ.

'Xenotime', https://dragos.com/blog/20180524Xenotime.html (Undated).

'Zero-day Vulnerability: What It is, and How It Works' (Undated), https://us.norton.com/internetsecurity-emerging-threats-how-do-zero-day-vulnerabilities-work-30sectech.html.

Index

A

Aaviksoo, Jaak 12, 14, 41, 42
Advanced Persistent Threats (APT) 7, 15, 55, 89–91, 109, 150, 186, 187, 227, 234, 235, 238, 249, 263, 299, 317, 321, 351, 377, 402, 415, 422
aerospace 99, 132, 178
Afghanistan 62, 264
Africa 87, 89
Ahmadinejad, Mahmoud 87. *See also* Iran
anonymity 316
anonymous 22, 42, 50, 51, 66, 126, 136, 144, 145, 210, 212, 214, 215, 218, 220, 225, 269, 336, 340, 343, 347, 349, 351, 356, 384, 385, 387, 390, 391, 393, 394, 396, 398. *See also* hacktivism
anti-virus 233, 246, 248, 292, 415
Apple 17, 23, 47, 233, 367, 368, 390, 402
Argentina 224
ARPANET 409
Arquilla, John 29, 50, 57, 116, 117, 401, 405, 424, 425
Artificial Intelligence (AI) 17, 18, 23, 44, 45, 198, 319, 320, 344, 423. *See also* machine learning (ML); singularity
Asia 91, 115, 157, 249, 299, 424
Asia-Pacific 107, 182, 404
Assange, Julian 363, 380, 398
Assante, Mike 154, 187, 202, 203, 219
Aurora 151–153, 202, 232, 402, 405

Australia 11, 80, 100, 158, 193, 402, 404, 413. *See also* Five Eyes
Austria 267, 279

B

Bangladesh 97, 101, 142
Barcelona 169, 210
Belarus 97
Bellingcat 302, 351
Biden, Joe 42, 105, 106, 110, 126, 191, 196, 197, 363. *See also* United States government
biometrics 40, 300, 304
blockchain 32, 51, 182, 249, 250
board-level decisionmaking 262, 413
Brantly, Aaron 117, 127, 307, 339, 340
Brenner, Joel 77, 127, 301, 306, 338, 340
Brown, Derren 289, 333
Byres, Eric 208, 264, 276, 280

C

California 169, 325, 349, 367, 390, 425
Canada 11, 80, 86, 107, 157, 193, 402. *See also* Five Eyes
Carlin, John 47, 427
CCDCOE. *See* Cooperative Cyber Defence Centre of Excellence
CERT. *See* Computer Emergency Response Teams
China (People's Republic of China) 2, 5, 6, 13, 14, 27, 28, 35, 41, 47, 49, 55, 66, 76–78, 80–82, 85, 87, 88, 94, 95, 97, 100, 102, 108, 110, 112, 113, 128–130, 154, 157, 162, 169, 179, 233, 238, 260, 263, 270, 278, 320, 337, 352, 353, 368, 370, 378–382, 387, 404, 413, 415–418, 420–423, 428, 429
People's Liberation Army (PLA) 78
CIA-triad 150, 263, 356
CISA. *See* Cybersecurity Information Sharing Act
Cisco Systems 31, 192, 261, 330
civil nuclear power 14, 66, 96–99, 117–120, 161, 168, 259
Clapper, James 356, 423, 429
Clarke, Richard 41, 47, 57–59, 73, 78, 116, 117, 120, 128, 201, 224, 404, 425
Clausewitz, Carl von 5, 8, 36, 38, 53–56, 58, 61–64, 77, 114, 115, 126, 127, 404, 421, 425
Clinton, Bill 322, 327
Clinton, Hillary 357, 380
cloud 18, 23, 197, 198, 238, 244, 245, 259, 282, 296, 325, 355, 368
Comey, James B. 365, 368
commercial-off-the-shelf-technologies (COTS) 173, 315
Computer Emergency Response Teams (CERTs) 183, 191, 322–324, 326, 417
Cooperative Cyber Defense Centre of Excellence (CCDCOE)

6, 70, 78. *See also* North Atlantic Treaty Organization (NATO); Tallinn process
counterespionage 330
counterintelligence 11, 93, 99, 104, 111, 284, 301, 330, 353, 419
counterterrorism (CT) 11, 62, 264, 305, 363, 366, 367, 419
cracking 8, 15, 245, 284, 298, 356. *See also* hackers and hacking
Creveld, Martin van 62, 118, 119
critical infrastructure (CI)
    chemical plants 88–90, 423
    civil aviation 175, 176, 179, 180, 405
    dams and reservoirs 159
        hydroelectric 2, 148, 159
    electricity generation and distribution 154
        electric grid 2, 27, 154–156, 410
    electricity substations 154
    financial services 20, 36, 81, 99, 101, 110, 148, 199, 201, 250, 253, 406, 409
    healthcare 20, 26, 99, 148, 188, 250, 352
    merchant shipping 170, 405
    oil, gas, and petroleum 1, 2, 28, 91, 148, 160, 162, 163, 166, 307, 374, 405
        oilfields and oil rigs 160, 162, 208
    owner-operators 79, 89, 91, 104, 106, 174, 180, 244, 251, 262, 283, 294, 327, 413
    pipelines 1, 20, 26, 28, 148, 160, 161, 170
    ports and logistics 169, 405
    rail 26, 28, 172, 174, 195, 405
    roads 173
    telecommunications 2, 9, 20, 89–91, 132, 191, 195, 238, 330, 359, 419
    transport 2, 19, 20, 26, 32, 159, 160, 172, 174, 175, 185, 235, 250, 256, 352, 376, 402, 405, 423
    water treatment and sanitation 148, 158, 405
CrowdStrike 9, 15, 39, 42, 90, 291, 378, 397
cryptocurrencies 32, 51, 98, 249, 250, 316, 371–373
cryptography 19
cyberattack 2–5, 8, 9, 11–14, 23, 25–29, 48, 54, 55, 58, 60, 65, 69–72, 74–78, 80, 81, 84–86, 88–90, 92, 93, 109–111, 113, 126, 141, 142, 148, 152, 155–157, 159–161, 167–169, 174, 176–181, 184–186, 188–190, 192, 197, 199, 200, 216, 229, 244, 255, 259, 309, 322, 373, 379, 380, 402–405, 407, 412, 414, 416, 421
cyber conference (CyCon) 170, 403. *See also* cyberwarfare; North Atlantic Treaty Organization (NATO); Tallinn process
cybercrime 6, 8, 9, 13, 24, 29, 33–35, 42, 54, 59, 64, 65,

72, 90, 91, 97–101, 112, 142, 150, 173–175, 185, 186, 190, 193, 197, 231, 235, 244, 250, 258, 261, 263, 270, 279, 283, 292, 293, 296, 301, 321, 324, 327, 331, 332, 351, 352, 354, 368–382, 393, 399, 406, 415, 416. *See also* cyber privateers; ransomware
  cybercrime prosecutions 81
  scammers and scamming 375
  state-sponsored 14, 29, 99, 148, 190, 193, 263, 327, 369, 371, 416, 417
cyberespionage 7, 8, 12, 13, 15, 23, 27–29, 34, 54–57, 59, 60, 65, 67, 72, 75, 79, 83, 87, 88, 90, 91, 93–97, 99, 100, 109, 132, 140, 148, 193, 194, 238, 240, 303, 370, 383, 404, 406, 407, 416–418, 422
  disinformation 25, 33, 56, 64, 102
cyber forensics 33
cyber insurance 7, 28, 120, 261
CYBERINT 285, 303, 358, 366, 367
cyber interdependencies 82
cyber privateers 381
cybersecurity 1, 3, 4, 7, 9, 11, 13, 15, 23, 29, 31, 33, 34
Cybersecurity Information Sharing Act (CISA) 105, 184, 191, 196, 307, 322, 323
cyberspace 7, 12, 13, 24, 30, 47, 55, 58, 59, 61, 65, 66, 68–70,
73, 77–79, 83, 92, 93, 103, 109, 110, 143, 144, 238, 263, 285, 302, 330, 331, 402–404, 408, 415, 418, 419, 422
cyberterrorism 185, 364
cyber threat intelligence (CTI) 198, 325, 329, 406, 413, 417
cyberwar
  cascading attacks 83
  cyber peacekeeping 86
  cyber Pearl Harbor 25, 27, 47, 73, 199, 401, 423
  cyber resilience 85
  peacetime non-aggression 423
  preparatory self-defense 60
  preparedness for 27
  prepositioning for 154
  proportionality 403. *See also* Just War Theory/tradition
  self-defense 65, 68, 71, 76, 79, 304, 403, 404, 406. *See also* Just War Theory/tradition
  undeclared war 60, 199
  zero-sum game 18, 55, 113, 233
cyberwarfare
  cyberdeterrence 37, 48, 113, 122, 127, 128, 144–146, 384, 390, 428
  cybergeddon 3, 199, 423
  decision-making 18, 35, 70, 77, 98, 107, 196, 262, 324, 418
  defense-in-depth 76
  Defensive Cyber Effects Operations (DCEO) 103
  demilitarized zones 94, 243, 356
  first-strike 110, 408
  Maginot Line 76

offensive cyber 5, 7, 11, 29, 30, 55, 56, 76, 79, 81, 83, 87, 93, 103, 105, 109, 110, 112, 198, 378, 404, 408, 420
Offensive cyber operations/Offensive Cyber Effects Operations (OCO/OCEO) 55, 103, 306, 405
politico-military decisions 23
proxy actors 378
rules of engagement (RoE) 6, 18, 65, 67, 103, 105, 112, 404
signaling 79, 82, 92, 379
Tallinn process 7, 60, 70, 403, 404. *See also* Cooperative Cyber Defense Centre of Excellence (CCDCOE); North Atlantic Treaty Organization (NATO); Schmidt, Michael N.

D

darknet 393
data breaches 9, 15, 140, 184, 293, 311, 319, 354
DEFCON (hacking conference) 9, 153, 171, 172, 391
defense-industrial base 102, 254, 265, 352, 354
Demchak, Chris 24, 47, 124, 171, 212, 422, 428
Democratic People's Republic of Korea
  Lazarus Group 98–101
  Sony hack 99

WannaCry 35, 100, 169, 185, 232, 246, 296, 376, 377, 382
disinformation, misinformation, malinformation 25, 31, 33, 51, 56, 64, 102, 196, 365, 386, 426. *See also* gray zone; hybrid warfare
Distributed denial-of-service (DDoS) attacks 15, 33, 92, 178, 190, 193, 194, 233, 351, 356
Dragos 91, 155, 156, 167, 182
drones 156, 176, 181, 314, 316, 364

E

election interference 64, 102, 110, 113, 406. *See also* cyberespionage; espionage
Electronic Intelligence (ELINT) 302, 303, 366
e-mail 15, 16, 31–33, 178, 227, 229, 231, 236, 238, 239, 242, 244, 248, 282, 283, 287, 292–294, 300, 355–357, 368, 373, 380, 411, 412, 414, 416
encryption 9, 17, 19, 31, 32, 35, 176, 185, 234, 238, 250, 252, 284, 365, 367, 368, 373, 376, 413
espionage 5, 6, 9, 11, 24, 27, 29, 58, 60, 72, 75–77, 79, 88, 91, 99, 144, 150, 163, 263, 281–284, 292, 299–304, 307, 310, 311, 320, 324, 328–330, 341, 354, 356,

362, 370, 372, 404, 406, 416, 417, 426
Estonia 6, 12, 14, 38, 50, 70, 113, 403, 406
European Union (EU) 7, 78, 157, 179–181, 184–186, 192, 198, 205, 218, 220, 262, 263, 321, 323, 324, 327, 376, 402
Europol 368, 371–373, 376

F

Facebook 245, 301, 353, 358, 359, 362, 365, 402
FireEye/FireEye-Mandiant 88, 100, 126, 141, 232, 270, 282, 325, 331
firewalls 16, 171, 176, 182, 227, 242, 243, 246, 248, 263, 292, 306, 356, 411, 415
firmware 16, 173, 231, 416
Five Eyes 11, 80, 128, 193, 402
France 107, 128, 233, 364

G

Galeotti, Mark 63, 64, 119, 379, 398, 406, 425
General Data Protection Regulation (GDPR) 180, 184, 185, 262, 263
Geneva Convention. *See also* international law; Just War Theory/tradition
Georgia (country) 10, 63, 113, 161, 406
GitHub 182, 237, 261, 314
globalization 4, 172, 410

Government Communications Headquarters (GCHQ) 9, 11, 79, 126, 221, 390, 402. *See also* Five Eyes; United Kingdom (UK)
gray zone 30, 58, 62, 101, 108. *See also* hybrid warfare
Greenlees, Colin 310–312, 341

H

hackers and hacking 355, 356
hacktivism 42, 98, 416
Hayden, Michael 57, 116
Holsti, Kalevi J. 60, 69, 118, 123, 421, 428
human intelligence (HUMINT) 284, 300, 302–304, 330, 356, 366
hybrid warfare 61, 62, 64, 108, 196, 407

I

Idaho National Laboratory (INL) 22, 130, 151, 154, 231, 316, 352, 405
India 53, 97, 100, 108, 157, 162, 168, 189, 233, 282, 321, 327, 376
Inkster, Nigel 14, 42, 300, 304
insider threats 7, 9, 38, 158, 177, 189, 281, 307–310, 330, 353–355, 392
Intel 18, 318
intellectual property theft 308, 309, 355
intelligence 6, 7, 9, 11, 12, 14, 27, 30, 33, 35, 39, 41, 55, 57,

58, 60, 63, 64, 67–69, 77, 79, 86, 87, 91, 93, 96–101, 104, 106, 109–111, 113, 123, 128, 129, 138, 140, 147, 151, 163, 168, 182–184, 189, 190, 192, 193, 195–198, 200, 232, 233, 238, 239, 244, 260, 263, 281–285, 287, 290, 299–306, 309–311, 317, 320, 323, 324, 329–331, 339, 347, 349, 351, 353–355, 357–362, 366–371, 376–379, 382, 383, 401, 404, 408, 411, 413–416, 418–420, 422
  tradecraft 302
international law
  Caroline test 71
  law enforcement 81, 261, 371
  Lawfare v
internet 2–4, 6, 9, 10, 15–17, 21, 22, 25, 28, 30–34, 40, 63, 64, 87, 95, 114, 130, 150, 151, 160, 167, 170, 173, 174, 176, 178, 187–189, 192, 196, 199, 229, 232, 235, 238, 240, 242, 244, 247, 248, 252, 254, 259, 261, 265, 285, 294, 296, 298, 299, 303, 305, 308, 353, 355–358, 364–370, 376, 381, 382, 409, 414, 422
  Internet of Things (IoT) 4, 7, 17, 22, 23, 85, 106, 147, 174, 187, 188, 232, 244, 315, 356, 358, 375, 410

Interpol 92, 376. *See also* cybercrime; police and policing
Iran 5, 6, 12, 56, 73, 75, 76, 80, 87–94, 108, 113, 132, 151, 154, 159, 162, 167, 178, 233, 286, 352, 366, 402, 405, 417
  Shamoon 88
Iraq 159, 162
Israel 87–89, 91, 93, 152, 167, 233

J

Jinping, Xi 418. *See also* China
Just War Theory/tradition 61, 62, 65, 66, 67, 71, 113, 193, 195, 379, 403. *See also* cyberwar; international law
  armed attack 65

K

Kant, Immanuel 424, 429
Kaspersky 130, 271, 325
Kim, Jong-un 94–96, 101
Knake, Robert 2, 3, 36, 41, 47, 58, 59, 73, 78, 117, 120, 128, 201, 224, 404, 425

L

leaktivism 6, 351
legacy systems 34, 147, 157, 171, 174, 180, 188, 252–259, 270, 315, 317, 377, 408, 409
LinkedIn 100, 245, 283, 311, 353, 358

M

Linux 182, 260, 297, 298, 414
Lithuania 190

machine learning (ML) 17, 18, 23, 244, 291, 306, 423
malware
  botnets 33, 193
  polymorphic malware 246, 247
  spyware 31, 314
  trojans 10, 31, 32, 177, 192, 193, 228, 230, 248, 300
  virus 7, 31, 32, 228. *See also* malware
  worms 10, 31, 32, 193
Maurer, Tim 14, 42, 378, 379, 397, 425
McAfee 7, 12, 22, 38, 41, 100, 188, 204, 219, 235, 262, 271, 279, 280, 295, 325, 372, 374, 376, 393, 395, 413, 426
Microsoft
  Microsoft Windows 232, 233, 248
  Windows XP 248, 260, 278, 375, 376
Milgram, Stanley 288, 298, 333, 337, 358
Mitnick, Kevin 16, 43, 291, 381
Mossad 202. *See also* Israel; Stuxnet
Mueller, Milton 10, 40, 51
Musk, Elon 190, 192

N

Nakasone, Paul M. 108, 110–112, 144–146. *See also* United States government, Department of Defense (DOD), National Security Agency (NSA)
National Cyber Security Centre (NCSC) UK 74, 104, 126, 143, 145, 183, 189, 199, 324, 347
Network and Information Security Directive (EU) 116–118, 167–168, 209
networks
  intranets 4, 370
  Local Area Network (LAN) 21, 237
  Wide Area Network (WAN) 21, 237
New Zealand 11, 80, 193, 365, 402, 404. *See also* Five Eyes
norms 60, 65, 66, 75, 82, 86, 108, 112, 289, 422, 423
North Atlantic Treaty Organization (NATO) 6, 7, 12, 17, 29, 30, 64, 69, 70, 78, 83, 107, 129, 170, 185, 190, 191, 195, 198, 403, 404, 421. *See also* Cooperative Cyber Defense Centre of Excellence (CCDCOE); cyber conference (CyCon); Tallinn process
Northrop-Grumman 370, 371
Norton security 232
NotPetya 35, 169, 194, 232, 376–378, 382, 397
nuclear weapons 18, 82, 83, 93, 94, 97, 99, 108, 118, 255
Nye, Joseph S. 58, 62, 82, 117, 129

O

Obama, Barack H. 1, 103, 104, 109, 110
Obama-Xi Agreement 238
Open Source Intelligence (OSINT) 16, 23, 33, 196, 283, 298, 301, 302, 304–306, 328, 353, 357, 358, 366, 381, 411, 416, 417
operational technology (OT) 4, 20–22, 106, 148, 161, 194, 234, 237, 244, 259, 317
Organisation for the Prohibition of Chemical Weapons (OPCW) 12, 120, 299, 300

P

Palo Alto Networks 325, 374
Pederson, Perry 151, 153, 202, 203
penetration testing 297, 310, 313, 316, 336, 372, 412, 417
Philippines 100, 162
phishing 15, 32, 88, 109, 132, 177, 184, 231, 235, 236, 240, 242, 245, 282, 283, 287, 292–295, 300, 325, 329, 357, 368, 372, 373, 375, 376, 380. *See also* social engineering (SE); spear phishing
Piper Alpha 163
police and policing 7, 183, 360, 362
psychology 282, 285, 287, 329, 330, 356, 360, 361, 382
Putin, Vladimir 56, 63, 108, 196–198, 417, 422

Q

quantum computing 17–19, 23, 423

R

ransomware 14, 23, 31, 32, 35, 89, 101, 148, 161, 169, 174, 177–179, 189, 190, 193, 194, 197, 230, 244, 258, 272, 283, 293, 296, 319, 352, 354, 373–377, 381, 406, 417. *See also* cybercrime
Republic of Korea 90, 94, 97, 99, 100, 102, 140, 162, 178, 404
Republic of Korea (South Korea) 102
responsible disclosure 180
Revolution in Military Affairs (RMA) debate 73, 82, 124, 406
Rid, Thomas 41, 49, 56, 58, 115, 117, 124, 136
risk management and mitigations 5, 110, 170, 181, 185, 239, 241, 248, 258, 265, 281, 285, 316, 330, 355, 413, 415, 423
Rogers, Michael S. 73, 102, 142, 186, 199, 200, 369, 401, 417
Ronfeldt, David 50, 57, 116
Russia
  cyber actors
    Cozy Bear (APT29) 422
    Energetic Bear 91, 130
    Fancy Bear (APT28) 192, 317, 377, 422

518 Index

Sandworm (GRU Unit 74455) 194
cyber campaigns
BlackEnergy 73–76, 86, 194, 195, 370, 402
Crashoverride 155
Dragonfly 86, 187, 253, 260, 283
Industroyer 75, 76, 155, 194, 195, 370, 402
Petya/NotPetya 35, 169, 376–378, 382
Gerasimov, Valery/Doctrine 62, 63, 65
intelligence agencies
Federal'naya Sluzhba Bezopasnosti 109, 110, 144, 196, 370
Glavnoye Razvedovatel'noye Upravlenie (GRU) 63, 169, 189, 194, 196, 299, 317, 377
Sluzhba Vneshnei Razvedki (SVR) 194, 196

S

sabotage 56, 58, 76, 97, 167, 187, 197, 307
San Francisco 174, 264, 325
Sanger, David E. 113, 136, 146, 195, 223, 225, 408, 425
satellites 7, 27, 107, 179, 181, 189–191, 211, 257, 302, 303
SATINT (Satellite Intelligence) 302, 366
Saudi Arabia 88–90, 161, 162, 167, 178, 189

Saudi Aramco 88, 161, 162, 168, 209
Schmitt, Michael N. 41, 49, 70, 71, 121, 123, 124, 126, 225, 393, 424. *See also* North Atlantic Treaty Organization (NATO); Tallinn process
Secret Intelligence Service (SIS/MI6) 302, 304, 306. *See also* United Kingdom (UK)
Self-Monitoring Analysis and Reporting Technology (SMART) 4, 37, 85, 147, 174, 176, 177, 187, 242, 244, 248, 253, 287, 356, 410
Shadow Brokers 297, 370, 377, 378
Shodan 22, 46, 74, 150, 229, 266
signals intelligence (SIGINT) 44, 96, 284, 302, 303
singularity 17. *See also* Artificial Intelligence (AI); machine learning (ML)
Snowden, Edward 11, 13, 79, 80, 103, 104, 305, 309, 355, 363, 367, 402, 408
social engineering (SE) 16, 32, 35, 97, 99, 177, 227, 234, 245, 281, 282, 284, 285, 289, 291, 300, 302, 307, 311, 312, 328–330, 355, 357, 362, 373, 381, 382, 411, 413–415
pretexting 90, 285, 311, 356, 357
Social Media Intelligence (SOCMINT) 302, 306
social network analysis (SNA) 306, 357–363, 382, 387

SolarWinds 198
SOLAS convention 170
Spain 169, 177, 189
spear phishing 90, 98, 101, 177, 231, 287, 292–296, 361, 373, 411, 413, 414
Stoltenberg, Jens 69, 123, 190. *See also* North Atlantic Treaty Organization (NATO)
Stuxnet 11, 12, 75, 76, 79, 93, 112, 113, 150–152, 202, 203, 232, 264, 286, 307, 370, 402, 405, 408
supercomputing 45
Supervisory Control and Data Acquisition Systems (SCADA) 19–21, 23, 74, 148–151, 155, 158, 159, 168, 172, 176, 181–183, 199, 204, 236, 243, 249, 251–254, 256, 257, 260–262, 264, 265, 270, 317, 331, 402, 405, 409
Symantec 187, 203, 325
Syria 364

T

tactics, techniques, and procedures (TTPs) 5, 25, 34, 90, 97, 98, 234, 241, 325, 409, 415
Taiwan 162, 376
Tallinn manual. *See* cyberwarfare; Tallinn process
Tallinn process 60, 70, 403, 404
Telegram 192, 362, 365, 372, 373
Texas 161, 325, 364
Thailand 97, 98

The Onion Router (TOR) 238, 367, 372, 382
Thucydides 62, 422
Trump, Donald J. 38, 56, 93, 104, 105, 107, 110, 191, 238, 256, 322, 420
Tzu, Sun 8, 26, 38, 48, 64, 114, 126

U

Ukraine 5, 25, 26, 28, 34, 56, 63, 64, 67, 69, 70, 73, 74, 76, 86, 91, 108, 112, 113, 125, 155, 157, 168, 189–198, 200, 220, 223, 227, 302, 376, 377, 402, 406–408, 422
United Kingdom (UK) 9, 14, 56, 68, 74, 79, 93, 107, 162, 178, 179, 183, 189, 192, 193, 199, 225, 233, 262, 264, 284, 289–291, 302, 325, 327, 328, 357, 364, 395, 413. *See also* Five Eyes; Government Communications Headquarters (GCHQ); National Cyber Security Centre (NCSC) UK Cyber Primer (UK Ministry of Defence) 43, 68, 122. *See also* international law
United Nations (UN) 12, 59, 60, 65, 67, 68, 76, 97, 172, 198, 403, 423
United Nations Commission on Human Rights (UNCHR) 95

United Nations Security Council 67
United States 2, 3, 4, 6, 7, 19, 20, 23, 27–30, 37, 40, 49, 55, 64, 66, 73, 74, 76, 79–87, 89–91, 93, 100–102, 104, 105, 107–109, 111, 112, 153, 156, 157, 161, 168, 177–179, 191–193, 197, 200, 233, 235, 236, 284, 289, 298, 309, 318, 320, 323, 325, 352, 356, 363, 364, 366–368, 376, 380, 391, 404, 408, 410, 413, 417–422, 424, 427, 428. *See also* Five Eyes
United States Air Force 111, 256, 277, 419
United States government
  department of commerce
    National Institute of Standards and Technology (NIST) U.S. 55, 183, 308, 327, 415
  Department of Defense (DOD)
    central intelligence agency (CIA) 21, 263, 356
    Defense Advanced Research Projects Agency (DARPA) 153
    Defense Intelligence Agency (DIA) 418, 419
    Department of Defense Information Network (DODIN) 85, 419
    Director of National Intelligence (DNI) 24, 96, 104, 256, 287, 356, 420
    National Security Agency (NSA) 11, 13, 73, 103, 153, 263, 356, 402, 418
    Pentagon 113
    PRISM 11, 13, 67, 360, 363, 367. *See also* Snowden, Edward
    U.S. Cyber Command (USCYBERCOM) 106, 111, 186, 192, 427
  Department of Energy (DOE) 104, 159, 191, 206
  Department of Homeland Security (DHS) 74, 105, 106, 151, 152, 256, 307, 322, 352, 354, 419
    National Cybersecurity and Communications Integration Center (NCCIC) 105, 307, 308, 322, 327, 346
    National Protection and Programs Directorate (NPPD) 105, 322
  Department of Justice (DOJ) 92, 167, 308, 317, 364, 419
  Federal Aviation Authority (FAA) 177–180, 256, 257
  Federal Bureau of Investigation (FBI) 11, 16, 34, 40, 101, 111, 186, 232, 264, 293, 301, 308, 309, 315, 352, 354, 356, 357, 365–369, 371, 372, 376
  Federal Emergency Management Agency (FEMA) 255, 323
United States Congress 363
White House 103, 357

National Security Council (NSC) 418
United States Senate 128, 144, 146

V

Verizon 9, 39, 51, 201, 238, 239, 241, 269, 270, 282, 283, 285, 294, 300, 332, 335, 354, 357, 373, 384
Verizon's Data Breach Investigations Report 39, 43, 50, 201, 265, 269, 270, 279, 282, 332, 335, 338, 354, 384, 394
Virtual Private Networks (VPNs) 39, 237, 245, 372, 382
virus. *See* malware

W

Wainwright, Rob 376. *See also* Europol; police and policing
Walzer, Michael 61, 65, 118, 120
Warner, Mark R. 174
Weiss, Joe 152, 153, 206
whistleblowing 309, 353, 363
Wi-Fi 21, 173, 181, 298, 317
Wikileaks 351, 363, 380

Y

Younger, Alex 302, 304, 306
YouTube 22, 223, 234, 291, 312, 337

Z

zero-days 16, 167, 231–235, 263, 299, 376, 415
Zetter, Kim 203, 206, 367, 385, 390

GPSR Compliance

The European Union's (EU) General Product Safety Regulation (GPSR) is a set of rules that requires consumer products to be safe and our obligations to ensure this.

If you have any concerns about our products, you can contact us on

ProductSafety@springernature.com

In case Publisher is established outside the EU, the EU authorized representative is:

Springer Nature Customer Service Center GmbH
Europaplatz 3
69115 Heidelberg, Germany

www.ingramcontent.com/pod-product-compliance
Lightning Source LLC
LaVergne TN
LVHW011005250326
834688LV00004B/75